Introductory mathematics

THROUGH SCIENCE APPLICATIONS

JOHN BERRY
Plymouth Polytechnic

ALLAN NORCLIFFE
Sheffield City Polytechnic

and

STEPHEN HUMBLE
Royal Military College of Science

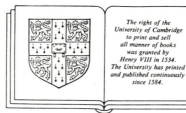

The right of the
University of Cambridge
to print and sell
all manner of books
was granted by
Henry VIII in 1534.
The University has printed
and published continuously
since 1584.

CAMBRIDGE UNIVERSITY PRESS

Cambridge

New York   New Rochelle

Melbourne   Sydney

Published by the Press Syndicate of the University of Cambridge
The Pitt Building, Trumpington Street, Cambridge CB2 1RP
32 East 57th Street, New York, NY 10022, USA
10 Stamford Road, Oakleigh, Melbourne 3166, Australia

© Cambridge University Press 1989

First published 1989

Printed in Great Britain at the University Press, Cambridge

British Library cataloguing in publication data

Berry, John 1947–
Introductory mathematics through science
applications
1. Mathematics – For science
I. Title   II. Norcliffe, Allan
III. Humble, Stephen
510'.245

Library of Congress cataloguing in publication data

Berry, J.S. (John Stephen), 1947–
Introductory mathematics through science applications/John
Berry, Allan Norcliffe, Stephen Humble.
    p. cm.
Includes index.
ISBN 0 521 24119 7. ISBN 0 521 28446 5 (paperback)
1. Mathematics—1961– 2. Science—Mathematics. I. Norcliffe,
Allan. II. Humble, Stephen. III. Title.
QA39.2.B475 1989
515—dc19   88-6983

ISBN 0 521 24119 7 hard covers
ISBN 0 521 28446 5 paperback

TM

CONTENTS

Preface   xiii

1   Basic functions   **1**
1.1   Scientific context   1
1.1.1   Loading a steel wire   1
1.1.2   Motion of a ball under gravity   2
1.2   Mathematical developments   4
1.2.1   Some basic terms   4
1.2.2   Polynomials of degree one: linear functions   6
1.2.3   Polynomials of degree two: quadratic functions   8
1.2.4   Complex numbers   12
1.2.5   Polynomials of degree n 14
1.2.6   Roots and factors of polynomials   15
1.3   Worked examples   17
1.4   Further developments   19
1.4.1   Linear interpolation and extrapolation   19
1.4.2   Approximate roots of polynomials   23
1.5   Further worked examples   27
1.6   Exercises   30

2   Functions for science 1: the exponential function   **34**
2.1   Scientific context   34
2.1.1   Orders of magnitude   34
2.1.2   Growth of a certain bacterium   35
2.2   Mathematical developments   37
2.2.1   Why e?   37
2.2.2   Power series representation of the exponential function e^x 40
2.2.3   Calculation of e^x 42
2.2.4   Approximate forms for e^x when x is small   42
2.2.5   Index laws   43
2.2.6   The graph of $y = e^x$ 45
2.3   Worked examples   46
2.4   Further developments   49
2.4.1   The decay function   49
2.4.2   Properties of the decay function   50
2.4.3   Hyperbolic functions   51
2.4.4   Properties of the hyperbolic functions   54
2.5   Further worked examples   58
2.6   Exercises   62

3	**Functions for science 2: trigonometric functions**	**66**
3.1	Scientific context	66
3.1.1	A swinging pendulum	66
3.1.2	Sound vibrations	68
3.1.3	Insect flight	68
3.2	Mathematical developments	69
3.2.1	Revision of trigonometric ratios for acute angles	69
3.2.2	Angles of any magnitude	74
3.2.3	Complete solutions of trigonometric equations	79
3.3	Worked examples	80
3.4	Further developments	81
3.4.1	Circular measure (radians)	81
3.4.2	Modelling with trigonometric functions	84
3.4.3	Small angles	87
3.4.4	Euler's formula for complex numbers	88
3.5	Further worked examples	92
3.6	Exercises	96
4	**Functions for science 3: inverse functions**	**101**
4.1	Scientific context	101
4.1.1	Bacterial growth revisited	101
4.1.2	Periodic systems	102
4.1.3	Obtaining scientific laws	103
4.2	Mathematical developments	106
4.2.1	Functions	106
4.2.2	Composite functions	108
4.2.3	Inverse functions	110
4.2.4	Logarithms	112
4.2.5	Inverse trigonometric functions	116
4.3	Worked examples	118
4.4	Further developments	120
4.4.1	Straightening curves	120
4.4.2	Inverse hyperbolic functions	123
4.5	Further worked examples	123
4.6	Exercises	128
5	**Other functions of science**	**132**
5.1	Scientific context	132
5.1.1	Equations of state	132
5.1.2	Vibrations	135
5.2	Mathematical developments	137
5.2.1	The function $f = 1/x$	137
5.2.2	Limits and continuity	139
5.3	Worked examples	143
5.4	Further developments	146
5.4.1	Asymptotes	146
5.4.2	Partial fractions	149
5.5	Further worked examples	154
5.6	Exercises	155

6   **Differentiation 1: rates of change**   **159**
6.1   Scientific context   159
6.1.1   Speed: a graphical model   159
6.1.2   Cell growth: an algebraic model   162
6.1.3   Potential energy and force: using x instead of t 163
6.2   Mathematical developments   164
6.2.1   Rates of change   164
6.2.2   Notation   167
6.2.3   Some basic functions and their derivatives   169
6.3   Worked examples   174
6.4   Further developments   176
6.4.1   Rules of differentiation   176
6.4.2   Higher derivatives   181
6.5   Further worked examples   182
6.6   Exercises   184

7   **Differentiation 2: stationary points**   **187**
7.1   Scientific context   187
7.1.1   Zero speed   187
7.1.2   Stability of a bead on a bent wire   189
7.2   Mathematical developments   190
7.2.1   Geometrical interpretation of the derivative   190
7.2.2   Stationary points   192
7.3   Worked examples   197
7.4   Further developments   200
7.4.1   Curve sketching   200
7.5   Further worked examples   204
7.6   Exercises   207

8   **Differentiation 3: approximation of functions**   **211**
8.1   Scientific context   211
8.1.1   Errors in measurement   211
8.1.2   Simplifying models   213
8.2   Mathematical developments   214
8.2.1   Taylor polynomials for small x 214
8.2.2   Taylor polynomials about $x = \alpha$ $(\alpha \neq 0)$ 218
8.3   Worked examples   220
8.4   Further developments   223
8.4.1   The Newton–Raphson method   223
8.5   Further worked examples   228
8.6   Exercises   230

9   **Integration 1: introduction and standard forms**   **233**
9.1   Scientific context   233
9.1.1   Speed–time graphs   233
9.1.2   Force–distance graphs   236
9.2   Mathematical developments   238
9.2.1   Definition of the integral as a summation   238

9.2.2	Evaluating integrals	242
9.3	Worked examples	245
9.4	Further developments	248
9.4.1	Numerical integration: Euler's method	248
9.4.2	The trapezoidal method	250
9.4.3	Simpson's method	251
9.5	Further worked examples	252
9.6	Exercises	254
10	**Integration 2: techniques of integration**	**259**
10.1	Scientific context	259
10.1.1	The centre of mass of a lamina	259
10.1.2	Chemical reactions	261
10.2	Mathematical developments	262
10.2.1	Direct substitutions	262
10.2.2	Indirect sustitutions	265
10.2.3	The substitution $t = \tan(x/2)$	268
10.3	Worked examples	269
10.4	Further developments	271
10.4.1	Use of partial fractions	271
10.4.2	Integration by parts	273
10.4.3	Reduction formulae	277
10.5	Further worked examples	278
10.6	Exercises	282
11	**Integration 3: further techniques**	**286**
11.1	Scientific context	286
11.1.1	Electric potential	286
11.1.2	Chemical reaction times	288
11.2	Mathematical developments	288
11.2.1	Infinite limits	288
11.2.2	Coping with singularities	289
11.3	Worked examples	291
11.4	Further developments	293
11.4.1	The length of a plane curve	293
11.4.2	Volumes of revolution	295
11.4.3	The centre of mass and moment of inertia of a solid of revolution	298
11.5	Further worked examples	300
11.6	Exercises	302
12	**First-order ordinary differential equations**	**305**
12.1	Scientific context	305
12.1.1	Chemical reactions	305
12.1.2	Newton's law of cooling	307
12.2	Mathematical developments	310
12.2.1	Classification of differential equations	310
12.2.2	Variables separable	311

12.2.3	Linear	313
12.2.4	Finding particular solutions	315
12.3	Worked examples	316
12.4	Further developments	320
12.4.1	A numerical solution	320
12.4.2	Improving the accuracy	324
12.5	Further worked examples	326
12.6	Exercises	328
13	**Second-order ordinary differential equations**	**334**
13.1	Scientific context	334
13.1.1	Cooling fins	334
13.1.2	Vibrations	337
13.2	Mathematical developments	338
13.2.1	Homogeneous equations	338
13.2.2	The method of solution	339
13.2.3	Interpreting the solution	344
13.3	Worked examples	349
13.4	Further developments	351
13.4.1	Non-homogeneous equations	351
13.4.2	The method of solution	351
13.4.3	Trial functions	352
13.5	Further worked examples	355
13.6	Exercises	359
14	**Statistics 1: frequency distributions and associated measures**	**364**
14.1	Scientific context	365
14.1.1	Radiation decay data	365
14.1.2	Velocity of light measurements	366
14.2	Mathematical developments	367
14.2.1	Organisation of data: frequency distributions	367
14.2.2	Presentation of data	370
14.2.3	Frequency curves and relative frequency curves	373
14.3	Worked examples	375
14.4	Further developments	380
14.4.1	Measures of centre	380
14.4.2	Measures of scatter	383
14.4.3	Some numerical considerations	386
14.5	Further worked examples	388
14.6	Exercises	391
15	**Statistics 2: probability and probability distributions**	**395**
15.1	Scientific context	395
15.1.1	Radiation decay data revisited	395
15.1.2	The alcohol in water experiment	396
15.2	Mathematical developments	398
15.2.1	The concept of probability	398

15.2.2	The probability scale	400
15.2.3	Simple rules	400
15.2.4	Permutations and combinations	401
15.2.5	General rules	402
15.2.6	Probability distributions	403
15.2.7	The concept of expectation	405
15.3	Worked examples	407
15.4	Further developments	409
15.4.1	The binomial distribution	409
15.4.2	The Poisson distribution	410
15.4.3	The Normal distribution	412
15.5	Further worked examples	420
15.6	Exercises	424
16	**Statistics 3: sampling, sampling distributions and hypothesis testing**	**430**
16.1	Scientific context	430
16.1.1	Estimating the value of the acceleration due to gravity	430
16.1.2	Alcohol in blood analysis	431
16.1.3	Iron alloy analysis	432
16.2	Mathematical developments	433
16.2.1	Definitions	433
16.2.2	The distribution of sample means	434
16.2.3	The distribution of the difference in two sample means	435
16.2.4	Best estimates	436
16.2.5	Sampling from a Normal distribution	439
16.2.6	The t-distribution	441
16.3	Worked examples	444
16.4	Further developments	447
16.4.1	The F-distribution	447
16.4.2	Elements of hypothesis testing	448
16.4.3	The t-test	453
16.4.4	The F-test	456
16.5	Further worked examples	458
16.6	Exercises	461
17	**Partial differentiation 1: introduction**	**466**
17.1	Scientific context	466
17.1.1	Van der Waals equation of state	466
17.1.2	A vibrating string	468
17.1.3	Potential fields	470
17.2	Mathematical developments	471
17.2.1	Functions of several variables	471
17.2.2	Definition of a partial derivative	472
17.2.3	Higher partial derivatives	476
17.2.4	Notation	477
17.2.5	Rules of partial differentiation	477
17.3	Worked examples	478

17.4	Further developments	482
17.4.1	Total changes	482
17.4.2	Error analysis	482
17.4.3	Differentials	484
17.4.4	The chain rule	487
17.5	Further worked examples	488
17.6	Exercises	493
18	**Partial differentiation 2: stationary points**	**498**
18.1	Scientific context	498
18.1.1	Positions of equilibrium	498
18.1.2	Optimum conditions	500
18.1.3	Making a clock	501
18.2	Mathematical developments	502
18.2.1	Taylor polynomials of second order	502
18.2.2	Stationary points	505
18.3	Worked examples	510
18.4	Further developments	510
18.4.1	Least squares analysis	513
18.5	Further worked examples	517
18.6	Exercises	521
	Answers to the exercises	525
	Index	543

PREFACE

The contents of this book consist of the basic mathematics taught in nearly all first-year service courses in mathematics for science and engineering students. It is based on the experience of the three authors of teaching such material for many years.

There is a growing awareness that we must not teach mathematics in isolation from its applications. In teaching mathematics to scientists, technologists and engineers, there is plenty of opportunity to provide applications as part of the syllabus and teaching approach. There are few textbooks which recognise this. One of the aims of writing this text has been to encourage the teaching of mathematics through its applications in science.

The importance of teaching mathematics through its applications is reflected in the format of this book. Each chapter starts with two or three examples setting the new techniques to be introduced in the context of the scientific world. The aim here is to answer the often posed question from science students: 'why do we need to learn this mathematics?'

Sections 2 and 4 of each chapter contain the basic mathematical techniques of each chapter, and sections 3 and 5 give worked examples showing the use of the new techniques introduced in the chapter.

Some of the material in the book will have been acquired by most students before entering higher education. For example, in the United Kingdom about half of the contents of the book appears listed in the syllabus of many GCE Advanced level and corresponding BTEC level 3 examinations. However, because of the varied success of students going on to higher education, we have taken the view that it is no bad thing to give a solid revision of the basic areas of mathematics on which the further ideas in universities and polytechnics are built. Students with a good A level or BTEC level 3 background should find the techniques of sections 2 and 4 of most chapters easy going but we hope that this book will be of value to them in showing the applications of mathematics to science and engineering. Our aim has been to provide a solid foundation on which to build in a later work.

The first five chapters of the book review the basic properties of functions. It is likely that some students will have met these ideas before, so that it will be necessary for them to dip into the relevant sections as a refresher. For other students all the material may be required.

Chapters 6–11 cover differentiation and integration, leading into two chapters on ordinary differential equations. The application of calculus in science is stressed throughout and the authors have been careful not to teach calculus for its own sake.

Statistics forms a major chunk of service courses in science and there are three chapters (14–16) introducing students to basic statistical ideas. Finally the book has two chapters on partial differentiation.

Although not mentioned explicitly, numerical methods are introduced in the progress of the book. For example, Chapter 1 has the bisection method for solving equations and Chapter 8 has Newton's method. This integration of the numerical and analytical methods reflects the authors' belief that numerical work should not be taught as a separate subject in isolation of the analytical work. Throughout the reading of the book it might prove useful to have a calculator handy.

Throughout the text we have also written simple illustrative programs. To cater for students' different tastes in high level languages we have deliberately written the programs in pseudocode, using the accepted constructs of structured programming. There should be no difficulty in implementing these programs in BASIC, FORTRAN, PASCAL or other high level languages.

The treatment of the mathematical analysis is not overtly rigorous although we have tried not to be sloppy in the presentation. Where appropriate we have avoided complicated proofs but have indicated how important results can be justified. The guiding principle has been to introduce the mathematics from a physical point of view deliberately avoiding strictly mathematical questions.

For the teacher one aim of this text is to provide a rich and varied source of applications to use in the classroom when teaching mathematics to scientists.

For the student the authors hope to have produced a textbook providing the opportunity for self-learning and an insight into the importance of learning and applying mathematics in their chosen major subject.

We are extremely grateful to Maggie Bedingham for her skilful and speedy typing of several of the chapters. We would also like to acknowledge Sheffield City Polytechnic for permission to use some past examination questions and to the Plymouth teaching staff whose exercise sheets have provided the basis of some of the exercises. Finally, we would like to thank the editorial staff of the Cambridge University Press for their patience and understanding in waiting for the manuscript.

John Berry
Allan Norcliffe
Stephen Humble

1

Basic functions

Contents: 1.1 scientific context – 1.1.1 loading a steel wire; 1.1.2. motion of a ball under gravity; 1.2 mathematical developments – 1.2.1 some basic terms; 1.2.2 polynomials of degree one: linear functions; 1.2.3 polynomials of degree two: quadratic functions; 1.2.4 complex numbers; 1.2.5 polynomials of degree *n*; 1.2.6 roots and factors of polynomials; 1.3 worked examples; 1.4 further developments – 1.4.1 linear interpolation and extrapolation; 1.4.2 approximate roots of polynomials; 1.5 further worked examples; 1.6 Exercises.

1.1 Scientific context

Many scientific laws are often expressed as relations between two or more physical quantities. In general these laws are obtained in one of two ways. Either the results of experiment are used directly to formulate empirical laws, or existing scientific knowledge is used, often together with mathematics, to arrive at new theories which can then be validated later by experiment. In formulating scientific laws we attempt to find a formula between the symbols representing the physical quantities of interest. Sometimes this is not possible and the relationship has to be expressed in the form of a table of values or a graph, for example.

If two quantities are related so that the value of one of them is uniquely determined when the other is known, then we say that there is a *functional relationship* between the variables. In these opening chapters we consider the basic mathematical *functions* which occur in science.

1.1.1 Example 1: loading a steel wire. Table 1.1 shows the results of an experiment to investigate how the length of a piece of mild steel wire changes when weights are attached to it. The unextended length of the wire is 2 m and its mean diameter is 1 mm.

It is clear that *l* increases with *W* but the nature of the relationship between the quantities *W* and *l* is more easily seen if we plot the points on a graph. As shown in Fig. 1.1, over the range of values of *W* considered, each point either lies on a straight line or is very close to it. Allowing for small random effects we may conclude that a simple straight-line or *linear* relationship exists between *W* and *l*.

Question: Is this to be expected?
   Answer: This experiment is a simple illustration of Hooke's law which states that for an elastic body (the wire in this case) the extension produced is proportional to the stretching force. We should therefore expect to obtain such a linear relationship.

Table 1.1. *Results of experiment 1: length of a steel wire for various loads*

Weight (W/newton)	Length (l/metre)
20	2.000 42
40	2.000 86
60	2.001 32
80	2.001 76
100	2.002 19

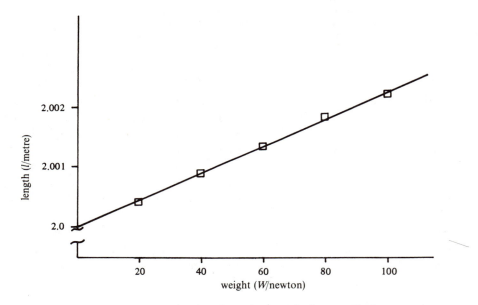

Fig. 1.1. Graph showing length against weight for a steel wire.

This experiment is typical of many experiments carried out in science in that we have ended up with a straight-line relationship. Not all experiments culminate in such relationships as our next example shows.

1.1.2 Example 2: motion of a ball under gravity. Suppose a ball is thrown upwards and a long-exposure photograph is taken of the ball under stroboscopic lighting as shown in Fig. 1.2. The resulting multiple images of the ball show its position every tenth of a second, which is the period of the stroboscopic light. Hence, by estimating the height of the ball from the position of its images on the photograph, we can make a plot of height h above the projection level against time t, as indicated by the squares in Fig. 1.3. The graph of h against t is clearly not a straight line.

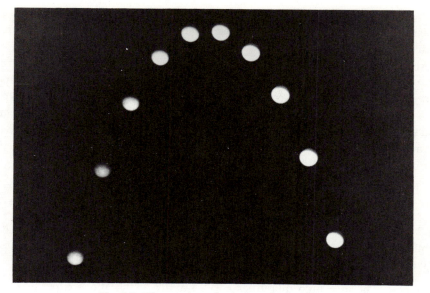

Fig. 1.2. The motion of a ball under gravity.

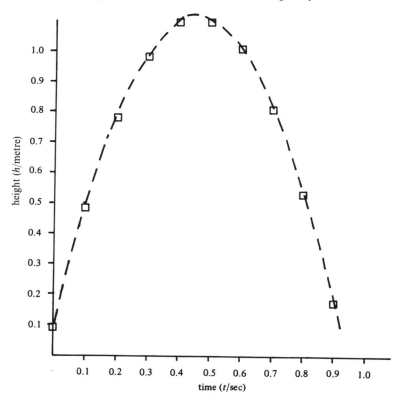

Fig. 1.3. Graph showing height of the ball against time.

Although initially the value of h increases as time increases, there comes a time when the ball stops rising and starts to fall and hence the values of h now decrease with time.

Although we have only measured h every tenth of a second it would seem reasonable to 'join up' the measured points in Fig. 1.3 by means of some smooth curve, as indicated. If air resistance was completely negligible, such a curve would be a parabola. The mathematics of parabolas, like that of straight lines, is simple and straightforward. Indeed, the line and the parabola that emerge from these two examples, whilst at first sight appearing totally different, are in fact related; they are both simple examples of more general functions called *polynomials*.

1.2 Mathematical developments

1.2.1 Some basic terms. Before looking at polynomials in any detail it will be helpful to first define several terms that we shall be using repeatedly when discussing polynomials and functions in general.

Variables. The measured quantities l, W, h and t in the above experiments are called *variables* because their values vary in the experiment. In each experiment we can choose a value of one variable (W or t), but having done this the corresponding value of the other variable (l or h respectively) is then determined. The chosen variable is usually called the *independent* variable, and the other variable is called the *dependent* variable because its value depends on the value chosen for the independent variable. In the first experiment, for example, we can identify the weight W as the independent variable since we are free to choose its value, at least within certain limits, while the resulting length of the wire l is clearly the dependent variable.

Functions. A *function* is a relationship between independent and dependent variables. The only restriction on such a relationship being called a function is that, for each possible value of the independent variable, there corresponds at most one value of the dependent variable.

In the first experiment we see from Fig. 1.1 that the length of the wire depends on the weight. Thus l is a function of W. We write 'l is a function of W' in mathematical shorthand as follows

$$l = f(W) \tag{1}$$

W, of course, cannot take on all values. In practice W can only be greater than zero or less than some maximum value W_{max} at which the wire breaks. We define the possible set of values of the independent variable to be the *domain* of the function, while the corresponding set of values of the dependent variable is called the *range* of the function.

When we perform an experiment, such as the one with the wire and weights, we are trying to identify two things. Firstly, we are trying to identify whether a functional relationship actually exists between the variables of interest that can be written as in (1) and, secondly, we are trying to identify if this relationship can be adequately represented, within the accuracy of the experiment, by some simple and easily identifiable function. In other words we are trying to see if we can *model* the relationship between the variables by some simple mathematical function.

Notation. Frequently, when there is no other obvious letter or variable name to use, the independent variable is denoted by the letter x, while the dependent variable is denoted by y. Thus

$$y = f(x)$$

is the shorthand way of writing 'y is a function of x', for x defined on some set of values. In some ways, then, the function f can be thought of as a devilish machine, as illustrated in Fig. 1.4, which takes an x value from this set, namely the domain, and turns it into a corresponding y value belonging to the range set. Having said that the independent variable in many cases is often represented by the single letter x, it is worth pointing out that there are functions of several independent variables. We will consider such functions later, but to begin with we will concentrate solely on those functions involving only one independent variable which are frequently used as models in science.

Polynomial functions. A polynomial function is defined to be a function of the form

$$f(x) = ax^n + bx^{n-1} + cx^{n-2} + \cdots + px + q \tag{2}$$

where $a, b, c, \ldots, p, q$ are constants and x is defined on some set of values. In this

Fig. 1.4. A 'devilish' function machine.

definition it is assumed that all powers of x that appear are non-negative integers. Since x^n is the highest power of x appearing in the expression we say that this is a polynomial of *degree n*.

We now start our study of polynomial functions by looking at polynomials of degree 1 and degree 2.

1.2.2 Polynomials of degree one: linear functions. In (2), when $n = 1$, we have

$$f(x) = ax + b \qquad (3)$$

If we draw a graph of $y = f(x)$ in this case for the set of x values for which it is defined, we obtain a straight line, as shown in Fig. 1.5, and $ax + b$ is called a *linear function* of x.

Any straight-line relationship connecting the independent variable x and the dependent variable y can always be written in the form $y = ax + b$ by choosing suitable values for the constants a and b. It is therefore pertinent to identify these constants in terms of the properties of the line.

To identify b, let us consider the value of the function when $x = 0$. From (3) we see that when $x = 0$, the corresponding y value is given by

$$y = f(0) = b$$

which is the *intercept* the straight line makes on the y axis, as seen in Fig. 1.5. Thus the constant b is the intercept. The intercept can, of course, take any value. If b is positive then the line intersects the y axis above the x axis, giving rise to a positive intercept. If the line crosses the y axis below the x axis, the value of b will be negative. A line through the origin (i.e. through the point $(0, 0)$), has zero intercept and consequently such a line has the equation $y = ax$.

The constant a is the value of the *slope* or *gradient* of the straight line. We can

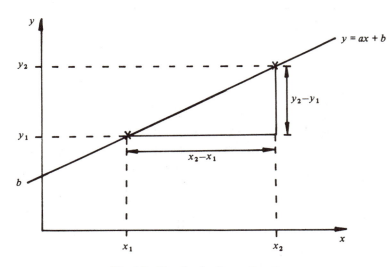

Fig. 1.5. Graph of a linear function.

easily show this to be the case as follows. The gradient is defined as the 'change in y value divided by the change in x value'. Thus, for example, when the value of x goes from x_1 to x_2, as shown in Fig. 1.5, y goes from the value $ax_1 + b$ to $ax_2 + b$, so that

$$\text{gradient} = (\text{slope} =)\frac{ax_2 + b - (ax_1 + b)}{x_2 - x_1} = \frac{a(x_2 - x_1)}{x_2 - x_1}$$

As x_1 and x_2 are assumed to be different values we can cancel the factor of $(x_2 - x_1)$ top and bottom to obtain the desired result that the gradient or slope is indeed equal to the constant a.

Example: What does the graph of a linear function look like if (i) a is negative and (ii) a is zero?
Solution: (i) If $a < 0$ the straight-line graph slopes downwards from left to right because as x increases, y decreases.
(ii) If $a = 0$ the relation $y = ax + b$ becomes simply $y = b$ for all values of x in the domain. This represents a straight line parallel to the x axis.

The roots of a linear function. In general the values of x which satisfy the equation

$$f(x) = 0 \tag{4}$$

for any function f are called the *roots* of that function.

If we were to plot a graph of $y = f(x)$ the roots would be those values of x where the graph crossed the x axis. For the linear function $f(x) = ax + b$, with $a \neq 0$, there is just one root, which is given by solving the equation

$$ax + b = 0$$

The root is therefore

$$x = -b/a$$

If we return to our first experiment, we can see that over the range of values of W, the linear relationship between l and W may be expressed as

$$l = \mu W + l_0$$

where l_0 is the unstretched length of the wire before any weights are added, i.e. $l = l_0$ when $W = 0$; and μ is the slope of the line in Fig. 1.1. The root of the linear function is therefore

$$W = -l_0/\mu$$

This corresponds to the value of W for which l is zero. Now $-l_0/\mu$ is clearly unphysical, corresponding to a negative value of W well outside the domain of validity of the function. It is tempting to identify this negative value as the compressive force necessary to shrink the length to zero. However, we must remember that Hooke's law only applies to an elastic body, and the wire would have ceased to be elastic well before the compressive force became so large. The linear function $f(W) = \mu W + l_0$ would certainly have ceased to be a reasonable model to represent the actual situation.

Before moving on to the next section it is worth noting that of all the functions we shall be studying in this book, linear functions are particularly important. They are easily recognised (i.e. as straight-line graphs) and their analysis is simple. In later chapters, particularly when discussing inverse functions, we shall consider how experimental data which are described by more complicated functional forms may sometimes be transformed to lie on simple straight-line graphs.

1.2.3 Polynomials of degree two: quadratic functions. In (2), when the degree of the polynomial is 2 we have

$$f(x) = ax^2 + bx + c \qquad (5)$$

This functional relationship, again defined over some set of x values, is characterised by three constants, a, b and c. As with the line, one of these constants, in this case c, is the intercept of the graph $y = ax^2 + bx + c$ on the y axis, i.e. the value of the function when $x = 0$. The other two constants, a and b, have no simple direct geometrical significance, although, of course, they do affect the shape of the graph.

In order to illustrate this fact let us consider the following examples:

(i) $y = x^2 - 4x + 3$;
(ii) $y = -2x^2 + 4x + \frac{5}{2}$;
(iii) $y = -x^2 + 2x - 1$;
(iv) $y = 3x^2 - x + 1$.

The graphs of these functions are shown in Fig.1.6. In each case the graph of the *quadratic function* has a typical symmetrical shape with only one turning point or 'hump', i.e. the points labelled A. This symmetrical shape is known as a *parabola*. The turning point corresponds to the minimum value of y when the constant a is positive, as in examples (i) and (iv), and corresponds to the maximum value of y when a is negative, as in examples (ii) and (iii). Thus the sign of constant a determines which way up the curve is. The reason for this is that for very large x values, the function is dominated by the term ax^2 and hence y has the same sign as a.

We see also that the graph will either cut the x axis in two places, x_1 and x_2, as in examples (i) and (ii), or will touch the x axis in one place, x_3, as in example (iii); or alternatively, will stay totally above the x axis, as in example (iv), or totally below the x axis (although we have not illustrated such a case). At the points x_1, x_2 or x_3, the value of y is zero and thus these values of x are the roots of the quadratic polynomial obtained by solving the equation

$$ax^2 + bx + c = 0 \qquad (6)$$

Roots of a quadratic. To find an expression for these roots let us add in and take out the quantity $b^2/4a$ so that (6) becomes

$$ax^2 + bx + \frac{b^2}{4a} - \frac{b^2}{4a} + c = 0$$

The first three terms $ax^2 + bx + (b^2/4a)$ can be written as $a[x + (b/2a)]^2$. (Check

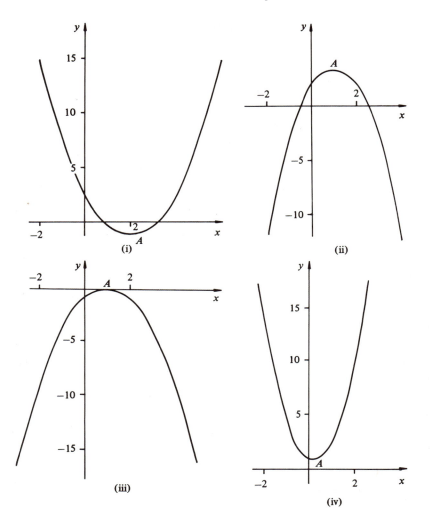

Fig. 1.6. Graphs of quadratic functions.

this for yourself.) Hence (6), after rearranging, can be written in the form

$$a\left(x + \frac{b}{2a}\right)^2 = \frac{b^2}{4a} - c$$

i.e. as

$$\left(x + \frac{b}{2a}\right)^2 = \frac{b^2 - 4ac}{4a^2}$$

where here we have taken the right-hand side over the common denominator $4a$, and divided both sides by a. Taking the square root of both sides (and remembering that if $X^2 = A^2$, then $X = +A$ or $-A$) we get

$$x + \frac{b}{2a} = \frac{\pm \sqrt{(b^2 - 4ac)}}{2a}$$

which we may rewrite in the well-known form as

$$x = \frac{-b \pm \sqrt{(b^2 - 4ac)}}{2a} \tag{7}$$

Provided $b^2 > 4ac$, so that we are taking the square root of a positive quantity, the formula in (7) gives rise to two distinct real roots x_1 and x_2 as indicated in Fig. 1.6. Therefore, in example (1), where $a = 1$, $b = -1$ and $c = 3$, we see that $x_1 = 1$ and $x_2 = 3$. Also, as in example (ii), where $a = -2$, $b = 4$, $c = \frac{5}{2}$, we see that $x_1 = -\frac{1}{2}$ and $x_2 = 2\frac{1}{2}$.

If $b^2 = 4ac$, as in example (iii), values given by (7) coincide, i.e.

$$x_1 = x_2 = -\frac{b}{2a}$$

so that the two roots converge to this single value and the graph just touches the x axis. In example (iii), with $a = -1$, $b = 2$, $c = -1$, $b^2 - 4ac = 0$ and $x_1 = x_2 = 1$.

If $b^2 < 4ac$ we end up trying to take the square root of a negative number in (7). This cannot be done in the real number system (try it on your calculator!) and hence we conclude that there are no actual values of x where the quadratic polynomial graph crosses or even touches the x axis. Example (iv) is an instance of this, since here with $a = 3$, $b = -1$, $c = 1$ we have that $b^2 - 4ac = -11$, which is clearly less than zero.

In connection with this latter case, if we introduce the symbol i to represent the square root of -1, i.e.

$$i = \sqrt{-1} \tag{8}$$

then we can *formally* write the two solutions of (7) when $b^2 < 4ac$ as

$$x = \frac{-b \pm \sqrt{(b^2 - 4ac)}}{2a}$$

$$= \frac{-b \pm \sqrt{[-1(4ac - b^2)]}}{2a}$$

$$= \frac{-b \pm i\sqrt{(4ac - b^2)}}{2a}$$

that is to say, the two roots x_1 and x_2 may be written as

$$x_1 = \alpha + i\beta, \quad x_2 = \alpha - i\beta \tag{9}$$

where

$$\alpha = -b/2a \quad \text{and} \quad \beta = \sqrt{(4ac - b^2)}/2a.$$

The two expressions involving i in (9) are examples of what are known as *complex numbers*. Complex numbers will be discussed in more detail in 1.2.4. At this stage it is appropriate simply to note that (9) is only a formal way of writing the roots of a quadratic function in the case $b^2 < 4ac$ when no *real* roots exist. The symbol i has no other meaning than that when multiplied by itself it produces the number -1.

Example: Evaluate the roots in terms of i in example (iv) where $a = 3$, $b = -1$ and $c = 1$.
Solution: The value of α will be $\frac{-(-1)}{2(3)}$ i.e. $\frac{1}{6}$. The value of β is $\sqrt{[4(3)(1)-(-1)^2]/2(3)} = \sqrt{\frac{11}{6}}$.
      The two complex number roots are therefore $x_1 = \frac{1}{6} + i\sqrt{\frac{11}{6}}$ and $x_2 = \frac{1}{6} - i\sqrt{\frac{11}{6}}$.

Turning points of a quadratic polynomial. Returning to the graph of the quadratic function, another value of x which is of interest is x_m, the value of x for which the function has its maximum or minimum value, as illustrated in Fig. 1.6. It may be seen by looking at the graph of the quadratic function that x_m lies midway between the two roots when they exist. From the expression for the two roots in (7) it follows that

$$x_m = -\frac{b}{2a} \qquad (10)$$

The corresponding value for y is thus

$$y_m = \frac{4ac - b^2}{4a}$$

The expressions in (7) and (10) are both important in identifying the shape of the graph of a quadratic function for particular values of the constants a, b and c. To illustrate this consider the second experiment in which the resulting graph is very similar to the quadratic curve shown in Fig. 1.6, example (ii). In fact it follows from Newton's laws of motion that, neglecting air resistance, we may write h as a quadratic polynomial of time t, namely

$$h = -\tfrac{1}{2}gt^2 + ut + h_0 \quad \text{(for } t \geqslant 0) \qquad (11)$$

where u is the initial vertical speed of the ball, g is the acceleration due to gravity (approximately $9.81 \, \text{m s}^{-2}$), and h is the height of the ball above the ground ($= h_0$ initially, i.e. at time $t = 0$). From what has been said it follows that the two roots of this quadratic function are

$$t = \frac{-u \pm \sqrt{(u^2 + 2gh_0)}}{-g}$$

and the maximum height is reached after a time $t_m = u/g$. We can also usefully note that if $h_0 = 0$, then the two roots are 0 and $2u/g$.

Example: What is the maximum height of the ball in our second experiment, and what is the time taken to reach this height, if $h_0 = 0.087$ m and $u = 4.48 \, \text{m s}^{-1}$?
Solution: From (10) the maximum of the quadratic function $h(t) = -\tfrac{1}{2}gt^2 + ut + h_0$ is given by $t_m = u/g = 4.48/9.81 = 0.46$ s. The maximum height is then worked out as the value of h at $t = 0.46$, which is 1.1 m.

We have deliberately spent some time on the quadratic function even though many of its properties are well known. A firm understanding of the shape of the graph of the quadratic function, as well as the ability to obtain the roots of a quadratic are essential to understanding a wide range of problems in the physical sciences. These include not only simple kinematical problems involving a constant gravitational force, as illustrated by our experiment 2, but also problems of oscillations, vibrations and other periodic motions as well as many problems in optimisation.

The discussion has also allowed us to introduce the idea of a complex number which is the subject of our next section.

1.2.4 Complex numbers. Quite simply, complex numbers are numbers that involve the square root of -1. They are not real numbers in the normal sense, but nevertheless, as we shall see later, they play an important role in the modelling of real systems in science, particularly those that have a tendency to vibrate or oscillate. Complex numbers are therefore important in electronics and in the study of wave motion and vibrations.

A complex number is usually called z and is written in the form

$$z = a + ib \tag{12}$$

where a and b are ordinary real numbers. The quantity a is known as the *real part* of z and b as the *imaginary part* of z. Thus in (9), if we take the complex number x_1, we see that the real part is α and the imaginary part is β. Note that the imaginary part of the complex number x_2 is $-\beta$.

Complex numbers, like ordinary numbers, can be used in calculations; that is to say they can be added, subtracted, multiplied, divided and so on. They therefore possess what is called an algebra, and the rules of this algebra are straightforward as we shall now see.

Addition and subtraction of complex numbers. If $z_1 = a_1 + ib_1$ and $z_2 = a_2 + ib_2$ are two complex numbers, then the complex numbers $z_1 \pm z_2$ are defined by

$$z_1 \pm z_2 = (a_1 \pm a_2) + i(b_1 \pm b_2)$$

Example: If $z_1 = 5 + 4i$ and $z_2 = 6 - 7i$ obtain the complex numbers $z_1 + z_2$ and $z_2 - z_1$.
Solution: The complex number $z_1 + z_2$ will be given by $(5 + 6) + i(4 - 7)$ which is $11 - 3i$.
　　　　The complex number $z_2 - z_1$ is $(6 - 5) + i(-7 - 4)$ which is $1 - 11i$.

Multiplication of complex numbers. The product $z_1 z_2$ is worked out exactly as we would expect. Thus if $z_1 = a_2 + ib_1$ and $z_2 = a_2 + ib_2$, their product is given by

$$
\begin{aligned}
z_1 z_2 &= (a_1 + ib_1)(a_2 + ib_2) \\
&= a_1 a_2 + ia_1 b_2 + ia_2 b_1 + i^2 b_1 b_2 \\
&= a_1 a_2 + i(a_1 b_2 + a_2 b_1) - b_1 b_2 \\
&= (a_1 a_2 - b_1 b_2) + i(a_1 b_2 + a_2 b_1)
\end{aligned}
$$

In general when multiplying complex numbers it is not necessary to remember this formula. All that need be remembered is how to multiply out brackets and that $i^2 = -1$.

Example: Obtain $(5 + 4i) \times (6 - 7i)$.
Solution: Multiplying these complex numbers together directly gives

$$(5 + 4i) \times (6 - 7i) = 30 - 35i + 24i - 28i^2$$
$$= 30 - 11i + 28$$
$$= 58 - 11i$$

Division of complex numbers. To do division we make use of the *complex conjugate* of a complex number. if $z = a + ib$ then the complex conjugate of z, written as $\bar{z}$, is $a - ib$. Note that the two roots of a quadratic polynomial, when $b^2 < 4ac$, are complex conjugates of one another being $\alpha \pm i\beta$. To work out z_1/z_2 and write the result in complex-number form we perform the division by actually evaluating $z_1\bar{z}_2/(z_2\bar{z}_2)$.

The wisdom of using the complex conjugate lies in the fact that $z\bar{z}$ always turns out to be a positive real number. In fact if $z = a + ib$ then $z\bar{z}$ will be $a^2 + b^2$. Since we know how to divide by a real number, we can form z_1/z_2 by multiplying the two complex numbers z_1 and $\bar{z}_2$ together and dividing the resulting product by the real number $z_2\bar{z}_2$.

Example: Obtain $(2 + 2i)/(1 + 3i)$.
Solution: We carry out the division by multiplying top and bottom by the complex conjugate of $1 + 3i$, which is $1 - 3i$.
Thus

$$\frac{2 + 2i}{1 + 3i} = \frac{(2 + 2i)(1 - 3i)}{(1 + 3i)(1 - 3i)}$$
$$= \frac{2 - 6i + 2i - 6i^2}{1^2 + 3^2}$$
$$= \frac{8 - 4i}{10}$$
$$= \frac{8}{10} - \frac{4}{10}i$$

Equality of complex numbers. With the above three rules it is possible to carry out calculations and analytical work with complex numbers. One very important property possessed by complex number expressions that are equal is that their respective real parts and imaginary parts are the same. Thus if $z_1 = a_1 + ib_1$ is equal to $z_2 = a_2 + ib_2$ then

$$a_1 = a_2 \quad \text{and} \quad b_1 = b_2$$

This additional rule is extremely important when complex numbers are part of some mathematical analysis.

Further properties of complex numbers will be developed later on, as and when they are needed. Let us now return to studying polynomials in general.

1.2.5 Polynomials of degree *n*. Consider the *n*th degree polynomial as given in (2), i.e.

$$f(x) = ax^n + bx^{n-1} + cx^{n-2} + \cdots + px + q$$

Polynomials which are of degree n ($n > 2$) occur as mathematical models of physical phenomena, as do linear and quadratic functions. For instance, cubic polynomials ($n = 3$) are used in the description of volumes; quartic polynomials ($n = 4$) arise in the modelling of black body radiation, in viscous flow of a fluid through a pipe and in the deformation of a heavy beam under its own weight; general polynomials describe the charge distributions around hydrogenic ions and give the angular distribution of particle-scattering cross-sections, etc. One important application of polynomials is in numerical analysis work where they are often used as simple expressions for approximating more complicated functions.

Example: What is the value of the cubic polynomial $f(x) = 2x^3 + 6x^2 + 3x + 17$ when $x = 2$,
and how many multiplication and addition operations are needed to calculate this
value?

Answer: Calculating the function value term by term we have

$$f(2) = 2 \times 2 \times 2 \times 2 + 6 \times 2 \times 2 + 3 \times 2 + 17 = 63$$

We see that the number of multiplications is six and the number of additions is
three.

Without thinking, most people would probably calculate the value of a polynomial term by term, adding up the result at the end, and at first sight this seems the obvious way to do the calculation. However, it is not the best way. Besides requiring intermediate values to be stored in some way, this method is also time consuming if n is large. It can be shown that in general it requires $\frac{1}{2}n(n+1)$ mutiplications and n additions, that is a total of $\frac{1}{2}n(n+3)$ operations, to calculate the value of an nth-degree polynomial.

What we should do in our calculations is to use the so-called *nested form* for a polynomial. The nested form for the polynomial in (2) is

$$f(x) = (\cdots((ax + b)x + c)x + \cdots + p)x + q$$

To work out the value we start by evaluating the innermost bracket first and work outwards. Thus the order of computation is

   (i) multiply a by x;
   (ii) add b to the result of (i) (which is ax);
   (iii) multiply the result of (ii) (which is $ax + b$) by x;
   (iv) add c to the result of (iii) (which is $ax^2 + bx$);
   (v) multiply the result of (iv) by x;
   (vi) add p to the result of (v);

(vii) multiply the result of (vi) by x;

$\vdots$

and so on.

The advantages of this nested multiplication over the term-by-term evaluation are that no stores are required and on a computer a simple looping procedure may be adopted. Also in general only n multiplications and n additions are required, i.e. a total of $2n$ operations to evaluate an nth degree polynomial.

Example: Write that part of a program that demonstrates this looping procedure and the fact that only n multiplications and n additions are required.

Solution: For convenience assume our nth-order polynomial is given by the function

$$f(x) = a_n x^n + a_{n-1} x^{n-1} + \cdots + a_1 x + a_0$$

Assuming that the constants $a_n, a_{n-1}, \ldots, a_0$ and the value of x have been input, the program, written in pseudocode, to obtain the value of the polynomial is as follows

begin
set polyvalue **to** a_n
for $i = 1$ **to** n **step** 1 **do**
    set polyvalue **to** polyvalue $* x + a_{n-i}$
endfor
print polyvalue
stop

The looping procedure is a simple *for* loop. Each time through the loop one multiplication and one addition is needed and since i ranges from 1 to n in steps of 1 there will be a total of n multiplications and n additions needed to obtain the final value of the polynomial.

Example: Evaluate $f(x) = 2x^3 + 6x^2 + 3x + 17$ for $x = 2$ and verify that only three multiplication and three addition operations are required.

Solution: Our polynomial in nested form is

$$f(x) = ((2x + 6)x + 3)x + 17$$

With $x = 2, f(2) = ((2 \times 2 + 6) \times 2 + 3) \times 2 + 17$, which is 63 as before. The number of multiplications is three and the number of additions is similarly three.

1.2.6 Roots and factors of polynomials. As we saw above, the graph of the quadratic polynomial $f(x) = ax^2 + bx + c$ cuts the x axis at the two points

$$x_1 = \frac{-b + \sqrt{(b^2 - 4ac)}}{2a}, \quad x_2 = \frac{-b - \sqrt{(b^2 - 4ac)}}{2a}$$

provided $b^2 > 4ac$. That is to say, x_1 and x_2 are the two roots of the polynomial, satisfying the equation $f(x) = 0$. It therefore follows that this quadratic function may be written in the form

$$f(x) = a(x - x_1)(x - x_2)$$

where it is now obvious that $f(x) = 0$ when $x = x_1$ or $x = x_2$ since one or other of the two brackets will be zero. The terms $(x - x_1)$, $(x - x_2)$ are called the *linear factors* of the function: 'linear' because they involve only the first power of x.

If $b^2 < 4ac$ the two roots are now the complex numbers $\alpha \pm i\beta$ given in (9) and $f(x)$ can be written as

$$f(x) = a(x - (\alpha + i\beta))(x - (\alpha - i\beta))$$

Since each linear factor has a single power of x there will be exactly n linear factors of an nth-degree polynomial provided that, as above, we may introduce factors containing complex roots if required. Note that, provided the coefficients of the polynomials are real, such complex roots always occur in 'complex conjugate' pairs, i.e.

$$x_1 = \alpha + i\beta \quad \text{and} \quad x_2 = \alpha - i\beta$$

so that

$$\begin{aligned}(x - x_1)(x - x_2) &= (x - \alpha - i\beta)(x - \alpha + i\beta) \\ &= (x - \alpha)^2 + \beta^2 \\ &= x - 2\alpha x + \alpha^2 + \beta^2 \end{aligned} \tag{13}$$

which is a quadratic polynomial with *real* coefficients.

If, however, we restrict ourselves to the real number system it follows that any polynomial function $f(x)$ may be written as a product of m linear factors, where $m \leqslant n$ (each factor corresponding to a real root of the polynomial) and a remaining multiplicative polynomial of degree $n - m$, i.e.

$$f(x) = a(x - x_1)(x - x_2) \cdots (x - x_m)r(x)$$

In turn this remaining multiplicative polynomial $r(x)$ may be written as a product of quadratic polynomials, each one like (13) corresponding to two complex conjugate roots of the original polynomial.

Sometimes it is necessary to factorise a polynomial function into its linear (and quadratic) factors. As we have seen, for a quadratic polynomial, this may be done using the quadratic 'formula' of (7) to identify the roots of the corresponding quadratic, and hence obtain the factors of the polynomial. For cubic and quartic polynomials there are corresponding formulae to determine the roots; however, these formulae are much more complicated than (7). For polynomials of degree $n > 4$ there is a theorem which shows that it is impossible, in general, to identify the roots by means of formulae involving only square roots, cube roots and in general nth roots of terms involving the coefficients of the polynomial. For these reasons, therefore, we must look for other methods to determine the roots and hence the factors of a polynomial function.

The most obvious method is the well-worn 'trial-and-error' approach in which we choose values for x and see if any values give a zero value for the function. Obviously in this approach we are restricted to fairly simple values of x. As an example, to factorise the polynomial $f(x) = 4x^3 - 24x^2 + 45x - 25$ we might use trial values of $x = 0, \pm 1, \pm 2$, say, to produce the following results:

x	$f(x)$
0	-25
1	0
-1	-98
2	1
-2	-243

Clearly $x = 1$ is a solution of the equation $4x^3 - 24x^2 + 45x - 25 = 0$ and therefore $(x - 1)$ is a factor of the polynomial. Having found this one factor we can now write

$$f(x) = (x - 1)r(x)$$

where $r(x)$ must be a polynomial of degree one less than the original, i.e. $3 - 1 = 2$, such that

$$r(x) = ax^2 + bx + c$$

Hence $f(x) = (x - 1)(ax^2 + bx + c) = ax^3 + (b - a)x^2 + (c - b)x - c$. For this to correspond to the polynomial $4x^3 - 24x^2 + 45x - 25$ for *all* values of x we must have $a = 4$, $(b - a) = -24$, $(c - b) = 45$, $-c = -25$. Since $r(x)$ is a quadratic polynomial we may use the quadratic formula (7) to obtain its two roots as $x_2 = 2.5$ and $x_3 = 2.5$. Thus we may write

$$r(x) = 4(x - 2.5)(x - 2.5)$$

and hence

$$f(x) = 4(x - 1)(x - 2.5)(x - 2.5)$$

When, as in this case, $f(x)$ has two equal roots, $x_2 = x_3 = 2.5$, we say this is a *repeated root* of the polynomial and $(x - 2.5)$ is a *repeated factor*. Of course, in real life, roots of polynomials may not be simple integers or fractions and the trial-and-error method adopted here may not be successful. In such cases there are *numerical methods* to evaluate the roots of the polynomial to any required level of accuracy. These will be discussed in 1.4.2 and later in Chapter 8.

1.3 Worked examples

Example 1. Hooke's law for the relationship between the length l and load W can be stated in the form

$$l = \mu W + l_0$$

For the experimental results shown in Table 1.1 find suitable values for μ and l_0 and then use the model to calculate the extension produced by a weight of 75 N.

Solution. The model $l = \mu W + l_0$ is a linear relationship and a plot of l against W is therefore a straight line of slope μ and intercept l_0.

The data in table 1.1 when plotted out give a straight line as shown in Fig.1.1. This line cuts the l axis at the point $l = 2.000\,00$ m and thus l_0 in our model is $2.000\,00$ m. This value matches up very nicely with the information we had about the original unextended length of the wire in 1.1.1, namely that it was 2 m.

Since μ is to be identified with the slope of the graph shown in Fig. 1.1, and noting that the point (100, 2.002 19) is also on the line, we have

$$\mu = \frac{2.002\,19 - 2.000\,00}{100 - 0} = 0.000\,021\,9\,\text{m N}^{-1}$$

The model relating l and W is thus

$$l = 0.000\,021\,9\,W + 2.000\,00$$

so that when $W = 75$ N, the value of l is 2.001 64 m. Since the original unextended length is 2 m it follows that the extension produced by a weight of 75 N will be 0.001 64 m, i.e. 1.64 mm.

Example 2. The thermo-emf E (measured in micro volts) corresponding to the temperature T (measured in °C) for a copper–iron thermocouple is modelled by a quadratic function of the form

$$E = aT^2 + bT$$

With $a = -0.0192$ and $b = 6.88$ the model is found to work well for temperatures between about 250 and 400 °C.

If the voltage indicated, when the thermocouple is used within this range, is $300\,\mu$V determine the temperature that is being measured. Are there likely to be any sensitivity problems when using the instrument over this range?

Solution. Over the range of temperatures given the model is

$$E = -0.0192\,T^2 + 6.88\,T$$

Thus E will be 300 when

$$-0.0192T^2 + 6.88T - 300 = 0$$

Using the quadratic formula the values of T satisfying this equation are

$$T = \frac{-6.880 \pm \sqrt{[6.88^2 - 4(-0.0192)(-300)]}}{2(-0.0192)}$$

$$= \frac{-6.880 \pm 4.929}{-0.0384}$$

$$= 307.5 \text{ or } 50.8$$

In the range of temperatures over which the thermocouple is used the temperature has to be 307.5 °C.

With regard to sensitivity problems, the thermocouple will be least sensitive when it is used to measure temperatures that are close to the minimum of the function $E(t) = -0.0192T^2 + 6.88T$. Close to the minimum of a quadratic function (see Fig. 1.6, for example) relatively large changes in the independent variable produce only small changes in the function. The minimum of our function will occur when

$$T_m = -b/2a = -6.88/-0.0384 = 179.2$$

i.e. at temperatures close to $180\,°C$. This is not inside the range over which the thermocouple is used, so sensitivity problems should not arise. For temperatures above $250\,°C$ it will be seen that E values begin to change quite rapidly.

1.4 Further developments

1.4.1 Linear interpolation and extrapolation. Let us return again to the wire loading experiment. Table 1.1 shows the results of the experiment for loads of $20\text{–}100\,\text{N}$ in steps of $20\,\text{N}$ and for convenience we have reproduced these results as points P_1, P_2, P_3, P_4, P_5 on a plot of l against W, as shown in Fig. 1.7.

Suppose that we want to predict the value of l for a load between $40\,\text{N}$ and $60\,\text{N}$; but suppose the only data we have to work with are the two points P_2 and P_3. It is reasonable to assume that the corresponding value of l will be between $2.000\,86$ and $2.001\,32\,\text{m}$. In fact since we know Hooke's law to be true we could join the points P_2 and P_3 with a straight line and read off the value of l corresponding to the value of W which lies on this line. This method of prediction is called the method of *linear interpolation*.

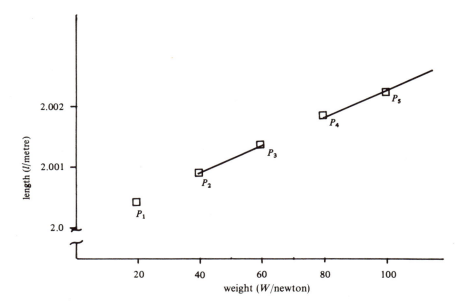

Fig. 1.7. Linear interpolation for finding the wire length given the weight.

Suppose now that the value of W in which we are interested lies outside of the given data, i.e. it is bigger than 100 N, but we do have the points P_4 and P_5 to work with. Again, because of Hooke's law, it may be reasonable to draw a straight line through the points P_4 and P_5 extending it to the value of W required and then read off the corresponding value of l. This method is called *linear extrapolation*.

Of course, using only two points does have its pitfalls as there could well be errors in the measured values of l and hence in the positions of the points. But off-setting this is the fact that a linear relationship does connect the variables involved. Ideally we should make predictions from more than just two data points. A more sophisticated method which allows for small inaccuracies in measurements and fits a 'best' straight line through N data points ($N \geqslant 2$) will be considered in Chapter 18.

Sometimes a set of data may lie on what appears to be a smooth curve as opposed to a straight line. In these circumstances it may still be adequate to use linear interpolation between two adjacent points in the data set. To illustrate this consider the results of experiment 2 which are given in Table 1.2. If we wish to estimate the height of the ball at, say, 0.25 s or at 0.275 s we could either fit the quadratic relationship $h = -\frac{1}{2}gt^2 + ut + h_0$ to these data or, more simply, but more approximately, use linear interpolation between appropriate adjacent points. To do this we might plot the two points $t = 0.2$ s, $h = 78.6$ cm and $t = 0.3$ s, $h = 99.0$ cm on a piece of graph paper, as shown in Fig. 1.8, and draw a straight line between them. From the graph we can then read off that at $t = 0.25$ s, $h = 88.8$ cm and at $t = 0.275$ s, $h = 93.9$ cm. We should note, however, that this

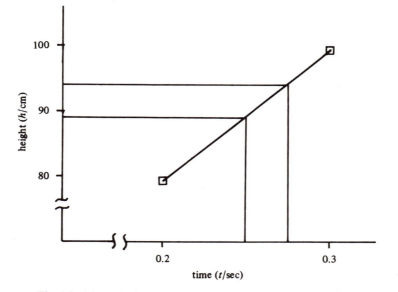

Fig. 1.8. Linear interpolation for finding the height of the ball.

Table 1.2. *Results of experiment 2: height of the ball at various times*

time t(s)	0.0	0.1	0.2	0.3	0.4	0.5	0.6	0.7	0.8	0.9
height of ball h(cm)	8.7	48.5	78.6	99.0	109.5	110.6	100.9	81.5	53.4	17.5

graphical method could be rather approximate, depending on the scale of the graph paper and how well the points have been plotted.

On the other hand, the assumption of a linear relationship between t and h in this *limited interval* does mean that we may *calculate* the interpolated values in a fairly straightforward manner. For example, since $t = 0.25$ s is half-way between $t = 0.2$ s and $t = 0.3$ s, the corresponding value of h must similarly be half-way between $h = 78.6$ cm and $h = 99.0$ cm, i.e. $h = 88.8$ cm. Also since $t = 0.275$ s is three-quarters of the way between $t = 0.2$ s and $t = 0.3$ s the corresponding value of h in this case must be three-quarters of the way from $h = 78.6$ cm to $h = 99.0$ cm, i.e. $h = 93.9$ cm. For more complicated intermediate values of t we shall need a mathematical formula to describe this fractional increase from one value of h to another. Such a formula is now obtained.

Interpolation and extrapolation formula. Let us suppose we are given two values x_1 and x_2 of some independent variable x, and two corresponding values y_1 and y_2 of the dependent variable y.

The *linear interpolation* value y_p of y corresponding to some intermediate value x_p of x can be worked out by noting that the distance $x_p - x_1$ is the fraction $(x_p - x_1)/(x_2 - x_1)$ of the distance $x_2 - x_1$. Hence the corresponding distance $y_p - y_1$ must be same fraction of the distance $y_2 - y_1$, i.e.

$$y_p - y_1 = \frac{x_p - x_1}{x_2 - x_1} \cdot (y_2 - y_1)$$

or

$$y_p = y_1 + \frac{x_p - x_1}{x_2 - x_1} \cdot (y_2 - y_1) \tag{14}$$

This same formula can be used for points x_p outside of the interval (x_1, x_2); i.e. it can also be used for *linear extrapolation*. In such cases the fraction $(x_p - x_1)/(x_2 - x_1)$ will take values greater than one (if $x_p > x_2$) or less than zero (if $x_p < x_1$).

Example: Using the data in Table 1.2 obtain the height of the ball at time $t = 0.77$ s by (i) linear interpolation, (ii) by linear extrapolation from the interval (0.6 s, 0.7 s) and (iii) by linear extrapolation from the interval (0.8 s, 0.9 s).

Solution: (i) Using (14) for the interval (0.7 s, 0.8 s) we have $y_1 = 81.5$, $y_2 = 53.4$, $x_1 = 0.7$, $x_2 = 0.8$ and $x_p = 0.77$. Thus

$$y_p = 81.5 + \frac{0.77 - 0.7}{0.8 - 0.7} \cdot (53.4 - 81.5)$$

$$= 61.83$$

The height of the ball is estimated to be 61.83 cm.

(ii) Here $y_1 = 100.9$, $y_2 = 81.5$, $x_1 = 0.6$, $x_2 = 0.7$ and $x_p = 0.77$. Thus

$$y_p = 100.9 + \frac{0.77 - 0.6}{0.7 - 0.6} \cdot (81.5 - 100.9)$$

$$= 67.92$$

The height of the ball is now estimated to be 67.92 cm.

(iii) Finally, we have $y_1 = 53.4$, $y_2 = 17.5$, $x_1 = 0.8$, $x_2 = 0.9$ and $x_p = 0.77$. Therefore

$$y_p = 53.4 + \frac{0.77 - 0.8}{0.9 - 0.8} \cdot (17.5 - 53.4)$$

$$= 64.17$$

The third estimate of the same height is now 64.17 cm.

When we use linear interpolation we are implicitly assuming that the function whose particular value we want to find is well approximated by a straight-line function over a restricted range of interest. This will be a reasonable assumption to make if the actual relationship between x and y is almost linear in this range. If the functional relationship between x and y is not linear it will always be more accurate to seek the nature of this relationship before interpolating. However, in many problems we may achieve reasonable accuracy with linear interpolation provided the data points are sufficiently close together.

With linear extrapolation, on the other hand, it is more likely that we will be in error using the linear approximation because of the added uncertainty of working outside the region of the data. Within the restricted range of interest the function

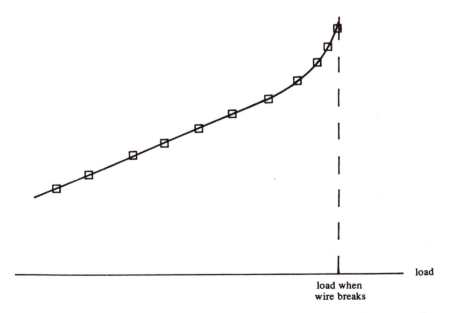

Fig. 1.9. Experiment shows that the graph becomes non-linear.

may well behave linearly (or almost so), but outside this range we cannot be sure that the function will still continue linearly as suggested. A good example of this is again provided by our first experiment in which a wire was stretched by adding weights to it. If we suppose that the functional relationship between the length l and the weight W is linear for all values of W, we can extend a line through the given points into a region for which no results have been obtained, i.e. for $W >$ 100 N, as indicated in Fig. 1.9. This implies that we can extend the wire to any length by adding a sufficiently large weight which, of course, is known not be the case. In fact, if we do continue the experiment the actual shape of the curve changes as shown in the diagram, and eventually the wire breaks. In conclusion, therefore, the method of linear interpolation must be used carefully and will often only give *approximate* values of y at intermediate values of x. Care is required, particularly in using linear extrapolation, and ideally we need some idea of the likely behaviour of the function beyond the range of given data values.

1.4.2 Approximate roots of polynomials. Frequently in science, given a functional relationship between variables of interest, such as $y = f(x)$, we wish to know the values of x corresponding to some particular value of y, e.g. $y = c$. That is to say, we need to solve the equation $f(x) = c$. This is equivalent to solving the equation

$$F(x) = f(x) - c = 0$$

i.e. to finding the roots of $F(x)$.

For example, in experiment 2, given that $h = -\frac{1}{2}gt^2 + ut + h_0$ with $h_0 = 0.087\,\text{m}$, $u = 4.480\,\text{m s}^{-1}$, $g = 9.81\,\text{m s}^{-2}$ we may wish to find the times when $h = 0.2\,\text{m}$. Thus we need to solve the equation

$$-4.905t^2 + 4.480t + 0.087 = 0.2$$

i.e. the equation

$$F(t) = -4.905t^2 + 4.480t - 0.113 = 0$$

This simple quadratic equation can be solved by means of formula (7). However, for more complicated functions we must either use the trial-and-error approach described in 1.2.6 or seek some other means. Below we describe two simple methods for the approximate solution of a polynomial equation. In fact it should be noted that the methods given in this section have a much wider applicability than just to polynomial functions. They will work for any function which is defined over a continuous range of x and whose graph is a *continuous* curve with no sudden 'jumps' or missing pieces. Functions that are discontinuous will be discussed later in Chapter 5.

The graphical method. In this method we simply calculate as many points of the function as possible, plot these points on graph paper, and then draw as accurately as possible a smooth curve through the points. We then read off from the graph the possible values of x where the curve crosses the x axis. For example, the graph

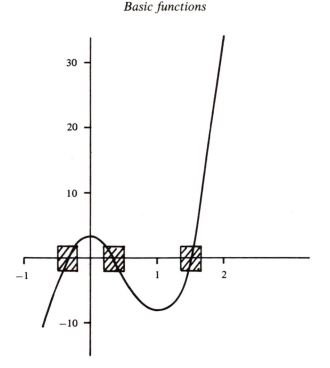

Fig. 1.10. The roots of the function $F(x) = 18x^3 - 27x^2 - 2x + 3$.

of $F(x) = 18x^3 - 27x^2 - 2x + 3$ is illustrated in Fig. 1.10. The zeros of this function are given approximately from this graph as -0.34, 0.34 and 1.5.

Clearly this graphical method has its disadvantages in that it depends on the accuracy of our drawing. We might improve the accuracy of our result, by one or two significant figures, by drawing an enlargement of the shaded regions shown in the diagram in which we expect that $F(x) = 0$. However, we really require not a graphical but a *numerical method* in which, by repeated application, we may find a root as accurately as we desire. Such a method is described below.

Method of interval bisection. If $F(x)$ is a polynomial function defined for all values of x between $x = a$ and $x = b$ and if $F(a)$ and $F(b)$ have opposite signs then, because $F(x)$ can be represented by a continuous curve, there must exist at least one solution of $F(x) = 0$ between $x = a$ and $x = b$. This is illustrated in Fig. 1.11 in the first graph. By drawing up a table of values for the function $F(x)$, provided $F(x)$ has an isolated real root, it is always possible to find an interval (a, b) such that $F(a)$ and $F(b)$ have opposite signs, i.e. either $F(a) < 0$, $F(b) > 0$ or else $F(a) > 0$, $F(b) < 0$.

Let us now investigate the sign of the function at a point mid-way between a and b, i.e. at $x = c = (a + b)/2$.

    (i) if $F(c) = 0$ (unlikely) then the root is $x = c$;

    (ii) if $F(c) > 0$ and $F(a) < 0$ then the root is between a and c as illustrated in the second graph of Fig. 1.11;

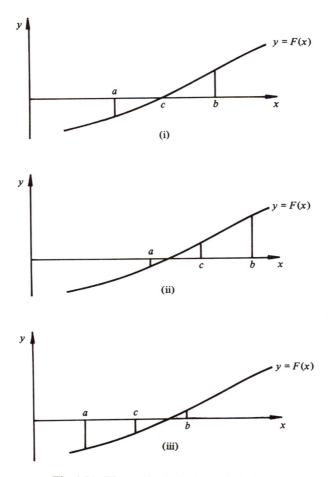

Fig. 1.11. The method of interval bisection.

(iii) if $F(c) < 0$ and $F(b) > 0$ then the root is between c and b as illustrated in the third graph of Fig. 1.11.

Hence either the root is $x = c$ or else we have reduced by one-half the width of the interval containing the root, i.e. from $(b - a)$ to either $(c - a)$ or $(b - c)$ depending on whether (ii) or (iii) holds. The process is now repeated using either a and c if (ii) holds or b and c if (iii) holds. Successive repetitions of this same process enable us to define smaller and smaller intervals in which the root must lie, until eventually we have defined an interval which is smaller than the required accuracy of the solution. This 'method of bisection' is an example of a process called *iteration*; that is, a method in which the same basic steps are repeated using data obtained from the previous step (or steps) to produce an even better approximation at each step.

Example: Correct to three decimal places, what is the value of the smaller positive root of the polynomial equation

$$F(x) = 18x^3 - 27x^2 - 2x + 3 = 0?$$

Solution: Firstly we must find an interval on the x axis which contains the smaller positive root. To do this we construct a table showing the values of the function at various values of x:

x	-1	0	1	2
$f(x)$	-40	3	-8	35

Clearly the three roots lie in the three intervals $(-1, 0)$, $(0, 1)$, $(1, 2)$. The smaller positive root therefore lies between 0 and 1. Using the bisection method as described we obtain the following table of results (it should be clear that the sign of $f(x)$ is important – not its value)

				Sign of	
a	b	c	$F(a)$	$F(b)$	$F(c)$
0	1	0.5	+	−	−
0	0.5	0.25	+	−	+
0.25	0.5	0.375	+	−	−
0.25	0.375	0.3125	+	−	+
0.3125	0.375	0.3475	+	−	−
0.3125	0.343 75	0.328 125	+	−	+
0.328 1 25	0.343 75	0.335 937 5	+	−	−
0.328 125	0.335 937 5	0.332 031 25	+	−	+
0.332 031 25	0.335 937 5	0.333 984 375	+	−	−
0.332 031 25	0.333 984 375	0.333 007 812 5	+	−	+
0.333 007 812 5	0.333 984 375	0.333 496 093 75	+	−	−

The root lies between 0.333 007 812 5 and 0.333 496 093 75 and to three decimal places these two values are identical, i.e. 0.333. Thus the smallest positive root to three decimal places is 0.333. In fact the exact value of this root is $\frac{1}{3}$.

This example shows that the method is rather slow and many iterations are needed unless $b - a$ is initially very small. However, the advantage of this method is that provided the function crosses the x axis at the root (and does not touch the x axis there) we can guarantee that it works. A program based on the method is easily written as our next example shows.

Example: Produce a program to solve $f(x) = 0$ accurate to some specified relative error e using the method of interval bisection.

Solution: Assuming initially that $f(x_1) < 0$ and $f(x_2) > 0$, then the following program, written in pseudocode, should provide the solution.

> **begin**
> **input** x_1, x_2, e
> **while** absolute value of $(1 - x_1/x_2) > e$ **do**
>   **set** x **to** $(x_1 + x_2)/2$
>   **if** $f(x) < 0$ **then**
>   **set** x_1 **to** x

```
        else
                set x₂ to x
        endif
endwhile
print x
stop
```

1.5 Further worked examples

Example 1. The following data show the results of an experiment involving the bending of a thin rod which is clamped so that its free length l varies from 0.30 m to 1.50 m and at the end of which is hung a 200-gm weight. The deflection y_{max} of the end of this rod is measured for each of the values of its length l with the following results:

l/m	0.30	0.50	0.70	0.90	1.10	1.30	1.50
y_{max}/m	0.001	0.006	0.015	0.032	0.059	0.097	0.148

(i) Estimate by means of linear interpolation and extrapolation the respective values of y_{max} when $l = 1.00$ m and $l = 1.60$ m.

(ii) If the relationship between y_{max} and l is in fact

$$y_{max} = cl^3$$

where $c = 0.044$ m^{-2}, test how good your linear interpolation and extrapolation values are.

Solution. Using the interpolation/extrapolation formula we can write

$$y_{max} = y_{max_1} + \frac{l - l_1}{l_2 - l_1}(y_{max_2} - y_{max_1})$$

where (l_1, y_{max_1}) and (l_2, y_{max_2}) are two adjacent readings from our table.

(i) For $l = 1.00$ m we therefore choose $l_1 = 0.90$ m and $l_2 = 1.10$ m with corresponding y_{max} values of 0.032 m and 0.059 m respectively. Hence an estimate of y_{max} corresponding to a length $l = 1.00$ m is given by linear interpolation as

$$y_{max} = 0.032 + \frac{1.0 - 0.9}{1.1 - 0.9} \cdot (0.059 - 0.032)$$

The estimate is thus 0.0455 m.

For $l = 1.60$ m we take $l_1 = 1.30$ m, $l_2 = 1.50$ m and using linear extrapolation we obtain a value of y_{max} corresponding to $l = 1.60$ m as

$$y_{max} = 0.097 + \frac{1.6 - 1.3}{1.5 - 1.3}(0.148 - 0.097)$$

The estimate is therefore 0.1635 m.

(ii) Assuming the relationship between y_{max} and l is actually $y_{max} = 0.044l^3$, we find that when $l = 1.00$ m then y_{max} is 0.044 m and when $l = 1.60$ m the value of y_{max} will be 0.180 m.

Therefore the percentage error in the interpolated value for l is

$$\frac{0.0455 - 0.0440}{0.0440} \times 100 = 3.41\%$$

whereas the percentage error in the extrapolated value is

$$\frac{0.1800 - 0.1635}{0.1800} \times 100 = 9.17\%.$$

This again indicates the care needed when attempting to extrapolate from a given data set.

Example 2. Values of the solubility s of a certain chemical in water at different temperatures T are given in the following table:

$T/°C$	20	25	30	35	40	45	50
$s/\text{mol m}^{-3}$	32.8	38.7	45.7	54.1	64.0	75.7	89.5

(i) Draw a suitable graph to show that, allowing for small errors in observation, there is a relationship between s and T of the form
$$s = 16 + aT + bT^3$$

(ii) From the graph estimate values of the constants a and b and hence obtain the value of the solubility at room temperature when $T = 15\,°C$.

(iii) Compare this result with the value obtained using linear extrapolation and the first two entries in the table.

(iv) Use the method of interval bisection to find the temperature T correct to two decimal places corresponding to a solubility of exactly $50.0\,\text{mol m}^{-3}$.

Solution. (i) A relationship of this type can be shown to describe the data if we first rearrange it into the form of a straight-line relationship. If $s = 16 + aT + bT^3$ describes the data then

$$s - 16 = aT + bT^3$$

i.e.

$$\frac{s - 16}{T} = a + bT^2$$

A plot of $(s - 16)/T$ versus T^2 should yield a straight line of intercept a and slope b with the data values given. From the table of original data we can draw up the following table:

$\dfrac{s - 16}{T} \Big/ \text{mol m}^{-3}\,°C^{-1}$	0.840	0.908	0.990	1.089	1.200	1.327	1.470
$T^2/°C^{-2}$	400	625	900	1225	1600	2025	2500

The graph of these data is shown in Fig. 1.12. We see that a straight line is obtained, indicating that the relationship $s = 16 + aT + bT^3$ is an acceptable model to take connecting s and T for the given data.

(ii) From the graph the value of a (the intercept) is given as $a = 0.72$ and b (the

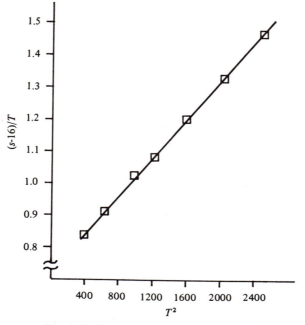

Fig. 1.12. Graph of $(s - 16)/T$ against T^2.

slope), using the two points indicated, is

$$b = \frac{1.2 - 0.84}{1600 - 400} = 0.0003$$

The model is therefore $s = 16 + 0.72T + 0.0003T^3$. When $T = 15\,^{\circ}\text{C}$ we see that $s = 16 + 0.72 \times 15 + 0.0003 \times 15^3$, which is $27.8\,\text{mol}\,\text{m}^{-3}$.

(iii) Returning to the original table and using linear extrapolation on the first two points we obtain an estimate of s of

$$s = 32.8 + \frac{15 - 20}{25 - 20}(38.7 - 32.8) = 26.9\,\text{mol}\,\text{m}^{-3}$$

The reason for this rather poor value compared to $s = 27.8\,\text{mol}\,\text{m}^{-3}$ is that the original data in the first table are related via a non-linear relationship and hence linear extrapolation will necessarily give only approximate results (notwithstanding the difficulties we have already stressed regarding extrapolation procedures in general).

(iv) Finally, regarding the value of T when $s = 50.0\,\text{mol}\,\text{m}^{-3}$. Substituting this value of s in the relationship given we see that we have to solve the cubic equation

$$bT^3 + aT + 16 = 50$$

i.e. the equation

$$F(T) = 0.0003T^3 + 0.72T - 34 = 0$$

From the original table we can see that the required value of T must lie somewhere between 30 and 35 °C. Substituting these values into $F(T)$ we see that $F(30) < 0$ and $F(35) > 0$. At $T = 32.5$ °C we have that $F(32.5)$ is negative, so that the root occurs between 32.5 and 35 °C. Continuing in this fashion and halving the appropriate interval each time we have

$$F(33.75) \quad \text{is positive}$$
$$F(33.125) \quad \text{is positive}$$
$$F(32.8125) \text{ is positive}$$
$$F(32.6563) \text{ is negative}$$
$$F(32.7344) \text{ is positive}$$
$$F(32.6953) \text{ is positive}$$
$$F(32.6758) \text{ is negative}$$
$$F(32.6856) \text{ is positive}$$
$$F(32.6807) \text{ is positive}$$

It follows that the root is between 32.6758 and 32.6807. Thus, to two places of decimals the temperature corresponding to a solubility of 50 mol m^{-3} is 32.68 °C.

1.6 Exercises

1. Find which of the following points lie on the line $y = 16.4x - 9.1$:
   (i) $(1.10, 8.94)$
   (ii) $(1.55, 15.32)$
   (iii) $(2.85, 37.64)$
   (iv) $(9.00, 137.6)$

2. Find the equation of the straight line in the form $y = ax + b$ for the following cases:
   (i) the line has slope 3.4 and intercept 0.3;
   (ii) the line is parallel to the line $y = 2x - 7$ and passes through the point $(7, 6)$;
   (iii) the line passes through the points $(4, 2)$ and $(6, 1)$;
   (iv) the line has intercept 4.3 and passes through the point $(5.9, 1.7)$.

3. Sketch the curves to describe the following polynomial functions:

   (i) $f(x) = x$ (ii) $f(x) = x^2$
   (iii) $f(x) = x^3$ (iv) $f(x) = x^4$
   (v) $f(x) = x^5$ (vi) $f(x) = (x - 1)^2$
   (vii) $f(x) = x^2 - 1$ (viii) $f(x) = x^2 - 3x + 2$
   (ix) $f(x) = 2x^2 - 3x + 1$ (x) $f(x) = x^2 - 2x^4$

   In each case find the roots and hence write each function as a product of factors.

4. Obtain the roots, by trial and error if necessary, of the following polynomials:

(i) $y = 7.6x - 2.4$

(ii) $y = x^2 - 16x + 60$

(iii) $y = x^3 + 2x^2 - x - 2$

(iv) $y = x^5 - x^4 - 3x^3 + x^2 + 2x$

5. Solve the following quadratic equations by completing the square:

(i) $x^2 - 5x + 6 = 0$ (ii) $x^2 + 8x + 7 = 0$

(iii) $x^2 + 4x + 3 = 0$ (iv) $x^2 + x - 6 = 0$

(v) $x^2 - 25 = 0$ (vi) $x^2 - 14 = 0$

(viii) $2x^2 + x - 1 = 0$ (viii) $6x^2 - x - 2 = 0$

(ix) $x^2 + 3x - 7 = 0$ (x) $2x^2 + 7x + 1 = 0$

6. Evaluate the following expressions for the given values of x, first putting each in an appropriate form for using nested multiplication:

(i) $x^3 + 3.14x^2 - 1.72x + 141$ when $x = 6.52$

(ii) $x^4 - 3.17x^3 + 5.29x^2 - 17.3$ when $x = 5.12$

(iii) $x^6 - 4x^5 - 3x^4 + 2x^3 - 3x^2 + x - 7.3$ when $x = 1.13$

Give the answers accurate to four decimal places.

7. Express the following in the form $a + bi$:

(i) $(2 + 3i) + (6 + 5i) - 3(7 - 2i) - (3 - 6i)$

(ii) $(2 + 3i)(2 - 5i)$

(iii) $(1 + i)/(1 + 2i)$

(iv) $(1 + i)/(1 - i)^2$

8. Find the values of x and y if

$$x + iy = (1 - 3i)(4 + i)/(2 + 3i)$$

9. Using the formula for the roots of a quadratic, solve the following equations, giving all solutions to two decimal places where appropriate:

(i) $p^2 - 16p + 100 = 0$

(ii) $4.1x^2 + 3.7x - 2.3 = 0$

(iii) $y^2 + y + 1 = 0$

(iv) $z^3 + 6z^2 - z = 0$

10. If the function $y = f(x)$ is approximately linear for values of x in the range $5.5 \leqslant x \leqslant 6.5$, and $y = 40.33$ when $x = 5.7$ and $y = 44.81$ when $x = 5.9$, use linear interpolation or extrapolation to obtain

(i) y when $x = 5.82$

(ii) y when $x = 5.65$

(iii) y when $x = 5.97$

11. Using the interval bisection method obtain the root of the equation

$$2x^3 + 3x - 3 = 0$$

correct to two decimal places, that occurs between $x = 0.5$ and $x = 1.0$.

12. Use the method of interval bisection to find all the roots of the equation

$$x^3 - 5x + 3 = 0$$

(three real roots) each accurate to 1 decimal place.

13. A simple root of the function $f(x) = 0$ is trapped in $(0, 1)$. How many bisections are necessary to locate it to an accuracy of three decimal places?

14. In the process of silver plating the number of grammes, m, of silver that is deposited on a metallic surface varies directly with the number of hours t. The electric current is kept steady throughout the process. Given that $m = 1.4$ when $t = 0.5$, find an equation giving m in terms of t. Hence find the number of grammes deposited in 2.5 hours.

15. The selling price P of a handtool consists of a fixed amount A plus a variable amount which is proportional to the square of the normal size S, so that $P = A + BS^2$ where A and B are constants. If a 20-mm tool costs 75 p and a 50-mm tool costs £3.50, what is the cost of a 60-mm tool?

16. Known standard frequencies are used to make a calibration curve for the dial of a signal generator. Angular position of the dial in terms of angle from a fixed reference point θ, together with the corresponding frequencies f, are as shown:

f /kHz	$\theta/°$
10	18.8
20	35.9
30	49.3
40	63.2
50	79.8

By drawing a graph determine the equation of the linear relationship connecting θ and f in the form $\theta = mf + c$, where m and c are constants. What is the value of f when $\theta = 25°$?

17. In the calibration of a mercury in glass thermometer, distances, L, from the bulb to the top of the mercury column are measured at various temperatures, T, with the results:

$T/°C$	5	20	35	50	80	105
L/mm	5.0	19.8	35.2	49.7	80.3	104.9

By drawing a suitable graph find the equation of the line $L = mT + c$, where m and c are constants, that describes these data, and hence predict where scale markings should be for 10 and 60 °C.

18. In order to see how quickly the heart returns to normal after running, four reasonably fit student volunteers exercised on a tread mill for about 12 min until their heart rate reached 190 beats m^{-1}. They then stopped exercising and their heart rate was monitored at regular intervals as it returned to normal. The average R of the four heart rates as a function of time after exercise t was as follows:

t/min	1	3	5	7	9	11
R/beats min^{-1}	170	136	111	95	86	80

Show by making a suitable plot that these data over the time interval of the experiment can be described by a model of the form

$$R = 190 + at + bt^2$$

where a and b are constants.

From your graph obtain values for a and b and hence determine the time indicated by the model for the heart rate R to become a minimum. What is this minimum predicted to be?

19. Using (i) linear interpolation and (ii) linear extrapolation estimate from the table of values given in Exercise 18 the average heart rate after $t = 9.5$ min. Compare your results with that predicted by the model $R = 190 + at + bt^2$. Why is the model $R = 190 + at + bt^2$ not likely to be a good model for large values of t?

20. The velocity of a certain chemical reaction is given by the equation

$$v = k(c_1 - x)(c_2 + x)$$

where c_1, c_2 and k are positive constants. Show that the maximum value of v is

$$k \left(\frac{c_1 + c_2}{2} \right)^2$$

provided that $c_1 \geqslant c_2$.

Sketch out the plot of v against x, and if v and x cannot be negative in the reaction, indicate that part of the plot that corresponds to reality.

21. In a particular circuit, involving resistors, an inductor and a capacitor, the following relationship exists between the quantities R, w, L, R_1, R_2, R_3 and C

$$\frac{R + iwL}{R_1} = \frac{R_2}{R_3 - i/wC}$$

where $i = \sqrt{-1}$. Find R and L in terms of R_1, R_2, R_3, w and C.

22. In solving the equation $f(x) = 0$, if the function changes sign between x_1 and x_2, show, using linear interpolation, that an estimate of the root is given by

$$x = x_1 - \frac{(x_2 - x_1)f(x_1)}{f(x_2) - f(x_1)}$$

Build this idea into an algorithm and write a program to obtain the solution of $f(x) = 0$ that is accurate to within a relative error e as specified by the user of the program.

2

Functions for science 1: the exponential function

Contents: 2.1 scientific context – 2.1.1 orders of magnitude; 2.1.2 growth of a certain bacterium; 2.2 mathematical developments — 2.2.1 why e?; 2.2.2 power series representation of the exponential function e^x; 2.2.3 calculation of e^x; 2.2.4 approximate forms for e^x when x is small; 2.2.5 index laws; 2.2.6 the graph of $y = e^x$; 2.3 worked examples; 2.4 further developments – 2.4.1 the decay function; 2.4.2 properties of the decay function; 2.4.3 hyperbolic functions; 2.4.4 properties of the hyperbolic functions; 2.5 further worked examples; 2.6 exercises.

2.1 Scientific context

2.1.1 Example 1: orders of magnitude. Many quantities in science when expressed in the usual SI units turn out to be either very large or very small numbers. For example, the speed of light c, to 10 significant figures, is $299\,792\,456.2\,\mathrm{m\,s^{-1}}$, and the Bohr radius a_0, which provides a measure of the size of the hydrogen atom, again to ten significant figures, is $0.000\,000\,000\,059\,217\,715\,81\,\mathrm{m}$. In order to write such quantities in a more compact form we often employ the scientific notation of writing the number as a multiple of a power of 10. For example,

$$c = 2.997\,924\,562 \times 10^8\,\mathrm{m\,s^{-1}}$$
$$a_0 = 5.921\,771\,581 \times 10^{-11}\,\mathrm{m}$$

Powers of 10 are useful in determining the relative orders of magnitude of quantities. For example, since the speed of sound in still air is approximately $3.4 \times 10^2\,\mathrm{m\,s^{-1}}$, the ratio of the speed of light to the speed of sound is roughly 10^6. We say, therefore, that the speed of light is approximately six *orders of magnitude* greater than the speed of sound. Note, however, that the quantities 1.09×10^4 and 9.3×10^3 would be considered to be of the same order of magnitude since their ratio $1.09 \times 10^4 / (9.3 \times 10^3)$ is approximately one.

This compact scientific notation has the advantage that we may easily approximate quantities, keeping only the number of significant figures we require for a particular calculation. For example, for most quick calculations such as that outlined in the above paragraph, we might take $c = 2.998 \times 10^8\,\mathrm{m\,s^{-1}}$ or even $3 \times 10^8\,\mathrm{m\,s^{-1}}$ depending upon how accurate we need our answer to be.

Further examples of physical quantities which are usually expressed in this notation, are the distance from the earth to the nearest star, Proxima Centauri, which is of the order of 4×10^{16} m, and the radius of the earth itself, which is about 6.4×10^6 m. Of course, it is unlikely that it is *exactly* 4×10^{16} m to Proxima Centauri, and we know that there are significant variations in the radius of the earth depending on whether we measure the radius at the equator or from pole to pole. However, it is convenient shorthand notation for estimates of very large (or very small) numbers.

In the above we have written numbers in terms of powers of 10, i.e. used the 'base 10' simply because of our ordinary decimal system of representing numbers. Sometimes, however, it is more convenient to use another base. For example, a simple binary code may consist of a string of 20 characters, each character being 0 or 1. A typical 20-bit 'word' in this code might be

$$10011010111001000110$$

Since each digit in this 'word' can be chosen in one of two ways, 0 or 1, there can be up to $2^{20} = 1\,048\,576$ 'words' or different character strings in this code, i.e. just over a million or 10^6 possibilities.

The reason for choosing a base 10 or 2, or for that matter any other positive number, is purely a matter of convenience or notational simplicity. In the above example 2^{20} is notationally simpler than $1.048\,576 \times 10^6$, though of course both notations describe the same number. This is demonstrated further in our next example.

2.1.2 Example 2: growth of a certain bacterium. Suppose a small number of bacterium *Escherichia coli* are placed on a microscope slide and maintained under conditions which allow growth. Every ten minutes the slide is examined and the number of cells of the bacterium is estimated. Suppose the results are produced as shown in Table 2.1. From these data we see that the number of cells doubles approximately every 20 min. In fact a better estimate is that on average they double every 21.3 min. (Although the origin of this number 21.3 may not be immediately apparent from the data, by the time you have read Chapter 4 and Chapter 18, you should be able to derive this value yourself.) Hence if we denote the number of cells at time t by $N(t)$ we have

$$N(21.3) \simeq 2 \times N(0)$$
$$N(42.6) \simeq 2 \times N(21.3) = 2^2 N(0)$$
$$N(63.9) \simeq 2 \times N(42.6) = 2^3 N(0)$$

and in general

$$N(t) \simeq N(m \times 21.3) = 2^m N(0)$$

Table 2.1. *Number of bacterium cells at 10-minute time intervals*

Time min^{-1}	0	10	20	30	40	50	60	70	80	90	100
Number of cells	7	10	14	20	27	39	54	76	108	151	213

where m is the number of 21.3 min intervals in time t, i.e. $m = t/21.3$. Therefore,

$$N(t) \simeq 2^{t/21.3} N(0)$$

This formula is a *model* for the growth of this type of bacterium. It allows us to deduce the approximate number of cells at any intermediate time t, i.e. to interpolate between readings; or to estimate how the bacterium will continue to grow after observations cease, i.e. to extrapolate the readings. However, the model is only an approximation since the graph of $N(t)$ against time is a continuous smooth curve, while the actual behaviour of the growth 'curve' will show numerous short discrete jumps as each individual cell divides. Nevertheless, the model does provide a good *average* description of the process as indicated in Fig. 2.1. Because of the

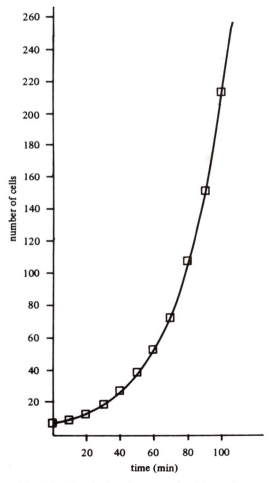

Fig. 2.1. Graph showing growth of bacterium.

way we have analysed the growth of the bacterium, the model we have come up with has involved a power of 2.

Question: Is there anything special about the number 2 in this case?
Answer: Not really. As well as doubling every 21.3 min, we see that the number of cells triples in a little over 30 min – in 33.8 min in fact. We can therefore write $N(t)$ as

$$N(t) \simeq 3^{t/33.8} N(0)$$

Thus the choice of the number 2 or 3 in these models is really quite arbitrary.

In fact we could have defined our model for $N(t)$ in terms of *any* positive number as our base. In practice, for reasons we discuss in the next section, we shall frequently choose a number somewhere between 2 and 3 called 'e'. The actual value of e is difficult to define explicitly, but to ten significant figures it is given by

$$e = 2.718\,281\,828$$

Thus we may write

$$N(t) = e^{ct} N(0)$$

where $1/c$ is the time scale for the bacterium to increase in size by the factor e ($\simeq 2.718\,281\,828$) i.e. approximately 29.3 min.

2.2 Mathematical developments

2.2.1 Why e? At first sight it would appear that the choice of this strange number e as a base upon which to build a model of a growth process is rather odd. We have stressed in the above paragraphs that we can choose *any* positive number as the base so long as the time scale is chosen accordingly. Why therefore should we introduce such an unlikely looking base as e? One answer lies in the underlying theoretical treatment of a growth process as we shall now explain.

Let us start from the quite reasonable assumption that the average rate of increase in the quantity $N(t)$, evaluated at time t, is directly proportional to the quantity present at time t. The quantity $N(t)$ need not necessarily represent the number of bacteria at time t; it could be the number of people in the world on any one day, or the money in a bank account subject to compound interest calculated daily, or indeed any quantity whose average rate of growth is proportional to the amount present at any given time.

If $N(0)$ is the quantity at time $t = 0$ then at some short time in the future τ the quantity is $N(\tau)$. Thus the average rate of increase in this time period is $(N(\tau) - N(0))/\tau$ which, by our assumption, is proportional to $N(0)$. Hence we can write

$$N(\tau) - N(0) = \tau c N(0) \tag{1}$$

where c is a positive constant of proportionality, i.e. a 'birth rate'. If we now consider several small intervals of time τ we have

$$N(\tau) - N(0) = \tau c N(0)$$
$$N(2\tau) - N(\tau) = \tau c N(\tau)$$
$$N(3\tau) - N(2\tau) = \tau c N(2\tau) \tag{2}$$

etc.

Thus rearranging each of the above equations in (2) gives

$$N(\tau) = N(0)(1 + c\tau)$$
$$N(2\tau) = N(\tau)(1 + c\tau) = N(0)(1 + c\tau)^2$$
$$N(3\tau) = N(2\tau)(1 + c\tau) = N(0)(1 + c\tau)^3$$

etc.

In general

$$N(n\tau) = N(0)(1 + c\tau)^n$$

Therefore if we wish to derive the quantity present at some time t, corresponding to n short time intervals τ, then since $t = n\tau$ we have that $\tau = t/n$, so

$$N(t) = N(0)\left(1 + \frac{ct}{n}\right)^n \tag{3}$$

Note that our assumption was that the average rate of increase of $N(t)$ at time t was proportional to $N(t)$. However, in (1) and (2) we assumed that the rate of increase *throughout* the time interval of length τ was proportional to $N(t)$. Of course this will not be entirely correct. As soon as we progress from time t, $N(t)$ will change and hence so should our rate of increase. In order for (3) to represent a good model for the quantity present at time t we should take the time interval between successive steps, $\tau = t/n$, as small as possible and hence the number of intermediate steps, n, as large as possible. Ideally therefore, instead of (3) we should write $N(t)$ as the value approached by (3) as n becomes very large. In mathematical notation we write

$$N(t) = \lim_{n \to \infty} \left[N(0)\left(1 + \frac{ct}{n}\right)^n \right] \tag{4}$$

i.e. $N(t)$ is defined as the limit of the function in square brackets as 'n tends to infinity'. The idea of the limit of a function is explored in some detail in Chapter 5. In the above expression we can never take n to be infinitely large. Nevertheless, by taking larger and larger values for n we may be able to deduce (to any given accuracy) what the value of this expression would be if we *could* let n go to infinity. For example, in Table 2.2 we list the values of

$$\left(1 + \frac{ct}{n}\right)^n$$

to ten significant figures for the particular cases $ct = 1$ and $ct = 2$ for various large values of n.

Table 2.2. *The values of* $(1 + ct/n)^n$ *for* $ct = 1$ *and* $ct = 2$

n	$\left(1 + \dfrac{1}{n}\right)^n$	$\left(1 + \dfrac{2}{n}\right)^n$
100	2.704 813 829	7.244 646 118
1000	2.716 923 932	7.374 312 390
10 000	2.718 145 927	7.387 578 632
100 000	2.718 268 237	7.388 908 321
1 000 000	2.718 280 469	7.389 041 321
10 000 000	2.718 281 692	7.389 054 621

From Table 2.2 we see that as n becomes larger more and more decimal places in these numbers become fixed; i.e. the numbers change less and less. We can imagine, therefore, that as $n \to \infty$ we arrive at constants for each of

$$\left(1 + \frac{1}{n}\right)^n \quad \text{and} \quad \left(1 + \frac{2}{n}\right)^n.$$

The limit of the expression

$$\left(1 + \frac{1}{n}\right)^n$$

is called 'e', and as we have said in 2.1.2 its value to ten significant figures is 2.718 281 828. The limit of

$$\left(1 + \frac{2}{n}\right)^n$$

as n tends to ∞ is 7.389 056 099 to ten significant figures, which is just the square of the quantity we have called e. Thus we define e as

$$e = e^1 = \lim_{n \to \infty}\left[\left(1 + \frac{1}{n}\right)^n\right] \tag{5}$$

and we define e^2 as

$$e^2 = \lim_{n \to \infty}\left[\left(1 + \frac{2}{n}\right)^n\right] \tag{6}$$

In general, therefore, we can define e^{ct} as

$$e^{ct} = \lim_{n \to \infty}\left[\left(1 + \frac{ct}{n}\right)^n\right] \tag{7}$$

Hence, returning to our expression (4) for the number of bacteria at time t, we see that we can express our model for $N(t)$ as

$$N(t) = N(0)e^{ct}$$

which answers the question 'why e?' that was posed at the beginning.

2.2.2 Power series representation of the exponential function e^x. The function $f(x) = e^x$ (or $\exp(x)$) is known as the *exponential function*. As we have just seen, by definition

$$e^x = \lim_{n \to \infty} \left[\left(1 + \frac{x}{n} \right)^n \right] \tag{8}$$

from which we can deduce values of e^x, at least to some required degree of accuracy, by considering sufficiently large values of n. However, there is another way of estimating e^x using this relation as we shall now illustrate. An alternative derivation will be given in Chapter 8, but here we will work with what is known as the *binomial expansion*.

By simply multiplying out brackets it is easy to show that

$$(1 + p)^2 = 1 + 2p + p^2$$
$$(1 + p)^3 = (1 + p)(1 + p)^2 = 1 + 3p + 3p^2 + p^3$$
$$(1 + p)^4 = (1 + p)(1 + p)^3 = 1 + 4p + 6p^2 + 4p^3 + p^4$$
etc.

and that, in general,

$$(1 + p)^n = 1 + np + \frac{n(n-1)}{2!} p^2 + \frac{n(n-1)(n-2)}{3!} p^3 + \cdots + p^n$$

$$= \sum_{k=0}^{n} \frac{n!}{(n-k)!k!} p^k$$

This last expression is known as the binomial expansion of $(1 + p)^n$ and we shall meet it on numerous occasions in later chapters. In writing out the binomial expansion we have in fact introduced some new notation. The expression $n!$, for example, reads as 'n factorial' and is mathematical shorthand for the product $n(n-1)(n-2)\cdots(2)(1)$. Thus $4!$ is $4 \times 3 \times 2 \times 1$ which is 24. Most scientific calculators have this factorial key on them and in the notation it is assumed that $0! = 1$. The *sigma notation* denoted by $\sum$ means 'add together'.

$$\sum_{k=0}^{n}$$

means add together terms of the form (in this case) $n!p^k/(n-k)!k!$ as k ranges from 0 to n. Thus the first term is when $k = 0$, the second when $k = 1$, the third when $k = 2$ and so on.

Example: What is the 5th term in the expansion of $(1 + p)^7$?
Solution: The general $(k + 1)$th term in $(1 + p)^n$ is

$$\frac{n!}{(n-k)!k!} p^k,$$

so the 5th term in $(1 + p)^7$ is when $k = 4$ and $n = 7$. The term is therefore given by

$$\frac{7!}{3!4!} p^4 = \frac{5040}{6 \times 24} p^4 = 35p^4$$

Returning, therefore, to our exponential function. If we let x/n in (8) be the quantity p, it follows from the binomial expansion that

$$\left(1 + \frac{x}{n}\right)^n = 1 + n\left(\frac{x}{n}\right) + \frac{n(n-1)}{2!}\left(\frac{x}{n}\right)^2 + \frac{n(n-1)(n-2)}{3!}\left(\frac{x}{n}\right)^3 + \cdots \qquad (9)$$

$$= \sum_{k=0}^{n} \frac{n!}{(n-k)!k!}\left(\frac{x}{n}\right)^k \qquad (10)$$

If n is assumed to be a very large number then in relative terms $(n-1)$ will be almost the same as n, $(n-2)$ will be almost the same as n and so on, so that we can write (10) approximately as

$$\left(1 + \frac{x}{n}\right)^n = 1 + n\left(\frac{x}{n}\right) + \frac{n^2}{2!}\left(\frac{x}{n}\right)^2 + \frac{n^3}{3!}\left(\frac{x}{n}\right)^3 + \cdots$$

$$= 1 + x + \frac{x^2}{2!} + \frac{x^3}{3!} + \cdots$$

(Of course for values of k close to n this approximation scheme will not be very good since we are trying to approximate

$$\frac{n!}{(n-k)!k!}\left(\frac{x}{n}\right)^k \quad \text{by} \quad \frac{x^k}{k!}$$

However these terms will be very small no matter how large x is provided k is sufficiently large.)

As n increases this approximation becomes more exact, so that in the limit as $n \to \infty$ we may write

$$e^x = \lim_{n \to \infty}\left[\left(1 + \frac{x}{n}\right)^n\right] = 1 + x + \frac{x^2}{2!} + \frac{x^3}{3!} + \cdots$$

$$= \sum_{k=0}^{\infty} \frac{x^k}{k!} \qquad (11)$$

where the sum extends over *all* non-negative integer values of k, i.e. it is an infinite sum called a *power series*. We should note that e^x is sometimes written as $\exp(x)$. This is purely a matter of convenience although this notation does have the advantage that it stresses that e^x is indeed a function of x. Also it is worth noting that, although you may not have recognised them as such, you have already met many examples of numbers which are expressible as an infinite sum of terms. One simple example is the fraction $\frac{1}{3}$. When expressed as a decimal we know that $\frac{1}{3} = 0.\dot{3}$ where the dot over the 3 indicates that the 3 recurs continually, i.e.

$$0.\dot{3} = 0.333\cdots$$

$$= 0.3 + 0.03 + 0.003 + 0.0003 + \cdots$$

$$= \frac{3}{10} + \frac{3}{100} + \frac{3}{1000} + \frac{3}{10000} + \cdots$$

$$= 3\sum_{k=1}^{\infty} 10^{-k}$$

It is interesting to note that although this number can never be calculated exactly we all know, or think we know, what we mean by the fraction $\frac{1}{3}$. One important difference however between the exponential function and a recurring decimal is that for a given value of x, the corresponding value of e^x is usually *not* a recurring decimal although it contains an infinite number of digits.

2.2.3 Calculation of e^x. Unlike the simple polynomial functions of the previous chapter, values of the function $f(x) = e^x$ cannot be calculated exactly. This follows because e^x is defined either as a limit as in (8) in which $n \to \infty$, or as a power series as in (11). Hence e^x can only be evaluated to *some given degree of accuracy*, and the results are presented in tables or on the display of a calculator or computer. For this reason e^x is sometimes called one of the 'special functions', although in science, as we shall see, such functions occur much more frequently than those that can be calculated exactly.

If we wish to evaluate e^x from (11) we need to know *two* numbers, the value of x *and* the accuracy required. The same is true even if we use a calculator. We key in the x value and the result is shown accurate to a certain number of significant figures. Usually the accuracy is to one or two significant figures more than the number of digits displayed. The calculator does not pluck out of its memory some appropriate value of e^x; it actually evaluates it by a process not dissimilar to adding together terms in an infinite series such as the one in (11).

Example: Write a program to evaluate e^x based on (11) in which the summation is terminated if each additional term becomes less than some specified value z.

Solution: At first sight we may be tempted to calculate each term from scratch, i.e. $1, x, x^2/2$, $x^3/6$, etc., checking each term to see if it is less than the specified value z and then adding them up. However, a much more efficient program emerges if we note that after the first term, every term is x/k times the previous one. An appropriate program in pseudocode is the following one, where $x \geqslant 0$.

```
begin
input x, z
set term to 1
set sum to 1
set k to 1
while term ⩾ z do
        set term to term * x/k
        set sum to sum + term
        set k to k + 1
endwhile
set result to sum
print result
stop
```

2.2.4 Approximate forms for e^x when x is small. Besides providing us with a means of calculating the function e^x for any value of x, the infinite series representation of e^x given by (11) also allows us to write down convenient expressions for e^x when x is small. This is useful for many quick calculations in science, but more importantly

Table 2.3. *Comparison of* e^x *with the linear and quadratic approximations*

x	$1 + x$	$1 + x + x^2/2$	e^x
0	1.0	1.0	1.0
0.05	1.05	1.0525	1.051 271
0.10	1.10	1.105 00	1.105 171
0.15	1.15	1.161 25	1.161 834
0.20	1.20	1.220 00	1.221 403
0.25	1.25	1.281 25	1.284 025
0.30	1.30	1.345 00	1.349 859
0.35	1.35	1.411 25	1.419 068
0.40	1.40	1.480 00	1.491 825
0.45	1.45	1.551 25	1.568 312
0.50	1.50	1.625 00	1.648 721
0.55	1.55	1.701 25	1.733 253

it provides us with simple and fairly accurate functional forms for e^x when x is small.

If x is small in comparison to 1, then successive powers of x rapidly become negligible. It is clear, therefore, that when x is very small we should not require to add on many terms of the series to obtain an accurate expression for e^x. From Table 2.3 we see that this is true. For example, we find that $1 + x$, i.e. the first two terms of the series in (11), is within 1.75% of the value of e^x for x values in the range 0 to 0.2. Similarly we see $1 + x + (x^2/2)$, which is the sum of the first three terms of (11), is within 1.85% of the value of e^x for x values of up to 0.55. Hence, if a simple polynomial expression is used to represent e^x then depending upon the accuracy we need, we may write

$$e^x \simeq 1 + x$$

or

$$e^x \simeq 1 + x + \frac{x^2}{2}$$

for *small* values of x.

2.2.5 Index laws. The definition of the exponential function has led us to the special number e and to the fact that $\exp(x)$ is 'e to the power x' or e^x. When we multiply or divide two expressions like e^{x_1} and e^{x_2} there are simple algebraic rules we can use to obtain the result and we do not necessarily have to appeal to the original definition of the exponential function. These rules are called the *index laws* and in fact apply to *any* power function and not just e^x.

Suppose, then, that a is any positive number and m and n are two integers. We interpret a^m in the usual way as

$$a^m = \underbrace{a \times a \times \cdots \times a}_{m \text{ times}}$$

For example, $a^3 = a \times a \times a$ and $a^6 = a \times a \times a \times a \times a \times a$. Similarly, we interpret a^n as

$$a^n = \underbrace{a \times a \times \cdots \times a}_{n \text{ times}}$$

The product, therefore, of a^m and a^n, written out in full, is

$$a^m \times a^n = \underbrace{(a \times a \times \cdots \times a)}_{m \text{ times}} \times \underbrace{(a \times a \times \cdots \times a)}_{n \text{ times}}$$

$$= \underbrace{a \times a \times \cdots \times a}_{m + n \text{ times}}$$

The product contains a total of $m + n$ a's, so we have the very important result that

$$a^m \times a^n = a^{m+n} \qquad (12)$$

If we now raise a^m to the power n, then writing out the result, again in full, we have

$$(a^m)^n = \underbrace{\underbrace{(a \times a \times \cdots \times a)}_{m \text{ times}} \times \underbrace{(a \times a \times \cdots \times a)}_{m \text{ times}} \times \cdots \times \underbrace{(a \times a \times \cdots \times a)}_{m \text{ times}}}_{n \text{ brackets}}$$

In total we have mn lots of a on the right-hand side so

$$(a^m)^n = a^{mn} \qquad (13)$$

Finally, if we consider a^m/a^n written out in full we have

$$\frac{a^m}{a^n} = \underbrace{\frac{(a \times a \times \quad \cdots \quad \times a)}{\underbrace{(a \times a \times \cdots \times a)}_{n \text{ times}}}}_{m \text{ times}}$$

Cancelling the n a's we see that there are $m - n$ a's left. Thus

$$\frac{a^m}{a^n} = a^{m-n} \qquad (14)$$

In this latter result, if the two powers m and n happen to be the same, then the left-hand side is clearly equal to 1 and the right-hand side is a^0. In fact, any non-zero number raised to the power zero is always equal to 1. One other consequence of (14), with $m = 0$, is the important result that

$$a^{-n} = \frac{1}{a^n} \qquad (15)$$

Example: Use the index laws as stated here to simplify the following expressions

    (i) $a^3 \times a^5/a^7$;
    (ii) $(x^4)^3 \times x^2$;
    (iii) $(b^6/b^9)^2$.

Solution: (i) $a^3 \times a^5 = a^8$ by (12) and by (14) we see that $a^8/a^7 = a$;
    (ii) $(x^4)^3 = x^{12}$ by (13) and $x^{12} \times x^2 = x^{14}$ by (12);
    (iii) $b^6/b^9 = b^{-3}$ by (14) and $(b^{-3})^2 = b^{-6}$ by (13) which by (15) may be written as $1/b^6$.

Although we have deduced the laws (12)–(15) for integer powers m and n, they are also valid for non-integer powers. So, for example,

$$(a^{0.7} \times a^{0.1})^{0.3} = (a^{0.8})^{0.3} = a^{0.24}$$

These index laws are of particular importance, as we have said, in manipulating expressions involving the exponential function. Thus, in addition to the properties of e^x that we have already considered we can add the following:

$$e^{x_1} \times e^{x_2} = e^{x_1 + x_2}$$
$$e^{x_1}/e^{x_2} = e^{x_1 - x_2}$$
$$(e^x)^k = e^{kx} \qquad (16)$$

2.2.6 The graph of $y = e^x$. The graph of $y = e^x$ is shown in Fig. 2.2. Because any non-zero number raised to the power zero is unity it follows that the graph passes through $y = 1$ when $x = 0$. The intercept is therefore 1 for the exponential function.

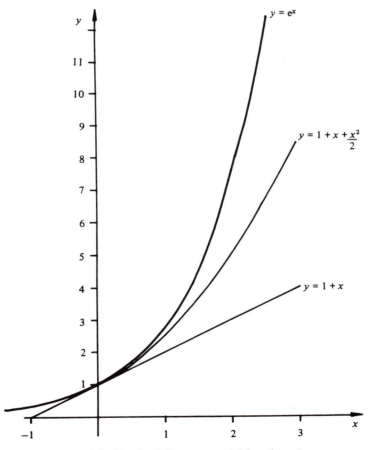

Fig. 2.2. Graph of the exponential function e^x.

Also from its series expansion in (11) we see that for $x > 0$, e^x is necessarily greater than 1 and increases as x increases. We have already seen that for small values of x, $e^x = 1 + x$ and, more accurately, $e^x = 1 + x + (x^2/2)$. These linear and quadratic small x approximations are shown on the diagram. However, as x increases the graph of $y = e^x$ quickly rises above these small x approximations. In fact it follows that the graph of e^x must eventually rise above *all* finite polynomial approximations for this function for sufficiently large x i.e. it 'rises to infinity faster than any finite power of x'.

Before moving on to the next section it is important to note that the exponential function is defined for *all* values of x and not just for values of x in the range $0 \leqslant x < \infty$. All the analysis of the previous sections applies for any value of x. Thus the series expansion of e^x in (11) can be used to obtain values of e^x for negative as well as positive values. The values of e^x for negative values are all less than 1, but never become zero as shown. The only place where extra care is needed when negative x values are involved is in any program to calculate e^x. Since terms in the series can now be negative we must pay particular attention to any condition that is used to terminate computation. In the pseudocode program of 2.2.3 the line

while term $\geqslant z$ **do**

must be altered to

while $- z \leqslant$ term $\leqslant z$ **do**

or, more conveniently, to

while absolute value of (term) $\geqslant z$ **do**

before the program is used to calculate e^x for a negative x input. The absolute value of (-3), for example, would be 3, as is the absolute value of (3). The mathematical shorthand for the 'absolute value of' involves the use of what are called modulus signs. Thus the absolute value of (x) is written as $|x|$.

2.3 Worked examples

Example 1. In a newly created wild-life preserve it is estimated that species A of animal will triple in number every 3.7 yr and species B will quadruple every 4.1 yr. If initially there are 46 of species A and 24 of species B, draw on the same diagram graphs of the growth of both species and hence estimate how long it will be before the numbers of the two species are equal.

Solution. Following along the lines of the discussions in 2.1.2 the mathematical *model* to take for species A would seem to be

$$N_A(t) = 46 \times 3^{t/3.7}$$

where t is the time measured in yr and N_A the number after t yr, and for species B

$$N_B(t) = 24 \times 4^{t/4.1}$$

In each case we note that the expressions $N_A(t)$ and $N_B(t)$ are only mathematical models of the real situation for the population of the two species. Animals come in discrete numbers – perhaps seasonally! From Fig. 2.3 the numbers of the two species are approximately equal after $15\frac{1}{2}$ years. In view of the nature of the cross-over of the two graphs, a non-graphical method of solving the equation

$$46 \times 3^{t/3.7} = 24 \times 4^{t/4.1}$$

may be more appropriate. One method is the method of interval bisection discussed in Chapter 1. Another method of solution will be given in Chapter 4.

Example 2. A mathematical model of the growth of a certain bacterium is given as

$$M = M_0 e^{0.4t}$$

where t is measured in minutes, and M_0 is the amount present at some time $t = 0$.

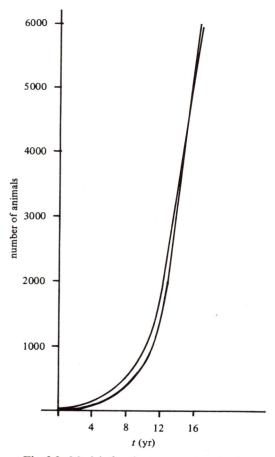

Fig. 2.3. Models for the two species of animal.

Use the small variable approximation to the exponential function to determine the time at which the mass has increased by 20%.

Solution. If M is 20% larger than M_0 then

$$\frac{M}{M_0} = \frac{120}{100} = 1.2$$

The infinite series representation of the exponential function is

$$e^{0.4t} = 1 + 0.4t + \frac{1}{2!}(0.4t)^2 + \frac{1}{3!}(0.4t)^3 + \cdots$$

Assuming that t is small enough that we may keep only the first two terms of this expansion, we get

$$e^{0.4t} = 1 + 0.4t$$

Thus the simplified model of the growth is

$$M = M_0(1 + 0.4t)$$

If $M/M_0 = 1.2$, then we require that

$$1.2 = 1 + 0.4t$$

Solving this we see that $t = (1.2 - 1)/0.4 = 0.5 \, \text{min}$.

To check if the simplified model of the growth process was a reasonable one, we can include the quadratic term in t so that

$$e^{0.4t} = 1 + 0.4t + \frac{1}{2!}(0.4t)^2$$

$$= 1 + 0.4t + 0.08t^2$$

and solve the equation

$$1.2 = 1 + 0.4t + 0.08t^2$$

i.e.

$$0.08t^2 + 0.4t - 0.2 = 0$$

The solutions of this quadratic equation are given by

$$t = \frac{-0.4 \pm \sqrt{[(0.4)^2 - 4 \times 0.8 \times (-0.2)]}}{2 \times 0.08} = -5.46 \quad \text{or} \quad 0.46 \, \text{min}$$

The negative solution $-5.46 \, \text{min}$ is not appropriate as t has to be positive. The only physical solution is $t = 0.46 \, \text{min}$, which is not too far from $t = 0.5 \, \text{min}$ obtained with the linear model. This suggests that a solution to the equation

$$1.2 = e^{0.4t}$$

is close to $t = 0.46 \, \text{min}$. To obtain a further improvement in the value of t, we could add in further terms of the exponential series and solve the resulting polynomial equation by the method of interval bisection.

2.4 Further developments

2.4.1 The decay function. So far we have considered the function e^{ct} as the function which describes the growth in various natural processes. We have seen that this model for the growth of physical quantities with time arises if the *rate of growth* is proportional to the amount present at time t.

There is, however, another class of processes in which the *average rate of decay* in some small time interval is proportional to the amount present at time t. One example of this is the decay of a radioactive substance. If there is a large amount of the substance present (so that one may average over the emission of individual quanta of mass, i.e. of individual particles) then the average rate of decay of the radioactivity of the substance may be thought of as being directly proportional to the amount still to decay at any time t. Another example is the cooling of a heated body. If the temperature of the body is much greater than the surrounding ambient temperature, its rate of cooling is greater than if its temperature is only just above the ambient temperature. Here again we may describe the *average rate of decrease*, in this case of the temperature difference, as being directly proportional to the temperature difference at any time.

The analysis of 2.2.1 can be applied to these decay situations. All that we have to do is replace the positive constant of proportionality c by some negative constant $-k$, say. The constant must be negative because we are describing a decrease and not an increase as previously.

Example: If the temperature T at any time t of a steel ingot cooling from $350\,°C$ to room temperature of $15\,°C$ is modelled by the expression

$$T(t) = 15 + 350\,e^{-kt}$$

where $k = 0.31\,\text{hr}^{-1}$, what is the temperature of the ingot at $t = 3.0\,\text{hr}$ and $t = 10.0\,\text{hr}$?

Solution: On substituting $t = 3$ into the formula for T we get

$$\begin{aligned} T(3.0) &= 15 + 350 \times \exp(-(0.31 \times 3)) \\ &= 15 + 350 \times 0.3946 \\ &= 153.1\,°C \end{aligned}$$

Similarly, putting $t = 10$, gives

$$\begin{aligned} T(10) &= 15 + 350 \times \exp(-(0.31 \times 10)) \\ &= 15 + 350 \times 0.0450 \\ &= 30.8\,°C \end{aligned}$$

Example: If the amount of each of the following substances present at time t is modelled by $N(t) = N(0)e^{-kt}$ determine how much of the substance is present after time T if

(i) for carbon 14, $k = 1.203 \times .10^{-4}\,\text{yr}^{-1}$, $N(0) = 3\,\text{gm}$ and $T = 5760\,\text{yr}$; and

(ii) for iodine 131, $k = 0.086\,66\,\text{day}^{-1}$, $N(0) = 5\,\text{gm}$ and $T = 8\,\text{day}$.

Give the answer to three decimal places.

Solution: Substituting for k, T and $N(0)$ in the given decay model we obtain

(i) $N(5760) = 3 \times \exp[-(1.203 \times 10^{-4} \times 5760)]$

     $= 3 \times \exp[-(0.6929)]$

     $= 1.500$ gm to three decimal places;

(ii) $N(8) = 5 \times \exp[-(0.086\,64 \times 8)]$

     $= 5 \times \exp[(0.6931)]$

     $= 2.50$ gm to three decimal places.

In this last example it is worth noting that in each case the value of $N(T)$ is one-half of its original value and that correct to three decimal places the value of kT is 0.693. In fact for any physical quantity that is modelled by the equation $N(T) = N(0)e^{-kt}$, when $kt = 0.693\,147$ (to six decimal places) the amount of the quantity will have been reduced to almost exactly one-half of its original value. This provides one way of defining the constant k in terms of the properties of the physical quantity.

We define T, the time for the quantity to be reduced to one-half of its original value, to be *the half-life* of the quantity and then

$$k = 0.693\,147/T$$

Having noted what the half-life is, in an experiment for example, the value of k for the substance can then be determined.

Example: In an experiment it is found that a sample of a substance contains 5 g of xenon 133. About 5 days later the amount of xenon 133 has reduced to 2.5 g. What are the values of the half-life of xenon 133 and the decay constant k?

Solution: The half-life of xenon 133 is clearly about 5 days. From our above discussions the value of k is about 0.693 147/5, i.e. about 0.14 day^{-1}.

2.4.2 Properties of the decay function. We shall now revert to our general notation of x for the independent variable and y for the dependent variable. Using a calculator we can work out values of $y = e^{-x}$ for a range of x values and plot out the shape of the decay curve as shown in Fig. 2.4. As before, for $x = 0$, $y = 1$, but now as x increases e^{-x} decreases, taking on smaller and smaller values and tending eventually to zero as $x \to \infty$.

The overall shape of the decay curve follows in fact from that of the growth curve. Since we obtain the decay function from e^x by replacing x by $-x$ it follows that the *reflection* of the curve $y = e^x$ in the y axis (i.e. the line $x = 0$) will therefore be the corresponding decay curve $y = e^{-x}$.

Also from the index laws it follows that

$$e^{-x} = \frac{1}{e^x}$$

so the decay function can be thought of as the reciprocal of the growth function.

The power series expansion of e^{-x} can be obtained from (11) by similarly

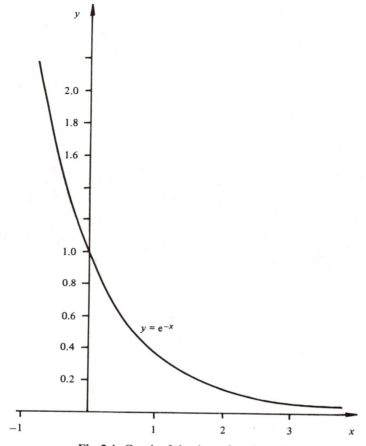

Fig. 2.4. Graph of the decay function e^{-x}.

replacing x by $-x$. Thus

$$e^{-x} = \lim_{n \to \infty}\left[\left(1 - \frac{x}{n}\right)^n\right]$$

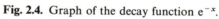

$$= 1 + (-x) + \frac{(-x)^2}{2!} + \frac{(-x)^3}{3!} + \cdots$$

$$= 1 - x + \frac{x^2}{2!} - \frac{x^3}{3!} + \cdots$$

$$= \sum_{k=0}^{\infty} \frac{(-x)^k}{k!} \tag{17}$$

2.4.3 Hyperbolic functions. By studying growth and decay processes we have seen how the separate functions e^x and e^{-x} arise as models. There are situations,

Fig. 2.5. The Tamar bridge from Plymouth.

however, in which special combinations of both e^x and e^{-x} occur as models and in the next two sections we make a study of functions involving such combinations.

To see the sort of combinations of e^x and e^{-x} that might arise, consider the example of a catenary. A catenary is the name given to the curve which a perfectly flexible cable adopts when held at two points and allowed to hang freely under its own weight. Fig. 2.5 shows a photograph of the cables supporting the Tamar bridge on the Devon/Cornwall border. The shape of each cable is an example of a catenary. Other examples in everyday life include telegraph wires, cables carried by pylons, a washing line, a skipping rope, etc.

The derivation of the equation of the shape of a catenary can be found in any book on statics. In its usual form the shape is described by the relation

$$y = \frac{c}{2}(e^{x/c} + e^{-x/c}) \tag{18}$$

where c is constant. This constant is called the parametric of the catenary and is related to w, the weight/unit length of the cable, and to T_0, the tension in the cable at the lowest point, by $c = T_0/w$. Some typical catenaries are shown in Fig. 2.6 for various values of c. Note that as the tension in the cable, and hence the value of c, increases the catenary becomes flatter. However, for no finite value of T_0 (and hence c) will the curve become a straight line. That is to say, there must always be some sag in a cable, allowed to hang freely under its own weight, no matter how tightly it is pulled.

It is also possible to show that the length s of the cable measured from its lowest

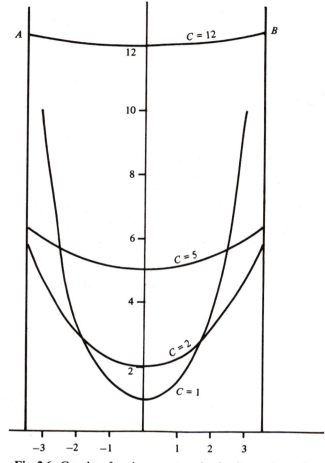

Fig. 2.6. Graphs of various cantenaries for four values of *c*.

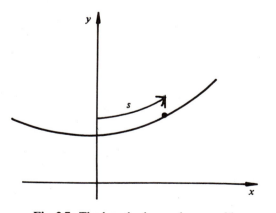

Fig. 2.7. The length along a heavy cable.

point as shown in Fig. 2.7 is given by

$$s = \frac{c}{2}(e^{x/c} - e^{-x/c}) \tag{19}$$

The expressions in (19) and (18) both involve combinations of the exponential growth and decay functions and can be written directly in terms of new functions called *hyperbolic functions*. The two basic hyperbolic functions are the so-called hyperbolic cosine and sine functions denoted, respectively, by $\cosh(x)$ and $\sinh(x)$. In terms of e^x and e^{-x} these functions are defined as follows:

$$\cosh(x) = \tfrac{1}{2}(e^x + e^{-x})$$
$$\sinh(x) = \tfrac{1}{2}(e^x - e^{-x}) \tag{20}$$

As their names suggest there is a correspondence between these functions and the trigonometric functions $\cos(x)$ and $\sin(x)$. Trigonometric functions are discussed in the next chapter and this correspondence will be investigated there. We see, for example, that the shape of a catenary can be described via the relation $y = c \cosh(x/c)$ and the length of the cable from its lowest point by $s = c \sinh(x/c)$.

There is a third hyperbolic function that is often used as a model in science called the hyperbolic tangent and denoted by $\tanh(x)$. This is defined in terms of $\sinh(x)$ and $\cosh(x)$ as follows

$$\tanh(x) = \frac{\sinh(x)}{\cosh(x)} = \frac{e^x - e^{-x}}{e^x + e^{-x}} \tag{21}$$

From their definitions in (20) and (21) it is clear that the properties of the hyperbolic functions are determined entirely by the properties of e^x and e^{-x}. We now look at these properties as well as considering a situation where the hyperbolic tangent function is used as a model.

2.4.4 Properties of the hyperbolic functions. The graphs of $y = \cosh(x)$, $y = \sinh(x)$ and how they relate to the curves $y = \tfrac{1}{2}e^x$ and $y = \pm\tfrac{1}{2}e^{-x}$ are all shown on one plot in Fig. 2.8. From this plot we can usefully note the following:

   (i) when $x = 0$ the values of $\sinh(x)$ and $\cosh(x)$ are given by $\sinh(0) = 0$ and $\cosh(0) = 1$;
  (ii) the graph of $y = \cosh(x)$ lies above the graphs of $y = \tfrac{1}{2}e^x$ and $y = \tfrac{1}{2}e^{-x}$ for all values of x;
  (iii) the graph of $y = \sinh(x)$ lies below the graph of $y = \tfrac{1}{2}e^x$ and above the graph $y = -\tfrac{1}{2}e^{-x}$ for all values of x; i.e. it lies between these graphs;
  (iv) both $\cosh(x)$ and $\sinh(x)$ tend to $\tfrac{1}{2}e^x$ as $x \to \infty$;
  (v) when $x \to -\infty$, $\cosh(x)$ tends to $\tfrac{1}{2}e^{-x}$ and $\sinh(x)$ tends to $-\tfrac{1}{2}e^{-x}$.

In the same way as the graphs of $\sinh(x)$ and $\cosh(x)$ may be determined from the graphs of $y = e^x$ and $y = e^{-x}$, so the power series representations of these hyperbolic functions may be determined from the power series representations of

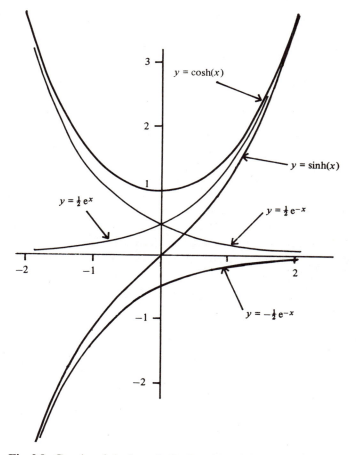

Fig. 2.8. Graphs of the hyperbolic functions sinh (x) and cosh (x).

e^x and e^{-x} given in (11) and (17), i.e.

$$\cosh(x) = \frac{1}{2}\left(1 + x + \frac{x^2}{2!} + \frac{x^3}{3!} + \frac{x^4}{4!} + \frac{x^5}{5!} + \cdots\right)$$

$$+ \frac{1}{2}\left(1 - x + \frac{x^2}{2!} - \frac{x^3}{3!} + \frac{x^4}{4!} - \frac{x^5}{5!} + \cdots\right)$$

$$= 1 + \frac{x^2}{2!} + \frac{x^4}{4!} + \cdots$$

$$= \sum_{k=0}^{\infty} \frac{x^{2k}}{(2k)!} \tag{22}$$

and

$$\sinh(x) = \frac{1}{2}\left(1 + x + \frac{x^2}{2!} + \frac{x^3}{3!} + \frac{x^4}{4!} + \frac{x^5}{5!} + \cdots\right)$$

$$-\frac{1}{2}\left(1 - x + \frac{x^2}{2!} - \frac{x^3}{3!} + \frac{x^4}{4!} - \frac{x^5}{5!} + \cdots\right)$$

$$= x + \frac{x^3}{3!} + \frac{x^5}{5!} + \cdots$$

$$= \sum_{k=0}^{\infty} \frac{x^{2k+1}}{(2k+1)!} \tag{23}$$

These power series representations allow us to calculate the value of these functions to any required degree of accuracy and for any value of x. Most scientific calculators now have these functions available as standard so there is no real need to have to use these power series for calculational purposes. However, just as with the exponential function, we do find that these series expansions can provide us with very useful polynomial approximations for the functions when values of x are small.

For small values of x we can approximate the hyperbolic sine and cosine

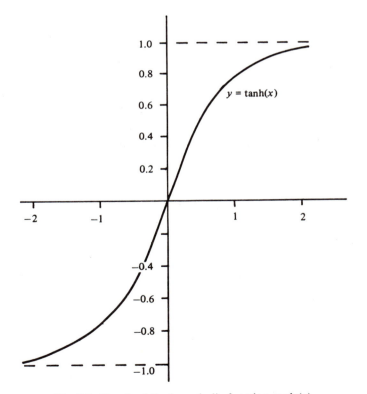

Fig. 2.9. Graph of the hyperbolic function $\tanh(x)$.

functions as follows

$$\sinh(x) \simeq x \text{ or } \sinh(x) \simeq x + \frac{x^3}{6}$$

and

$$\cosh(x) \simeq 1 \text{ or } \cosh(x) \simeq 1 + \frac{x^2}{2} \tag{24}$$

With regard to properties of the hyperbolic tangent function, the graph of $y = \tanh(x)$ may be drawn from values of the function obtained using a calculator. The resulting graph is shown in Fig. 2.9. Note that the value of $\tanh(x)$ when $x = 0$ is $\tanh(0) = 0$. Also as $x \to \infty$, we see that the values of $\tanh(x)$ tend to 1 'from below'; that is to say, for large positive values of x, $\tanh(x)$ is close to unity, but always slightly below it. The line $y = 1$ is said to be an *asymptote* of the graph $y = \tanh(x)$, and we say that $\tanh(x)$ tends asymptotically to the value 1 as x tends to infinity. As $x \to -\infty$, on the other hand, we see that $\tanh(x) \to -1$ 'from above'. The line $y = -1$ is also another asymptote of the graph. With reference to the exponential function we now see that the x axis, i.e. the line $y = 0$, is an asymptote of the graph $y = e^x$.

Finally, before moving on to the Further Worked Examples, it is worth considering one important area where the hyperbolic tangent function is potentially very useful as a model.

Magnetic hysteresis. When a magnetic field of strength H is applied to a ferromagnetic specimen, the magnetic intensity, M, induced in the specimen in the direction of the applied field, varies from zero to the saturation level M_s as shown by the curve OA in Fig. 2.10. As the field is reduced back to zero and then reversed in direction, the induced magnetic intensity also decreases and changes sign and reaches its minimum value $-M_s$ as indicated by the curve ABC in the diagram. A second reversal of the applied field induces another reversal of the magnetic intensity as indicated by the curve CDA, to form a closed loop in the M–H plane, called the *hysteresis loop*. These hysteresis loops are symmetric about the origin and are reproducible, at least after the first few cycles.

The work done by taking the specimen through this cycle of magnetisation and demagnetisation is given by the area enclosed by the hysteresis loop. It is clearly desirable, therefore, to be able to provide a simple mathematical model (i.e. two simple functions) for the two parts of the loop, which can be simply expressed and from which the properties of the hysteresis loop may be deduced without recourse to experimental labour.

Unfortunately, there is no *simple* mathematical description of the hysteresis curves provided by the underlying theory of the phenomenon (as was the case for the catenary). However, one possible simple model description for a hysteresis loop which may be applicable, to some degree of approximation for certain ferromagnetic substances, is provided by the following two functions based on the

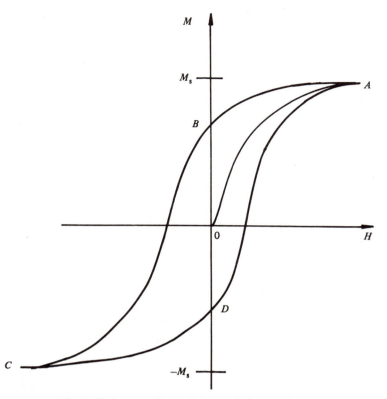

Fig. 2.10. An experimental magnetic hysteresis curve.

hyperbolic tangent function

$$\text{for the curve } ABC\colon M = M_s \tanh\left(\alpha(H + H_c)\right)$$
$$\text{for the curve } CDA\colon M = M_s \tanh\left(\alpha(H - H_c)\right)$$

where $\pm H_c$ are the values of the field H at which the magnetic intensity is zero, and α is some scaling factor which is related to how fast M approaches its limiting values $\pm M_s$.

This model is shown in Fig. 2.11. Note that in the model, M approaches its limiting values $\pm M_s$ as H tends to $\pm \infty$ respectively. However, its distance away from this asymptotic value, and hence the distance apart of the two curves is so small for large values of $\pm H$ that for all practical purposes these functions may well describe an experimentally observed hysteresis curve.

2.5 Further worked examples

Example 1. The luminous intensity I at a depth of x m in clean sea water is given by

$$I(x) = I_0 e^{-2x}$$

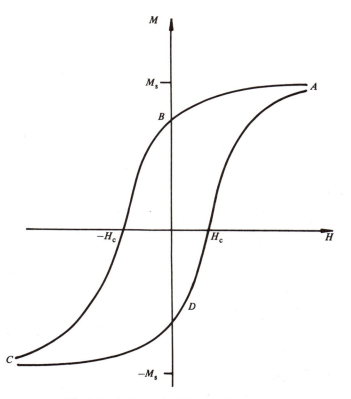

Fig. 2.11. A theoretical hysteresis curve.

where I_0 is the intensity at the surface of the sea. Show that, relative to I_0, the intensity decreases by 48.7% in the first $\frac{1}{3}$ m, but only decreases by a further 37.8% in the next $\frac{2}{3}$ m.

Solution. Substituting $x = \frac{1}{3}$ m into the expression for the luminosity, we find that

$$I(x) = I_0 e^{-2x} = 0.513\,42\,I_0$$

Thus at a depth of $\frac{1}{3}$ m the intensity has been reduced to 51.34% of the intensity at the surface. The intensity has decreased by $100 - 51.34 = 48.66\%$ in the first $\frac{1}{3}$ m. To one decimal place this is 48.7%. Substituting $x = 1$ m into the expression for I, we see that

$$I = I_0 e^{-2x} = 0.135\,34\,I_0$$

The difference in intensities between this value and the value at $x = \frac{1}{3}$ m is therefore

$$0.513\,42\,I_0 - 0.135\,34\,I_0 = 0.378\,08\,I_0$$

The intensity has therefore decreased by a further 37.8% of its original value on the surface.

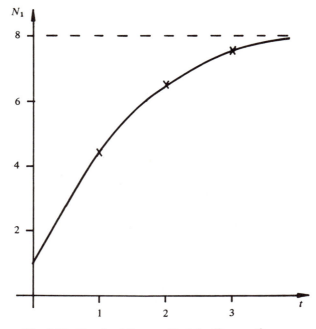

Fig. 2.12. Graph of the von Bertalonffy growth curve.

Example 2. The von Bertalonffy growth curve (Fig. 2.12) is given by the expression

$$N_1 = (a - be^{-kt})^3$$

where a, b and k are constants with $a > b$.

(i) Sketch the behaviour of this curve for $0 \leqslant t < \infty$ assuming values for a, b, k of $2, 1, 1$ respectively.

(ii) Show that for small values of t this growth curve has the same behaviour as the curve

$$N_2 = (a - b)^3 \exp\left(\frac{3bkt}{a - b}\right)$$

and discuss how the two curves differ for larger values of t.

Solution. (i) With $a = 2$, $b = 1$, $k = 1$ we have

$$N_1 = (2 - e^{-t})^3$$

Since the graph of e^{-t} for $t \geqslant 0$ starts at a value of 1 and slowly decreases to zero the function $2 - e^{-t}$ will start at $t = 0$ with a value of $2 - 1 = 1$ and gradually increase to 2. The cube of this function, therefore, will start at 1^3 and rise to 2^3 as $t \to \infty$. The sketch of this function is shown above. To aid drawing we have calculated N_1 at three typical values of t, i.e. $t = 1, 2, 3$:

$$N_1(t = 1) = (2 - e^{-1})^3 = 4.3477$$
$$N_1(t = 2) = (2 - e^{-2})^3 = 6.4834$$
$$N_1(t = 3) = (2 - e^{-3})^3 = 7.4173$$

Note that the curve gets to within 8% of its asymptotic value of $N_1 = 8$ by $t = 3$.

(ii) For small values of t we may approximate the exponential function by the first two terms of its power series expansion. That is

$$N_1 = (a - be^{-kt})^3 = (a - b(1 - kt))^3$$
$$= ((a - b) + bkt)^3$$

If we now expand the cubic expression using the binomial expansion we have

$$N_1 = (a - b)^3 + 3(a - b)^2 bkt + \cdots$$

Note that since we only kept the linear term in t in the exponential expansion, there is no point in keeping higher-order terms in the binomial expansion.

If we now look at the expansion of

$$N_2 = (a - b)^3 \exp\left(\frac{3bkt}{a - b}\right)$$

for small t values we have

$$N_2 = (a - b)^3 \exp\left(\frac{3bkt}{a - b}\right) = (a - b)^3 \left(1 + \frac{3bkt}{(a - b)} + \cdots\right)$$
$$= (a - b)^3 + 3(a - b)^2 bkt + \cdots$$

and hence we find that N_1 and N_2 are the same for *small* values of t, where terms involving t^2, t^3, etc, may be neglected.

However, for larger values of t, since

$$\frac{3bk}{a - b}$$

is greater than zero, N_2 will increase exponentially and tend to $+\infty$ as $t \to \infty$, while, as we have seen, N_1 tends to a finite limit, namely a^3, as $t \to \infty$.

Example 3. We have already seen in 2.4.3 that the equation for the shape of a hanging cable is given by

$$y = c \cosh\left(\frac{x}{c}\right)$$

where the origin of x and y coordinates ($x = 0$, $y = 0$) is a distance c below the lowest point of the curve, and the length of the cable measured from its lowest point is given by

$$y = c \sinh\left(\frac{x}{c}\right)$$

Use these two equations to show that if a cable is suspended between two points at the same level, 200 m apart and has a sag of 20 m at its mid-point, then the length of cable is approximately 205.4 m.

Solution. At one end of the cable $y = c + 20$, and $x = 100$. Therefore

$$c + 20 = c \cosh\left(\frac{100}{c}\right)$$

In order to solve this equation for c, let us suppose that c is much larger than 100, so that we may approximate the hyperbolic cosine function by the first two terms in its power series expansion, i.e.

$$c + 20 = c\left[1 + \frac{1}{2}\left(\frac{100}{c}\right)^2\right]$$

$$= c + \frac{5000}{c}$$

Therefore cancelling the c from both sides, we see that

$$20 = 5000/c, \text{i.e. } c = 250.$$

This is consistent with our assumption that $c > 100$. However, to see the size of error our assumption has induced let us consider the next term in the power series of

$$\cosh\left(\frac{100}{c}\right)$$

i.e.

$$\frac{1}{4!}\left(\frac{100}{250}\right)^4 = 0.0011$$

which is indeed a relatively small quantity compared with the previous two terms, 1 and

$$\frac{1}{2}\left(\frac{100}{250}\right)^2 = 0.08.$$

Hence our approximation scheme seems to be satisfactory.

Substituting this value of c into the second of the given expressions we find that the length of the cable s from its lowest point to one of its supports is

$$s = 250 \sinh\left(\frac{100}{250}\right) = 102.7\,\text{m}$$

Hence the total length of cable required is twice this value, which is 205.4 m. Since c is not exact then it is likely that this figure is only approximate.

2.6 Exercises

1. Use the laws of exponents to simplify the following using positive exponents only:

   (i) $3^2 \times 3^4$ (ii) $x^5 \times x^2$ (iii) $x^{-3} \times x^{-2}$

   (iv) $\dfrac{10^3}{10^5}$ (v) $\dfrac{a^5}{a^7}$ (vi) $\dfrac{x^{-2}}{x}$

   (vii) $\dfrac{b^3}{b^{-2}}$ (viii) $(10^5)^3$ (ix) $(x^{-4})^2$

   (x) $(b^{-4})^{-2}$ (xi) $(4y)^3$ (xii) $\left(\dfrac{a^3 a^{-2}}{b^{-3} b^4}\right)^3$

   (xiii) $\left(\dfrac{2a^3 b^{-3}}{5abc}\right)^{-1}$

2. Using the index laws simplify the following:

   (i) $(a^{0.6}/a^{-1.3})^{1.6}$

   (ii) $((x^{0.5})^3 (x^{-1})^{1.5})^2$

   (iii) $((2b)^{3.5}/(4b)^{1.5})^{0.5}/\sqrt{2}$

3. Expand the following using the binomial expansion:

   (i) $(1 + x)^7$

   (ii) $(1 - x)^7$

   (iii) $(1 - 2p)^5$

   (iv) $(x + y)^n$

   N.B. In (iv) take x^n out as a factor first and then use the binomial expansion to expand $(1 + y/x)^n$.

4. The possible energy levels, E_n, of the electron in a hydrogen atom are given by the formula

   $$E_n = \frac{-e^4 m 2\pi^2}{h^2 n^2}, \quad n = 1, 2, 3, \ldots$$

   where $e = 1.602\,189\,2 \times 10^{-19}$ is the electronic charge in coulombs, $m = 9.109\,534 \times 10^{-31}$ is the rest mass of the electron in kilogrammes, and $h = 6.626\,176 \times 10^{-34}$ is Planck's constant in joule seconds. Obtain the value of the ground-state energy level, i.e. the value of E_1.

5. Use a calculator (or microcomputer) where necessary to write down the values of the following as decimals correct to three significant figures:

   (i) e^2 (ii) $e^{3/2}$ (iii) e^0

   (iv) $1/e$ (v) 10^0 (vi) 10^3

   (vii) 10^{-3} (viii) $\sqrt{e}$ (ix) $e^2 e^{-2}/10$

6. A function of great importance in statistics is one of the form $f(x) = \exp(-x^2)$. Sketch this function for values of x in the range $-3 \leqslant x \leqslant 3$.

7. How many terms in the power series expansion of e^x are needed to obtain the sum
   (i) correct to three decimal places;
   (ii) accurate to within 0.5% when $x = 1$?

8. From the basic definitions of $\sinh(x)$ and $\cosh(x)$ in terms of e^x and e^{-x}, show that
   (i) $\sinh(x \pm y) = \sinh(x)\cosh(y) \pm \cosh(x)\sinh(y)$
   (ii) $(\cosh(x) + \sinh(x))^n = \cosh(nx) + \sinh(nx)$
   (iii) $1 - \tanh^2(x) = \text{sech}^2(x)$ where $\text{sech}(x) = 1/\cosh(x)$

9. Prove that $e^{2x} = (1 + \tanh(x))/(1 - \tanh(x))$

10. By approximating the exponential function by a quadratic function, determine the values of x satisfying
    (i) $e^x = 1.1$
    (ii) $e^{-x} = 0.8$
    (iii) $e^x + e^{-x} = 2.3$

11. If $e^{2x} = 3$ show that $4\tanh(x) = \coth(x)$, where $\coth(x) = 1/\tanh(x)$.

12. If $3\sinh(x) - \cosh(x) = 1$ find the possible values of e^x.

13. Use the method of interval bisection to solve the following equations correct to three decimal places
    (i) $x = e^{-x}$
    (ii) $x^2 + x = e^{2x}$

14. For the exponential growth process given by $M = M_0 e^{kt}$, where t is the time measured in years, obtain the percentage growth increase every month, every six months, and each year when
    (i) $k = 0.1 \text{ yr}^{-1}$
    (ii) $k = 0.2 \text{ yr}^{-1}$

15. In a certain kind of growth process the amount y after time t is given by the expression

    $$y = \frac{k_1 a[\exp(-k_1 t) - \exp(-k_2 t)]}{(k_2 - k_1)}$$

    where a, k_1 and k_2 are positive constants.
    Assuming $a = 1$ and $k_1 = 0.2 \text{ yr}^{-1}$, sketch the plot of y against t in the following cases:
    (i) $k_2 = 0.1$
    (ii) $k_2 = 0.5$
    (iii) $k_2 = 1.0$

    From each plot estimate the time taken for y to reach its maximum value.

16. The excess temperature of a cooling body, which is initially 120 °C above its surroundings, is observed to be about 96 °C after 10.5 minutes and 77 °C after 21 min. Obtain a mathematical model to describe this cooling process and use it to predict the excess temperature after half an hour.

17. In an exponential decay process given by $M = M_0 e^{-kt}$ the original amount M_0 has been reduced by a factor 16 in 321 dy. How many days did it take to be reduced by a factor of 2? What is the value of k?

18. In an experimental growth process given by $N = N_0 e^{kt}$ the value of k is $0.152\,\mathrm{hr}^{-1}$. Find the time taken for N to double.

    Given that 10^6 is approximately 2^{20}, calculate how long it will take for N to be a million times greater than N_0.

19. A wire hangs in the catenary $y = 200 \cosh(0.005x)$. Find the length of wire and the sag at the middle if the points of suspension are at the same height and 100 m apart.

20. An electrical transmission line is stretched between two identical towers 300 m apart on level ground. If the tension is equal to the weight of 1500 m of cable, find the sag midway between the two towers.

3

Functions for science 2: trigonometric functions

Contents: 3.1 scientific context – 3.1.1 a swinging pendulum; 3.1.2 sound vibrations; 3.1.3 insect flight; 3.2 mathematical developments – 3.2.1 revision of trigonometric ratios for acute angles; 3.2.2 angles of any magnitude; 3.2.3 complete solutions of trigonometric equations; 3.3 worked examples; 3.4 further developments – 3.4.1 circular measure (radians); 3.4.2 modelling with trigonometric functions; 3.4.3 small angles; 3.4.4 Euler's formula for complex numbers; 3.5 further worked examples; 3.6 exercises.

3.1 Scientific context

3.1.1 Example 1: A swinging pendulum. The photograph in Fig. 3.1 shows the motion of a swinging pendulum. The multiple images of the pendulum, obtained using stroboscopic lighting, represent its position at the times when the stroboscope flashed which was every tenth of a second. In Fig. 3.2 we have drawn a graph of the horizontal displacement of the pendulum bob from its central position (i.e. when the string is vertical).

None of the functions we have considered so far provides a model for the way the displacement changes with time and this is obvious if we draw the graph of the displacement over a much longer time interval. This graph is shown in Fig. 3.3. As the bob swings backwards and forwards its displacement varies from -0.076 m to $+0.076$ m in a periodic way.

Polynomials and exponential functions are not periodic and so it is clear that a new class of functions is going to be needed if we are to describe the kind of motion shown. This wave-type behaviour is a very common feature of many periodic systems in science.

Question: What kinds of system and physical phenomena exhibit this behaviour?
  Answer: Systems giving rise to waves are obvious candidates. A stretched string that is plucked, waves on water, the motions of tides and planets, light waves, radio waves, alternating currents in electricity, are all examples of wave motion. Vibrations give rise to this wave-like behaviour as does a study of biorythms and so on. This list is really as long as we care to make it.

A simple experiment to illustrate the wave nature of vibrations is illustrated in the next example.

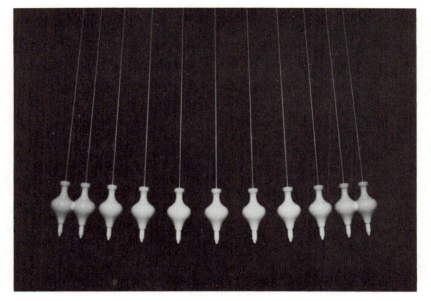

Fig. 3.1. A swinging pendulum.

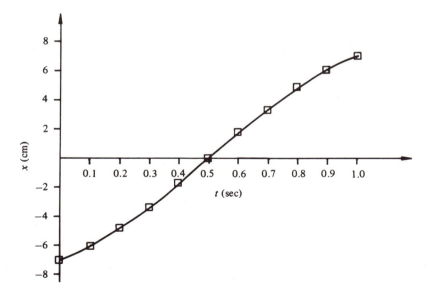

Fig. 3.2. Graph of pendulum bob displacement against time over a short interval.

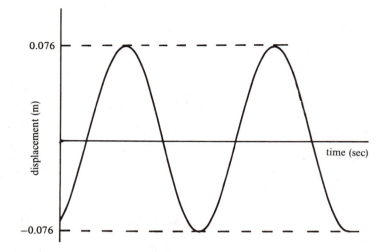

Fig. 3.3. Graph of pendulum bob displacement against time over a longer interval.

3.1.2 Example 2: sound vibrations. In our second example we consider the sound generated by the vibrations of a tuning fork. Suppose that a tuning fork is held close to a microphone, which is connected in a circuit to an oscilloscope. As the tines of the tuning fork vibrate the air pressure near them rises and falls, and hence the microphone which is sensitive to changes in air pressure transmits a varying electric current to the oscilloscope. The oscilloscope is set up to produce a picture on its screen of these current variations as a function of time. The changes in current are proportional to the changes in air pressure. The resulting image is very similar to that shown in Fig. 3.3 for the pendulum. The air pressure continually rises and falls as the tines vibrate.

Oscillatory functions in mathematics which have graphs similar to the one shown in Fig. 3.3 have wide application as models in science. The function shown in Fig. 3.3 is one of the basic *trigonometric functions* called the *sine function*. Your introduction to sine was probably through the ratio of the lengths of the 'opposite side' to the 'hypotenuse side' of a right-angled triangle and it may be a surprise to be told that it is in fact a *basic oscillatory function of science*. However, this alternative interpretation of the sine function, and indeed of the trigonometric functions in general, as periodic functions is but a simple extension of their definitions as properties of right-angled triangles as we shall show in this chapter.

3.1.3 Example 3: insect flight. In this third example we look at the motion of the wings of an insect. Birds and insects are able to fly by a complicated process of flapping their wings. The graph in Fig. 3.4 represents the oscillations of the wings of a locust in flight. Along the vertical axis is measured the angle of the wing between the actual wing position and the horizontal. Along the horizontal axis is

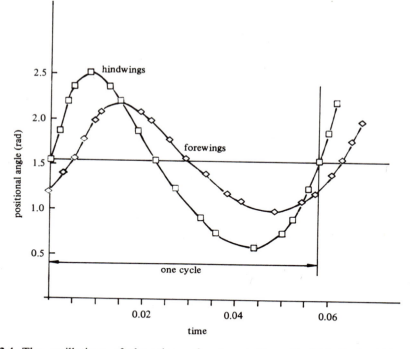

Fig. 3.4. The oscillations of the wings of a locust (from T. Weis-Fogh, 1953, *Proc. Roy. Soc.* B).

measured the time in hundredths of a second. Once again we see that we have an example of oscillatory type motion.

We see from the diagram that in this example the graphs of the functions are more complicated than the previous two. As you might expect the curves are no longer perfectly symmetric. Also the curve for the hindwings appears to lag behind the curve for the forewings. However, we shall see that we can model these vibrations by using the basic trigonometric functions that are introduced in the next section.

3.2 Mathematical developments

3.2.1 Revision of trigonometric ratios for acute angles. We begin with a brief revision of the definitions of the trigonometric ratios sine, cosine and tangent and their values for acute angles (i.e. angles less than 90°). Fig. 3.5 shows a right-angled triangle. The three angles are denoted by A, B, C and are measured in degrees. The lengths of the sides are denoted by a, b, c. The angle C is the right-angle, i.e. $C = 90$ and the side opposite this angle is called the hypotenuse. The sine and cosine of the angle A, denoted by $\sin(A)$ and $\cos(A)$, respectively, are then defined

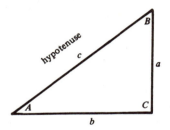

Fig. 3.5. A right-angled triangle.

in the usual way as ratios involving the lengths a, b and c as follows:

$$\sin(A) = a/c \tag{1}$$

$$= \frac{\text{length of side opposite to } A}{\text{length of hypotenuse}}$$

$$\cos(A) = b/c \tag{2}$$

$$= \frac{\text{length of side adjacent to } A}{\text{length of hypotenuse}}$$

These two functions of angle are the basic trigonometric functions and all other trigonometric functions can be defined in terms of them. Thus the tangent of the angle A, denoted by $\tan(A)$, is defined as follows:

$$\tan(A) = \frac{\sin(A)}{\cos(A)} = \frac{a}{c} \div \frac{b}{c} = \frac{a}{b} \tag{3}$$

$$= \frac{\text{length of side opposite to } A}{\text{length of side adjacent to } A}$$

Since the formulas in (1), (2) and (3) are defined as the ratio of two lengths they do not have any units attached to them. In other words, $\sin(A)$ is just a number with no units (and similarly for $\cos(A)$ and $\tan(A)$).

Example: Determine the formulas defining $\sin(B)$, $\cos(B)$ and $\tan(B)$.
Solution: Using the definitions above we have

$$\sin(B) = b/c$$
$$\cos(B) = a/c$$
$$\tan(B) = b/a$$

At first sight the two trigonometric functions sine and cosine may appear to be quite independent; but by using some of the properties of the right-angled triangle in Fig. 3.5 we can find simple relations connecting them.

Property 1: complementary angle formulas. Since the angles of a triangle add up to 180° and $C = 90$ it follows that $B = 90 - A$. Hence using definitions in (1) and (2) as they apply to A and B, we have

$$\sin(B) = \sin(90 - A) = b/c = \cos(A)$$
$$\cos(B) = \cos(90 - A) = a/c = \sin(A)$$

Thus we have the important complementary angle formulas, namely

$$\sin(90 - A) = \cos(A) \quad \text{and} \quad \cos(90 - A) = \sin(A) \tag{4}$$

Property 2: sum of squares formula. If we apply Pythagoras' theorem to the triangle in Fig. 3.5 we obtain a formula relating $\sin(A)$ and $\cos(A)$. We have

$$a^2 + b^2 = c^2$$

Now dividing each side of this equation by c^2 gives

$$(a/c)^2 + (b/c)^2 = 1$$

and using the definitions of $\sin(A)$ and $\cos(A)$ we can write

$$\sin^2(A) + \cos^2(A) = 1 \tag{5}$$

Note that in this expression we have written $\sin^2(A)$ instead of $(\sin(A))^2$; this is the conventional way of writing it. You should be careful with this notation when first meeting it because $\sin^2(A)$ means the square of $\sin(A)$ not sine of 2 times A or sine of the sine of A. Property 2 is extremely useful in manipulating expressions involving trigonometric functions. Furthermore, it shows that we only really require a set of values for one of the trigonometric functions because the values of the others can be found from this one. For example, suppose that we know the values of $\sin(A)$ for all angles A between 0 and 90, then all the values of $\cos(A)$ can be obtained using Property 2 and the values of $\tan(A)$ can then be found from the ratio $\sin(A)/\cos(A)$. We should note, of course, that a calculator can be used to obtain the values of these trigonometric functions, so we need not concern ourselves unduly with the intricacies of calculating their values. However having just said this, it is often very useful in science to know the values of certain trigonometric ratios in closed form when they exist as· such. The numerical values of the trigonometric functions for the angles 30°, 45° and 60° can be obtained relatively easily by considering the two special triangles shown in Fig. 3.6.

Figure 6(a) shows an equilateral triangle divided into two equal halves, so that $A = 60$ and $B = 30$. The lengths of the sides of the equilateral triangle are two units so that the lengths of the sides of the triangle ABC are 2, 1 and $\sqrt{3}$ as shown. Using the definition of sine we find that $\sin(60°) = \frac{\sqrt{3}}{2}$ and $\sin(30°) = \frac{1}{2}$. We thus see that the sine value for each of these angles is a simple closed form expression. These closed form expressions are invaluable in crystallography where angles of 60° and 30° (as well as 90° and 45°) occur quite naturally.

Table 3.1. *Some values of sine, cosine and tangent*

A	$\sin(A)$	$\cos(A)$	$\tan(A)$
0	0	1	0
30	$\frac{1}{2}$	$\frac{\sqrt{3}}{2}$	$\frac{1}{\sqrt{3}}$
45	$\frac{1}{\sqrt{2}}$	$\frac{1}{\sqrt{2}}$	1
60	$\frac{\sqrt{3}}{2}$	$\frac{1}{2}$	$\sqrt{3}$
90	1	0	∞

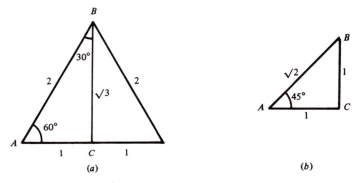

Fig. 3.6. Two special triangles (*a*) with angles 60° and 30° (*b*) with angles 45°.

Example: Fig. 3.6(*b*) shows an isosceles triangle in which angles A and B both equal 45°. What are the values of the sine and cosine of 45°?

Solution: Using the definitions (1)–(3) we have

$$\sin(45°) = \tfrac{1}{\sqrt{2}}$$

and

$$\cos(45°) = \tfrac{1}{\sqrt{2}}.$$

Note that an angle of 45° is the only acute angle for which the sine and cosine are equal.

Table 3.1 summarises the above results. The table also includes the tangents of the angles and the values for 0 and 90°. The table shows the exact values of the trigonometric functions for these rather special angles. More generally as we have already mentioned, we would use our calculator (or microcomputer) to find the values of sine, cosine and tangent. When the value of a trigonometric function is computed, a method similar to that for the exponential function is used, in that the value of the required ratio is obtained progressively until the desired accuracy required has been achieved. Power series representations of the sine, cosine and tangent functions, which enable us to obtain their values to any desired level of accuracy, are given in 3.4.3. These series expansions, however, do not give insight into the very important property possessed by these functions – namely the fact

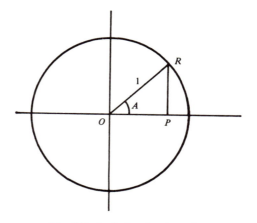

Fig. 3.7. A circle of unit radius.

that they are periodic. For the time being, then, it will be instructive to stick with the geometrical approach that we have adopted. The values of the sine function and particularly the way these values change with angle can easily be seen as follows. Fig. 3.7 shows a circle of radius 1 unit. R is any point on the circumference of this circle in the first quadrant. If we draw in the right-angled triangle ORP the value of $\sin(A)$ is simply the length of the side RP since the hypotenuse has length 1. As we take different positions of the point R we can build up a set of lengths RP for each position and hence we can build up a table of values for $\sin(A)$ as A varies between 0 and 90°. Using such a technique and the exact values from Table 3.1 we can sketch the graph of $\sin(A)$. This is shown in Fig. 3.8, together with the graphs of $\cos(A)$ and $\tan(A)$ whose values are obtained from $\sin(A)$ as described for each angle.

Before leaving this section and considering angles outside of the range 0–90°, it is worth listing the other trigonometric functions along with their definitions and important properties. The cotangent, secant and cosecant functions of A,

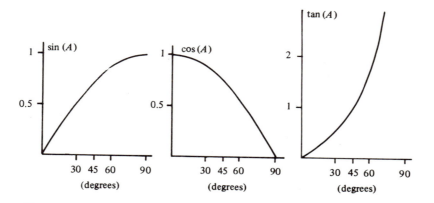

Fig. 3.8. Graphs of sine, cosine and tangent for angles between 0° and 90°.

denoted by cot(A), sec(A) and cosec(A), respectively, are given by

$$\cot(A) = \cos(A)/\sin(A)$$
$$\sec(A) = 1/\cos(A)$$
$$\operatorname{cosec}(A) = 1/\sin(A)$$

It is easily seen that

$$\cot(A) = 1/\tan(A)$$
$$\sec^2(A) = 1 + \tan^2(A)$$
$$\operatorname{cosec}^2(A) = 1 + \cot^2(A)$$

3.2.2 Angles of any magnitude. In the last subsection we defined the trigonometric formulas for angles between 0 and 90°; now we extend these ideas for angles outside this range. We begin by defining positive and negative angles graphically.

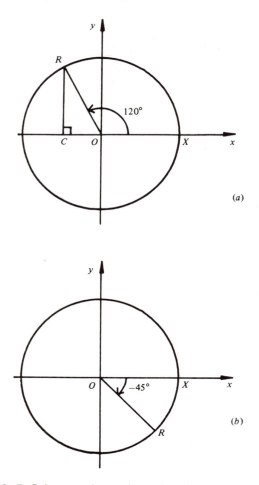

Fig. 3.9. Defining negative angles and angles greater than 90°.

Fig. 3.9 shows a circle of unit radius whose centre at O is the origin of a Cartesian coordinate system (in other words the usual x–y system of axes). Fig. 3.9(*a*) shows the conventional method of defining positive angles; such angles are measured in an anti-clockwise direction starting from the positive x axis. The angle shown is $+120°$. Notice that the position reached by the line OR will be the same if it is rotated from the initial line OX by $120°$, or $(120 + 360)°$, or $(120 + 720)°$, and so on. Thus the angle described by the lines OX and OR may be written in general as

$$A = 120° + 360°n \quad \text{for any integer } n.$$

If an angle is measured from OX in a clockwise direction then it is defined as a negative angle; fig. 3.9(*b*) shows the angle $-45°$. The definitions of the trigonometric formulas as ratios make less geometrical sense for angles greater than $90°$; so we use the idea of a rotation to extend the definitions to such angles.

Consider the angle shown in Fig. 3.9(*a*). To arrive at the point R the radius OX has swept out an angle of $120°$. Now the Cartesian coordinates of the point R are $x = -\frac{1}{2}$ and $y = \frac{\sqrt{3}}{2}$. These values are defined to be the cosine and sine of $120°$ respectively. In a sense this is consistent with the ratio definitions given earlier. The value of y is the ratio of the lengths of the lines RC and OR, and x is minus the ratio of the lengths of the lines OC and OR. However, using the definitions (1) and (2) the values of the ratios themselves are the sine and cosine of $60°$. To avoid any confusion with minus signs we need some clear rule for working out the sine and cosine values when A is an arbitrary angle and not restricted to the first quadrant.

Definition of sin(A) and cos(A) for an arbitrary angle. Suppose, then, that A is any arbitrary angle and we rotate a unit length of the x axis through the angle A as shown in Fig. 3.10.

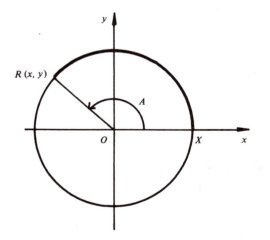

Fig. 3.10. Finding sine and cosine for angles greater than $90°$.

Then we define

$$\sin(A) = \text{the } y \text{ coordinate of the point } R \text{ obtained}$$
$$\cos(A) = \text{the } x \text{ coordinate of the point } R \text{ obtained}$$

With these definitions we can specify and evaluate the sine and cosine of any angle between 0 and 360°.

Example: What are the values of $\cos(180°)$, $\sin(240°)$, $\cos(330°)$?
Solution: Fig. 3.11 shows the three angles given. From Fig.3.11(a) the coordinates of the point R are $x = 1$ and $y = 0$. Thus $\cos(180°) = -1$. From Fig. 3.11(b) the coordinates of the point S are $x = -\frac{1}{2}$ and $y = \frac{-\sqrt{3}}{2}$. Thus $\sin(240°) = -\frac{\sqrt{3}}{2}$. From Fig. 3.11(c) the coordinates of the point T are $x = \frac{\sqrt{3}}{2}$ and $y = -\frac{1}{2}$. Thus $\cos(330°) = \frac{\sqrt{3}}{2}$.

Fig. 3.12 shows the graphs of $\sin(A)$ and $\cos(A)$ for this extended range of angles. The following features of the graphs should be noted:

(i) The graph of $\cos(A)$ is just that of $\sin(A)$ but displaced by 90°. In other words

$$\cos(A) = \sin(90 + A)$$

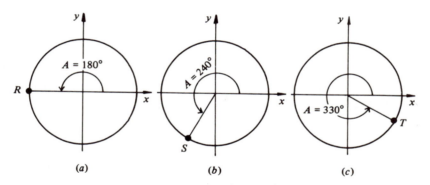

Fig. 3.11. The angles 180°, 240° and 330°.

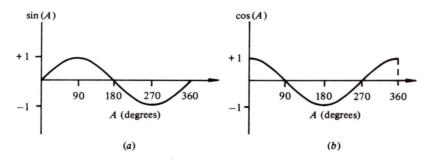

Fig. 3.12. Graphs of sine and cosine for angles between 0° and 360°.

If the angle somehow measures time, as in the oscillations of a pendulum for example, we say that the cosine function 'leads' the sine function by 90°.

(ii) For each function the graph can be built up from the portion between 0 and 90° by suitable combination of repositioning and reflections.

(iii) Each graph lies between the values +1 and −1.

For negative angles the coordinates of the point R again give the values of the sine and cosine function. Fig. 3.13 shows the angle $-A$. The coordinates of the point R, x and y are then the values of $\cos(-A)$ and $\sin(-A)$ respectively. From this diagram we can see that

$$\cos(-A) = \cos(A)$$

and
$$\sin(-A) = -\sin(A)$$

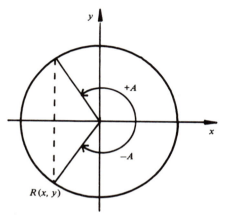

Fig. 3.13. Finding sine and cosine for negative angles.

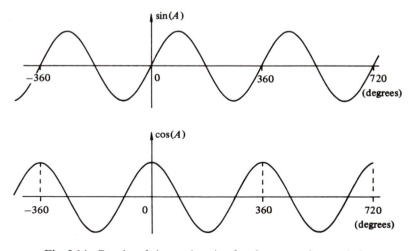

Fig. 3.14. Graphs of sine and cosine for three complete periods.

Thus, for example $\cos(-45°) = \frac{1}{\sqrt{2}}$ and $\sin(-45°) = -\frac{1}{\sqrt{2}}$ (see Fig. 3.9(b)). Finally, for angles greater than 360° we note that the addition of any integral multiple of 360° to an angle does not alter the position of the point R in Fig. 3.10. Thus the angle 785° is just the same as 65° so that the sine and cosine of 785° are, respectively, the sine and cosine of 65°. Thus the graphs of sine and cosine are repeats of those shown in Fig. 3.12 for the range 0–360°. We say that these functions have a *period of 360°*. Because they repeat their values over a fixed period we call them *periodic functions*. Fig. 3.14 shows the general graphs of sine and cosine.

The symmetry of these graphs greatly assists in the evaluation of trigonometric functions and in the solution of trigonometric equations (which are discussed in 3.2.3).

Example: What are the values of $\sin(140°)$ and $\cos(210°)$?
Solution: From the graph of $\sin(A)$ (see Fig. 3.14(a)) it is clear from the symmetry about the line $A = 90$ that the value of $\sin(140°)$ is the same as the value of $\sin(40°)$. Now evaluating $\sin(40°)$ using a calculator we have

$$\sin(140°) = \sin(40°) = 0.6428 \quad \text{(to four decimal places)}.$$

Similarly, from Fig. 3.14(b) the shape of the cosine curve reveals that the value of $\sin(210°)$ must be of the same magnitude as $\sin(30°)$ but with opposite sign. Thus $\sin(210°) = -\sin(30°) = -\frac{1}{2}$.

Finally, the tangent of any angle is the ratio of the sine of the angle to the cosine. To assist us in evaluating the tangent of any angle, the graph of $\tan(A)$ over an extended angle range is also important. The graph of $\tan(A)$ is shown in Fig. 3.15. Note that this graph is periodic with period 180°, but it is a discontinuous function. Because it possesses these discontinuities the tangent function is seldom used as a model in the description of wave-like phenomena.

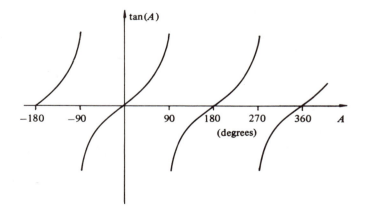

Fig. 3.15. Graph of the tangent function.

The trigonometric functions of sine, cosine and tangent satisfy far more relations than the ones we have written down so far. Formulas that are useful for manipulating equations involving sine, cosine and tangent are given in the short appendix to this chapter.

We complete our initial study of trigonometric functions by considering some of the problems that arise in solving equations involving trigonometric functions.

3.2.3 Complete solutions of trigonometric equations. When solving equations involving trigonometric functions there are in general many solutions. For example, consider the equation

$$\cos(x) = 0.5$$

Using the $\cos^{-1}$ key on our calculator we obtain the single answer $x = 60°$ (the $\cos^{-1}$ function is an example of an inverse function and on most calculators you have to press the inverse key to access it. Inverse functions will be considered in detail in Chapter 4.) But note, by inspecting the graph of cosine (see Fig. 3.14) we see that there are very many angles that also have 0.5 as their cosine value. You can check, for example, that $x = -60°, 300°, 420°, 660°, \ldots$ are also solutions of $\cos(x) = 0.5$. In general we can write the solution as

$$x = \pm 60° \pm 360°n \quad \text{for } n = 0, 1, 2. \ldots$$

Example: Obtain all the solutions of the equations $\tan A = 2.7$.
Solution: Using the $\tan^{-1}$ key on a calculator we have $A = 69.7°$. However from Fig. 3.16 we see that other possible solutions are:

$$A = 69.7 + 180, \quad 69.7 + 360, \ldots$$

as well as

$$A = 69.7 - 180, \quad 69.7 - 360, \ldots$$

In general we can write the solution as

$$A = 69.7 \pm 180n, \quad \text{for } n = 0, 1, 2, \ldots$$

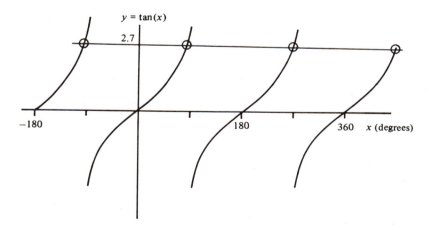

Fig. 3.16. Solutions of the equation $\tan(A) = 2.7$.

These two examples illustrate the care that is needed in solving equations which involve trigonometric functions. We really need to have some idea of the size of the angles before we begin; usually we need to know in which of the following quadrants the angle belongs $(0, 90)$, $(90, 180)$, $(180, 270)$, $(270, 360)$. As we have said we will investigate these ideas further in Chapter 4 where we discuss inverse functions.

3.3 Worked examples

Example 1. Fig. 3.17 shows a graph of the horizontal displacement from the vertical position x of the bob of a simple pendulum as a function of time t. The period of the motion is $15\,\mathrm{s}$ and the largest displacement of the bob is $1\,\mathrm{m}$. Find a formula for x in terms of t using a sine or cosine function. Find the smallest value of t for which $x = -0.75$.

Solution. If we compare the graph in Fig. 3.17 with the graphs of sine and cosine shown in Fig. 3.14, then the displacement could be represented by a sine function because x values only vary between $+1$ and -1. We may therefore write

$$x = \sin(A(t))$$

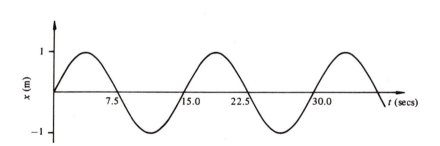

Fig. 3.17. The displacement of the bob of a simple pendulum.

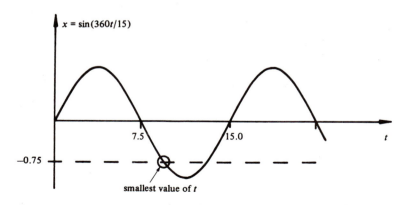

Fig. 3.18. A solution of the equation $\sin(360t/15) = -0.75$.

where the 'angle' A is a function of t. Clearly, the curve in the diagram has a period of 15 s and the sine curve has a period of 360. Thus the function $A(t)$ is given by

$$A(t) = 360t/15$$

and the formula relating x and t is

$$x = \sin(360t/15).$$

This relationship is shown graphically in Fig. 3.18. There are many values of t for which $x = -0.75$ the smallest value is shown. The smallest positive angle whose sine is -0.75 is $(180 + 48.6) = 228.6$. Thus the smallest value of t is $15 \times 228.6/360 = 9.53$ s.

Example 2. The voltage across a component in an electrical circuit is given by $V = 7\sin(18\,000t)$ while the current flowing through the component is $I = 5\cos(18\,000t)$. Derive an equation relating V to I which does not involve the variable t. Is Ohm's law satisfied in this component? If t is measured in seconds determine by how many seconds the current 'leads' the voltage in this component.

Solution. We can eliminate t from the two expressions for V and I using the sum of squares formula in (5). We have that

$$\sin^2(18\,000t) + \cos^2(18\,000t) = 1$$

so that

$$(V/7)^2 + (I/5)^2 = 1$$

Thus multiplying through by $7^2 \times 5^2$, i.e. by 1225, we see that

$$25V^2 + 49I^2 = 1225$$

i.e.

$$V^2 = (1225 - 49I^2)/25$$

This is clearly not a linear relationship between V and I and so Ohm's law, which states that the current flowing in a conductor is proportional to the applied voltage, does not apply for the component in question. We should note that with alternating currents, Ohm's law quite often does not describe the resulting behaviour in an electrical component. Since the cosine function 'leads' the sine function by 90° it follows that in degrees the current leads the voltage by this amount. Therefore we must have $18\,000t = 90$ for the number of seconds t by which the current leads the voltage. The lead is therefore $\frac{1}{200}$ s.

3.4 Further developments

3.4.1 Circular measure (radians). In Section 3.2 we used degrees as the unit of angle, but degrees are not the only units in which angles are measured. An alternative to degrees, which is used very widely in science, and which in many ways is a more natural measure of angle, is a measure known as *radian or circular* *measure*. The basis on which this measure is defined is the angle in a sector of a circle, thus the name circular measure.

Fig. 3.19 shows a sector of a circle of radius *a* containing an angle θ. The angle determines the shape of the sector so that two sectors of circles of different radii having the same angle will be similar in shape. This is illustrated in Fig. 3.20. Of course the arc length of the larger sector is longer than that of the smaller sector but the ratio of arc length to radius is the same in each case. This ratio is fixed by the angle θ in the sector. This provides an alternative method of defining the way we measure angles.

Radian measure.

The measure of an angle in radians is defined by the ratio

$$\theta = \frac{\text{arc length}}{\text{radius}} \qquad (6)$$

For example, if the radius of the circle in Fig. 3.19 is 4 cm and the arc length is 2 cm, the angle of the sector is $\frac{2}{4}$ rad or 0.5 rad. Of course, as well as providing an alternative unit of measure for angles, the definition gives a formula for the

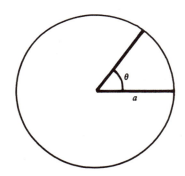

Fig. 3.19. A sector of a circle containing angle θ.

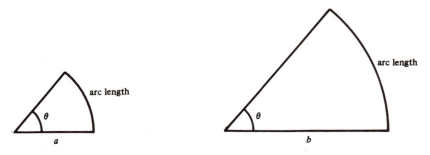

Fig. 3.20. Sectors of different circles with equal angles are similar.

arc length in terms of the radius. We have

$$\text{arc length} = \text{radius} \times \text{angle}$$

where the angle is measured in radians.

There is a simple relation between the units of degrees and radians, and there are many ways of getting at this relation. For instance, we could use a right-angled sector, a half-turn or a complete circle. We shall use the last. For a circle of radius a, the length of the circumference of the circle is $2\pi a$ and the angle turned through is $360°$. Hence in terms of radians we have

$$A = \frac{\text{arc length}}{\text{radius}} = \frac{2\pi a}{a} = 2\pi$$

Thus 2π rad $= 360°$.

To convert an angle from radians to degrees we use the formula

$$1 \text{ rad} = (360/2\pi)° = 57.295 \cdots \text{deg}$$

Alternatively if we wanted to convert an angle given in degrees to radians we would use the formula

$$1 \text{ degree} = (2\pi/360)\text{rad} = 0.017\,453\,2 \cdots \text{rad}$$

Example: What are the following angles when measured in radians: $0°$, $30°$, $45°$, $60°$, $90°$, $120°$, $180°$?

Solution: The following table shows the radian form of these common angles:

angle measured in degrees	angle measured in radians
0	0
30	$\pi/6$
45	$\pi/4$
60	$\pi/3$
90	$\pi/2$
120	$2\pi/3$
180	π

It is important when evaluating the trigonometric functions to ensure that your calculator knows exactly which units of measure you are using. There is usually a key or a switch to define degrees (often written DEG) or radians (often written RAD). To illustrate the care that is needed the following is the value of $\sin(1)$ in each unit of measure:

$$\sin(1 \text{ rad}) = 0.841\,47 \quad \text{and} \quad \sin(1°) = 0.017\,45.$$

They are quite different. When using a computer the trigonometric functions available in most high-level languages require that the angle is measured in radians.

Radian measure will be used almost exclusively throughout the rest of this text.

It is, as we have said, a more natural measure, certainly in science. Degrees are mostly only used in geometry.

You may have noticed on your calculator a unit written as GRA; this is a unit used in some European countries and is short for GRADE. One grade is one hundredth of a right-angle. Thus

$$1 \text{ grade} = 0.9 \text{ deg} = 0.007\,854 \text{ rad}$$

3.4.2 Modelling with trigonometric functions. In Section 3.1 we introduced three examples of physical phenomena, the properties of which could be described by graphs similar to those of the sine and cosine functions. Further, in worked example 1, we used the sine function to represent the displacement of a pendulum. There are many phenomena in science that can be modelled using the trigonometric functions and in this subsection we consider certain parameters that will help in this modelling. Consider the function of x given by

$$f(x) = A \sin(kx + c) \tag{7}$$

Usually this function, for appropriate values of A, k and c, can be made to 'fit' most wave-like disturbances. The parameter A is called the *amplitude* of the function and is a key parameter in modelling the 'size' of the disturbance. The quantity c is known as the *phase* of the function and is needed in order to bring the function exactly into synchronisation with the disturbance. The period of oscillation of the

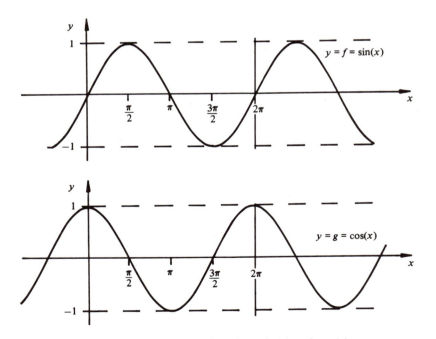

Fig. 3.21. Graphs of the functions $\sin(x)$ and $\cos(x)$.

wave-like disturbance may be exactly the same as the function in (7), but slightly *out of phase*. The parameter c will correct for this. We see, for example in Fig. 3.21, that the functions $f(x) = \sin(x)$ and $g(x) = \cos(x)$ are both periodic with period 2π. As we have noted before, close observation of the two curves shows that the cosine curve is identical in shape to the sine curve but is displaced from it by $\pi/2$ rad. Sine and cosine are therefore out of phase by $\pi/2$ rad. However, we can model the cosine function using the sine function if we use a phase, c, of $\pi/2$. Explicitly we have $\cos(x) = \sin(x + \pi/2)$. The third parameter k is determined by the *period* of the disturbance being modelled. The period of the sine function in (7) is 2π rad. If the angle $kx + c$ changes by 2π, then x must have changed by $2\pi/k$. The quantity $2\pi/k$ is known as the period of the function $f(x)$ in (7). Note that the period relates to x, not to the angle $kx + c$ (its period is 2π rad). Once we know the period of the wave-like disturbance, then the value of k in (7) can be fixed. In connection with the size of the oscillations, the amplitude A is half the difference between the largest and smallest values of the disturbance. For the simple functions shown in Fig. 3.21 the largest value is $+1$ and the smallest value is -1. The amplitude of these functions is 1 (i.e. $(+1 - (-1))/2 = 1$). The function $A\sin(kx + c)$ clearly has amplitude A because its values oscillate between $+A$ and $-A$.

The following table summarises these definitions

Name	Symbol	Meaning
Amplitude	A	half the difference between the largest and smallest values of a periodic function
Period	$2\pi/k$	the value of x over which the graph of a periodic function repeats iself
Phase angle	c	amount by which the graph is out of phase with a similar function

It is worth pointing out that our choice of a sine function in (7) was really quite arbitrary. All that has been said for the function $f(x) = A\sin(kx + c)$ applies equally well to the cosine function $g(x) = A\cos(kx + c)$. The choice of a sine or cosine is often one of personal preference.

Example: Sketch a graph of the function $2 + 3\sin(0.5x - \pi/4)$ showing one complete cycle, and write down the values of the amplitude and period.

Solution: The graph of the function is shown in Fig. 3.22. The period of the function is $2\pi/0.5 = 4\pi$. The largest and smallest values of the function are $+5$ and -1. Hence the amplitude is

$$(+5 - (-1))/2 = 3.$$

Now with these ideas we can 'work backwards' and use the trigonometric functions to model physical phenomena for which we know the form of the graph

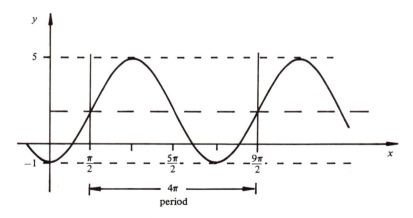

Fig. 3.22. Graph of the function $f(x) = 2 + 3\sin(0.5x - \pi/4)$.

of some property. Consider again the simple pendulum of 3.1.1. The amplitude is 0.076 and the period of the motion is 14 s. The dependent variable in this case is time t (this is usually the case in modelling oscillations as we describe changes with time), so take as a model for the graph in Fig. 3.23 the function $x = A\sin(kt + c)$. The amplitude A is 0.076 and setting the period $2\pi/k$ equal to 14 we have $k = \pi/7$.

Finally we can find the phase angle c using the intercept $t = 5$ when $x = 0$. We then have $5k + c = 0$ so that $c = -5k = -5\pi/7$. The motion of the simple pendulum

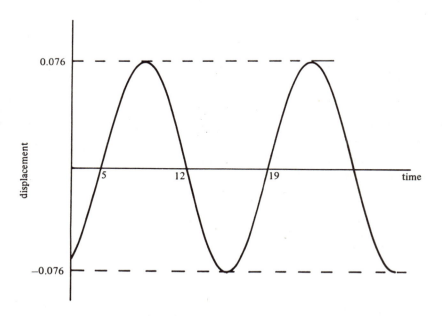

Fig. 3.23. Graph of the displacement of the simple pendulum of section 3.1.1.

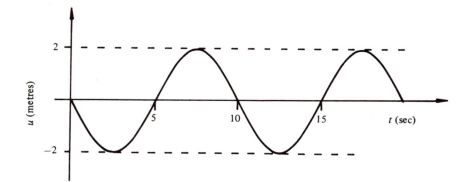

Fig. 3.24. Graph of a periodic function with amplitude 2 metres and period 10 seconds.

is described by

$$x = 0.076 \sin (\pi t/7 - 5\pi/7)$$
$$= 0.076 \sin \pi(t - 5)/7$$

Example: From the graph of a periodic function shown in Fig. 3.24, find a functional relation between the variables u and t.

Solution: From the diagram we see that the period is 10 s, the amplitude of the oscillations is 2 m and the graph is 5 s out of phase with the graph of sine. So we can model the relation between u and t by

$$u = A \sin (kt + 5)$$

Hence the amplitude $A = 2$. The period $2\pi/k = 10$, hence $k = 2\pi/10 = 0.2\pi$. Thus the formula

$$u = 2 \sin (0.2\pi t + 5)$$

is a model for this problem.

3.4.3 Small angles. In Chapter 2 we showed that the function $f = e^x$ could be expressed as a power series in x which was particularly useful for obtaining approximate functional forms for e^x when x was small. In the same way we can find series expansions for the trigonometric functions. The derivations of these formulas are given in Chapter 8; here we quote the series without proof.

Series expansions of sine, cosine and tangent. Provided x is measured in radians, we can write

$$\sin (x) = x - \frac{x^3}{3!} + \frac{x^5}{5!} - \frac{x^7}{7!} + \cdots \tag{8}$$

$$\cos (x) = 1 - \frac{x^2}{2!} + \frac{x^4}{4!} - \frac{x^6}{6!} + \cdots \tag{9}$$

$$\tan (x) = x + \frac{x^3}{3} + \frac{2x^5}{15} + \frac{17x^7}{315} + \cdots \tag{10}$$

For sufficiently small values of x the first term in each of these expansions gives a good approximation to the value of the functions. We then have

$$\sin(x) \simeq x \tag{11}$$

$$\cos(x) \simeq 1 \tag{12}$$

$$\tan(x) \simeq x \tag{13}$$

The following table gives some idea of the accuracy of these approximations:

x (degrees)	x (radians)	$\sin(x)$	$\cos(x)$	$\tan(x)$
0	0	0	1	0
1	0.017 453 3	0.017 452 4	0.999 847 7	0.017 455 0
2	0.034 906 7	0.034 899 4	0.999 390 8	0.034 920 7
3	0.052 359 9	0.052 335 9	0.998 629 5	0.052 407 7
4	0.069 813 2	0.069 756 4	0.997 564 1	0.069 926 8
10	0.174 532 9	0.173 648 1	0.984 807 7	0.176 326 9

This table shows that the errors are very small for angles up to 4° and that even for 10° the errors are less than 2%. Of course, better approximations are obtained by taking the first two terms in each series. Thus simple and accurate functional forms, involving only a few powers of x, can easily be obtained to represent the trigonometric functions. This is often extremely convenient in analytical work where trigonometric functions are involved.

3.4.4 Euler's formula for complex numbers. In Chapter 1 we introduced complex numbers and we saw how to add, multiply and divide two complex numbers. Adding two complex numbers is reasonably easy, all we do is to add the real and imaginary parts. For example,

$$z_1 + z_2 = (a_1 + ib_1) + (a_2 + ib_2)$$
$$= (a_1 + a_2) + i(b_1 + b_2).$$

Multiplication and division is more tedious with the complex numbers in this form. In this section we introduce a different way of denoting a complex number which allows multiplication and division to be carried out quickly and which establishes an important and fundamental link between the exponential function and the trigonometric functions cosine and sine. This link is often referred to as *Euler's formula*.

We begin with a geometric representation of a complex number on a special diagram called the *Argand diagram*. Because the Argand diagram has x and y axes it is perhaps more natural to refer to a complex number as $z = x + iy$ instead of $a + ib$. The complex number $z = x + iy$ is then represented by the point P with

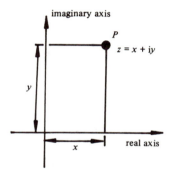

Fig. 3.25. Argand diagram showing the complex number $z = x + iy$.

Cartesian coordinates x and y as shown in Fig. 3.25. In this diagram the Ox axis is called the *real axis* and the Oy axis is called the *imaginary axis*. The representation $z = x + iy$ is called the *Cartesian form* of the complex number.

Fig. 3.26, however, shows an alternative way of specifying the point P using polar coordinates r and θ. It is Euler's formula that provides the link between the two representations. From the geometrical definitions of cosine and sine, we see that

$$z = x + iy = r\cos(\theta) + ir\sin(\theta)$$
$$= r(\cos(\theta) + i\sin(\theta))$$

Now the expression $\cos(\theta) + i\sin(\theta)$ can be related to the exponential function by using the series expansions for $\sin(\theta)$ and $\cos(\theta)$. We have

$$\cos(\theta) + i\sin(\theta) = \left(1 - \frac{\theta^2}{2!} + \frac{\theta^4}{4!} + \cdots\right) + i\left(\theta - \frac{\theta^3}{3!} + \frac{\theta^5}{5!} + \cdots\right)$$
$$= 1 + i\theta + \frac{(i\theta)^2}{2!} + \frac{(i\theta)^3}{3!} + \frac{(i\theta)^4}{4!} + \cdots$$

The expression on the right-hand side is just the series expansion for the exponential

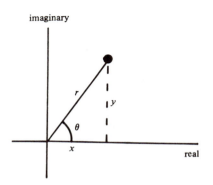

Fig. 3.26. The polar coordinate form of a complex number.

function $e^{i\theta}$. So we have

$$\cos(\theta) + i\sin(\theta) = e^{i\theta} \tag{14}$$

This is *Euler's formula* and it is a very useful link between the exponential and the trigonometric functions. The formula allows us to write $\cos(\theta)$ and $\sin(\theta)$ in terms of $e^{\pm i\theta}$ and vice versa. For example, we can now write a complex number z in *polar form*. We have

$$z = re^{i\theta} \tag{15}$$

The quantity r is called the *modulus* of z and θ is called the *argument* of z. Clearly, from the Argand diagram $r = \sqrt{(x^2 + y^2)}$ and $\tan(\theta) = y/x$.

Example: Put the following complex numbers in polar form:

$$z_1 = 1 + i \quad \text{and} \quad z_2 = 1 - i$$

Solution: On the Argand diagram below the two complex numbers are represented by the points $P(z_1)$ and $Q(z_2)$ (Fig. 3.27). We see that in each case $r^2 = 1^2 + 1^2 = 2$; so the modulus of each complex number is $\sqrt{2}$. (Note that the modulus is always taken as the positive square root.) For z_1 the angle θ is clearly $\pi/4$. Thus $z_1 = 1 + i = \sqrt{2}e^{i\pi/4}$. For z_2 we have to be more careful. According to the formula for θ we have $\tan(\theta) = y/x = -1$. Now the $\tan^{-1}$ key on a calculator will give $\theta = -\pi/4$. However, the angle θ must lie between 0 and 2π. Clearly, the angle we are looking for is $2\pi - \pi/4 = 7\pi/4$. Thus $z_2 = 1 - i = \sqrt{2}\exp(i7\pi/4)$.

The polar form of a complex number makes multiplication and division of complex numbers an easy operation. If $z_1 = r_1\exp(i\theta_1)$ and $z_2 = r_2\exp(i\theta_2)$ then

$$z_1 z_2 = r_1 r_2 \exp i(\theta_1 + \theta_2)$$

so that the product of a complex number of modulus r_1 argument θ_1 and a complex number of modulus r_2 argument θ_2 is just a complex number of modulus $r_1 r_2$ argument $\theta_1 + \theta_2$. The sum and product of two complex numbers can be illustrated on the Argand diagram using the Cartesian and polar representations respectively (Fig. 3.28).

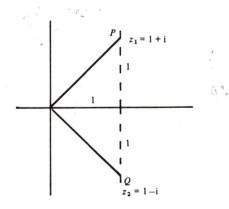

Fig. 3.27. The complex numbers $z_1 = 1 + i$ and $z_2 = 1 - i$.

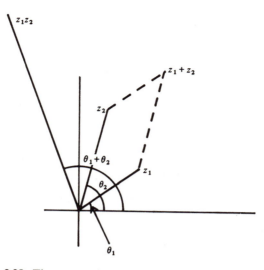

Fig. 3.28. The sum and product of two complex numbers.

Example: Evaluate $(2e^{i\pi/2} \times 4e^{i7\pi/4})/3e^{i\pi}$ and express the answer in Cartesian form.
Solution:

$$2e^{i\pi/2} \times 4e^{i7\pi/4} = 8\exp\left[i\left(\frac{\pi}{2}+\frac{7\pi}{4}\right)\right] = 8e^{i9\pi/4}$$

$$8e^{i9\pi/4}/3e^{i\pi} = \frac{8}{3}\exp\left[i\left(\frac{9\pi}{4}-\pi\right)\right] = \frac{8}{3}e^{i5\pi/4}$$

The Cartesian form can be found using Euler's formula. We have

$$\frac{8}{3}e^{i5\pi/4} = \frac{8}{3}\left(\cos\frac{5\pi}{4}+i\sin\frac{5\pi}{4}\right)$$
$$= -1.886 - 1.886i$$

This example shows how much easier is multiplication and division of complex numbers when we use the polar form.

Complex numbers are important in many branches of physics and engineering. For example, alternating current circuits are important because our usual domestic supply of power is an AC supply. Because of the way power is generated the voltage V can be represented by a cosine function $V = V_0\cos(wt)$ where V_0 is the constant peak voltage, w is the angular frequency of the generator and t is the time. This alternating current is often represented as the projection on the x axis of what is called a *phasor* (see Fig. 3.29). The phasor representing the voltage is the rotating line of length V_0 shown, whose angular velocity is w. Clearly, the projection on the x axis is $V_0\cos(wt)$. The important thing is that the phasor can be represented as a complex number moving in the complex plane. The phasor is $z = V_0e^{iwt}$ and then V is the real part of z. An alternating voltage supply as input

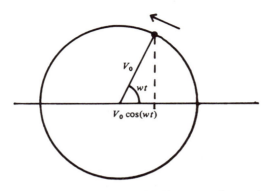

Fig. 3.29. The projection of the phasor on the *x*-axis.

to any electrical device will cause an alternating output of the same frequency. Thus any output can also be represented as some complex phasor. Analysing the system using complex numbers saves a lot of work and is easier. If the manipulation of these phasors involves multiplication or division then the polar form is the easiest to work with. If the manipulation involves addition then we can easily change the complex number to Cartesian form, find the sum and convert back to polar form. The important rule that provides the link and that we need later in Chapter 12 and 13 is Euler's formula

$$e^{i\theta} = \cos(\theta) + i\sin(\theta).$$

3.5 Further worked examples

Example 1. A person's blood pressure $P(t)$ (in mb) at time t (in s) is modelled by the function

$$P = 95 + 25\cos(6t).$$

Using this model find the values of the maximum and minimum pressures. (These are called the systolic and diastolic pressures respectively.) What is the period of time between consecutive maxima in the pressure?

Solution. The function $P(t)$ is a periodic function with amplitude 25 and period $2\pi/6$. Thus the largest pressure is $95 + 25 = 120$ mb, and the smallest pressure is $95 - 25 = 70$ mb.

The extreme values are repeated in a time equal to the period. Thus the extreme values are achieved every $(2\pi/6) = 1.047$ s.

Example 2. Figure 3.30 shows graphs of the angular displacement of the forewings and hindwings of an insect. The period of each motion is 0.06 s. The amplitude of the motions of the forewings and hindwings are 0.5 rad and 1 rad respectively. Find models to describe each of these graphs.

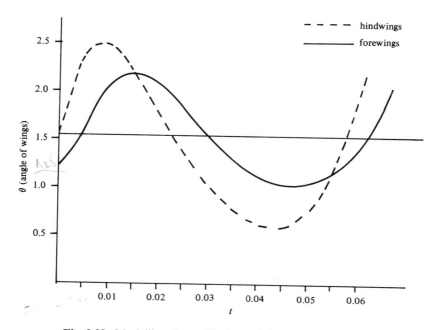

Fig. 3.30. Modelling the oscillations of the wings of a locust.

Solution. It is clear from the graphs that a sine or cosine curve will not fit these graphs exactly. That a known mathematical function does not describe reality exactly is often the case in modelling. However, each graph exhibits many of the features of a sine function so we may usefully proceed to find models in terms of this function.

So for the hindwings, suppose we denote the angle of the wings by h, and let

$$h = H + a \sin(kt + c)$$

for constants H, a, k and c. For the hindwings the amplitude a is 1 and $H = 1.5$. The graph passes through the origin so that $c = 0$. The period of the motion is 0.06; thus

$$2\pi/k = 0.06 \text{ and then } k = 2\pi/0.06$$

The model of the angular displacement for the hindwings is then

$$h = 1.5 + \sin((2\pi/0.06)t)$$

For the forewings, denoting the angle by f, we let

$$f = F + a \sin(kx + c)$$

for constants F, a, k and c. For the forewings the amplitude a is 0.5 and $F = 1.5$. The period of the motion is 0.06; thus $k = 2\pi/0.06$, as before. The graph of the forewings' angle leads the hindwings' by 0.005 s so that the phase angle is -0.005, i.e.

$$c = -0.005$$

For the forewings the model of the angular displacement is

$$f = 1.5 + 0.5 \sin((2\pi/0.06)t - 0.005).$$

Example 3. A pendulum swings so that after t s the displacement of its bob from the vertical is given by the equation

$$x = 2.4 \sin(t) - 0.7 \cos(t)$$

Express x in the form $A \sin(t - a)$ evaluating the constants A and a. Find the smallest value of t for which $x = 1.5$ and $x = -1.5$.

Solution. Assume that x is written as $x = A \sin(t - a)$. Expanding this expression using the appropriate addition formula given in the appendix, we have

$$x = A \sin(t)\cos(a) - A \cos(t)\sin(a)$$

If we compare this formula with the given one for x, i.e.

$$x = 2.4 \sin(t) - 0.7 \cos(t)$$

we must have

$$A \cos(a) = 2.4 \quad \text{and} \quad A \sin(a) = -0.7.$$

Solving for A we get

$$(A \cos(a))^2 + (A \sin(a))^2 = A^2 = 2.4^2 + 0.7^2$$

i.e.

$$A = 2.5$$

Eliminating A we have

$$\cos(a) = 2.4/2.5 \quad \text{and} \quad \sin(a) = 0.7/2.5$$

Solving for a gives $a = 0.284$ (in radians). Thus we can write

$$x = 2.5 \sin(t - 0.284)$$

The graph of this function is shown in Fig. 3.31. The smallest values of t for which $x = 1.5$ and $x = -1.5$ are also shown on this graph. When $x = 1.5$ we have $\sin(t - 0.284) = 0.6$. The angle whose sine is 0.6 is 0.644, so that $t = 0.644 + 0.284 = 0.928$. When $x = -1.5$ we have $\sin(t - 0.284) = -0.6$. The angle whose sine is -0.6 is 3.785, so that $t = 3.785 + 0.284 = 4.069$.

Example 4. In the field of alternating current theory the voltage response of an element in a circuit is related to an alternating current input through the equation $V = IZ$, where I and V are the phasors representing the current and voltage, respectively, and Z is a complex number called the impedance. For a current of frequency w the form of Z for a resistance R, a capacitance C, and an inductance L, is given by R, iwL and 1/iwC respectively. In a circuit in which a resistance,

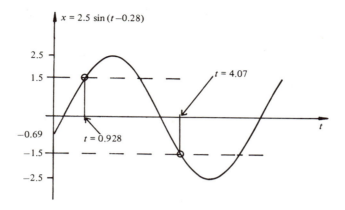

Fig. 3.31. Graph of the function $x = 2.5 \sin(t - 0.28)$.

capacitance and inductance are connected in series the impedance Z is the sum of the individual impedances. If the actual current in such a circuit is $i = 3 \cos(2t)$ find the overall voltage v.

Solution. First we express the actual current as a phasor I. In polar form we can write

$$i = 3 \cos(2t) = \text{real part of } 3e^{2ti}$$

so that the phasor is

$$I = 3e^{2ti}$$

The impedance of the circuit is given by the sum of the individual impedances:

$$Z = R + iwL + 1/iwC$$

where $w = 2$. Writing $1/iwC$ as $-i/wC$ (since $1/i = -i^2/i = -i$) the form of Z is

$$Z = R + i(wL - 1/wC)$$

The phasor representing the voltage is related to I and Z by $V = IZ$, so on substituting for I and Z we have

$$V = 3e^{2ti}(R + i(2L - 1/2C))$$

The actual voltage v is the real part of the phasor V. Thus

$$v = \text{real part of } 3e^{2ti}(R + i(2L - 1/2C))$$
$$= \text{real part of } 3(\cos 2t + i \sin 2t)(R + i(2L - 1/2C))$$

Carrying out the multiplication and choosing the real part we have

$$v = 3R \cos 2t - (6L - 3/2C) \sin 2t$$

3.6. Exercises

1. Write each of the following in terms of angles between 0 and 360°:
   (i) $\sin(140°)$ (ii) $\cos(400°)$ (iii) $\sin(630°)$
   (iv) $\tan(720°)$ (v) $\cos(3360°)$ (vi) $\sin(-840°)$
   (vii) $\cos(-2160°)$ (viii) $\tan(-300°)$

2. Express the following in terms of the sine of an angle in the first quadrant:
   (i) $\sin(636°)$
   (ii) $\sin(-342°)$
   (iii) $\cos(542°)$
   (iv) $\cos(-124°)$

3. Express the following in terms of the cosine of an angle in the first quadrant:
   (i) $\sin(420°)$
   (ii) $\cos(1000°)$
   (iii) $\cos(-123°)$
   (iv) $\sin(350°)$

4. Use the triangle definitions of sine, cosine and tangent in the following:
   (i) $\sin(A) = \frac{1}{3}$ find $\cos(A)$ and $\tan(A)$
   (ii) $\cos(B) = \frac{3}{5}$ find $\sin(B)$ and $\tan(B)$
   (iii) $\tan(C) = \frac{4}{3}$ find $\sin(C)$ and $\tan(C)$
   (iv) $\sin(x) = (a^2 - b^2)/(a^2 + b^2)$ find $\cos(x)$ and $\tan(x)$

5. Obtain the angles between 0 and 360° that satisfy the following equations:
   (i) $2\sin(x) = \sqrt{3}$
   (ii) $\cos^2(x) = 0.75$
   (iii) $2\sin^2(x) + \sin(x) = 1$
   (iv) $3\tan^2(x) - 2\sqrt{3}\tan(x) - 3 = 0$

6. Use graphs of $\sin(A)$, $\cos(A)$ and $\tan(A)$ to find all the solutions of the following equations:
   (i) $\sin(A) = -0.5$
   (ii) $\cos(2A) = \frac{\sqrt{3}}{2}$
   (iii) $\tan(3A) = -1$

7. Prove the following:

   (i) $\dfrac{\sin(4A)(1 - \cos(2A))}{\cos(2A)(1 - \cos(4A))} = \tan(A)$

   where $\cos(2A) \neq 0$ and $1 - \cos(4A) \neq 0$

   (ii) $\dfrac{\cos(A) - 1}{\sec(A) + \tan(A)} + \dfrac{\cos(A) + 1}{\sec(A) - \tan(A)} = 2(1 + \tan(A))$

   (iii) $\dfrac{\sin(3A)\sin(6A) + \sin(A)\sin(2A)}{\sin(3A)\cos(6A) + \sin(A)\cos(2A)} = \tan(5A)$

8. Write the expression $4\sin(2x) + 3\cos(2x)$ in the forms $A\sin(2x + \alpha)$ and $B\cos(2x + \beta)$ where A and B are positive and α and β are acute angles.

9. Write the expression $4\cos(A) + \sin(A)$ in terms of t where $t = \tan(A/2)$ and hence solve the equation $4\cos(A) + \sin(A) = 1$ for values of A between 0 and 360°.

10. Express the following angles, given in degrees, as angles measured in rad:
    (i) 15.5°   (ii) $-69.4°$
    (iii) 291.6° (iv) 596.7°

11. Express the following angles, given in terms of radians, as angles measured in degrees between 0 and 360°:
    (i) 0.87 (ii) 2.7
    (iii) 5.24 (iv) 0.25

12. Put into polar form the following complex numbers
    (i) 2   (ii) 3i   (iii) -2
    (iv) $-2i$ (v) $3 + 4i$ (vi) $3 + \sqrt{3}i$
    (vii) $2 + 2i$ (viii) $\sqrt{3} - i$

13. Multiply or divide the following complex numbers by first writing them in polar form:
    (i) $(2 + 2i)(3 + 4i)$ (ii) $(3 + \sqrt{3}i)(\sqrt{3} - i)$
    (iii) $2/(\sqrt{3} - i)$ (iv) $3i/(3 + 4i)$

    (v) $\dfrac{(\sqrt{3} - i)(2 + 2i)(3 + \sqrt{3}i)}{3i(3 + 4i)}$

14. Find the values of a and b if

    (i) $a + bi = \dfrac{3}{\cos(A) + i\sin(A)}$

    (ii) $a + bi = \dfrac{(2 - i)}{1 + \cos(A) - i\sin(A)}$

15. Calculate the amplitude, period and phase of the following expressions:
    (i) $f(x) = 2.3\sin(7x + \pi/2)$
    (ii) $f(x) = 5\cos(2x + 1.5)$
    (iii) $f(x) = 3\sin(2x) + 4\cos(2x)$
    (iv) $f(x) = 4.5\sin(3x)\cos(3x)$

16. Solve the following equations for values of x lying between 0 and 2π:
    (i) $\sin(2x) = \tan(x)$
    (ii) $\cos(4x) + \cos(2x) - \sin(4x) + \sin(2x) = 0$
    (iii) $2\cos^2(x) = 1 + \sin(x)$

17. Calculate the approximate values of $\sin(x)$, $\cos(x)$ and $\tan(x)$ for $x = 0.1$ and 0.2 rad by including only the first three terms in their power series expansions. What are the respective percentage errors?

18. If x is a very small acute angle, use the second degree polynomial expansions for sine, cosine and tangent to solve the equations
    (i) $3\sin(x) + 2\cos(x) + \tan(x) = 2.05$
    (ii) $\cos(x) = 0.95$

19. Assuming that $\sin(A) = A - kA^3$ (where k is a constant) is a good

approximation to the value of $\sin(A)$ when A is small, use the formula $\sin(3A) = 3\sin(A) - 4\sin^3(A)$ to show that $k = \frac{1}{6}$.

20. Use the method of interval bisection to find the smallest root of $\sin(x) = 1/2x$ to two decimal places.

21. Use the method of interval bisection to solve

$$x - \pi/2 = \sin(x)$$

correct to three decimal places.

22. Find, correct to two decimal places, the root of $\tan(x) = x$ which is near 4.5 rad.

23. The displacement x (measured in m) of a mechanical system is given as a function of time t by the expression

$$x = \text{Re}(X)$$

where $X = 0.8e^{i(2t + A)}$ where $A = -2.2143$ rad.
Obtain the explicit expression for x in terms of t.
Find the amplitude, period, the phase of the oscillations and the distance travelled between $t = 1$ and $t = 2$ s.

24. The mean monthly temperature in Marseilles peaks in August at about 72 °F and has a minimum in February of about 45° F. Assuming the variation in temperature is roughly sinusoidal, obtain a mathematical model to represent the mean monthly temperature, and use this model to predict the mean monthly temperature in June and in November.

25. Show that when two wave forms of roughly the same amplitude but slightly different periods are added together the result is a wave of roughly the same period but with an amplitude that varies with a frequency equal to the difference in frequencies of the two waves.

26. In the Bristol Channel there is an unusually high rise and fall in the tides. At a typical spring tide there is a difference between high and low water of as much as 13 m. If the time between successive high tides is 12.4 hr, obtain the values of A and w in the model

$$y = A\sin(wt + \delta)$$

which gives the height y in m of the water above mean sea level as a function of the time t measured in hr.
If, at the time of a spring tide, high water happens to occur at midnight (i.e. $t = 0$) on a given day, find the value of δ. Use the model to determine the height above mean sea level of the water at midnight on the following three successive nights.

27. The colour of the fiddler crab is governed by two biological rhythms. Chromatophore activity is locked into the solar day (period 24 hr) and the tidal cycle (period about 12.4 hr). The crab darkens during daylight hours and is more susceptible to darkening at low tide. If the variation in chromatophore activity, due to the solar day, is modelled by the expression

$$E_s = a\sin(2\pi t - \pi/2)$$

where a is a positive constant and t is the time measured in (solar) days, show that the effect is greatest at midday, i.e. when $t = 0.5$ (or $1.5, 2.5, \ldots$) The effect, E_T, due to tides is modelled by the expression

$$E_T = b \sin(1.935\,48\pi t - 0.145\,16\pi)$$

Show that E_T is largest on the first day at 8 a.m., i.e. $t = \frac{1}{3}$. Show also that the colour of the crab should be darkest (i.e. $E_s + E_T$ should be largest) on midday of the 6th day.

28. Data on sunspots shows that the sunspot number waxes and wanes with a more or less regular solar cycle time which repeats every 11 yr. If there was a minimum in the sunspot number in 1744 of about 10 and a maximum of about 90 in 1750, use the model

$$n = n_0 + a \sin(wt + \delta)$$

with appropriate values for n_0, a, w and δ, to predict the sunspot number n during the years 1752, 1754, 1756 and 1758.

In the cycle that occurred around 1950, the peak (of about 150) occurred in 1948. The value of the following minimum sunspot number was about 20. Estimate the value of the sunspot number in 1951.

Appendix: some useful trigonometric formulas

$$\sin(90 + A) = \cos(A)$$
$$\sin(90 - A) = \cos(A)$$
$$\cos(90 + A) = -\sin(A)$$
$$\cos(90 - A) = \sin(A)$$
$$\cos(180 + A) = -\cos(A)$$
$$\cos(180 - A) = -\cos(A)$$

Addition formulas

$$\sin(A + B) = \sin(A)\cos(B) + \cos(A)\sin(B)$$
$$\sin(A - B) = \sin(A)\cos(B) - \cos(A)\sin(B)$$
$$\cos(A + B) = \cos(A)\cos(B) - \sin(A)\sin(B)$$
$$\cos(A - B) = \cos(A)\cos(B) + \sin(A)\sin(B)$$
$$\tan(A + B) = \frac{\tan(A) + \tan(B)}{1 - \tan(A)\tan(B)}$$
$$\tan(A - B) = \frac{\tan(A) - \tan(B)}{1 + \tan(A)\tan(B)}$$

Double angle formulas

$$\sin(2A) = 2\sin(A)\cos(A)$$
$$\cos(2A) = \cos^2(A) - \sin^2(A)$$
$$\tan(2A) = \frac{2\tan(A)}{1 - \tan^2(A)}$$

Factor formulas

$$\sin(A) + \sin(B) = 2\sin((A+B)/2)\cos((A-B)/2)$$
$$\sin(A) - \sin(B) = 2\cos((A+B)/2)\sin((A-B)/2)$$
$$\cos(A) + \cos(B) = 2\cos((A+B)/2)\cos((A-B)/2)$$
$$\cos(A) - \cos(B) = -2\sin((A+B)/2)\sin((A-B)/2)$$

4

Functions for science 3: inverse functions

Contents: 4.1 scientific context – 4.1.1 bacterial growth revisited; 4.1.2. periodic systems; 4.1.3 obtaining scientific laws; 4.2 mathematical developments – 4.2.1 functions; 4.2.2 composite functions; 4.2.3 inverse functions; 4.2.4 logarithms; 4.2.5 inverse trigonometric functions; 4.3 worked examples; 4.4 further developments – 4.4.1 straightening curves; 4.4.2 inverse hyperbolic functions; 4.5 further worked examples; 4.6 exercises.

4.1 Scientific context

In the first three chapters, by looking at appropriate examples from science, we have been able to motivate the study of polynomials, exponential functions and trigonometric functions. For example, by looking at the growth of a certain kind of bacterium and asking how large the colony would be after time t, we arrived quite naturally at the exponential function. However, had we looked at this growth problem from a different standpoint and tried to discover how long it would be before the colony reached a certain size, we would have been led to a quite different but not unrelated function. In our study of the growth process the independent variable was the time t and the dependent variable was the number N of the bacterium. Had we adopted this different standpoint, the role of the variables would have been reversed. N would now be the independent variable and t the dependent one. The function we would have been led to would have been the so-called *inverse function* of the exponential function. Many functions have inverses and these inverse functions often play an important role in mathematical modelling. In this chapter we look at those inverse functions that are of importance in science.

4.1.1 Example 1: bacterial growth revisited. The models of growth and decay that we have produced so far have all been geared to telling us how large something might be after a certain period, or what the temperature of a cooling body is after a certain time interval, and so on. As well as the usual questions of 'how big?' 'how strong?', 'how hot?', 'how cool?', that are all important in science, there is the very important class of questions beginning with 'when?': 'when will the number of bacterium cells double?', or 'when will the steel ingot have cooled to $100\,°C$?' or 'when is the intensity of radiation going to be safe?' and so on. As we have said, when we ask these questions we are effectively wanting to change the roles of the variables involved and trying to 'invert' the functional relationship that specifies the amount or the temperature in terms of the time t.

Table 4.1. *Number of bacterium cells, at 10-min time intervals*

time (min)	0	10	20	30	40	50	60	70	80	90	100
N	7	10	14	20	27	39	54	76	108	151	213

Consider again the growth of the bacterium *Escherichia coli* cells on the microscope slide that we considered first of all in 2.1.2. The data that we worked with is reproduced for convenience in Table 4.1 and shows the number of cells N after t min have elapsed. If we are indeed interested in the time taken for the colony size to reach a certain number, then it would seem sensible to plot out the data on a graph whose independent variable was N and whose dependent variable was t. Such a plot is shown in Fig. 4.1. From the graph we can easily read off the time taken for the colony size to reach a given number. For example, the time taken for N to reach 100 is about 77.5 min. The graph is clearly that of the function that 'undoes' the functional relationship that specifies N in terms of t. We say it represents the *inverse function* of this relationship.

The mathematical function whose shape is the same as the graph in Fig. 4.1 is the logarithm function. There are many logarithm functions as we shall see. If we choose to model N in terms of t using powers of e, then the most convenient logarithm function to invert this relationship will be the so-called natural logarithm function written as $\log_e$ or ln.

If we choose powers of 2, then the corresponding logarithm is $\log_2$ and so on. The mathematical properties of logarithm functions are considered in 4.2.4.

4.1.2 Example 2: periodic systems. Our study of trigonometric functions in Chapter 3 was motivated by looking at periodic systems and wave-like phenomena and in

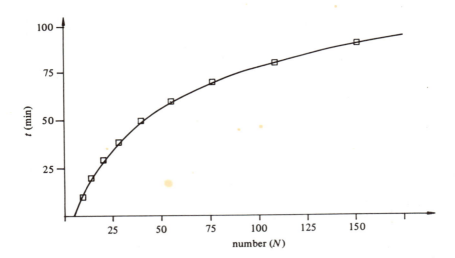

Fig. 4.1. Graph of time against number of cells for *Escherischia cola*.

3.4.2 we put forward general trigonometric expressions for modelling these periodic systems. In general, models provided by these trigonometric functions enable us to specify in a convenient way 'where' a periodic system might be as a function of time. They do not, however, readily enable us to ask 'when' the system will be in a certain position or configuration. But such questions are important in science and we often need to know when high tide is, or when lighting-up time is and so on. The fact that these times can be published well in advance, and with accuracy too, is because the periodic motion of the tides and the planets is well understood and the functions that specify the mean height of the sea, or the amount of daylight, as a function of time t, can be *inverted* to give us t as the dependent variable. It follows, therefore, that if we are to be able to answer 'when-type' questions in connection with periodic systems, described by the simple trigonometric functions of Chapter 3, we are going to need to know how to invert such functions. The mathematical properties of *inverse trigonometric functions* are discussed in 4.2.5.

4.1.3 Example 3: obtaining scientific laws. Inverse functions, as we have just seen, provide mathematical models of appropriate situations in science. However, certain inverse functions, particularly logarithms, are especially useful as a means of transforming scientific data for the purposes of establishing empirical laws connecting the variables comprising the data. The key lies in transforming the data to straight-line form. This idea is not new and we met it in Chapter 1. There, for example, we faced the problem of finding the law relating the load W and the length l of a steel wire of unextended length 2 m from the results of an experiment which are shown again in Table 4.2. Plotting out the values of length against load leads to the graph shown in Fig. 4.2. Recognising this plot as a straight-line graph enables us immediately to postulate Hooke's law, i.e.

$$1 = \mu W + l_0$$

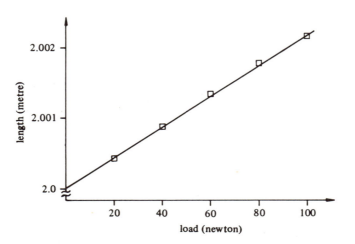

Fig. 4.2. Graph showing length against load for a steel wire.

Table 4.2. *Lengths of a steel wire for various loads*

load W (newton)	length l (metre)
20	2.000 42
40	2.000 86
60	2.001 32
80	2.001 76
100	2.002 19

Table 4.3. *Mean distances and periods of the planets*

planet	distance ($\times 10^6$ km)	period (days)
Mercury	57.9	88
Venus	108.2	225
Earth	149.6	365
Mars	227.9	687
Jupiter	778.3	4 329
Saturn	1427.0	10 753
Uranus	2870.0	30 660
Neptune	4497.0	60 150
Pluto	5907.0	90 670

where μ and l_0 are constants. In fact, as we showed in Chapter 1, the actual model is

$$l = 0.000\,021\,9\ W + 2$$

Because these data already lie on a straight line there is little or no difficulty in deducing Hooke's law. But as we know data do not always lie conveniently on straight lines.

Consider how the sidereal periods of the planets (i.e. the times taken for the planets to go round the sun) are related to the distances from the sun. These values are shown in Table 4.3. A plot of period as a function of distance is shown in Fig. 4.3. Clearly, any assumption of a straight line is not sensible here and the law relating sidereal period and distance is not linear. The fact that the points lie on a smooth curve does suggest however that there might be a functional relation between the variables that is mathematically simple.

Question: How can we find this relationship?
  Answer: Clearly, a detailed study of Newtonian mechanics applied to the motion of the planets is one possible answer! However, a simpler answer lies in the use of logarithms.

The properties of logarithms enable us to extract a straight-line relationship from the data and from this we can deduce the required law that relates period

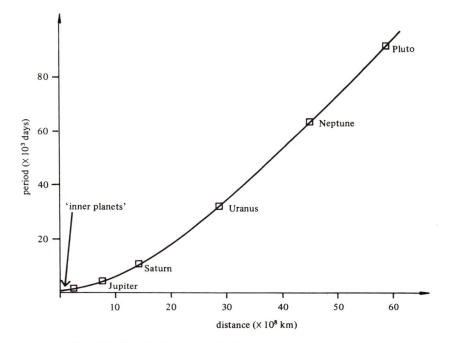

Fig. 4.3. Graph of the period of the planets against distance.

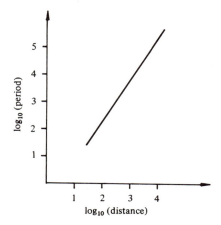

Fig. 4.4. Log–log graph for the period of the planets.

and distance. We shall see in Section 5 that if we draw a graph of log(period) against log(distance) the graph in Fig. 4.4 is obtained. This graph strongly suggests a linear relationship between log(period) and log(distance) and from this we can easily deduce the form of the original relationship between period and distance. Although logarithms have been used we will see that this original relationship does not involve logarithms. The use of logarithms in this way is an important application in science.

4.2 Mathematical developments

4.2.1 Functions. Although we have already defined what we mean by a function and introduced several important functions of relevance in science, it is perhaps opportune to formalise our ideas before proceeding to a study of logarithms and other inverse functions.

Essentially a function consists of three parts

    (i) a *rule*;

    (ii) a set of inputs called its *domain*;

    (iii) a set containing its possible values or outputs called the *codomain* or *target*.

We have seen the idea of two variables being connected by an equation. For instance, the temperature scales of Fahrenheit and Centigrade are connected by the equation $C = 5(F - 32)/9$ where C and F represent a temperature in degrees Centigrade and Fahrenheit respectively. The equation tells us that C depends on F and it enables us to calculate the value of C as soon as we are given the value of F. When two variables are related by an equation of this kind we say that C is *a function of F*; the equation is the *rule*. The rule provides a set of processing instructions for getting from the *independent variable* to the *dependent variable*, in this case from F to C.

When talking about functions in general where no definite rule is given explicitly, we write

$$y = f(x)$$

where x is the independent variable and y is the dependent variable. We say that *y is a function of x*. One important requirement of the rule that defines the function is that it must be unambiguous. Given any value of the independent variable x, from the domain, the rule must uniquely specify just one corresponding value of the dependent variable y in the target. For instance, we must be careful with defining $y = x^{1/2}$ as a function. The possible values of $x^{1/2}$ are $+\sqrt{x}$ and $-\sqrt{x}$; for example, $4^{1/2}$ takes the value $+2$ and -2. On a calculator the 'square root function' works via the unambiguous rule $y = +\sqrt{x}$, i.e. all square root values on a calculator are restricted to non-negative values.

We can describe a rule connecting x and y pictorially or geometrically by drawing its graph. This can provide a visual test as to whether the rule specifies a functional relationship between y and x. A curve in the x–y plane is the graph of a function if and only if a vertical line cuts or touches the curve at no more than one point. Fig. 4.5 shows two curves and the 'vertical line test' shows that the first is a function and that the second is not.

As well as being restricted to give only non-negative values, the square root function on a calculator has another restriction applied to it; the values of x must

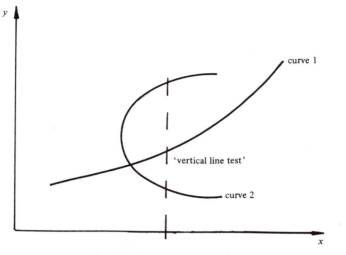

Fig. 4.5. The vertical line test.

all be positive so that the square root is a real number. The set of values that the independent variable may take in a functional relationship is called *the domain of the function*. This is the set of x on which the rule acts. For the square root function the domain is the set of x values in the range $0 \leqslant x < \infty$.

Example: Is this really the domain of the square root function on a calculator?
Solution: Strictly speaking the answer is no. The square root function is really two functions in one. One of them is the function we have just specified whose domain is the set of x values in the range $0 \leqslant x < '\infty'$ where '∞' is now the largest number the calculator can handle, and the other is a function which if it receives a negative x value as input returns the one value of 'error' each time. Implemented as a program the square root function on a calculator is the following

```
begin
input x
if x ⩾ 0 then
    set output to √x
    print output
else
    set output to 'error'
    print output
end if
step
```

Such a program represents a function whose domain is the set of *all* real numbers capable of being processed by a calculator.

The set of values of the dependent variable is called *the image set of the function*. This is the set of y that the rule produces by acting on the set of values in the domain. It is sometimes called the *range* set. For the square root function defined by the rule $y = +\sqrt{x}$ the image is the set of y values satisfying $0 \leqslant y < \infty$. For

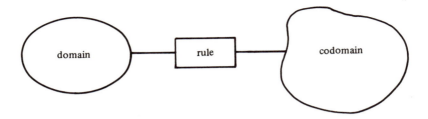

Fig. 4.6. The three elements of a function.

the square root function on a calculator it is the set of outputs satisfying $0 \leqslant$ output $<$ '∞' together with the value 'error'. When mathematical functions are used as models in science their domains and ranges may well be restricted due to physical constraints and common-sense requirements. For instance, the temperature of water is restricted by a lower value at which it freezes and an upper value at which it boils. So the function

$$C = 5(F - 32)/9$$

has domain $(32, 212)$, i.e. the allowable values of F and image set $(0, 100)$, i.e. the resulting values of C. In this example it is clear how to work out the image set, but this is not always the case. It is usually sufficient just to specify any set which contains the image set. This larger set is called *the codomain or target of the function* and is the third element making up a function. Sometimes it turns out to be convenient to specify a larger set containing the domain known as the *source* of the function. Restrictions on the source of the type just described then give rise to the domain. Diagrammatically we can represent the three elements that make up a function as in the Fig. 4.6. Mathematical modelling then, using functions, is concerned with defining the function rule in terms of well-understood mathematical expressions, together with establishing the domain and range of the functions involved.

4.2.2 Composite functions. In this subsection we introduce a way of combining functions to give new functions. Many functions that we encounter in science are combinations of other functions. The most familiar ways of combining functions algebraically are by addition, multiplication and division.

Another important way of combining two functions is by the successive application of the two rules. For instance, consider the two rules $f = x^2$ and $g = x + 2$. We can express the rule for $f(x) = x^2$ as $f = x^2$, i.e. the f value produced from x is given by $f = x^2$. The rule for $g(x)$ is therefore $g = x + 2$. With $x = 2$, if we apply the rule for f we get $2 \mapsto 2^2 = 4$. The notation '$\mapsto$' means 'maps to' or 'goes to' and is standard notation in connection with functions. If we now apply the rule for g to 4 we get $4 \mapsto 4 + 2 = 6$. Similarly, if we begin with $x = 3$ then

Fig. 4.7. Forming a composite function.

successive application of the two rules gives

$$3 \mapsto 3^2 = 9$$
$$9 \mapsto 9 + 2 = 11$$

Diagrammatically the process is shown in Fig. 4.7. Such a combination is called *a composite function* or more often a *function of a function*. The latter reminds us of the process of successive applications of two rules. In symbols we write such a combination as

$$y = g(f(x)) \tag{1}$$

Example: For the two rules $f = +\sqrt{x}$ and $g = \sin(x)$, find the values of $g(f(x))$ and $f(g(x))$ for $x = 2$ and $x = 4$.

Solution: For $g(f(x))$ we begin with the 'inner function' $f(x) = +\sqrt{x}$, evaluate this and then substitute the answer into the rule for g. For $x = 2$, $f(2) = 2^{1/2} = 1.414$ (to 3 d.p.); then

$$g(f(2)) = g(1.414) = \sin(1.414)$$

We should note here that it is assumed that x is measured in rad, so the value of $g(f(2))$ is 0.988 to three decimal places. For $x = 4$, $f(4) = +\sqrt{4} = 2$; then

$$g(f(4)) = g(2) = \sin(2) = 0.909$$

Consider now $f(g(x))$; here we evaluate the rule for g first. For

$$x = 2,\ g(2) = \sin(2) = 0.909 \text{ (to 3 d.p.)}$$

then

$$f(g(2)) = f(0.909) = +\sqrt{(0.909)} = 0.953$$

For

$$x = 4,\ g(4) = \sin(4) = -0.757$$

then

$$f(g(4)) = f(-0.757) = +\sqrt{(-0.757)} = ?$$

In this example the last composite function cannot be formed because the value for $g(x)$ is not in the domain of the function f.

Care must be taken with the domains and codomains when evaluating composite functions. For a composition to be possible, the image set of the first function must be in the domain of the second function. This is summarised pictorially in Fig. 4.8. Thus for the composite functions of the last example to work as functions we require that the domain of the function $g(f(x)) = \sin(+\sqrt{x})$ is the set of x values satisfying $0 \leqslant x < \infty$. The image set is then the set of numbers in the interval $(-1, +1)$. For the function $f(g(x)) = +\sqrt{(\sin(x))}$, the domain is the set of x values satisfying $2n\pi \leqslant x \leqslant (2n+1)\pi$ for any integer n; i.e. those values of x for which $\sin(x)$ is non-negative and the image set is the set of numbers $(0, 1)$.

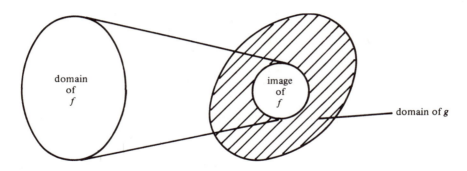

Fig. 4.8. Image of f is in the domain of g.

4.2.3 Inverse functions. So far in our formal treatment of functions we have started with an x value in the domain of a function and proceeded to obtain the value of the function in the image set.

Now we look at a slightly different problem. Given a rule $f(x)$ and a number in the image set of f, what is the corresponding value of x? For example, if $f(x) = 2x - 1$ and the value of f in the image set happens to be 3, what is the value of x? This is quite an easy problem to solve. If this particular x value is 'a' we know that with the function rule $a \mapsto 3$, so a is the solution of the equation $2x - 1 = 3$, i.e. $a = 2$.

But now consider the function $f(x) = x^2 - 1$. Suppose again we choose a value in the image set that happens to be 3. The particular x value a in the domain that results in a function value of 3 is therefore the solution of the equation $x^2 - 1 = 3$. In this case there are two possible values for a of ± 2.

These two examples serve to contrast the two different types of functions involved. For $f(x) = 2x - 1$ each member of the domain has *one* member in the image set. However, for $f(x) = x^2 - 1$, there are *two* members in the domain for each member of the image set.

We classify functions as one of two types. If for each member of the domain there is a different member of the image set then the function is said to be *one-to-one*. Otherwise the function is *not one-to-one*. So a function f is one-to-one means that

$$\text{if } a \neq b \text{ then } f(a) \neq f(b) \text{ or if } f(a) = f(b) \text{ then } a = b \tag{2}$$

The function $f(x) = 2x - 1$ is an example of a function that is one-to-one, whereas $f(x) = x^2 - 1$ is not one-to-one.

In this subsection the 'slightly different problem' we have been looking at has involved us going backwards from the image set to the domain. We now define a new function which takes any member from the image set and finds the member of the domain from which it comes. This new function is called *the inverse function*.

Consider the two functions $f(x) = 3x - 1$ and $g(t) = (t + 1)/3$ where the domain and image set of each function is the set of real numbers. If we form the composite

$g(f(x))$ then

$$g(f(x)) = g(3x - 1)$$
$$= ((3x - 1) + 1)/3$$
$$= x$$

In a sense then, the function g 'undoes' the effect of the function f and so g is the inverse function of f. In general, for any function f, the inverse function is the function g such that

$$g(f(x)) = x \text{ for all } x \text{ in the domain of } f \tag{3}$$

We adopt a special notation for g, we write *the inverse function of f as f^{-1}*, so

$$g = f^{-1}$$

$$\tag{4}$$

Note that this does not mean one over f.

Example: Find the inverse function for each of the following
    (i) $f(x) = 2x + 7$ where the domain is the set of all real numbers;
    (ii) $f(x) = x^2 + 2$ where the domain is the set of positive numbers.
Solution: In each case we need to find the rule for a function g such that $g(f(x)) = x$.
    (i) The formula $t = 2x + 7$ determines t uniquely when x is given. Rearranging the formula, if we are given t then we can find x from $x = (t - 7)/2$. Consider the function g with this rule. Forming the composite $g(f(x))$ we get

$$g(f(x)) = g(2x + 7) = ((2x + 7) - 7)/2 = x$$

as required. So $f^{-1}(x) = (x - 7)/2$.
    (ii) Following a similar approach, the rule for g is $x = \pm \sqrt{(t - 2)}$. We choose the positive square root because the domain of f (i.e. the positive x values) is the set of positive numbers. Forming the composite function $g(f(x))$ we get

$$g(f(x)) = g(x^2 + 2)$$
$$= + \sqrt{[(x^2 + 2) - 2]}$$
$$= x$$

Thus

$$f^{-1}(x) = + \sqrt{(x - 2)}$$

The original function f has to be one-to-one for an inverse function to exist; a function that is not one-to-one does not have an inverse. Because the inverse function acts on the images of the original function, the domain of the inverse function is the image set of the original function. The image set of the inverse function is thus the domain of the original function.

Finally in the same way as we can use the rule of a function to find its inverse, we can use the graph of the original function to draw the graph of the inverse function. For example, consider the function $f(x) = x^2 + 2$, with domain $(0, 2)$ and therefore image set $(2, 6)$. The inverse function is then $g(x) = + \sqrt{(x - 2)}$ with domain $(2, 6)$ (i.e. the image set of f) and image set $(0, 2)$ (i.e. the domain of f). Fig. 4.9 shows the graphs of these two functions. The graph of g is just the reflection of the graph of f in the line $y = x$. This is in general true for any function and its

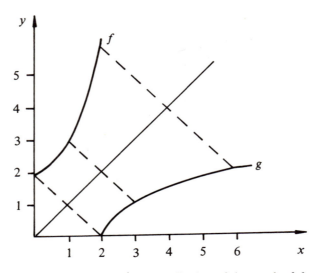

Fig. 4.9. Graph of f^{-1} is the reflection of the graph of f.

inverse. When a function possesses an inverse then *the graph of* f^{-1} *is the reflection in* $y = x$ *of the graph of* f.

We now complete this section by considering two special classes of inverse function of importance in science, namely the logarithm function and inverse trigonometric functions.

4.2.4 Logarithms. In 4.2.1 we said that the logarithm function was that function which would 'undo' the functional relationship connecting N and t in an exponential growth process. We also said that there were many logarithm functions. In our discussion of logarithm functions we start by looking at *natural logarithms*. We define the natural logarithm as the inverse function of the exponential function. For $x > 0$ the natural logarithm of x is the number y which satisfies the equation

$$x = \exp(y) \text{ or } x = e^y$$

We write $y = \log_e(x)$ or $y = \ln(x)$ as this natural logarithm and thus

$$\text{if } x = e^y \text{ then } y = \ln(x) \tag{5}$$

Seeing the expression $y = \ln(x)$ or $y = \log_e(x)$ written down simply means that $x = e^y$. The subscript e in the $\log_e$ notation is the *base* of the natural logarithm function of importance in science, namely the logarithm function and inverse remind us that the logarithm is the natural logarithm associated with the exponential function. We shall adopt the notation $\ln(x)$ throughout this text, although you may find the notation $\log_e(x)$ in others. On calculators, for example, you will find both notations used.

Two important identities can be written down immediately from the definition given in (5). Since exp and ln are inverse functions we can form the composite functions exp(ln) and ln(exp) which do not change the numbers they act on. In symbols we have

$$\exp(\ln(x)) = x \quad \text{or} \quad e^{\ln(x)} = x$$

and

$$\ln(\exp(x)) = x \quad \text{or} \quad \ln(e^x) = x$$

(6)

The first of these identities holds the key to how we can actually evaluate the natural logarithms of a positive number. Before we look at this problem it will be instructive to first develop some important properties of natural logarithms.

Some important properties of natural logarithms. Suppose that y_1 and y_2 are any two numbers and let $\exp(y_1) = x_1$ and $\exp(y_2) = x_2$. From the properties of the exponential function it follows that x_1 and x_2 will both be positive numbers. Using either (5) or (6) we can write

$$y_1 = \ln(x_1) \text{ and } y_2 = \ln(x_2)$$

(7)

The two numbers y_1 and y_2 are the powers to which the quantity e is raised to give x_1 and x_2, respectively, and so logarithms are simply powers or indices. The properties of logarithms should therefore follow from the index laws we established in Chapter 2, Section 2.2.5.

Consider the product $x_1 x_2$. In terms of the powers of e this is given by

$$x_1 x_2 = \exp(y_1) \exp(y_2)$$

which, by the first index law, is $\exp(y_1 + y_2)$ Thus

$$x_1 x_2 = \exp(y_1 + y_2)$$

From (5) or (6) and using (7) it follows that

$$\ln(x_1 x_2) = y_1 + y_2 = \ln(x_1) + \ln(x_2)$$

Thus we obtain the following important property of natural logarithms, namely

$$\ln(x_1 x_2) = \ln(x_1) + \ln(x_2)$$

(8)

Next, by considering the quotient x_1/x_2 and using the index law for quotients, we obtain

$$\ln(x_1/x_2) = \ln(x_1) - \ln(x_2)$$

(9)

Finally by looking at x_1^k, i.e. at $(\exp(y_1))^k$ and using the fact that $(\exp(y_1))^k = \exp(y_1 k)$ we obtain

$$\ln(x_1^k) = k \ln(x_1)$$

(10)

These three properties are very useful and enable us to do complicated arithmetic calculations with moderate ease. If we want the value of $19.432^{2.76}$, for example, then all we need do is find the natural logarithm of 19.432, multiply this by the power 2.76 and the result is the natural logarithm of the required value. The exponential function can then be used to furnish the result. This may sound complicated, but it is more or less how such a calculation would be done using the x^y key on a calculator.

The logarithmic series. In the above calculation we said that what we need first is the natural logarithm of 19.432. As yet we have not said how we calculate $\ln(x)$ values. Obviously we use a calculator, but how are the displayed values computed? Values of the natural logarithm can be obtained to any desired degree of accuracy from a power series expansion of $\ln(1 + x)$, known as the *logarithmic series*. The logarithmic series is easily obtained by looking at the identity

$$\exp(\ln(1 + x)^n) = (1 + x)^n$$

Because of property (10) the left-hand side is $\exp(n\ln(1 + x))$ and when this is expanded as a power series in $n\ln(1 + x)$ we have

$$\exp(n\ln(1 + x)) = 1 + n\ln(1 + x) + \frac{(n\ln(1 + x))^2}{2!} + \cdots$$

The right-hand side can be expanded by the binomial expansion to give

$$(1 + x)^n = 1 + nx + \frac{n(n - 1)x^2}{2!} + \frac{n(n - 1)(n - 2)x^3}{3!} + \cdots$$

Now, if n is *not* a positive integer power, the binomial expansion never terminates because the coefficients in the expansion never become zero. The binomial expansion becomes a power series expansion. The snag then is that the power series expansion is only valid (i.e. is a proper representation of $(1 + x)^n$) if $|x| < 1$, i.e. if $-1 < x < 1$.

However, what we can deduce from all of this is that for *any* n the following identity must hold provided $|x| < 1$:

$$1 + n\ln(1 + x) + \frac{(n\ln(1 + x))^2}{2!} + \cdots$$

$$= 1 + nx + \frac{n(n - 1)x^2}{2!} + \frac{n(n - 1)(n - 2)x^3}{3!} + \cdots \tag{11}$$

Equating coefficients of n in (11) we see that

$$\ln(1 + x) = x - \frac{x^2}{2} + \frac{x^3}{3} - \cdots$$

$$= \sum_{k=1}^{\infty} \frac{(-1)^{k+1}x^k}{k} \tag{12}$$

This power series expansion is the logarithmic series and despite the restrictions on x it can be used to compute the natural logarithm of any number.

Example: How can we compute (i) $\ln(10)$ (ii) $\ln(18.2)$?
Solution: (i) We can usefully note from (10) with $k = -1$ that $\ln(x_1^{-1}) = \ln(1/x_1) = -\ln(x_1)$. Thus $\ln(10) = -\ln(1/10) = -\ln(0.1)$. The value of $\ln(0.1)$ can be calculated from the series as $\ln(1 - 0.9)$, that is as $\ln(1 + x)$ with x equal to -0.9
    (ii) For $\ln(18.2)$ we can note that 18.2 is equal to 10×1.82. Thus
$$\ln(18.2) = \ln(10 \times 1.82) = \ln(10) + \ln(1.82)$$

$\ln(10)$ is calculated as in (i) and $\ln(1.82)$ is calculated as $\ln(1 + x)$ with $x = 0.82$

The graph of $y = \ln(x)$. Because $y = \ln(x)$ is the inverse function of $y = e^x$ its graph can be sketched out as the reflection of the graph of $y = e^x$ in the line $y = x$. Alternatively, we can draw the graph from values obtained using a calculator. The plot of $y = \ln(x)$ is shown in Fig. 4.10. We see that the domain of $\ln(x)$ is the set of positive real numbers and is the same as the range or image set of e^x. The range of $\ln(x)$ is the same as the domain of e^x and is just the set of real numbers. It follows that we cannot obtain the natural logarithm of a negative number as negative x values are not in the domain of $\ln(x)$. If we take the natural logarithm of a negative number on a calculator, the value 'error' is returned.

Example: Write the program that corresponds to the implementation of the ln function on a calculator.

Solution: In pseudocode such a program would be the following

```
begin
input x
if x > 0 then
    set output to ln (x)
    print output
else
    set output to error
    print output
endif
stop
```

Logarithms to other bases. Natural logarithms, i.e. logarithms with a base e, are particularly convenient for undoing or inverting expressions involving the

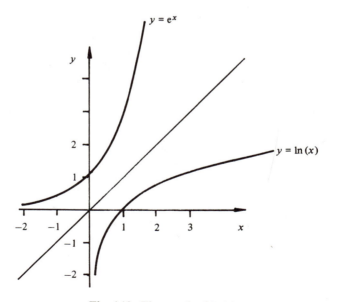

Fig. 4.10. The graph of $\ln(x)$.

exponential function. If powers of some number other than e are involved, it is often more useful to use logarithms with a base equal to that number. There is thus a whole range of different logarithm functions.

In general, for any positive number a, we can define the corresponding $\log_a$ function as follows

$$\text{if} \quad x = a^y \quad \text{then} \quad y = \log_a(x) \tag{13}$$

The three properties expressed in (8),(9) and (10) all hold for $\log_a$ functions and the graph of $y = \log_a(x)$ is the reflection in $y = x$ of $y = a^x$. Bases in common use, besides e, are base 10 and base 2. Logarithms to the base 10 are called common logs and the common log of a number x is usually written as $\log_{10}(x)$ or simply $\log(x)$. Where 2 is an important number as in genetics, digital communications and information theory, logarithms to the base 2 arise. The log of a number in such a case would be written as $\log_2(x)$. Logarithms to any base can be evaluated from the values of natural logarithms using the 'change of base rule'. Suppose we require to evaluate $\log_a(x)$. Let y be this value, then $x = a^y$. Let a, as a power of e, be given by $a = e^c$, so that we can write $x = e^{cy}$. Grouping all this information together it follows that

$$c = \log_e a \quad \text{and} \quad cy = \log_e x$$

Thus we have

$$y = \log_a x = \frac{1}{c} \log_e x = \frac{\log_e x}{\log_e a} = \frac{\ln(x)}{\ln(a)}$$

The formula

$$\log_a x = \frac{\log_e x}{\log_e a} = \frac{\ln(x)}{\ln(a)} \tag{14}$$

is known as *the change of base formula* and allows us to compute any logarithm value in terms of the corresponding natural logarithm value.

4.2.5 Inverse trigonometric functions. In the same way as we need 'the number whose exponential is ...', there are many situations in science when we need to find 'the number whose sine is ...'. In other words suppose we know the number y such that $y = \sin(x)$, then what is the angle x for this identity to hold? Similar questions can be asked for the other trigonometric functions cosine, tangent, etc. Such questions bring us naturally to the inverse trigonometric functions.

Example: Solve the equation $0.5 = \sin(x)$.
Solution: The easiest method of solving this equation is to use the graph of the sine function shown opposite. From the graph (Fig. 4.11) we see that there are many solutions to our problem. In radians some of the solutions are

$$-11\pi/6, \, -7\pi/6, \, \pi/6, \, 5\pi/6, \, 13\pi/6, \, 17\pi/6.$$

In fact there is an infinite number of solutions given by

$$x = 2\pi n + \pi/6 \quad \text{and} \quad x = 2\pi n + 5\pi/6 \quad \text{for} \quad n = 0, \pm 1, \pm 2, \pm 3, \ldots$$

Table 4.4. *The range of the inverse trigonometric functions*

Function	Inverse function	Range
$y = \sin(x)$	$x = \arcsin(y)$	$-\pi/2 \leqslant x \leqslant \pi/2$
$y = \cos(x)$	$x = \arccos(y)$	$0 \leqslant x \leqslant \pi$
$y = \tan(x)$	$x = \arctan(y)$	$-\pi/2 \leqslant x \leqslant \pi/2$

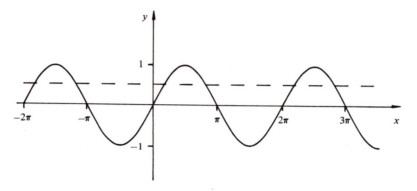

Fig. 4.11. Solutions of the equation $0.5 = \sin(x)$.

The solution to this example shows that the sine function is certainly not a one-to-one function. To define an inverse we must restrict the domain of $\sin(x)$ somehow to ensure that it is one-to-one and define an inverse over this restricted set. We note from the graph of the sine function that it is one-to-one for x values satisfying $-\pi/2 \leqslant x \leqslant \pi/2$ and for this set of x values we can define a function called *arcsin* that is the inverse of $\sin(x)$. Thus $\arcsin(y)$ is the angle x whose sine is y where $-\pi/2 \leqslant x \leqslant \pi/2$. This set of x values is called *the principal value range* of $\arcsin(y)$. Thus for this range we can say

$$\text{if} \quad y = \sin(x) \quad \text{then} \quad x = \arcsin(y) \tag{15}$$

The notation $\sin^{-1}(y)$ is sometimes used for $\arcsin(y)$. Thus if we wanted to solve the equation $0.5 = \sin(x)$, as in the last example, the value of x in the range $-\pi/2 \leqslant x \leqslant \pi/2$ would be $\arcsin(0.5)$. This is the angle $\pi/6 = 0.5236$ rad and is the value given by a calculator using the $\sin^{-1}$ key. All other values satisfying $0.5 = \sin(x)$ have then to be obtained using the symmetry and periodic properties of the sine function.

The trigonometric functions $\cos(x)$ and $\tan(x)$ also have inverse functions whose ranges are restricted. They are listed in Table 4.4 along with the inverse sine function for completeness.

It is worth investigating inverse trigonometric functions on your calculator. You should find that the values given are all within the principal value ranges given in the definitions. Inverse trigonometric functions can be evaluated to any degree of accuracy from the following power series expansions:

$$\arcsin(x) = x + \frac{x^3}{2.3} + \frac{1.3x^5}{2.4.5} + \frac{1.3.5x^7}{2.4.6.7} + \cdots$$

$$\arccos(x) = \pi/2 - \arcsin(x)$$

$$\arctan(x) = \begin{cases} x - \dfrac{x^3}{3} + \dfrac{x^5}{5} - \dfrac{x^7}{7} + \cdots & \text{for} \quad |x| \leqslant 1 \\ \dfrac{\pi}{2} - \dfrac{1}{x} + \dfrac{1}{3x^3} - \dfrac{1}{5x^5} + \cdots & \text{for} \quad |x| > 1 \end{cases}$$

4.3 Worked examples

Example 1. When sound passes through a glass window its intensity is reduced according to the formula

$$I_0 - I_1 = I_0(1 - 10^{-k})$$

where I_0 and I_1 are the outside and inside sound intensities respectively and k is a constant. When the intensity of loud street traffic is $6.5 \times 10^{-5}\,\mathrm{W\,m^{-2}}$ and of ordinary conversation is $5.0 \times 10^{-6}\,\mathrm{W\,m^{-2}}$, find the value of k that a glass window should have in order that the traffic noise is reduced inside a room to 'conversation levels'.

Solution. Starting from the law

$$I_0 - I_1 = I_0(1 - 10^{-k})$$

we rewrite this formula as

$$10^{-k} = 1 - (I_0 - I_1)/I_0$$

Substituting in values for $I_0 = 6.5 \times 10^{-5}$ and $I_1 = 5 \times 10^{-6}$ we have

$$10^{-k} = 1 - 0.923 = 0.077$$

Taking logs to base 10 of either side

$$-k = \log_{10}(0.077) = -1.1135$$

Hence the value of k is 1.1135 to four decimal places.

Example 2. The decay of radioactivity satisfies the law $N(t) = N(0)e^{-kt}$ where k is a constant depending on the element. The following table gives some typical values of k:

$$\text{carbon 14} \quad k = 1.203 \times 10^{-4}$$
$$\text{iodine 131} \quad k = 8.664 \times 10^{-2}$$
$$\text{xenon 133} \quad k = 1.386 \times 10^{-1}$$

(i) Find the half-life in each case.
(ii) Find how long it takes for the amount of radiation to reduce to $\frac{1}{4}$ and $\frac{3}{4}$ of the original amount.

Solution (i) The half-life of a substance was introduced in Chapter 2. It is the time taken for the radiation to reduce to half its original amount. So we require the value of t for which $N(t) = N(0)/2$. From the radioactivity decay law we have

$$N(0)/2 = N(0)e^{-kt}$$

i.e.

$$0.5 = e^{-kt}$$

Using natural logarithms we have

$$\ln(0.5) = -kt$$

Using a calculator and working to eight figures we have $\ln(0.5) = -0.693\,147\,1$. Thus in each case the half-life T is given by

$$T = \ln(0.5)/(-k) = 0.693\,147\,1/k$$

Substituting for the k values we have

$$\begin{array}{lll} \text{carbon 14} & k = 1.203 \times 10^{-4} & T = 5762\,\text{yr} \\ \text{iodine 131} & k = 8.664 \times 10^{-2} & T = 8\,\text{yr} \\ \text{xenon 133} & k = 1.386 \times 10^{-1} & T = 5\,\text{yr} \end{array}$$

(ii) For the radiation to reduce to a fraction D of its original amount, we have

$$D = e^{-kt}$$

and using natural logarithms we obtain

$$t = \ln(D)/(-k)$$

For $D = \frac{1}{4}$ and $D = \frac{3}{4}$, we have $\ln(\frac{1}{4}) = -1.386\,294\,4$ and $\ln(\frac{3}{4}) = 0.287\,682$. The following table shows the results for each substance:

Substance	Time for $\frac{1}{4}$	Time for $\frac{3}{4}$
carbon 14	11524 years	2391 years
iodine 131	16 years	3.3 years
xenon 133	10 years	2.08 years

Example 3 The magnitude of earthquakes is defined in terms of the Richter Scale where the scale number R is related to the relative intensity of the shock I by $R = \log_{10}(I)$.

(i) Rewrite this law in exponential form.
(ii) For the Italian earthquake in November 1980 the Richter Scale number was 7 and for the San Francisco earthquake of 1979 the Richter Scale number was 6. How many times more intense was the Italian earthquake?

Solution

(i) The logarithm function $\log_{10}$ is the inverse function for the exponential function 10^x. Thus we can write

$$I = 10^R$$

(ii) For the Italian earthquake we have $I = 10^7$ and for the San Francisco earthquake $I = 10^6$.

So the Italian earthquake was ten times more intense than the San Francisco earthquake.

Example 4. The displacement x (in metres) of the bob of a simple pendulum from its vertical position is modelled to an excellent approximation by the following function of time t (in seconds)

$$x = 0.16 \cos(1.4t)$$

This displacement results from the pendulum bob being drawn back 16 cm from its position when the pendulum is vertical and released at time $t = 0$. Find all the subsequent times at which the bob is 10 cm from the central position when the pendulum is vertical.

Solution. The equation we have to solve, for times t, is

$$0.10 = 0.16 \cos(1.4t)$$

In other words we have to solve $\cos(1.4t) = \frac{0.1}{0.16} = 0.625$. The value of arccos (0.625) is $0.895\,665\ldots$ rad. The first time that the displacement is 10 cm from vertical will be the t value satisfying

$$1.4t = 0.895\,665$$

This time is 0.640 s. The time period of the pendulum is $2\pi/1.4$ s, i.e. 4.488 s, so the next time the displacement is 10 cm will be after a time of $4.488 - 0.640 = 3.848$ s. All subsequent times will be integer multiples of the period added onto these two times. Thus the times are

$$0.640 + 4.488n \text{ s}$$

and

$$3.848 + 4.488n \text{ s}$$

for $n = 0, 1, 2 \ldots$

4.4 Further developments

4.4.1 Straightening curves. Right at the beginning of Chapter 1 we pointed out that scientific laws usually arise in one of two ways. Either the results of experiments are used directly to formulate empirical laws or existing scientific knowledge is used to develop new theories that are then tested later by experiment. In both these approaches logarithms can play an important role. We know that if we make a plot of one variable of interest against another and obtain a straight-line graph, then we can formulate the nature of the mathematical relationship connecting the two variables. If there is no theory about the situation we are studying then the straight-line graph enables us to formulate an empirical law; if on the other hand existing theory had suggested a linear relationship, then the straight-line plot actually verifies or confirms this theory.

Now we saw in 4.1.3 how logarithms could be used to transform data that originally did not satisfy a linear relationship, to values which, when plotted out, then lay on a straight line. What we have just said about formulating empirical laws or verifying existing theories applies here, but now to the transformed variables. Unfortunately, however, not all curves can be transformed to straight-line form, so the approach we used in 4.1.3 does have its limitations. Nevertheless, many important relationships can be transformed to straight-line form using logarithms. We now consider two important ones below.

A power law. Suppose that the relationship between two variables is thought to be a power law of the form $y = ax^b$. Then a typical graph of y against x will be of the form shown in Fig. 4.12 – clearly not a straight line. Now if we take natural

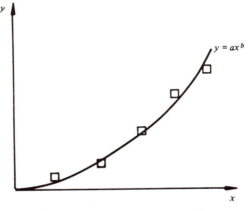

Fig. 4.12. The power law $y = ax^b$.

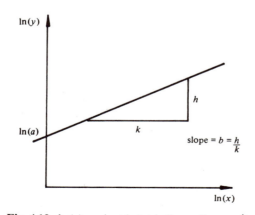

Fig. 4.13. $\ln(y)$ against $\ln(x)$ is linear if $y = ax^b$.

logarithms of each side we get

$$\ln(y) = \ln(a) + \ln(x^b)$$
$$= \ln(a) + b\ln(x)$$

Thus if we plot $\ln(y)$ against $\ln(x)$ we should obtain a straight-line graph. The slope and intercept of this graph can then be used to find the values of b and $\ln(a)$, respectively, as shown in Fig. 4.13.

An exponential law. Suppose that the relationship between two variables is thought to be of the form $y = ae^{bx}$. A typical graph of y against x (for b positive) will be as shown in Fig. 4.14. If we take natural logarithms of each side we get

$$\ln(y) = \ln(a) + bx$$

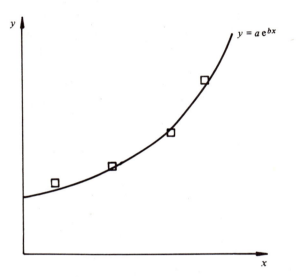

Fig. 4.14. The exponential law $y = ae^{bx}$.

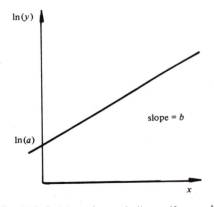

Fig. 4.15. $\ln(y)$ against x is linear if $y = ae^{bx}$.

Table 4.5. *Domain and range of the inverse hyperbolic functions*

Function	Inverse function	Domain	Range
$y = \sinh(x)$	$x = \sinh^{-1}(y)$	$-\infty < y < \infty$	$-\infty < x < \infty$
$y = \cosh(x)$	$x = \cosh^{-1}(y)$	$1 \leqslant y < \infty$	$0 \leqslant x < \infty$
$y = \tanh(x)$	$x = \tanh^{-1}(y)$	$-1 < y < 1$	$-\infty < x < \infty$

Now if we plot $\ln(y)$ against x we should obtain a straight-line graph and the slope and intercept can be used to find b and a (see Fig. 4.15).

The examples in Section 4.5 illustrate this theory in action. In this subsection we have suggested how the appropriate relation between two variables can be found. To establish a criterion for selecting the 'best' straight line is left to Chapter 18. At this stage we draw the best line by eye.

4.4.2 Inverse hyperbolic functions. The hyperbolic functions $\sinh(x)$, $\cosh(x)$ and $\tanh(x)$ were introduced in Chapter 2 and defined in terms of the exponential functions e^x and e^{-x}. The inverse functions of these are defined in the obvious way. For example, if $y = \sinh(x)$ then $x = \sinh^{-1}(y)$ is 'the number whose sinh is y'. Table 4.5 shows the inverse functions and their range. To calculate the value of x corresponding to a value of y we can write each inverse function in terms of natural logarithms. Consider $x = \sinh^{-1}(y)$. Now $y = \sinh(x) = (e^x - e^{-x})/2$. Multiplying through by $2e^x$ we have

$$2e^x y = e^{2x} - 1$$

Denoting e^x by s, we see that s satisfies the quadratic equation

$$s^2 - 2ys - 1 = 0$$

Solving for s we get

$$s = y \pm \sqrt{(1 + y^2)}$$

Since $s = e^x$ we must choose the positive root of the quadratic. So using the definition of logs we obtain

$$x = \ln(y + \sqrt{(1 + y^2)})$$

A similar analysis for $\cosh^{-1}(y)$ gives

$$x = \ln(y + \sqrt{(y^2 - 1)})$$

and for $\tanh^{-1}(y)$

$$x = 0.5 \ln((1 + y)/(1 - y))$$

Most scientific calculators nowadays have inverse hyperbolic functions available.

4.5 Further worked examples

Example 1. The table below gives the values of the atmospheric pressure, expressed as a percentage of its sea-level value, at various altitudes:

Altitude x (km)	Pressure y (% of sea level value)
0	100.0
5	53.0
10	26.0
14	14.0
20	5.4
24	2.9
30	1.2

(i) Draw the graphs y against x, $\ln(y)$ against x, $\ln(y)$ against $\ln(x)$.

(ii) From these graphs decide on the form of the law relating y and x in this case. Hence find a possible scientific law relating pressure and altitude.

Solution. Figs. 4.16(i), (ii) and (iii) show graphs of y against x, $\ln(y)$ against x, and $\ln(y)$ against $\ln(x)$, respectively, where x is the altitude and y is the pressure. From the diagram we can deduce that for this range of data there is a linear relation between ln (pressure) and altitude. The equation of the straight line in Fig. 4.16(ii) gives

$$\ln(\text{pressure}) = 4.6 - 0.147\,(\text{altitude})$$

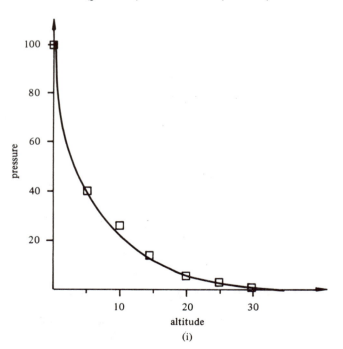

(i)

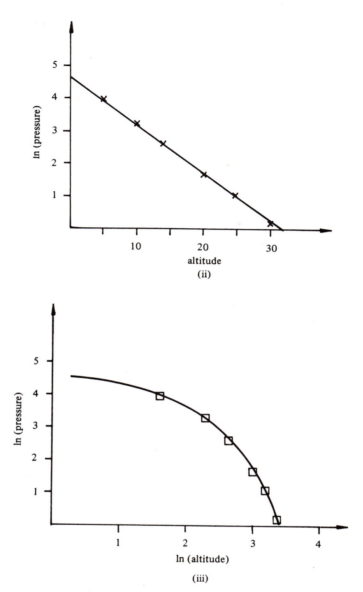

Fig. 4.16. (i) Graph of atmospheric pressure against altitude. (ii) Graph of ln (atmospheric pressure) against altitude (iii) Graph of ln (atmospheric pressure) against ln (altitude).

Writing this in terms of the exponential function we have

$$\text{pressure} = \exp(4.6 - 0.147x) = 99.5 \exp(0.147 \text{ (altitude)})$$

Example 2. The data in the table below shows the distance of each of the planets from the sun (measured in millions of kilometers) and the time (measured in days) that it takes each planet to travel round the sun once, this time is called the sidereal period:

Planet	Distance R ($\times 10^6$ km)	Period T (days)
Mercury	57.9	88
Venus	108.2	225
Earth	149.6	365
Mars	227.9	687
Jupiter	778.3	4329
Saturn	1427	10 753
Uranus	2870	30 660
Neptune	4497	60 150
Pluto	5907	90 670

(i) Draw the graphs T against R, $\log(T)$ against R, $\log(T)$ against $\log(R)$.

(ii) From these graphs decide on the form of the law relating R and T in this

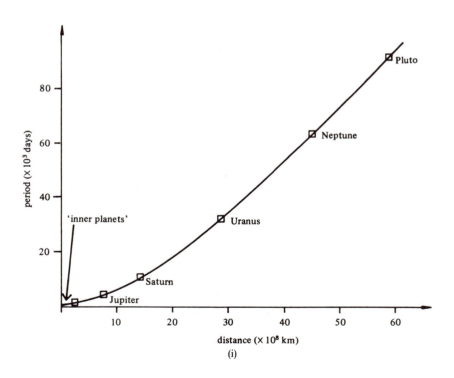

(i)

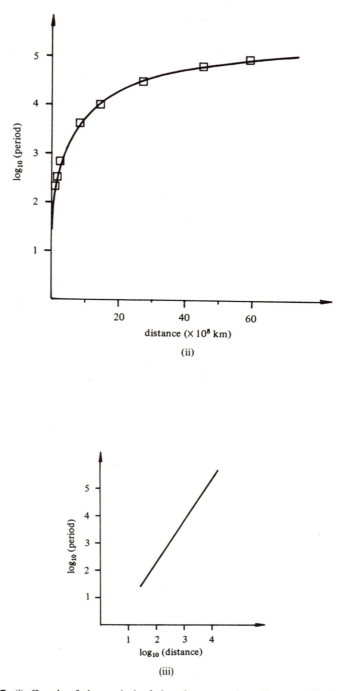

Fig. 4.17. (i) Graph of the period of the planets against distance. (ii) Graph of log(period) against distance. (iii) Graph of log(period) against log(distance).

case. Hence find a possible scientific law relating the sidereal period and the distance.

Solution. Figure 4.17 shows graphs of T against R, $\log(T)$ against R, and $\log(T)$ against $\log(R)$ where R is the distance and T is the sidereal period. In this case the third figure suggests a straight-line graph. From the equation of the line we have

$$\log(\text{period}) = -0.7 + 1.5\log(\text{distance})$$

From this equation we obtain the law

$$\text{period} = 0.20\,(\text{distance})^{1.5}$$

This relation was first found by Kepler in the seventeenth century and is called Kepler's third law. It says that the square of the sidereal period is proportional to the cube of the distance of the planet from the sun.

4.6 Exercises

1. Evaluate
   - (i) $\ln(7)$
   - (ii) $\ln(5)$
   - (iii) $\ln(3^5)$
   - (iv) $\ln(0.7)$
   - (v) $\log(2)$
   - (vi) $\log(8)$
   - (vii) $\log_9(27)$
   - (viii) $\log_7(343)$
   - (ix) $\log_4(8)$
   - (x) $\log_4(\sqrt{8})$
   - (xi) $\log_{10}(10)$
   - (xii) $\log_a(a)$
   - (xiii) $\ln(e)$
   - (xiv) $\ln(e^2)$
   - (xv) $\ln(e^2)\ln(1/e)$

2. If $\ln(y) = 3\ln(x) - 0.5\ln(x+2) + \ln(x^{\frac{3}{2}})$ show that

   $$y^2 = \frac{x^9}{x+2}$$

3. If $e^{2x} - 2e^x - 1 = 0$, show that $x = \ln(1 + \sqrt{2})$.

4. Solve the following equations:
   - (i) $3\sinh(x) - \cosh(x) = 1$
   - (ii) $4\tanh(x) = \coth(x)$
   - (iii) $3\cosh(2x) = 3 + \sinh(2x)$

5. Evaluate the following (quoting the principal values only):
   - (i) $\arcsin(\frac{1}{\sqrt{2}})$
   - (ii) $\arcsin(-\frac{1}{\sqrt{2}})$
   - (iii) $\arccos(\frac{1}{\sqrt{2}})$
   - (iv) $\arccos(\frac{-1}{\sqrt{2}})$
   - (v) $\arctan(1)$
   - (vi) $\arctan(-1)$
   - (vii) $\arcsin(-1)$
   - (viii) $\arccos(\frac{-\sqrt{3}}{2})$
   - (ix) $\text{arcsec}(-2)$
   - (x) $\arcsin(0.4)$
   - (xi) $\arccos(0.2)$
   - (xii) $\arcsin(-0.6)$
   - (xiii) $\arctan(\pi/4)$

6. Show that the following equations hold:
   - (i) $\arctan(\frac{1}{3}) + \arcsin(\frac{1}{\sqrt{5}}) = \pi/4$
   - (ii) $\arctan(x) + \text{arccot}(x) = \pi/2$

7. Solve the following equations giving general formulae for all possible solutions:

   (i) $4\cos(x) + 5 = 6\sin^2(x)$

   (ii) $3\tan^3(x) - 3\tan^2(x) = \tan(x) - 1$

   (iii) $6\sin^2(x) + 5\cos(x) = 7$

   (iv) $3\sin^4(x) + 2\sin^2(x) - 1 = 0$

8. In each of the following, write the left-hand side as $R\sin(x + A)$ where R is positive and A is an acute angle and hence solve the equations:

   (i) $\sin(x) - 2\cos(x) = 1$

   (ii) $3\sin(x) + 4\cos(x) = 5$

   (iii) $\sin(x) + \cos(x) = 1$

9. What is the domain and range of the function arcsine? What is the domain and range of the function on your calculator that implements the arcsine function?

10. Find all values of x in the range $(0° \leqslant x \leqslant 360°)$ which satisfy

$$\tan(x) + \sec(x) = 2\cos(x)$$

11. Solve the following equations for all values of x in the range $0 \leqslant x \leqslant 2\pi$:

   (i) $4^{x+1} = 8^{x^2}$

   (ii) $\tan(x) - 2\cot(x) = 1$

12. The displacement x of a linear oscillator at time t may be written as

$$x = 2\sin(2t) + \cos(2t)$$

Obtain the amplitude and phase of the oscillator and determine the smallest positive t value for which $x = 1.5$.

13. Given the function $f(x) = (x^2 + 1)/(2x^2 + 1)$ for $x \geqslant 2$, find an expression for $f^{-1}(x)$ and show that $f^{-1}(f(x)) = x$.

14. Consider the functions $f(x) = e^{2x}$ and $g(x) = \sqrt{(1 - 4x^2)}$ for values of x between -0.5 and 0. Find expressions for the inverse functions $f^{-1}(x)$ and $g^{-1}(x)$ and write down the domain and range of each of these two inverse functions.

15. The function $f(x) = x^2 + 1$ has domain D given by $\{x : x > -1\}$. State the range of the function and show that an inverse function f^{-1} does not exist. Suggest an interval such that if f is restricted to this interval then an inverse function f^{-1} does exist.

16. In a particular second-order chemical reaction the time t and concentration x of a product are related via the equation

$$kt = \frac{1}{(a - b)}\ln\left(\frac{b(a - x)}{a(b - x)}\right)$$

where a, b and k are constants. Show, by rearranging the equation, that

$$x = \frac{ab[\exp((a - b)kt) - 1]}{a\exp((a - b)kt) - b}$$

17. In a radioactive decay process, one-third of the original mass happens to disintegrate in 70 dy; calculate correct to the nearest day, the time

required for the radioactive material to be reduced to half its original mass.

18. The heat of combustion H(joule mol^{-1}) for a petroleum hydrocarbon of molecular mass M is given in the following table

H	213	373	530	688	845	1002	1159
M	16	30	44	58	72	86	100

Show, using logarithms and a suitable graph, that a relationship of the form $H = kM^n$ exists between H and M where k and n are constants. From your graph determine values for k and n and find the value of H when $M = 50$.

19. The velocity constant k of the reaction

$$2N_2O_5 \rightarrow 2N_2O_4 + O_2$$

varies with the Celsius temperature t as follows:

$t(°C)$	0	25	45	65
$k(min^{-1})$	0.000 047 2	0.002 03	0.0299	0.292

Show that these data are consistent with the law

$$k = A e^{-Q/RT}$$

where A, Q and R are constants; and $T = t + 273$ is the thermodynamic temperature. Obtain the expression relating T to k and determine the value of T when k is 0.012. The value of the gas constant R may be taken as 8.314 $JK^{-1} mol^{-1}$.

20. If the amount of glass produced by the glass industry is increasing on average by 8% yr^{-1}, find the value of the rate constant k if the growth of output is described by the model $a = a_0 e^{kt}$ where t is the time in years.

21. The following data relate to the cooling of a steel ingot:

time t(hr)	1	2	3	4	5	6
temperature $T(°C)$	370	274	200	150	112	83

Show that $T = T_0 e^{-kt}$, where T_0 and k are constants, is a reasonable model to describe the relationship between temperature and time. When will the ingot have cooled down to 60 °C?

22. N_0 atoms of uranium I decay to give uranium X by α emission. After 44×10^9 yr only half of the original concentration will remain. The number of atoms N remaining at any time t is given by $N = N_0(1 - e^{-\lambda t})$ where λ is a constant. Find the value of λ.

23. The following data show the variation with temperature of the solubility S of calcium acetate in water:

temperature (°C)	0	20	40	60	80
solubility (mol dm^{-3})	36.4	34.9	33.7	32.7	31.7

By plotting a suitable graph verify that the data are described by a relationship of the form $S = A e^{a/T}$ where A and a are constants and T is the thermodynamic temperature (0 °C = 273 K). From your plot obtain values for A and a and find the solubility at 27 °C.

24. An experiment on the decomposition of a certain gas yielded the following values for the velocity constant k at various temperatures T:

T(K)	1005	1058	1069	1105
k(mol min^{-1})	0.447	2.00	2.52	6.31

If these experimental values are consistent with a law of the form $k = Ce^{-A/RT}$ where A, C and R are constants, find graphically a value for A in terms of R, the gas constant.

5

Other functions of science

Contents: 5.1 scientific context – 5.1.1 equations of state; 5.1.2 vibrations; 5.2 mathematical developments – 5.2.1 the function $f = 1/x$; 5.2.2 limits and continuity; 5.3 worked examples; 5.4 further developments – 5.4.1 asymptotes; 5.4.2 partial fractions; 5.5 further worked examples; 5.6 exercises.

5.1 Scientific context

In the first four chapters we have introduced various important functions and investigated their properties. In the last chapter we presented some of the more formal aspects of the properties of a function, such as its domain and image, and considered how to build up composite functions by combining basic functions. In this chapter we continue to investigate the properties of functions and introduce one more important function namely the function $f(x) = 1/x$.

5.1.1 Example 1: equations of state. Pure substances can exist in various phases, namely gas, liquid and solid, and whether the phase is a gas, a liquid or a solid is determined essentially by the values of the pressure, volume and temperature of the substance. Relationships between p, V and T lead to different models called *equations of state*. To investigate these models fully we obviously need to understand the properties of functions of more than one variable. Functions of several variables are discussed in Chapter 17, but to simplify matters, suppose we look at the changes that can take place in a substance when the temperature remains constant. (Such changes are called isothermal changes.)

Imagine then that an experiment is carried out where the volume of gas is measured at different pressures, the temperature remaining constant. Suppose the results of the experiment are those shown in Table 5.1. In a first attempt to produce a model we might plot a graph of the pressure against the volume, as shown in Fig. 5.1. This graph suggests that there is a functional relationship between p and V but it is not linear. Without knowing anything about the kinematic theory of gases, we might try drawing graphs of $\ln(p)$ against V, or $\ln(p)$ against $\ln(V)$, to try to 'straighten the curve'.

But suppose instead we introduce the function $1/V$ and draw the graph of p against $1/V$, as shown in Fig. 5.2. From this graph we see that it is reasonable to propose a linear relationship as the model connecting p and $1/V$. From the graph we obtain

$$p = -1.0 + 4.71(1/V)$$

This relation between p and V has been obtained from the properties of the graph

Table 5.1. *Pressure and volume for a gas at constant temperature*

pressure p (Mpa)	16	21	26	31	36
volume V (m^3)	0.31	0.22	0.18	0.15	0.13

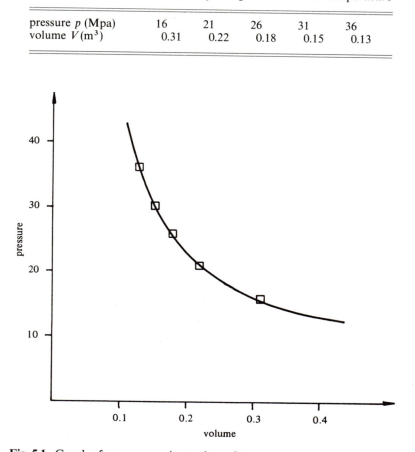

Fig. 5.1. Graph of pressure against volume for a gas at constant temperature.

using the experimental data. Now using physical arguments we could suggest that the graph should pass through the origin because a given mass of gas is likely to have zero pressure as the volume increases indefinitely. We would then have the relation

$$p = 4.66(1/V)$$

This model of the relation between p and V, namely that pV is a constant, is called Boyle's Law and can be written generally as $p = a/V$, where a is the constant. The law introduces another important function in science

$$f(x) = 1/x.$$

This rather simple 'one-over x' formula can be used to form more general composite functions; for example, $f(x) = 1/(x-4)^3$.

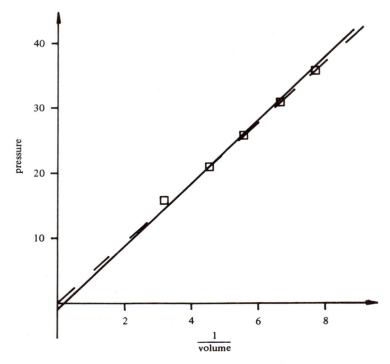

Fig. 5.2. Graph of pressure against 1/volume.

Returning to the model of the state equation for a gas, an important composite function occurs as a more general equation of state called van der Waal's equation,

$$p = \frac{RT}{(V - b)} - \frac{a}{V^2}$$

Functions of this type have important properties which are different to functions such as e^x and $\sin(x)$ already discussed.

We have just used the property that $1/x$ becomes zero as x becomes very large to justify drawing the pV graph through the origin. An important question to be asked now is what happens to $1/x$ as x gets very small? Investigating this introduces two new ideas (i) the idea of *the limit* of a function and (ii) the idea of *a discontinuous function*.

Functions such as polynomials, exponentials and the trigonometric functions, sine and cosine, are continuous in that we can draw graphs of them without any breaks. However, a graph of the function $1/x$ has a break or a discontinuity at the origin. It is not only at the origin that problems can occur. If we draw a graph of the function $\tan(x)$ between $x = 0$ and $x = 2\pi$ (see Fig. 5.3), there are breaks at $x = \pi/2$ and $x = 3\pi/2$. So $\tan(x)$ is said to be discontinuous at these values of x. In fact $\tan(x)$ is discontinuous for all odd multiples of $\pi/2$.

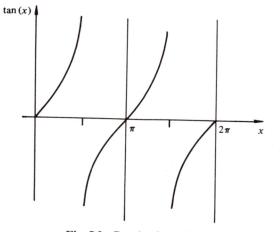

Fig. 5.3. Graph of tan (x).

5.1.2 Example 2: vibrations. The idea of a limit of a function has many applications. Models of complicated physical systems often consist of combinations of more basic functions and in Chapter 13 we shall introduce a model which is used to describe vibrating systems. For a damped system which is forced to oscillate, the mathematical function describing the motion consists of terms involving exponential and trigonometric functions together with 'amplitude' terms depending on the frequency of the forcing motion. Knowing how these functions and terms behave in various limits enables us to predict how the system might behave. For example, an electrical network consisting of an inductance (1 henry), a resistance (100 ohm) and a condenser (10^{-4}Farad) connected in series can be forced to provide damped oscillating currents by applying an oscillating voltage $E(t) = a \sin(wt)$ where a and w are constants. The general solution describing the value of the charge on the condenser plates as a function of time t is

$$q = A(w)\cos(wt + b(w)) + Be^{-50t}\cos(50\sqrt{3}t + c)$$

where B and c are constants and the amplitude term $A(w)$ is most conveniently written as

$$A(w) = 1/f(w)$$

i.e. as a 'one-over' type of function. The function $b(w)$ is just a phase that varies with w. This solution is made up of two parts:

and
$$q_p = A(w)\cos(wt + b(w))$$
$$q_c = Be^{-50t}\cos(50\sqrt{3}t + c)$$

These functions are shown in Fig. 5.4. We see that the function q_c dies out as time increases, this is called the *transient response* of the system. The function q_p does not approach zero, this is called the *steady state response*.

The behaviour of a vibrating system is important in many applications. For

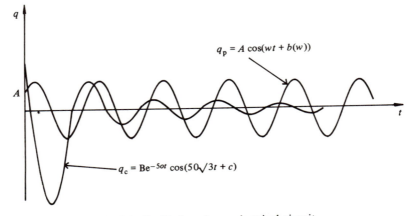

Fig. 5.4. Oscillations in an electrical circuit.

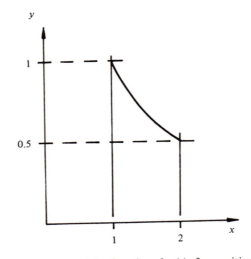

Fig. 5.5. Graph of the function $f = 1/x$ for positive x.

example, systems controlling the operation of engineering structures often involve mechanical or electrical components which oscillate. If the response of these components grows with time, for whatever reason, the system is unstable and could lead to the collapse of some structure, or some other catastrophe. Even if a system is 'stable' as in our electrical system above (q_c dies out as time increases) catastrophic effects can occur if the term $A(w)$ in the steady state response becomes large. This will only happen if $f(w)$ in our 'one-over' function becomes small. This can happen near the so called resonant frequency of the system. To predict the ultimate behaviour of a system requires not only a knowledge of the limit, as t increases indefinitely, of the basic functions such as the exponential and trigonometric functions, but also a knowledge of the behaviour of the $1/x$ type of function as x becomes small. In some cases the graphs of the functions provide the answer but in other cases an analytical approach is often easier.

5.2 Mathematical developments

5.2.1 The function $f = 1/x$. Fig. 5.5 shows the graph of the function $f = 1/x$ in the interval between $x = 1$ and $x = 2$. The function gradually decreases from 1 to 0.5. Now the behaviour of the function $f = 1/x$ in this interval causes no problems. But what about the interval between $x = -1$ and $x = 1$? As x gets smaller from the value $x = 1$ the function grows quite rapidly. The following table illustrates this growth:

x	$f(x)$
1	1
0.1	10
0.01	100
0.001	1000
0.0001	10 000

As x gets closer to the value 0, the value of the function increases indefinitely; we say that the function 'tends to infinity as x tends to zero' and this is written as

$$1/x \to \infty \quad \text{as} \quad x \to 0+$$

The notation $0+$ tells us that we approach zero from positive values of x. Now consider the behaviour of f when x is negative and approaches zero. Again the size of f grows indefinitely but this time to a large negative number. This is illustrated in the table below:

x	$f(x)$
-1	-1
-0.1	-10
-0.01	-100
-0.001	-1000
-0.0001	$-10 000$

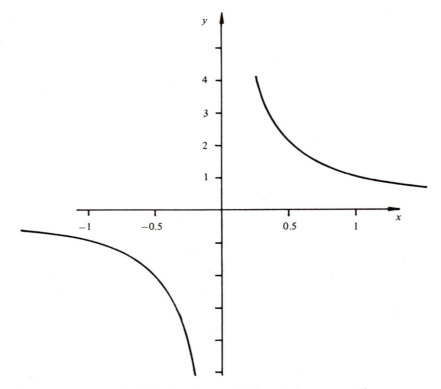

Fig. 5.6. Graph of the function $f = 1/x$ for positive and negative x.

Now we say that the function 'tends to minus infinity as x tends to zero' and we write

$$f(x) \to -\infty \quad \text{as} \quad x \to 0-$$

The notation $0-$ tells us that we approach zero from negative values of x. Figure 5.6 shows the graph of $f = 1/x$ for the interval between $x = 1$ and $x = -1$. We have not drawn the graph at the origin; the gap we have left suggests the peculiar behaviour at $x = 0$. This behaviour of the function $f = 1/x$ illustrates two important concepts in the theory of functions, (i) the limit of a function and (ii) the idea of continuity. If there is a break in the graph of the function, we say that it is *discontinuous*; if, as in this case, the graph 'goes to infinity' we say that the function has a *singular point* or $x = 0$ is a *singularity* of the function.

Another property of $f = 1/x$ which is significant in science is the behaviour as x increases indefinitely. Consider the values in Table 5.2. This table suggests that as x increases indefinitely the function $f = 1/x$ tends to zero. We write

$$1/x \to 0 \quad \text{as} \quad x \to \infty$$

So we have introduced the basic properties of $f = 1/x$ and at the same time introduced the ideas of limit and continuity. We now discuss these ideas further.

Table 5.2. $1/x$ *becomes small*
as x increases

10	0.1
100	0.01
1000	0.001
10 000	0.0001
100 000	0.000 01
1 000 000	0.000 001

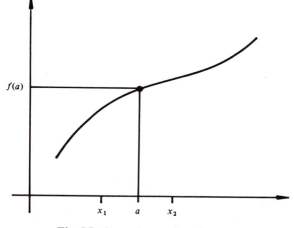

Fig. 5.7. A continuous function.

5.2.2 Limits and continuity. Consider a general function $f(x)$ and some interval
between $x = x_1$ and $x = x_2$ which contains the point a. Now as x approaches the
value a, finding the corresponding value of the function is often easy; we substitute
in $x = a$ and calculate the value of f. For example, if $f = x^3 + 1$ and $a = 2$ then we
have $f(2) = 2^3 + 1 = 9$. However, the value of $f = (x^2 - 1)/(x - 1)$ is not so obvious
when $a = 1$. Direct substitution gives $0/0$ which has no meaning. In fact writing
$x^2 - 1$ as $(x + 1)(x - 1)$ and dividing top and bottom by $(x - 1)$ gives $f(x) = (x + 1)$.
Now setting $x = 1$ we have $f(1) = 2$. In each case the function does equal a unique
value when x is taken to be the value a. Formally we say that

$$f(x) \text{ tends to } f(a) \text{ as } x \text{ tends to } a.$$

The value $f(a)$ is called the *limit of f as* $x \to a$ and the notation used is

$$\lim_{x \to a} f(x) = f(a).$$

What this means is that if we choose a value of x very close to a, then the function
value will be very close to the value $f(a)$. Figure 5.7 illustrates this idea. It does

not matter from which direction we approach the value $x = a$. If we approach $x = a$ 'from below', i.e. starting at $x = x_1$ on the left side of $x = a$ and choose values of x between x_1 and a, the limit is $f(a)$; and we write

$$\lim_{x \to a-} f(x) = f(a - 0)$$

If we approach $x = a$ 'from above', i.e. starting at x_2 on the right of $x = a$ and choose values of x between x_2 and a, the limit is again $f(a)$; and we write

$$\lim_{x \to a+} f(x) = f(a + 0)$$

In this case $f(a - 0)$ is the same value as $f(a + 0)$ so that

$$f(a - 0) = f(a) = f(a + 0)$$

A function for which this property holds is said to be *continuous* at the point a.

We naturally think of a continuous curve as one without any breaks, for example, gaps or jumps, so that we can draw such a curve without removing the pen or pencil from the paper. For a function to be continuous in some interval between $x = x_1$ and $x = x_2$, it must be continuous at each point within the interval. For example, $f(x) = x^3 + 1$ is continuous for all points in the interval between $x = -1$ and $x = 1$ whereas $f(x) = 1/x$ is not continuous at the origin $x = 0$, as we have seen; $f = 1/x$ is an example of a *discontinuous function*.

It is not necessary for a function to become indefinitely large for it to be discontinuous. Consider the function shown in Fig. 5.8. In this case there is a discontinuity at the point $x = 2$. We have $f(2 - 0) = 1$ and $f(2 + 0) = 4$. There is a 'jump' in the value of the function as x increases through the value 2. Such a function may appear to be rather arbitrary and you might expect real functions of science not to exhibit finite jumps of this type.

However, consider the function

$$f = \begin{cases} -x & x < 0 \\ x & x > 0 \end{cases}$$

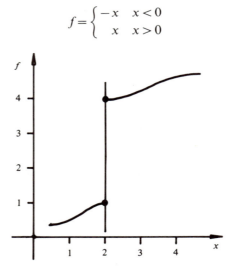

Fig. 5.8. A discontinuous function.

shown in Fig. 5.9. This function is continuous but consider the slope of the function. The direction of the curve changes abruptly at the origin. If we denote the direction by the angle θ, then θ changes from $3\pi/4$ to $\pi/4$ when $x = 0$. So a function describing the direction is given by

$$g = \begin{cases} \pi/4 & x < 0 \\ 3\pi/4 & x > 0 \end{cases}$$

There is a finite jump in the value of g at the origin, so g is a discontinuous function. In this case the function g is not defined at the origin. We will see that the slope of a function provides a very powerful tool in mathematics when we introduce the idea of calculus in Chapter 6. Often in science the function modelling some physical property or situation is continuous but its slope is discontinuous.

A special discontinuity that we must look out for in our scientific models is one where the function tends to infinity. Such a point is called *a singular point* and we say that the function has a *singularity* there. For example, $x = 1$ is a singular point of the function $f(x) = 3/(1-x)$. A function often has a singularity if the denominator becomes zero for some value of the independent variable.

Most functions in science occur as combinations of simple functions such as the polynomial, exponential and trigonometric functions. To compute the limits of

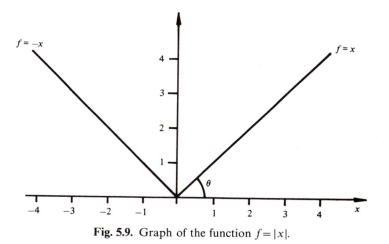

Fig. 5.9. Graph of the function $f = |x|$.

Table 5.3. *Limit theorems for composite functions*

Suppose that $\lim_{x \to a} f(x) = f(a)$ and $\lim_{x \to a} g(x) = g(a)$:
1. if k is a constant, then $\lim_{x \to a} kf(x) = kf(a)$
2. if r is a positive constant, then $\lim_{x \to a} (f(x))^r = (f(a))^r$
3. $\lim_{x \to a} (f(x) + g(x)) = f(a) + g(a)$
4. $\lim_{x \to a} (f(x) - g(x)) = f(a) - g(a)$
5. $\lim_{x \to a} f(x)g(x) = f(a)g(a)$
6. if $g(a) \neq 0$, then $\lim_{x \to a} f(x)/g(x) = f(a)/g(a)$

such combinations of functions we use the so-called limit theorems (rules) shown in Table 5.3. Although our discussion has been illustrated with finite values of a, the rules can be applied when a is indefinitely large provided the limit can be interpreted. For example if Rule 6 gives 0/0 as $x \to \infty$ we must seek another method to find the limit of the ratio of two functions.

Example: Use the limit theorems to evaluate the limits of the following functions as $x \to a$ for the values stated. Are there any singular points in the respective functions?

   (i) $f(x) = x^4$, $a = 2$;
   (ii) $f(x) = 4x^3 - 7$, $a = 3$;
   (iii) $f(x) = (x^2 + x + 1)/(2x + 1)$, $a = 0$;
   (iv) $f(x) = (x^2 - 4)/(x + 2)$, $a = -2$;
   (v) $f(x) = (e^x - 1)/x$, $a = 0$;
   (vi) $f(x) = e^{-x}/x$, $a = \infty$.

Solution: (i) If $f(x) = x^4$ then using Rule 2,

$$\lim_{x \to 2} f = 2^4 = 16$$

(ii) If $f(x) = 4x^3 - 7$ then using Rules 1, 2 and 3 we have

$$\lim_{x \to 3} f = 4 \times 3^3 - 7 = 101$$

(iii) For $f(x) = (x^2 + x + 1)/(2x + 1)$ we consider the limits of the functions $(x^2 + x + 1)$ and $(2x + 1)$ and then apply Rule 6. We have

$$\lim_{x \to 0} x^2 + x + 1 = 1 \text{ and } \lim_{x \to 0} 2x + 1 = 1$$

so that

$$\lim_{x \to 0} f = 1/1 = 1$$

The point $x = -0.5$ is a singular point since $\lim_{x \to -\frac{1}{2}} f = \frac{(3/4)}{0}$

(iv) For the function $f(x) = (x^2 - 4)/(x + 2)$ we consider the limits of the functions $(x^2 - 4)$ and $(x + 2)$ and apply Rule 6. We have

$$\lim_{x \to -2} x^2 - 4 = 0 \text{ and } \lim_{x \to -2} x + 2 = 0$$

In this case we have 0/0 which we cannot interpret. However, if we write $x^2 - 4$ as the product $(x - 2)(x + 2)$ then the function f becomes $f = x - 2$. Now we can write $\lim_{x \to -2} f = \lim (x - 2) = -4$. The function has a finite value when $x = -2$ so that it is continuous; in this case $x = -2$ is not a singular point.

(v) For $f(x) = (e^x - 1)/x$ consider the functions $e^x - 1$ and x. When $x = 0$ we have

$$\lim_{x \to 0} e^x - 1 = 0 \text{ and } \lim_{x \to 0} x = 0$$

so that if we try to apply Rule 6 we obtain 0/0 which we cannot interpret. However, if we use the power series expansion for e^x we have

$$(e^x - 1)/x = ((1 + x + x^2/2 + \cdots) - 1)/x$$
$$= 1 + x/2 + \cdots$$

The limit of this expression as $x \to 0$ is thus 1. Thus $\lim_{x \to 0} (e^x - 1)/x = 1$ and $x = 0$ is not a singular point.

(vi) For the function $f = e^{-x}/x$ consider the product of e^{-x} and $1/x$. We have $\lim_{x \to \infty} e^{-x} = 0$ and $\lim_{x \to \infty} (1/x) = 0$; so applying Rule 5 we have

$$\lim_{x \to \infty} f = \lim (e^{-x})(1/x) = 0 \times 0 = 0$$

The function is singular at the origin $x = 0$.

5.3 Worked examples

Example 1. The focal length of a convex lens is 0.25 m. When an object is placed at a distance u from the lens the image is at a distance v from the lens. In an experiment the following results are obtained.

object distance (u m)	0.3	0.4	0.5	0.6	0.7	0.8	0.9	1.0
image distance (v m)	1.5	0.67	0.5	0.43	0.39	0.36	0.35	0.33

By drawing an appropriate graph show that the law relating u and v is

$$\frac{1}{u} + \frac{1}{v} = 4$$

Solution. In this example the model is suggested, so we draw a graph of $1/v$ against $1/u$ (Fig. 5.10). The graph suggests a linear relation between $1/u$ and $1/v$. From the intercept and slope we obtain

$$\frac{1}{v} = 4 - \left(\frac{1}{u}\right)$$

Allowing for experiment errors this confirms the law given in the question as $1/u + 1/v = 4 \, (= 1/\text{focal length})$.

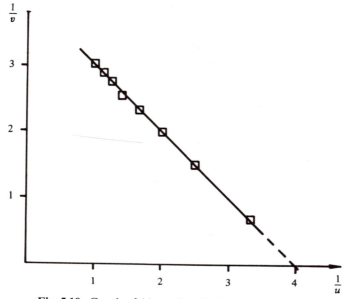

Fig. 5.10. Graph of $1/u$ against $1/v$ for a convex lens.

Example 2. The power delivered into the load R of a simple class A amplifier of output resistance Q is given by

$$\text{power} = \frac{E^2}{(Q+R)^2}R$$

where E is the output voltage. Find the limit of this expression for the power (i) when $R = 0$ and (ii) when R becomes indefinitely large.

Solution.

    (i) In this case we consider the two functions R and $1/(Q+R)^2$ and use Rule 5 for the limit of products:

$$\lim_{R \to 0} R = 0 \text{ and } \lim_{R \to 0} (1/(Q+R)^2) = 1/Q^2$$

Thus

$$\lim_{R \to 0} \text{power} = E^2 \times 0 \times (1/Q^2) = 0$$

    (ii) If we adopt the same approach as in (i) the product in the limit $R \to \infty$ would be $\infty \times 0$, which we cannot interpret. In this case we take R^2 out of the denominator to give

$$\text{power} = \frac{E^2}{R(Q/R+1)^2}$$

Now we consider the fuctions $1/R$ and $((Q/R)+1)^2$ and apply Rule 6 for quotients. We obtain

$$\lim_{R \to \infty} (1/R) = 0 \text{ and } \lim_{R \to \infty} ((Q/R)+1)^2 = 1$$

Thus

$$\lim_{R \to \infty} \text{power} = 0/1 = 0.$$

A graph of the function looks like that shown in Fig. 5.11.

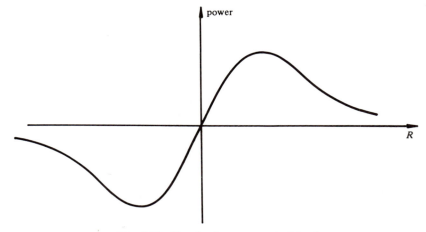

Fig. 5.11. Graph of power against load.

Example 3. Van der Waal's equation of state for a gas is given by

$$p = \frac{RT}{(V-b)} - \frac{a}{V^2}$$

If T is held fixed, find the singular points of this function. Are there any restrictions that we must place on the function $p(V)$?

Solution. The function is singular for zeros of the common denominator on the right-hand side. Clearly this occurs when $V = 0$ and $V = b$. A physical restriction that we must place on the function $p(V)$ is that for a real gas the pressure and volume must certainly be positive. In fact there are other restrictions on V which are discussed later in Chapter 17. Since the volume of a gas must be positive we restrict V to positive values. When V is small (and such that $0 < V < b$) each term on the right-hand side is negative so that p is negative, which is not allowed. We conclude that $V > b$. When $V = b_+$, p is large and positive. When V becomes indefinitely large each term becomes zero so that $\lim_{V \to \infty} p = 0$. Furthermore, p behaves like RT/V for large V, i.e. since T is being held fixed it behaves according to Boyle's law, so that $p \to 0_+$. Now mathematically the function $p(V)$ could become negative between $V = b$ and $V = \infty$. The graph of the function crosses the V axis at the roots given by $p(V) = 0$, which leads to the following equation in V

$$RTV^2 - aV + ab = 0$$

Solving for V we have

$$V = \frac{a \pm \sqrt{(a^2 - 4RTab)}}{2RT}$$

If $a < 4RTb$ then there are no real roots and p is always positive (for $V > b$). However, if $a > 4RTb$ there are two real roots and the function $p(V)$ takes negative values between the two roots. So for the simplistic physical reasons outlined above, in this case we restrict the values of V to

$$b < V < \frac{a - \sqrt{(a^2 - 4RTab)}}{2RT}$$

and

$$\frac{a + \sqrt{(a^2 - 4RTab)}}{2RT} < V < \infty$$

Example 4
(i) A forced damped oscillating system is described by the following equation:

$$f = A \cos(wt + b) + B e^{-ct} \sin(w_0 t + d)$$

where A, b, B, c, d, w and w_0 are constants, with c being positive. Find the form of the steady state vibrations of the system.

(ii) Regarded as a function of w, A can be written as $A_0 / \sqrt{((k - mw^2)^2 + (aw)^2)}$ where A_0, k, m and a are positive constants. Find the value of A when $w = \sqrt{(k/m)}$ and find the limit of A for this value of w when $a \to 0$.

Solution

(i) The trigonometric functions cosine and sine always lie between -1 and $+1$ and we cannot deduce definite values for $\lim_{t \to \infty} \cos(wt + b)$ and $\lim_{t \to \infty} \sin(w_0 t + d)$. However, we can assert that the limits are finite. Now, for positive c, $\lim_{t \to \infty} e^{-ct} = 0$. Using Rule 5 for the limit of products of functions we can write

$$\lim_{t \to \infty} B e^{-ct} \sin(w_0 t + d) = B \times 0 \times L$$

where $-1 < L < 1$. Thus

$$\lim_{t \to \infty} B e^{-ct} \sin(w_0 t + d) = 0.$$

The steady state vibrations of the system are described by the equation

$$f = A \cos(wt + b).$$

(ii) As a function of w, A is given by

$$A = \frac{A_0}{\sqrt{[((k - mw^2)^2 + (aw)^2)]}}$$

When $w = \sqrt{(k/m)}$, the value of A is thus A_0/aw, i.e. it is

$$\frac{A_0}{a} \sqrt{\left(\frac{m}{k}\right)}$$

As a tends to 0 $\lim_{a \to 0} A = \infty$. The amplitude thus increases indefinitely with decreasing a at this w value. The quantity a relates directly to the amount of damping in the system and $w = \sqrt{(k/m)}$ is in fact the value w_0 appearing in the transient part of the expression for f.

5.4 Further developments

5.4.1 Asymptotes. In Section 5.2 we saw that some functions are not defined for all values of x; for example, $f = 1/x$ is not defined when $x = 0$ and $f = \sqrt{x}$ cannot take real values when x is negative.

Fig. 5.12 shows a graph of the function $f = 1/x$ as x increases indefinitely. As x increases the graph gets closer to the x axis but never crosses this axis. We often say that the graph 'touches' the x axis at 'infinity'. The graph of $f = 1/x$ 'straightens out' and for large x approaches a straight line. We call this line an *asymptote*. In this case, the y axis (i.e. the line $x = 0$) is also an asymptote.

An asymptote is defined in the following way:

> *An asymptote of a function f is a line which is a tangent to the graph of f at an infinite distance from the origin*

Fig. 5.13 illustrates this definition. The most common asymptotes are the horizontal lines $y = $ constant and vertical lines $x = $ constant. This is the case for $f = 1/x$. To find asymptotes which are parallel to the axes is a relatively easy task. For horizontal

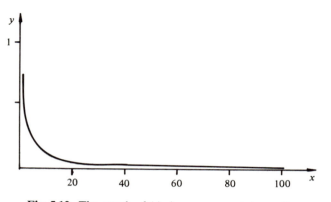

Fig. 5.12. The graph of $1/x$ has an asymptote $y = 0$.

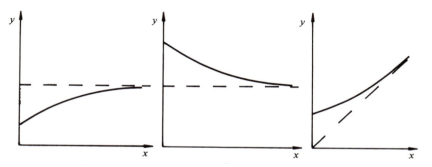

Fig. 5.13. Geometrical interpretation of asymptotes.

asymptotes of a graph we calculate the limits

$$\lim_{x \to \infty} f(x) \text{ and } \lim_{x \to -\infty} f(x)$$

If either limit exists, L say, then its value determines a horizontal asymptote $y = L$. For a vertical asymptote we look for singular points, i.e. values of x which give division by zero. For example, $f = 1/(x + 2)$ has a vertical asymptote $x = -2$.

Example: Find the vertical and horizontal asymptotes of the function

$$f = 2 + \frac{x}{1 + x}$$

Solution: We first evaluate the limit of f as $x \to \infty$ and as $x \to -\infty$:

$$\lim_{x \to \infty} f = 3, \text{ and } \lim_{x \to -\infty} f = 3$$

Hence $y = 3$ is a horizontal asymptote. If we let $x = -1$ then we get division by zero, so $x = -1$ is a vertical asymptote.

Sometimes a graph of a function will approach a straight line which is neither

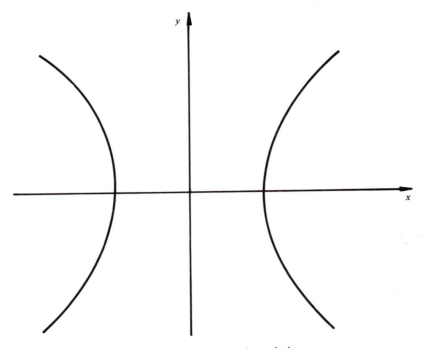

Fig. 5.14. Graph of an hyperbola.

vertical nor horizontal. One such graph is shown in Fig. 5.14. This is the graph of a curve called a hyperbola. Points on this graph satisfy the equation

$$\frac{x^2}{a^2} - \frac{y^2}{b^2} = 1$$

To investigate the asymptotes of the function describing the hyperbola, we are interested in what happens to the graph in Fig. 5.14 a long way from the origin. Consider the points of intersection of the hyperbola and the straight line $y = mx + c$. Substituting for y we have

$$\frac{x^2}{a^2} - \frac{(mx + c)^2}{b^2} = 1$$

which can be written as a quadratic equation in $1/x$ as

$$\frac{a^2(c^2 + b^2)}{x^2} + \frac{2a^2mc}{x} + a^2m^2 - b^2 = 0$$

From the definition of an asymptote, the line $y = mx + c$ is an asymptote of the hyperbola if it meets the hyperbola at an infinite distance from the origin, i.e. when $x = \infty$.

If $a^2m^2 - b^2 = 0$ and $2a^2mc = 0$, then the equation in $1/x$ has two zero roots $x = -\infty$ and $x = \infty$, and the line $y = mx + c$ meets the hyperbola 'at infinity' as

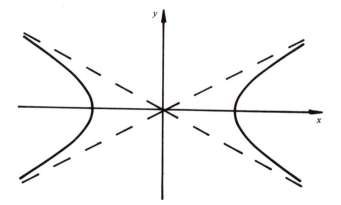

Fig. 5.15. Asymptotes of a hyperbola.

required. (Note that if $2a^2mc \neq 0$ there exists a finite point of intersection which is not allowed for an asymptote of a hyperbola.) Solving for m and c we have

$$m = b/a \text{ or } m = -b/a \text{ and } c = 0.$$

Thus the two straight lines, passing through the origin,

$$y = bx/a \text{ and } y = -bx/a$$

touch the hyperbola at an infinite distance from the origin and so are asymptotes of the hyperbola. The asymptotes are shown in Fig. 5.15 and in this case are equally inclined to the x axis.

5.4.2 Partial fractions. In this chapter we have specifically chosen to look at the function $f(x) = 1/x$ and the composite functions that can arise from this 'one-over-x' formula. One very important class of functions that at first sight appears to have no connection with the function $f(x) = 1/x$ is the class of functions known as *rational functions*. Rational functions are easily formed by dividing one polynomial function by another. So, for example, the following function:

$$f(x) = \frac{2x^3 - 4x^2 + 6x - 1}{x^2 - 2x + 7}$$

is a rational function. Note also that the expression for the pressure p given by van der Waal's equation, namely

$$p(V) = \frac{RTV^2 - aV + ab}{V^3 - bV^2} \tag{1}$$

is also a rational function.

Rational functions do occur naturally in science, and not only in connection with gases and equations of state. However, where they do occur it is often the case that when we have to manipulate rational functions it is far easier to deal with them if they have been decomposed first of all into what are called *partial fractions*. For example, if we consider the expression for the pressure again in van der Waal's equation, we are far happier dealing with the expression

$$p(V) = \frac{RT}{(V - b)} - \frac{a}{V^2}$$

than the form given in (1). This latter expression is the partial fraction decomposition of the rational function we had to begin with, and the expressions $RT/(V - b)$ and a/V^2 are the partial fractions in this decomposition. We see therefore that partial fractions are themselves rational functions, but we see now that they have more in common with the 'one-over x' function, which is the subject of this chapter. One other very important property that partial fractions possess is that they are *proper* rational functions, in that the degree of the polynomial in the numerator of the partial function is less than the degree of the polynomial in the denominator. This latter property holds the key to how we set about decomposing a rational function into partial fractions. To obtain the partial fractions to associate with a rational function we make use of a theorem (which we will not prove!) which states the following:

> If $f(x) = p(x)/q(x)$ is a proper rational function (i.e. one polynomial divided by another, with the degree of the denominator being greater than that of the numerator) then $p(x)/q(x)$ can be expressed as a sum of partial fractions (i.e. proper rational functions) having denominators that are the factors of $q(x)$.

Thus, for example, if we wanted to express $f(x) = 2x/(x^2 - x - 2)$ in terms of partial fractions we would first note that the denominator $(x^2 - x - 2)$ was the product of the factors $(x + 1)$ and $(x - 2)$. The denominators of the two partial fractions involved would therefore be $(x + 1)$ and $(x - 2)$. The numerators of the partial fractions would each have to be polynomials having degrees less than the polynomials $(x + 1)$ and $(x - 2)$. The numerators in each case would have to be polynomials whose degrees were less than 1. In other words, they would be polynomials of degree zero, i.e. constants. In establishing the numerators we obviously try any polynomial having degree one less than the degree of the polynomial in the denominator. Thus if a factor in the numerator of our original rational function was a quadratic term, the corresponding partial fraction would be a linear polynomial divided by this quadratic.

Returning to our example, then, we would express $f(x) = 2x/(x^2 - x - 2)$ as the sum

$$\frac{A}{(x + 1)} + \frac{B}{(x - 2)}$$

These two expressions are the partial fractions of $f(x)$. The problem is how to obtain the values of A and B. We can determine A and B as follows:

step 1: $f(x)$ and its partial fractions form are equivalent, so

$$\frac{2x}{(x^2 - x - 2)} = \frac{A}{(x + 1)} + \frac{B}{(x - 2)}$$

Multiplying through by the product $(x + 1)(x - 2)$ gives:

step 2: $2x = A(x - 2) + B(x + 1)$

This identity holds for all values of x. So choosing any two different values of x will create two equations for the two unknowns A and B. We can easily find A by letting $x = -1$, for then the coefficient of B is zero:

step 3: $2(-1) = A(-1 - 2) + B(0)$

Hence

$$A = -2/-3 = 2/3$$

Similarly, setting $x = 2$, we can obtain B:

step 4: $2(2) = A(0) + B(2 + 1)$

So

$$B = 4/3$$

The following example illustrates the method further.

Example: Find the partial fraction expansions for

    (i) $f(x) = 1/(x^2 + x)$;
    (ii) $g(x) = (x + 1)/(x^2 + x - 2)$.

Solution: In each case we must begin by writing the denominator as a product of its factors.

    (i) Now $x^2 + x = x(x + 1)$, so we decompose f into the sum

$$\frac{1}{x^2 + x} = \frac{A}{x} + \frac{B}{(x + 1)}$$

Multiplying through by $x^2 + x$ we have

$$1 = A(x + 1) + Bx$$

Let $x = 0$ then

$$1 = A(1) + B(0)$$

and

$$A = 1$$

Now let $x = -1$ then

$$1 = A(0) + B(-1)$$

So

$$B = -1$$

The partial fraction expression for $f(x) = 1/(x^2 + x)$ is $(1/x) - 1/(x + 1)$.

    (ii) For g, the factors of $x^2 + x - 2$ are $(x + 2)(x - 1)$, so we seek values of A and B such that

$$\frac{x + 1}{x^2 + x - 2} = \frac{A}{(x + 2)} + \frac{B}{(x - 1)}$$

Multiply through by $(x + 2)(x - 1)$ to give

$$x + 1 = A(x - 1) + B(x + 2)$$

Let $x = 1$ then

$$2 = A(0) + B(3)$$

So

$$B = 2/3$$

Let $x = -2$ then

$$-1 = A(-3) + B(0)$$

So

$$A = 1/3$$

The partial fraction expansion for $f(x) = (x + 1)/(x^2 + x - 2)$ is
$[(1/3)/(x+2)] + [(2/3)/(x - 1)]$.

The above examples are straightforward; however there are complications that can occur.

Suppose that the denominator of our original rational function is of a lower degree than the numerator. How can we use our theorem now to work out the partial fractions of $f(x) = (2x^3 + x + 6)/(x^2 + x - 2)$, for example? All that we need to do to use our theorem, which relates only to proper rational functions, is to divide out our (improper) rational function to obtain a polynomial plus a proper rational function. This is obviously best demonstrated by looking at an example, so let us consider how we determine the partial fractions of $f(x) = (2x^3 + x + 6)/(x^2 + x - 2)$. In such cases we must first divide the numerator by the denominator until the remainder is less than the degree of the denominator. For example, dividing the denominator into the numerator by long division gives

$$
\begin{array}{r}
2x - 2 \\
x^2 + x - 2 \overline{\smash{)}2x^3 + x + 6} \\
2x^3 + 2x^2 - 4x \\
\hline
-2x^2 + 5x + 6 \\
-2x^2 - 2x + 4 \\
\hline
7x + 2
\end{array}
$$

and the remainder $7x + 2$ is of degree 1 which is lower than the degree of the denominator. So

$$\frac{2x^3 + x + 6}{x^2 + x - 2} = 2x - 2 + \frac{7x + 2}{x^2 + x - 2}$$

Now to write f in partial fraction form we need to find A and B so that

$$\frac{7x + 2}{x^2 + x - 2} = \frac{A}{(x - 1)} + \frac{B}{(x + 2)}$$

Multiply through by $(x - 1)(x + 2)$, then

$$7x + 2 = A(x + 2) + B(x - 1)$$

Let $x = 1$, then $9 = A(3) + B(0)$; so $A = 3$. Let $x = -2$, then $-12 = A(0) + B(-3)$;

so $B = 4$. In partial fraction form, we have

$$f(x) = \frac{2x^3 + x + 6}{x^2 + x - 2} = 2x - 2 + \frac{3}{(x-1)} + \frac{4}{(x+2)}$$

A second special case occurs if the denominator contains repeated factors. For instance, consider the function

$$f(x) = \frac{1}{(x-1)^2(x+2)}$$

We might well try the following decomposition (which is quite correct):

$$\frac{1}{(x-1)^2(x+2)} = \frac{Ax + B}{(x-1)^2} + \frac{C}{(x+2)}$$

but as we shall see in the later chapters on calculus, the first partial fraction is not all that convenient to handle from a mathematical standpoint. It is far better to write this as

$$\frac{Ax + B}{(x-1)^2} = \frac{Ax - A + A + B}{(x-1)^2} = \frac{A(x-1)}{(x-1)^2} + \frac{A+B}{(x-1)^2}$$

which, after cancelling the $(x-1)$ factor, is the same as

$$\frac{A}{(x-1)} + \frac{B'}{(x-1)^2}$$

where $B' = A + B$, and use as the partial fractions for $f(x)$ the following:

$$\frac{1}{(x-1)^2(x+2)} = \frac{A}{(x-1)} + \frac{B'}{(x-1)^2} + \frac{C}{(x+2)}$$

In general it is worth noting that the partial fractions associated with the repeated terms $(x + a)^n$ are

$$\frac{A}{(x+a)} + \frac{B}{(x+a)^2} + \cdots + \frac{P}{(x+a)^n}$$

where $A, B, \ldots, P$ are constants.

Returning to the example we can obtain A, B' and C without too much difficulty as follows. If we multiply through by $(x-1)^2 (x+2)$ we obtain

$$1 = A(x-1)(x+2) + B'(x+2) + C(x-1)^2$$

We can find B' and C as before. Let $x = -2$, then $1 = A(0) + B'(0) + C(9)$, so $C = \frac{1}{9}$. Let $x = 1$, then $1 = A(0) + B'(3) + C(0)$, so $B' = \frac{1}{3}$.

We have found the values of C and B' easily by choosing values of x so that two terms on the right-hand side are zero. To find the third constant A we choose x equal to *any* value except -2 and 1, and use the calculated values of B' and C. Let $x = 0$ say, then $1 = A(-2) + B'(2) + C$; substituting for B' and C and solving we have $A = -\frac{1}{9}$.

The decomposed form of f into partial fractions is

$$\frac{1}{(x-1)^2(x+2)} = \frac{-1/9}{(x-1)} + \frac{1/3}{(x-1)^2} + \frac{1/9}{(x+2)}$$

At the final stage of finding A, any other value of x will give the same value for A but in this example $x = 0$ is easiest to work with.

5.5 Further worked examples

Example 1. Determine the equations of the asymptotes for van der Waal's equation of state

$$\left(p + \frac{a}{V^2}\right)(V - b) = RT$$

Solution. Rewriting van der Waal's equation in the form $p(V)$ we have

$$p = \frac{RT}{(V-b)} - \frac{a}{V^2}$$

The horizontal asymptotes (if they exist) are given by

$$p = \lim_{V \to \infty} p(V).$$

Each term in the function $p(V)$ becomes zero in the limit so that the V axis, $p = 0$, is one asymptote. For vertical asymptotes we look for values of V which give division by zero. Clearly, $V = 0$ and $V = b$ are such values, so that the vertical lines $V = 0$ (i.e. the p axis) and $V = b$ are asymptotes. A graph of the function $p(V)$ is shown for various values of the ratio $a/4RTb$ (Fig. 5.16). (To sketch this graph requires the techniques of Chapter 7.) For physically realistic situations, the pressure and volume must be positive so that only the solid lines are useful models.

Example 2. An expression for the acceleration of a body moving with velocity v m s^{-1} in a viscous medium at time t s is

$$f(v) = -v - kv^2$$

where k is a positive constant which depends on the properties of the fluid. To find an expression for the velocity requires the expansion of $1/f(v)$ in partial fractions (and the techniques of integration of Chapters 9–11). Find the partial fractions of $1/f(v)$ in this case.

Solution. The factors of $v + kv^2$ are v and $(1 + kv)$ so we must find constants A and B such that

$$\frac{1}{v + kv^2} = \frac{A}{v} + \frac{B}{(1 + kv)}$$

Multiply through by $v(1 + kv)$ to give

$$1 = A(1 + kv) + Bv$$

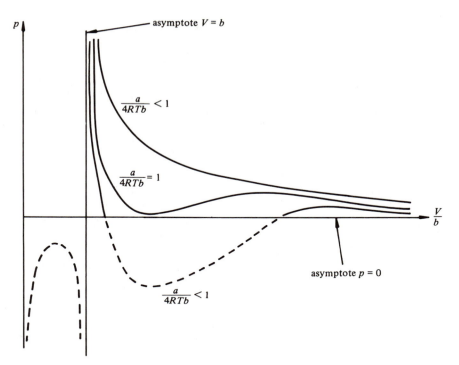

Fig. 5.16. Van der Waal's equation of state.

Putting $v = 0$ we get

$$1 = A(1 + 0) + B(0)$$

and

$$A = 1$$

Putting $v = -1/k$ we get

$$1 = A(0) + B(-1/k)$$

so that

$$B = -k$$

Finally, we can write

$$\frac{1}{f(v)} = \frac{-1}{v + kv^2} = \frac{-1}{v} + \frac{k}{(1 + kv)}$$

5.6 Exercises

1. Evaluate the following limits:

(i) $\lim\limits_{n \to \infty} \dfrac{4}{5n - 1}$

(ii) $\lim\limits_{n \to \infty} \dfrac{3n^2 + 2n + 7}{n^3 - n + 2}$

(iii) $\lim\limits_{n \to \infty} \dfrac{1 + 2 + 3 + 4 \cdots\cdots + n}{n^2}$

(iv) $\lim\limits_{x \to 1} \dfrac{x^3 - 1}{x - 1}$

(v) $\lim\limits_{x \to \infty} \dfrac{\sqrt{5x - 4}}{x + 2}$

(vi) $\lim\limits_{x \to 0} \dfrac{\sqrt{1 + x} - \sqrt{1 - x}}{4x}$

(vii) $\lim\limits_{n \to 1/2} \dfrac{6n^2 - n - 1}{2n^2 + 5n - 3}$

(viii) $\lim\limits_{x \to 0} \dfrac{\sin(x)}{x}$

(ix) $\lim\limits_{x \to \infty} x \sin(1/x)$

(x) $\lim\limits_{x \to 0} \dfrac{\cos(x)}{x}$

2. State the points at which the following functions are singular:

(i) $\dfrac{1}{x + 3}$

(ii) $\dfrac{x + 2}{(x^2 - 9)}$

(iii) $\dfrac{\cos(x)}{x - 1}$

(iv) $\dfrac{3x + 7}{1 - \sin(x)}$

(v) $\dfrac{x^2 - 2x + 1}{x^2 + 2x + 2}$

(vi) $\cot(x)$

(vii) $\dfrac{1}{x^4 - 15x^2 + 36}$

3. Write the following in partial fraction form:

(i) $\dfrac{2}{(x - 1)(x + 2)}$

(ii) $\dfrac{1}{(x - 2)(x + 3)}$

(iii) $\dfrac{(x + 1)}{x^2 - 3x + 2}$

(iv) $\dfrac{2 - 3x}{3x^2 - 4x + 1}$

(v) $\dfrac{(x + 3)}{x^3 - 6x^2 + 8x}$

(vi) $\dfrac{x^2}{(x - 1)(x + 2)}$

(vii) $\dfrac{x^2 - 2x + 1}{x^2 + 2x + 1}$

(viii) $\dfrac{x^3 + 3x}{x^2 + 3x + 2}$

(ix) $\dfrac{(4 - x)}{(x - 3)^2}$

(x) $\dfrac{1}{(x - 2)(x + 1)^3}$

4. The frequency f kHz of radio waves is inversely proportional to their wavelength λ m. Given that $f = 375$ when $\lambda = 800$ find the relation between f and λ. Hence

(i) calculate the wavelength of BBC Radio 2 whose frequency is 200 kHz,

(ii) calculate, to the nearest unit, the frequency of BBC Radio 1 whose wavelength is 284 m.

5. Two electrical charges attract one another with a force P units which varies inversely as the square of the distance, x units, between them. If $P = 5.4$ when $x = 6$, find P when $x = 9$.

6. Find the equations of the asymptotes for each of the following functions:

(i) $\dfrac{1}{(x + 3)}$
    (ii) $\dfrac{2}{(x - 1)(x + 5)}$

(iii) $\dfrac{(x + 1)}{(x - 1)}$
    (iv) $\dfrac{x^2 - 2x + 1}{x^2 + 2x + 2}$

(v) $\dfrac{3x + 7}{5x - 1}$
    (vi) $\dfrac{2x - 3}{x - \sin(x)}$

7. A parachutist of mass m drops from an aircraft and falls towards the earth. The parachutist opens the chute immediately and subsequently the velocity of the parachutist is given by

$$v(t) = \frac{mg}{k}[1 - \exp(-kt/m)]$$

where k is a constant. Find the limiting velocity or terminal velocity of the parachutist.

8. In a multiple decay process a type A atom decays into a type B atom and then decays into a type C atom. The number of type B atoms at any time t is given by

$$N_B = \frac{N_0 T_B}{(T_A - T_B)}[\exp(-at/T_A) - \exp(-at/T_B)]$$

where N_0 is the initial number of type A atoms, a is a constant and T_A and T_B are the half-lives of type A atoms and type B atoms respectively. Find the limiting number of type B atoms.

9. The inside of a solid sphere of radius R is initially at zero temperature. The sphere is dropped into a heat bath which is maintained at temperature T_0. Subsequently the temperature at radius r and time t is (approximately) given by

$$T = T_0\left[1 - \frac{2R}{\pi r}\exp(-\pi^2 kt/R^2)\sin(\pi r/R)\right]$$

where k is a constant. By choosing smaller and smaller values of x, use your calculator to show that $\lim_{x \to 0} \sin(x)/x = 1$.

Hence find an expression for the temperature at the centre of the sphere.

10. The following expression occurs in fracture mechanics:

$$K = \frac{cX^2}{(1 + X^2)^2}$$

where c is a constant. Find the value of $\lim_{x \to \infty} K$.

11. A large tank contains L litres of water. A brine solution of salt water flows into the tank at a constant rate of α litres min^{-1}. The mixture in the tank is stirred continuously and is drawn off from the tank at a rate of β litres min^{-1}. If the concentration of salt in the brine solution entering the tank is 1 kilogram litre^{-1} then the concentration of salt in the tank as a function of time is

$$c(t) = \frac{\alpha}{(\beta + 1)(\alpha - \beta)} \left(1 - \frac{1}{[1 + (\alpha - \beta)t/L]^{\beta + 1}} \right)$$

Find where this function is singular and interpret your solution physically. If the input is greater than the output, find the limit of the concentration after a long time period.

12. In population dynamics an important mathematical model which describes the size of many populations as a function of time is the logistic equation

$$P(t) = \frac{a}{b + (a/P_0 - b)\exp(-at)}$$

where a and b are constants. Find the equilibrium population size.

6

Differentiation 1: rates of change

Contents: 6.1 scientific context – 6.1.1 speed: a graphical model; 6.1.2 cell growth: an algebraic model; 6.1.3 potential energy and force: using x instead of t; 6.2 mathematical developments – 6.2.1 rates of change; 6.2.2 notation; 6.2.3 some basic functions and their derivatives; 6.3 worked examples; 6.4 further developments – 6.4.1 rules of differentiation; 6.4.2 higher derivatives; 6.5 further worked examples; 6.6 exercises.

6.1 Scientific context

We now begin several chapters dealing with *calculus*, which is a very important technique of analysis used in all branches of science. In the opening chapters we shall concentrate on working out rates of change and on locating where the maximum and minimum values occur in variables of interest. The technique used is one branch of calculus called *differentiation*.

Speed is a familiar concept; we are all used to travelling, whether on foot, by bicycle or by car, and often talk about the speed at which we are travelling. If we travel by car 170 miles in 4 hours then we say that our *average speed* for the journey is $42\frac{1}{2}$ m.p.h. We calculate average speed for a journey by working out the ratio

$$\text{average speed} = \frac{\text{distance travelled}}{\text{time taken}}$$

But the average speed is almost certainly not what the speedometer reading gives. That reading will vary from 0 when the car is at rest to perhaps 70 m.p.h. when cruising along an open stretch of road. The reading that the speedometer gives is the speed at an instant or the *instantaneous speed*. Average speed and instantaneous speed are both examples of *rates of change*.

The rate of change of a variable is an important idea in science; for instance the rate of change of the electric charge on the plates of a battery in a circuit tells us about the current that flows in the circuit. Not all rates of change need to be with respect to time, however. We are often interested in finding rates of change with respect to other quantities such as distance. For example, the rate of change with distance of the potential energy stored in a spring equals the force exerted on the spring.

To motivate the mathematical study of rates of change let us now consider three different examples.

6.1.1 Example 1 speed: a graphical model. Consider, once again, experiment 2 in

Chapter 1. A ball is thrown upwards and a photograph is taken of the motion of
the ball. From this photograph we can obtain a graph showing the position of the
ball at any instant. This is shown in Fig. 6.1.

Suppose that we want to find the vertical speed of projection of the ball. We
could start by finding the average speed of the ball over the first 0.1 s of its flight.
Fig. 6.2 shows that the ball rises 39.8 cm in this time interval so the average vertical

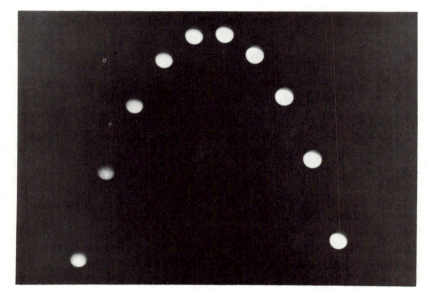

Fig. 6.1. The motion of a ball under gravity.

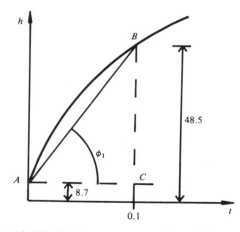

Fig. 6.2. Finding the average speed over 0.1 seconds.

speed of the ball is

$$\frac{39.8}{0.1} = 398 \,\text{cm s}^{-1}$$

This average speed, which is the ratio of the distance travelled over time taken, is the slope of the straight line or chord joining A to B, i.e.

$$\text{average speed} = \frac{BC}{AC} = \tan(\phi_1)$$

For the first 0.05 s we get a different value for the average speed. When $t = 0.05$ s the ball has travelled through a height $(30.5 - 8.7)$ cm as shown in Fig. 6.3. The average speed over this time interval of 0.05 s is therefore

$$\frac{30.5 - 8.7}{0.05} = 436 \,\text{cm s}^{-1} = \tan(\phi_2)$$

which is somewhat different from the value over 0.1 s. The diagram shows that ϕ_2 is greater than ϕ_1, i.e. the slope of the chord is steeper for this second time interval. As B moves closer and closer to A the chord AB becomes parallel to the tangent to the curve at A. The actual value of $\tan(\phi)$ for the tangent is called the *gradient of the curve at A* and equals the *instantaneous speed* of the ball at A in a vertical direction, i.e. at the instant of projection. The process of letting B move closer to A is a complicated one. The calculation for average speed has to be carried out carefully because the numerator and denominator of the formula for average speed are both getting smaller although the ratio is tending to a definite value. This process is a *limiting process* and the instantaneous speed is the *limit* of the average speed as the time interval becomes zero. In our example, if we work carefully with the plot in Fig. 6.2, the instantaneous speed of projection in a vertical direction

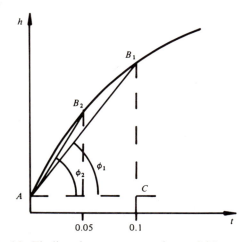

Fig. 6.3. Finding the average speed over 0.05 seconds.

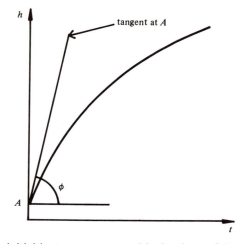

Fig. 6.4. The initial instantaneous speed is the slope of the tangent at A.

will be found to be about 448 cm s^{-1}, which is much closer to our second average speed than the first. The gradient of the tangent at A is shown in Fig. 6.4.

Question: This does not seem to be a very accurate way of finding the instantaneous speed. Is there not a better way?

Answer: In the absence of a suitable model, i.e. function that tells us what the height h is in terms of t, the answer has to be no. With such a model, however, we can actually obtain a mathematical expression for this speed.

Our next example involves a situation where such a model is known.

6.1.2 Example 2 cell growth: an algebraic model. The yeast culture, *Saccharomyces Cerevisiae*, is very important in the making of wine and beer – it turns sugar into alcohol. It is known that the increase in the number of yeast cells depends on many factors such as the sugar content and the temperature. It is important to build up a strong yeast colony quickly so that the wine or beer does not spoil. An important question then is, how fast is the yeast colony growing?

Under ideal conditions suppose that the number of cells of the yeast culture doubles every 60 min. If at time $t = 0$ there are 10 yeast cells, then after 60 min there are 20; after 120 min there are 40 yeast cells and so on (a similar example was discussed in 2.1.2). Under these conditions the growth of the population of yeast cells can be expressed in terms of an equation relating the number of yeast cells N and the time t as follows:

$$N = 10 \times 2^{t/60}$$

However, it is not always easy to count the number of cells in a fermenting liquid, so a biologist usually measures the amount of yeast in terms of its biomass. Suppose, then, that the biomass M of our colony of *Saccharomyces Cerevisiae* is modelled

Table 6.1. *Average growth rate for different time intervals*

Start	Interval end	Length	Average growth rate of M/K $\min^{-1}$
300	360	60	0.533 33
330	360	30	0.624 84
345	360	15	0.678 84
350	360	10	0.698 25
355	360	5	0.718 41
360	365	5	0.761 13
360	370	10	0.783 76
360	375	15	0.807 28
360	390	30	0.883 66
360	420	60	1.066 67

as a function of time by

$$M = K \times 2^{t/60}$$

where K is a constant proportional to the average mass of the yeast cell.

The biomass grows continuously because each yeast cell grows continuously before splitting in two. What is of interest to a biologist is the growth rate of the yeast colony at a particular instant of time, in other words the instantaneous growth rate.

For the example here, consider the time 360 min after the start of the experiment. Table 6.1 shows the average growth rates of the biomass defined in the following way for different intervals of time. We say that the average growth rate over a certain time interval is the ratio of 'the increase in the number of cells during the time interval' over 'the time interval'. In the table each interval either starts or ends at $t = 360$ and the formula $M/K = 2^{t/60}$ has been used to compute the appropriate values of M/K at the times quoted. As with the example on average speed in 6.1.1 we see that the smaller we take the interval from 360 min, then the smaller is the difference between the average growth rates. This is true whether we approach 360 from above or below. Again we have a ratio for which both the numerator and denominator are becoming smaller but the ratio itself is remaining a finite number. The average speed is the rate of change of distance over a time interval and the average growth rate is the rate of change in the biomass over a time interval. It is what happens to these rates of change as the time interval is reduced to zero that we investigate in this chapter. In Section 6.2 we shall introduce a straightforward way of calculating the instantaneous rate of change for known algebraic functions.

6.1.3 Example 3: potential energy and force: using x instead of t.

Our two examples so far have used t as the independent variable; however we can find the rate of change with respect to other variables. Consider experiment 1 in Chapter 1. We stretched a piece of mild steel wire by suspending weights from the end, and plotted

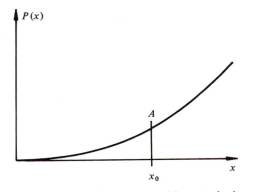

Fig. 6.5. Potential energy stored in a steel wire.

a graph of the length of the wire against the weight. The work done in stretching the wire is stored in the wire in the form of *potential energy P* which is given by the formula

$$P(x) = \tfrac{1}{2}kx^2$$

where x is the extension of the wire and k, a constant called the *stiffness*, is a property of the material which measures how easily the wire is stretched. For the steel wire in experiment 1, $k = 45 \times 10^4 \, \mathrm{N \, m^{-1}}$. Fig. 6.5 shows the graph of $P(x)$ against x. The rate of change of $P(x)$ for some extension x_0 say, i.e. the slope of the tangent to the curve at A, equals the tension in the wire when it is stretched by an amount x_0. So an algebraic formula for the rate of change of $P(x)$ with respect to x gives us information about the tension in the wire. Clearly, it is more convenient to have this rate of change as a formula than to have to draw tangents to the graph of $P(x)$. In 6.2 we will see how to work out the rate of change in the function $\tfrac{1}{2}kx^2$.

6.2 Mathematical developments

6.2.1 Rates of change. Let s denote a variable which depends on t, where t is the time. We have obtained the average rate of change of s over a time interval $t_2 - t_1$ using the ratio $(s_2 - s_1)/(t_2 - t_1)$ where s_1 and s_2 are the values of s at times t_1 and t_2 respectively.

We shall now see what happens to the average rate of change as the interval $t_2 - t_1$ is reduced in size and t_1 is held constant in a particular example where the functional form of s is known.

Example: If $s = 4t + 5t^2$, what are the average rates of change in s for $t_1 = 2$ and $t_2 = 6, 4,$ 3 and 2.5?

Solution: The average rates of change are shown in Table 6.2.

Table 6.2. *Average rates of change of s*

t_1	t_2	s_1	s_2	$\left[\dfrac{s_2 - s_1}{t_2 - t_1}\right]$
2	6	28	204	44
2	4	28	96	34
2	3	28	57	29
2	2.5	28	41.25	26.5

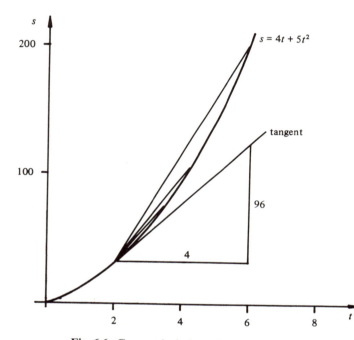

Fig. 6.6. Geometrical view of rates of change.

Fig. 6.6 shows graphically what is happening in the example above. One obvious question we might ask concerning this example is can we let $t_2 = 2$? The answer, algebraically, appears to be no. There is a problem here because if $t_2 = 2$, then $s_2 = 28$ and the average rate of change becomes 0/0 which we know is meaningless. However, on the graph in Fig. 6.6 the chords are gradually tending to the tangent to the curve at $t = 2$, so let us take the table of values further by reducing t_2 towards the value 2, but without it actually becoming equal to 2 (Table 6.3). It seems that as the value of t_2 gets closer to 2, the value of s_2 gets closer and closer to 28, which is what we would expect. But it also appears that the ratio $(s_2 - s_1)/(t_2 - t_1)$ is getting closer to a definite value although both numerator and denominator are

Differentiation 1: rates of change

Table 6.3. *Average rates of change of s as the time interval is reduced*

t_1	t_2	s_1	s_2	$\dfrac{s_2 - s_1}{t_2 - t_1}$
2	2.1	28	30.45	24.5
2	2.01	28	28.2405	24.05
2	2.001	28	28.024 005	24.005
2	2.0001	28	28.002 400 05	24.0005

tending to zero. It looks as if the limiting value of $(s_2 - s_1)/(t_2 - t_1)$ as t_2 approaches 2 should be 24. Certainly, the graph gives this value for the slope of the tangent at $t = 2$. We can investigate mathematically whether this *is* the value by letting $t = 2 + \tau$ and seeing what happens when τ tends to zero. When $t_1 = 2$ and $t_2 = 2 + \tau$ we get

$$s_2 = 4(2 + \tau) + 5(2 + \tau)^2$$
$$= 28 + 24\tau + 5\tau^2$$

so that

$$\frac{s_2 - s_1}{t_2 - t_1} = \frac{28 + 24\tau + 5\tau^2 - 28}{2 + \tau - 2} = \frac{24\tau + 5\tau^2}{\tau} = 24 + 5\tau$$

Now if we let τ get smaller and smaller, t_2 tends closer and closer to the value 2, and indeed the ratio $(s_2 - s_1)/(t_2 - t_1)$ gets nearer to 24. If we actually let $\tau = 0$, the limit of $(s_2 - s_1)/(t_2 - t_1)$ equals 24. We say that the ratio $(s_2 - s_1)/(t_2 - t_1)$ tends to the limit 24 as τ tends to 0, or as t_2 tends to 2.

The limit of the ratio that we have found is called the *instantaneous rate of change* at $t = 2$. It tells us how the variable s is changing at a particular value of the variable t, instead of over an interval for t.

Using the same notation here for a limit as we used before we write

$$\lim_{t_2 \to t_1} \{(s_2 - s_1)/(t_2 - t_1)\} \tag{1}$$

to stand for the limit of the ratio $(s_2 - s_1)/(t_2 - t_1)$ as the value of t_2 tends to the value t_1. Note that we must take care in the evaluation of the limit. The analysis carried out on the ratio $(s_2 - s_1)/(t_2 - t_1)$ must be for t_2 *not* equal to t_1, i.e. we let $t_2 - t_1 = \tau$ where τ is *not* zero. We cannot put $t_2 = t_1$ in the expression for the limit because that would give 0/0. Therefore *first* we divide by τ in the denominator and numerator, and *then* we can let $\tau = 0$. Notice how in the example above, this procedure leaves an expression whose limit is easier to find.

Example: If $s = t^2 - 3t + 1$ what is the value of $\lim_{t_2 \to t_1} \{(s_2 - s_1)/(t_2 - t_1)\}$ when $t_1 = 1$?
Solution: The first step is to let $t_2 - t_1 = t_2 - 1 = \tau$ where τ is a small but non-zero number. Then

$$s_1 = t_1^2 - 3t_1 + 1 = -1 \text{ when } t_1 = 1$$

and
$$s_2 = (1 + \tau)^2 - 3(1 + \tau) + 1 = \tau^2 - \tau - 1$$

The ratio $(s_2 - s_1)/(t_2 - t_1)$ becomes

$$\frac{(\tau^2 - \tau - 1) - (-1)}{\tau} = \frac{\tau^2 - \tau}{\tau}$$

The next step is to divide numerator and denominator by τ so that $(s_2 - s_1)/(t_2 - t_1) = \tau - 1$. Now we let τ tend to zero giving

$$\lim_{t_2 \to t_1} \{(s_2 - s_1)/(t_2 - t_1)\} = \lim_{\tau \to 0}(\tau - 1) = -1 \text{ when } t_1 = 1$$

6.2.2 Notation. The last section introduced the idea of small changes in a variable. In calculus we adopt a special notation for these small changes.

Suppose that we make a 'small change in the variable t' then we write this small change as δt. This does not mean δ times t; it is one of the occasions where we write two symbols next to each other that does not mean multiplication. So

$$\delta t \text{ means a small change in } t \qquad (2)$$

The small change δt is sometimes called an '*increment*' in the value of t'. Writing $t_2 = t_1 + \delta t$ implies that t_1 is close to t_2. In the examples above, when we made a small change in t, we obtained a corresponding small change in s which was a function of t. So that if $s = s(t)$ we can then write $s_2 = s_1 + \delta s$ or in general $s(t + \delta t) = s(t) + \delta s$ as shown in Fig. 6.7. This means that the ratio $(s_2 - s_1)/(t_2 - t_1)$ for $t_1 = t$ can be written as $\delta s/\delta t$ for small changes in the variable t.

The following example brings together the notation of 'a small change' and the 'limit' and introduces the process of *differentiation*.

Example: If $s = 2t^3 + 2t - 1$ what is the value of $\lim_{\delta t \to 0} (\delta s/\delta t)$?
Solution: $\delta s/\delta t$ means the value of $(s_2 - s_1)/(t_2 - t_1)$ for $t_1 = t$ and $t_2 = t_1 + \delta t = t + \delta t$.

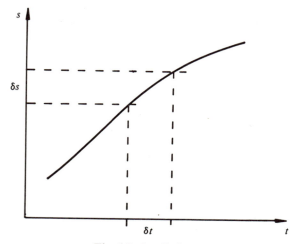

Fig. 6.7. Small changes.

Now the value of s_1 is

$$s(t_1) = s(t) = 2t^3 + 2t - 1$$

and

$$s_2 = s(t_2) = s(t + \delta t) = 2(t + \delta t)^3 + 2(t + \delta t) - 1$$

Thus the value of $\delta s = s_2 - s_1 = s(t + \delta t) - s(t)$ which is

$$2(t + \delta t)^3 + 2(t + \delta t) - 1 - (2t^3 + 2t - 1)$$

Simplifying this expression for δs we obtain

$$\delta s = (6t^2 + 1)\delta t + 6t\,\delta t^2 + 2\delta t^3$$

The value of $\delta s/\delta t$ is

$$(6t^2 + 2) + 6t\,\delta t + 2\delta t^2$$

To find the limit as $\delta t \to 0$, we let δt get smaller and smaller in this expression and obtain

$$\lim_{\delta t \to 0} \left(\frac{\delta s}{\delta t}\right) = 6t^2 + 2$$

Thus the instantaneous rate of change of s with respect to t is the function $6t^2 + 2$.

The process of finding the instantaneous rate of change is called *differentiation* and $\lim_{\delta t \to 0}(\delta s/\delta t)$ is known as the *derivative* of s with respect to t. We adopt a special notation for this limit and write

$$\frac{\mathrm{d}s}{\mathrm{d}t} = \lim_{\delta t \to 0} \left(\frac{\delta s}{\delta t}\right) \tag{3}$$

We should note that $\mathrm{d}s/\mathrm{d}t$ is *one* symbol meaning the instantaneous rate of change of s with t. It should not be confused with the ratio $\delta s/\delta t$. The value of $\mathrm{d}s/\mathrm{d}t$ tells us how quickly s is changing with t.

Average rates of change and the derivative of a function of *any* variable (and not just t) can be defined in the same way. Suppose that a variable y is given as a function of a variable x. Then we define the average rate of change in y as x changes from x_1 to x_2 to be $(y_2 - y_1)/(x_2 - x_1)$ where y_1 and y_2 are the corresponding values of y. If $x_1 = x$ and $x_2 = x + \delta x$ then this average rate of change becomes $\delta y/\delta x$ where $\delta y = y(x + \delta x) - y(x)$. In the limit $\delta x \to 0$ we obtain the derivative $\mathrm{d}y/\mathrm{d}x$. Thus

$$\frac{\mathrm{d}y}{\mathrm{d}x} = \lim_{\delta x \to 0} \left(\frac{\delta y}{\delta x}\right) \tag{4}$$

As well as defining the rate of change of variables we can easily extend the idea of differentiation to functions rather than just variables.

The derivative of a function $f(x)$ with respect to x is defined as the limit given by

$$\lim_{\delta x \to 0} \left[\frac{f(x + \delta x) - f(x)}{\delta x}\right] \tag{5}$$

and is denoted by the symbol $f'(x)$. This follows on naturally from what we have just said above; for if x and y are variables such that $y = f(x)$ then

$$\lim_{\delta x \to 0} \left[\frac{f(x + \delta x) - f(x)}{\delta x}\right] = \lim_{\delta x \to 0} \left(\frac{\delta y}{\delta x}\right) = \frac{\mathrm{d}y}{\mathrm{d}x}$$

Example: If $f(x) = x^2 + 2x + 1$, what is $f'(x)$?

Solution: Using the definition in (5) we have

$$f'(x) = \lim_{\delta x \to 0} \left[\frac{(x + \delta x)^2 + 2(x + \delta x) + 1 - (x^2 + 2x + 1)}{\delta x} \right]$$

$$= \lim_{\delta x \to 0} \left[\frac{(2x + 2)\delta x + \delta x^2}{\delta x} \right] = \lim_{\delta x \to 0} (2x + 2 + \delta x)$$

$$= 2x + 2$$

This last example shows indeed that the derivative of a function is itself a new function of x. To remind us of this fact we often call $f'(x)$ the *derived function* of f. It is rather tedious having to use the formula

$$f'(x) = \lim_{\delta x \to 0} \left[\frac{f(x + \delta x) - f(x)}{\delta x} \right]$$

for every example. Therefore we work out the derivatives of the basic functions that occur frequently in science and apply these results in specific problems as we need them. We shall also need rules for differentiating combinations of basic functions, e.g. sums, products, etc.

6.2.3 Some basic functions and their derivatives. *The derivative of ax^n (for positive integer values of n).* For $f(x) = ax^n$, where a is a constant, we have

$$f(x + \delta x) - f(x) = a(x + \delta x)^n - ax^n$$

Using the binomial expansion to expand the term $(x + \delta x)^n$ we obtain

$$f'(x) = \lim_{\delta x \to 0} \left[\frac{a(x^n + nx^{n-1}\delta x + (n(n-1)/2)x^{n-2}\delta x^2 + \cdots + \delta x^n) - ax^n}{\delta x} \right]$$

$$= \lim_{\delta x \to 0} \left[a[nx^{n-1} + (n(n-1)/2)x^{n-2}\delta x + \cdots + \delta x^{n-1}]] \right]$$

$$= anx^{n-1}$$

The derivative of ax^n is anx^{n-1}; thus

$$\text{if } f(x) = ax^n, \quad f'(x) = anx^{n-1} \tag{6}$$

This result, in fact, holds for any value of n, for example $n = -2, \frac{1}{2}$, etc., and not just positive integers.

Special cases of the result in (6) that are of note occur when $n = 1$ and $n = 0$. If $n = 1$, then $f(x) = ax$. The derivative $f'(x)$ is therefore $a \cdot 1 \cdot x^0$ i.e. a. This result can be appreciated graphically because the graph of $y = ax$ is a straight line of slope a as shown in Fig. 6.8. The chord joining any two points on it has the same slope as the tangent. In fact for this case the chord is the tangent at every point. If $n = 0$, i.e. $f(x) = a$, then $f'(x) = a \cdot 0 = 0$. *The derivative of a constant is zero.* For this case the line $y = a$ is a line parallel to the x axis as shown in Fig. 6.9. The angle between this line and the x axis is zero so its slope is also zero – which is the derivative.

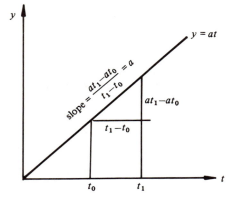

Fig. 6.8. Slope of a linear function.

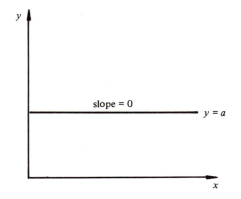

Fig. 6.9. The derivative of a constant is zero.

The derivative of a sum $g(x) + h(x)$. In the last section we showed that if $f(x) = x^2 + 2x + 1$ then $f'(x) = 2x + 2$ and we did it by using the definition of $f'(x)$ as a limit. Notice, however, that the derivative of x^2 is $2x$, of $2x$ is 2 and of 1 is 0. So $f'(x)$ is the sum of the derivatives of the functions forming $f(x)$.

In general the derivative of a sum of functions is the sum of the derivatives of the individual functions:

$$\text{if } f(x) = g(x) + h(x), \quad \text{then} \quad f'(x) = g'(x) + h'(x) \tag{7}$$

To show this result we use the definition of $f'(x)$ as a limit. If $f(x) = g(x) + h(x)$ then

$$\frac{f(x + \delta x) - f(x)}{\delta x} = \frac{g(x + \delta x) + h(x + \delta x) - (g(x) + h(x))}{\delta x}$$

$$= \frac{g(x + \delta x) - g(x)}{\delta x} + \frac{h(x + \delta x) - h(x)}{\delta x}$$

So taking the limit $\delta x \to 0$ on each side of this equation we have

$$\lim_{\delta x \to 0} \left[\frac{f(x + \delta x) - f(x)}{\delta x} \right] = \lim_{\delta x \to 0} \left[\frac{g(x + \delta x) - g(x)}{\delta x} \right] + \lim_{\delta x \to 0} \left[\frac{h(x + \delta x) - h(x)}{\delta x} \right]$$

i.e.

$$f'(x) = g'(x) + h'(x)$$

establishing the required result.

Polynomials of degree n. We can use this last property to find the derivative of a polynomial of degree n, of which the function $f(x) = x^2 + 2x + 1$ is a special case. In Chapter 1 we defined the nth-degree polynomial as a sum of $(n + 1)$ terms in the following way:

$$f(x) = ax^n + bx^{n-1} + cx^{n-2} + \cdots + px + q$$

Using the 'derivative of a sum' rule we have

$$f'(x) = anx^{n-1} + b(n-1)x^{n-2} + c(n-2)x^{n-3} + \cdots + p + 0 \tag{8}$$

The derivative of e^x. The exponential function e^x can be differentiated as follows. If we let $f(x) = e^x$ then

$$f(x + \delta x) - f(x) = e^{x + \delta x} - e^x$$
$$= e^x e^{\delta x} - e^x$$
$$= e^x (e^{\delta x} - 1)$$

and hence

$$f'(x) = \lim_{\delta x \to 0} \left[\frac{e^x (e^{\delta x} - 1)}{\delta x} \right]$$
$$= e^x \lim_{\delta x \to 0} \left[\frac{(e^{\delta x} - 1)}{\delta x} \right]$$

Using a calculator we can get an idea of what this limit is. Table 6.4 gives the value of the ratio for various values of δx. The table suggests that as δx gets smaller the ratio $(e^{\delta x} - 1)/\delta x$ approaches the value 1. The value of the limit of the ratio $(e^{\delta x} - 1)/\delta x$, in fact, can be shown to be 1 by using the power series expansion of

Table 6.4. *Finding the limit of* $(e^{\delta x} - 1)/\delta x$

δx	$\dfrac{e^{\delta x} - 1}{\delta x}$
0.1	1.051 71
0.01	1.005 017
0.001	1.000 500
0.0001	1.000 050

the exponential function (see below). So we have the important result that

$$\text{if } f(x) = e^x, \quad \text{then} \quad f'(x) = e^x \tag{9}$$

The derivative of e^x is the same function and this is a special feature of the exponential function. There is no other non-zero function for which $f'(x) = f(x)$.

Using the series expansion, then, to investigate the limit $(e^{\delta x} - 1)/\delta x$ as δx tends to zero we have

$$e^{\delta x} - 1 = \left(1 + \delta x + \frac{\delta x^2}{2!} + \cdots \right) - 1$$

$$= \delta x + \frac{\delta x^2}{2!} + \cdots$$

so that

$$(e^{\delta x} - 1)/\delta x = 1 + \delta x/2! + \cdots$$

In the limit as $\delta x \to 0$, the right-hand side equals 1, as we suggested above.

For the more general function $f(x) = e^{cx}$, where c is a constant, we get

$$f'(x) = \lim_{\delta x \to 0} e^{cx} \left[\frac{e^{c\delta x} - 1}{\delta x} \right]$$

$$= e^{cx} \lim_{\delta x \to 0} \left[c + \frac{c^2 \delta x}{2!} + \cdots \right]$$

$$= c e^{cx}$$

Thus

$$\text{if } f(x) = e^{cx}, \quad \text{then} \quad f'(x) = c e^x \tag{10}$$

The derivative of a^x. The general exponential function is the function $f(x) = a^x$ where a is a positive constant, i.e. $a > 0$. To differentiate $f(x) = a^x$ we use the identity

$$a^x = \exp[\ln(a^x)] = \exp[x \ln(a)]$$

Using result (10) we see that the derivative of $\exp[x \ln(a)]$ will be $\ln(a) \exp[x \ln(a)]$ i.e. $\ln(a)a^x$. Thus

$$\text{if } f(x) = a^x, \quad \text{then} \quad f'(x) = \ln(a)a^x \tag{11}$$

Example: What is the value of $f'(x)$ for $f(x) = 2^x$ when $x = 1$?
Solution: Using (11) we have $f'(x) = \ln(2)2^x$ when $x = 1$, this has the value 0.6931×2.
Thus $f'(1) = 1.3862$.

The derivative of $\sin(x)$. Let $f(x) = \sin(x)$ then

$$f(x + \delta x) - f(x) = \sin(x + \delta x) - \sin(x)$$

$$= \sin(x)\cos(\delta x) + \cos(x)\sin(\delta x) - \sin(x)$$

$$= \sin(x)(\cos(\delta x) - 1) + \cos(x)\sin(\delta x)$$

Hence $f'(x)$ is given by

$$f'(x) = \lim_{\delta x \to 0} \left[\sin(x) \left(\frac{\cos(\delta x) - 1}{\delta x} \right) + \cos(x) \left(\frac{\sin(\delta x)}{\delta x} \right) \right]$$

Table 6.5. *Finding the limits of*
$(\cos(\delta x) - 1)/\delta x$ *and* $\sin(\delta x)/\delta x$

δx	$\dfrac{\cos(\delta x) - 1}{\delta x}$	$\dfrac{\sin(\delta x)}{\delta x}$
1	$- 0.459\,70$	$0.841\,47$
0.1	$- 0.049\,958$	$0.998\,33$
0.01	$- 0.005$	$0.999\,98$
0.001	$- 0.0005$	$0.999\,999\,8$

We need, therefore, to investigate the limits of $(\cos(\delta x) - 1)/\delta x$ and $\sin(\delta x)/\delta x$ as $\delta x \to 0$. Use of a calculator will give us some idea of these limits. Table 6.5 shows the values of these ratios for decreasing values of δx. The values in Table 6.5 suggest that

$$\lim_{\delta x \to 0}\left[\frac{\cos(\delta x) - 1}{\delta x}\right] = 0$$

and

$$\lim_{\delta x \to 0}\left[\frac{\sin(\delta x)}{\delta x}\right] = 1$$

Using these limits in our expression for $f'(x)$ we see that

$$f'(x) = \sin(x) \times 0 + \cos(x) \times 1$$

Thus

$$\text{if } f(x) = \sin(x), \quad \text{then,} \quad f'(x) = \cos(x) \tag{12}$$

We can, in fact, obtain the above limits using the power series expansions for $\sin(x)$ and $\cos(x)$

$$\cos(\delta x) - 1 = \left(1 - \frac{\delta x^2}{2!} + \frac{\delta x^4}{4!} - \cdots\right) - 1$$

$$= -\frac{\delta x^2}{2!} + \frac{\delta x^4}{4!} - \cdots$$

So

$$\frac{\cos(\delta x) - 1}{\delta x} = -\frac{\delta x}{2!} + \frac{\delta x^3}{4!} - \cdots$$

As $\delta x \to 0$, the right-hand side becomes zero, as was suggested above. Also

$$\sin(\delta x) = \delta x - \frac{\delta x^3}{3!} + \frac{\delta x^5}{5!} - \cdots$$

so that

$$\frac{\sin(\delta x)}{\delta x} = 1 - \frac{\delta x^2}{3!} + \frac{\delta x^4}{5!} - \cdots$$

The limit of the right-hand side is now 1, as was again suggested by Table 6.5.

The derivative of $\cos(x)$. Let $f(x) = \cos(x)$ then

$$f(x + \delta x) - f(x) = \cos(x + \delta x) - \cos(x)$$
$$= \cos(x)\cos(\delta x) - \sin(x)\sin(\delta x) - \cos(x)$$
$$= \cos(x)(\cos(\delta x) - 1) - \sin(x)\sin(\delta x)$$

Hence

$$f'(x) = \lim_{\delta x \to 0}\left[\cos(x)\left(\frac{\cos(\delta x) - 1}{\delta x}\right) - \sin(x)\left(\frac{\sin(\delta x)}{\delta x}\right)\right]$$

and using the values of the limits obtained above, we have

$$f'(x) = \cos(x) \times 0 - \sin(x) \times 1$$

Thus

$$\text{if } f(x) = \cos(x), \quad \text{then,} \quad f'(x) = -\sin(x) \tag{13}$$

The derivative of $cg(x)$. Let $f(x) = cg(x)$, where c is a constant and $g(x)$ is any arbitrary function of x. We have, therefore, that

$$f(x + \delta x) - f(x) = cg(x + \delta x) - cg(x) = c(g(x + \delta x) - g(x))$$

Hence

$$f'(x) = \lim_{\delta x \to 0}\left[\frac{c(g(x + \delta x) - g(x))}{\delta x}\right]$$
$$= c \lim_{\delta x \to 0}\left[\frac{g(x + \delta x) - g(x)}{\delta x}\right]$$
$$= cg'(x)$$

Thus

$$\text{if } f(x) = cg(x), \quad \text{then,} \quad f'(x) = cg'(x) \tag{14}$$

Example: If $f(x) = 5\cos(x) + 3e^{2x}$ what is $f'(x)$?
Solution: We need to use the results of (7), (10), (13) and (14). Putting together the results of (13) and (14) we have

$$\frac{d}{dx}(5\cos(x)) = 5(-\sin(x)) = -5\sin(x)$$

Then using (10) and (14) we have

$$\frac{d}{dx}(3e^{2x}) = 3(2e^{2x}) = 6e^{2x}$$

Adding these results according to (7) we obtain

$$f'(x) = -5\sin(x) + 6e^{2x}$$

6.3 Worked examples

Example 1. A stone is thrown upwards so that its height in metres is given by $h(t) = 5 + 21t - 5t^2$, where t is in seconds.

(i) Find the speed of the stone when $t = 1, 2, 3, 4$.

(ii) What is the average speed for the intervals (0 to 1) and (1 to 2) seconds?

(iii) For what range of values of t is the stone
  (a) climbing
  (b) falling?

Solution. The speed v of the stone at any time t is given by the equation $v = h'(t)$. We can differentiate $h(t)$, which is a polynomial function of degree 2. We have

$$v = h'(t) = 0 + 21 - 10t$$

(i) The following table shows the values of v when $t = 1, 2, 3$ and 4

t s	1	2	3	4
v m s^{-1}	11	1	-9	-19

(ii) The average speed over the interval (t_1 to t_2) is given by the expression $(h(t_2) - h(t_1))/(t_2 - t_1)$. For the interval (0 to 1) we have

$$\text{average speed} = (21 - 5)/(1 - 0) = 16 \, \text{m s}^{-1}$$

For the interval (1 to 2) we have

$$\text{average speed} = (27 - 21)/(2 - 1) = 6 \, \text{m s}^{-1}$$

(iii) The stone comes to rest when $v = 0$. Now $v = 0$ when

$$21 - 10t = 0, \quad \text{i.e. when} \quad t = 2.1 \, \text{s}$$

  (a) Initially the value of v is positive and the stone is climbing. It climbs for times t satisfying $t < 2.1$ s.
  (b) For times $t > 2.1$ s the value of v is negative and hence the stone is falling.

Example 2. The air pressure P (in atmospheres) changes with altitude h (in km) according to the formula

$$P = (0.885)^h$$

Find the rate of change of pressure with height at the earth's surface (i.e. when $h = 0$) and at heights $2, 4, 6$ and 8 km. What does the negative sign mean in each of the values?

Solution. The rate of change of P with h is given by

$$P'(h) = \ln(0.885)(0.885)^h$$

At the earth's surface $h = 0$, so that

$$P'(0) = \ln(0.885) \times 1$$
$$= -0.122 \, \text{atm km}^{-2}$$

The following table shows the rate of change of P at the other values of h:

h km	2	4	6	8
$P'(h)$ atm km^{-1}	-0.096	-0.075	-0.059	-0.046

The negative sign means that the pressure is decreasing with increasing altitude, which we know to be the case.

Example 3. A mathematical model of the growth of a certain bacterium is given as

$$M = M_0 e^{0.4t}$$

where t is the time and M_0 is the mass of bacteria present at time $t = 0$. Find the instantaneous growth rate of the bacterium at time t, and show that it is proportional to the mass at that time.

Solution. With $M_0 = M_0 e^{0.4t}$ we have $M'(t) = M_0(0.4 e^{0.4t})$. The instantaneous growth rate at time t is therefore

$$M'(t) = 0.4 M_0 e^{0.4t}$$

In terms of M we see that this is simply 0.4 M. Thus the growth rate is directly proportional to M.

6.4 Further developments

6.4.1 Rules of differentiation. We now know how to differentiate some of the basic functions that occur in science, and we have seen that the derivative of a sum of two functions is just the sum of the derivatives of the two functions. We can, of course, construct more complicated functions by, for instance, taking products and quotients of our basic functions. We could, for example, form the function $x^2 \sin(x)$. We might then want to know what the derivative of $f(x) = x^2 \sin(x)$ was.

We can always start from our definition of the derivative as a limit and obtain the answer that way. Approaching the problem this way we have

$$
\begin{aligned}
f(x + \delta x) - f(x) &= (x + \delta x)^2 \sin(x + \delta x) - x^2 \sin(x) \\
&= (x^2 + 2x\delta x + \delta x^2) \sin(x + \delta x) - x^2 \sin(x) \\
&= (2x\delta x + \delta x^2) \sin(x + \delta x) + x^2 (\sin(x + \delta x) - \sin(x))
\end{aligned}
$$

Hence

$$
\begin{aligned}
f'(x) &= \lim_{\delta x \to 0} \left[(2x + \delta x) \sin(x + \delta x) + x^2 \left(\frac{\sin(x + \delta x) - \sin(x)}{\delta x} \right) \right] \\
&= 2x \sin(x) + x^2 \cos(x)
\end{aligned}
$$

because the expression

$$\lim_{\delta x \to 0} \left[\frac{\sin(x + \delta x) - \sin(x)}{\delta x} \right]$$

is the derivative of $\sin(x)$.

So we have an interesting answer. The derivative of $x^2 \sin(x)$ is just $(x^2 \times$ derivative of $\sin(x)) + ($derivative of $x^2 \times \sin(x))$. For a function built out of our basic functions x^2 and $\sin(x)$, then the derived function is built up out of the functions and derived functions of x^2 and $\sin(x)$. This process works because there are simple rules for differentiating combinations of functions which we shall now formulate.

The product rule. Suppose that $f(x)$ is a product of two functions g and h; i.e. let $f(x) = g(x)h(x)$. Then

$$f(x + \delta x) - f(x) = g(x + \delta x)h(x + \delta x) - g(x)h(x)$$

Now we expect to have derivatives of $g(x)$ and $h(x)$ in the expression for $f'(x)$ so we introduce the terms $g(x + \delta x) - g(x)$ and $h(x + \delta x) - h(x)$. To do this we rewrite the above expression as

$$f(x + \delta x) - f(x) = (g(x + \delta x) - g(x))h(x + \delta x) + g(x)(h(x + \delta x) - h(x))$$

Hence

$$f'(x) = \lim_{\delta x \to 0} \left[\left(\frac{g(x + \delta x) - g(x)}{\delta x} \right) h(x + \delta x) + g(x) \left(\frac{h(x + \delta x) - h(x)}{\delta x} \right) \right]$$

$$= g'(x)h(x) + g(x)h'(x)$$

In general, then

$$\text{if } f(x) = g(x)h(x), \quad \text{then} \quad f'(x) = g'(x)h(x) + g(x)h'(x) \tag{15}$$

Example: If $f(x) = (x^3 + 4x)\cos(x)$ find $f'(x)$.
Solution: The function $f(x)$ is the product of two functions

$$g(x) = (x^3 + 4x) \quad \text{and} \quad h(x) = \cos(x)$$

Noting that $g'(x) = 3x^2 + 4$ and $h'(x) = -\sin(x)$ we have, using the product rule, that

$$f'(x) = (3x^2 + 4)\cos(x) + (x^3 + 4x)(-\sin(x))$$
$$= (3x^2 + 4)\cos(x) - (x^3 + 4x)\sin(x)$$

The product rule has to be used repeatedly to differentiate a function involving the product of three or more functions. This usually presents no difficulties.

The quotient rule. Suppose now that $f(x)$ is a quotient of the functions g and h so that $f(x) = h(x)/g(x)$. Then

$$f(x + \delta x) - f(x) = \frac{h(x + \delta x)}{g(x + \delta x)} - \frac{h(x)}{g(x)}$$

$$= \frac{h(x + \delta x)g(x) - h(x)g(x + \delta x)}{g(x + \delta x)g(x)}$$

The numerator can be rewritten to include terms $h(x + \delta x) - h(x)$ and $g(x + \delta x) - g(x)$ in the following way:

$$f(x + \delta x) - f(x) = \frac{(h(x + \delta x) - h(x))g(x) - (g(x + \delta x) - g(x))h(x)}{g(x + \delta x)g(x)}$$

Hence, dividing by δx and taking the limit $\delta x \to 0$ we obtain

$$f'(x) = \lim_{\delta x \to 0} \left[\frac{(\{h(x + \delta x) - h(x)\}/\delta x)g(x) - (\{g(x + \delta x) - g(x)\}/\delta x)h(x)}{g(x + \delta x)g(x)} \right]$$

$$= \frac{h'(x)g(x) - g'(x)h(x)}{g(x)^2}$$

Thus we obtain the following quotient rule:

$$\text{if} \quad f(x) = h(x)/g(x), \quad \text{then} \quad f'(x) = [h'(x)g(x) - g'(x)h(x)]/g(x)^2 \quad (16)$$

Example: If $f(x) = \tan(x)$ obtain $f'(x)$.
Solution: $f(x) = \sin(x)/\cos(x)$; so letting $h(x) = \sin(x)$ and $g(x) = \cos(x)$. Therefore the derivative of $\tan(x)$ is given by

$$f'(x) = (\cos(x)\cos(x) - (-\sin(x))\sin(x))/\cos^2(x)$$
$$= (\cos^2(x) + \sin^2(x))/\cos^2(x)$$
$$= 1/\cos^2(x) = \sec^2(x)$$

The reciprocal rule. A special case of the quotient rule is when the numerator is 1, i.e. $h(x) = 1$ for all x. So $h'(x) = 0$ and therefore

$$\text{if} \quad f(x) = 1/g(x), \quad \text{then} \quad f'(x) = -g'(x)/g(x)^2 \quad (17)$$

An important example of this rule is when $f(x) = x^{-n}$, i.e. when $f(x) = 1/x^n$. Using the reciprocal rule we see that $f'(x) = -nx^{n-1}/x^{2n} = -nx^{-n-1}$. This shows that the rule for differentiating $f(x) = x^n$ thus extends to the case of negative powers as we stated earlier.

The function of a function rule. We have seen how to differentiate basic functions such as $\sin(x)$, e^x, etc., but what about functions of more complicated expressions such as $\sin(4x)$, e^{7x}, or even $\sin(2x^2 + 4)$?

Example: Using the definition of differentiation as a limit, what is $f'(x)$ when $f(x) = \sin(4x)$?
Solution: If $f(x) = \sin(4x)$, then

$$f(x + \delta x) - f(x) = \sin(4(x + \delta x)) - \sin(4x)$$
$$= \sin(4x)\cos(4\delta x) + \cos(4x)\sin(4\delta x) - \sin(4x)$$

Hence dividing by δx and taking the limit $\delta x \to 0$ we obtain

$$f'(x) = \lim_{\delta x \to 0} \left[\sin\left(\frac{\cos(4\delta x) - 1}{\delta x}\right) + \cos(4x)\left(\frac{\sin(4\delta x)}{\delta x}\right) \right]$$

We know the values of

$$\lim_{\delta x \to 0} \left[\frac{\cos(\delta x) - 1}{\delta x} \right] \quad \text{and} \quad \lim_{\delta x \to 0} \left[\frac{\sin(\delta x)}{\delta x} \right]$$

but here we have $\sin(4\delta x)$ and $\cos(4\delta x)$. Suppose we let $4\delta x = h$ then $\delta x = h/4$ and as $\delta x \to 0$, $h \to 0$ also. So we can write

$$\lim_{\delta x \to 0} \left[\frac{\sin(4\delta x)}{\delta x} \right] = \lim_{h \to 0} \left[\frac{\sin(h)}{h/4} \right] = 4$$

and

$$\lim_{\delta x \to 0} \left[\frac{\cos(4\delta x) - 1}{\delta x} \right] = \lim_{h \to 0} \left[\frac{\cos(h) - 1}{h/4} \right] = 0$$

Thus with these limits we obtain $f'(x) = 4\cos(4x)$

In this last example we should note that whilst the derivative of $\sin(x)$ is $\cos(x)$, the derivative of $\sin(4x)$ is $4\cos(4x)$. The sine becomes a cosine as before but the $(4x)$ introduces a 4 multiplying the cosine.

Sin$(4x)$ is an example of a *function of a function* or a composite function. If $u = 4x$, then $\sin(4x)$ is the function $\sin(u(x))$ which is of the form $g(h(x))$. To find a rule to differentiate the function $f(x) = g(h(x))$ we go back to our basic idea of differentiation.

If we make a small change δx in the variable x then there is a small change $\delta f = f(x + \delta x) - f(x)$ in the function of x, and the ratio $\delta f / \delta x$ equals the derivative $f'(x)$ in the limit as $\delta x \to 0$. For the function $g(h(x))$ let the variable u be $h(x)$ and let the variable v be $g(h(x))$, i.e. the function $g(u)$. Suppose there is a small change δx in x. This gives rise to a small change δu where $\delta u = h(x + \delta x) - h(x)$, and this small change in u produces a small change δv in v where $\delta v = g(u + \delta u) - g(u)$.

The quantities δu, δv and δx are non-zero, so we can apply the normal rules of algebra to them. We require the ratio $\delta v / \delta x$, and its limit as $\delta x \to 0$. Now

$$\frac{\delta v}{\delta x} = \frac{\delta v}{\delta u} \cdot \frac{\delta u}{\delta x} = \left(\frac{g(u + \delta u) - g(u)}{\delta u} \right) \left(\frac{h(x + \delta x) - h(x)}{\delta x} \right)$$

The derivative $v'(x)$ is defined by

$$\lim_{\delta x \to 0} \left(\frac{\delta y}{\delta x} \right),$$

so that

$$\lim_{\delta x \to 0} \left(\frac{\delta y}{\delta x} \right) = \lim_{\delta x \to 0} \left(\frac{\delta y}{\delta u} \cdot \frac{\delta u}{\delta x} \right) = \lim_{\delta x \to 0} \left[\left(\frac{g(u + \delta u) - g(u)}{\delta u} \right) \left(\frac{h(x + \delta x) - h(x)}{\delta x} \right) \right]$$

and since $\delta u \to 0$ as $\delta x \to 0$ we have

$$\frac{dy}{dx} = \frac{dy}{du} \cdot \frac{du}{dx} = g'(u)h'(x) = f'(x) \tag{18}$$

This result is sometimes called the *chain rule* and is used frequently when finding derivatives.

Example: Find $f'(x)$ in the following cases:
    (i) $f(x) = \cos(x^2 + 1)$;
    (ii) $f(x) = (x^3 + 2x + 1)^5$.

Solution:

    (i) Using the chain rule we let $u = x^2 + 1$ and $v = \cos(u)$. It follows that $du/dx = 2x$ and $dy/du = -\sin(u)$. The derivative $f'(x)$ is therefore given by

$$f'(x) = \frac{dy}{dx} = \frac{dy}{du} \cdot \frac{du}{dx} = -\sin(u) \cdot 2x$$
$$= -2x \sin(x^2 + 1)$$

(In terms of the composite function idea, we could let $h(x) = x^2 + 1$ and $g(u) = \cos(u)$ so that $\cos(x^2 + 1) = g(h(x))$. The derivative of $g(h(x))$ is then $g'(u)h'(x)$, which is $-\sin(u) \cdot 2x$, i.e. $-2x \sin(x^2 + 1)$ as before.)

    (ii) Let $u = x^3 + 2x + 1$ and $v = u^5$, so that

$$f'(x) = \frac{dy}{dx} = \frac{dy}{du} \cdot \frac{du}{dx}$$
$$= 5u^4 \cdot (3x^2 + 2)$$
$$= 5(x^3 + 2x + 1)^4 (3x^2 + 2)$$

The derivative of inverse functions. Suppose we have the function $g(x)$. Its inverse, if it exists, is the function we write as $g^{-1}(x)$. As we showed in Chapter 4, the inverse function is such that $g(g^{-1}(x)) = x$. Starting from this identity and regarding x as the composite function $g(g^{-1}(x))$, we can see how to differentiate $g^{-1}(x)$. Using rule (18) on the composite function $x = g(g^{-1}(x))$ we see that

$$1 = g'(u) \cdot g^{-1\prime}(x)$$

In other words

$$g^{-1\prime}(x) = 1/g'(u) \text{ where } u = g^{-1}(x) \tag{19}$$

We will now use this formula on the important inverse functions we have considered so far.

The derivative of $\ln(x)$. Let $g^{-1}(x) = \ln(x)$, so that $g(x) = e^x$. We know that $g'(x) = e^x$, so the derivative of $f(x) = \ln(x)$ is given by

$$f'(x) = \frac{1}{e^u} \text{ where } u = \ln(x)$$

$$= \frac{1}{x}$$

Thus we have the important result that

$$\text{if } f(x) = \ln(x) \quad \text{then } f'(x) = 1/x \tag{20}$$

The derivative of arc sin (x). Let $g^{-1}(x) = \arcsin(x)$, so that $g(x) = \sin(x)$. We have that $g'(x) = \cos(x)$, so the derivative of the function $f(x) = \arcsin(x)$ is given by

$$f'(x) = \frac{1}{\cos(u)} \text{ where } u = \arcsin(x)$$

$$= \frac{1}{\sqrt{[1 - \sin^2(u)]}}$$

If $u = \arcsin(x)$ then $\sin(u) = x$, thus

$$\text{if } f(x) = \arcsin(x), \quad \text{then } f'(x) = \frac{1}{\sqrt{(1 - x^2)}} \tag{21}$$

The derivatives of arc cos (x) *and* arc tan (x). In a similar manner it is easily shown that

$$\text{if } f(x) = \arccos(x), \quad \text{then } f'(x) = \frac{-1}{\sqrt{(1 - x^2)}} \tag{22}$$

and

$$\text{if } f(x) = \arctan(x), \quad \text{then } f'(x) = \frac{1}{1 + x^2} \tag{23}$$

Parametric differentiation. There are many occasions in science when we have a relation between two variables x and y, which might be $y = f(x)$ say, where x and y are both functions of a third variable t say $x = x(t)$ and $y = y(t)$. For example, the horizontal and vertical positions of a projectile are given by the formulae

$$x(t) = x_0 + u_0 t$$
$$y(t) = y_0 + v_0 t - gt^2/2$$

where t represents the time and the other quantities are constants. Now, if we eliminate t between these two equations we obtain the equation of the path of the projectile, namely $y = y(x)$. At any instant of time the direction of motion of the projectile is the derivative of y with respect to x, i.e. dy/dx.

The problem of evaluating this derivative is often easier if we carry out the derivatives of $x(t)$ and $y(t)$ with respect to t and then express dy/dx in terms of dx/dt and dy/dt. This method of obtaining dy/dx is called 'parametric differentiation', where t is the *parameter* in question. The method follows from the chain rule. Suppose that $y = f(x)$, $x = g(t)$ and $y = h(t)$, then the chain rule gives

$$\frac{dy}{dt} = \frac{dy}{dx} \cdot \frac{dx}{dt}$$

and providing dx/dt is non-zero we can write

$$\frac{dy}{dx} = \frac{dy}{dt} \bigg/ \frac{dx}{dt} \quad \text{or} \quad f'(x) = h'(t)/g'(t) \tag{24}$$

Example: What is $f'(x)$ in terms of t when $x = g(t) = c(t - \sin(t))$ and $y = h(t) = c(1 - \cos(t))$ and c is a constant?

Solution: $h'(t) = c(\sin(t))$ and $g'(t) = c(1 - \cos(t))$.
Thus using (24) we have

$$f'(x) = h'(t)/g'(t) = c\sin(t)/c(1 - \cos(t))$$
$$= \sin(t)/(1 - \cos(t))$$

6.4.2 Higher derivatives. The need for the derived function of $f'(x)$ is apparent from the relation between speed, acceleration and distance. The speed is the rate of change of distance with time, i.e. the derivative of the distance function with respect to time. Similarly, the acceleration is the rate of change of speed with time, i.e. it is the derivative of the speed with respect to time. This means that in obtaining the acceleration we have to differentiate the distance function *twice*.

If $s = f(t)$ is the distance travelled by a body at time t then the speed v is $ds/dt = f'(t)$ and the acceleration a is dv/dt. Hence

$$a = \frac{dv}{dt} = \frac{d}{dt}\left(\frac{ds}{dt}\right)$$

We write the right-hand side in the form d^2s/dt^2, so that

$$\frac{d}{dt}\left(\frac{ds}{dt}\right) = \frac{d^2s}{dt^2}$$

This is called the *second derivative* of s with respect to t.

Example: If $s = t^2 + 2t + 1$ find d^2s/dt^2.
Solution: We have that $ds/dt = 2t + 2$ so that $d^2s/dt^2 = 2$.

The second derivative of a function $y = f(x)$ is thus written in exactly the same way, namely

$$\frac{d}{dx}\left(\frac{dy}{dx}\right) = \frac{d^2 y}{dx^2} \tag{25}$$

Example: If $y = 2e^x + 4\sin(x)$ find $d^2 y/dx^2$.
Solution: $dy/dx = 2e^x + 4\cos(x)$, and hence

$$\frac{d^2 y}{dx^2} = 2e^x - 4\sin(x).$$

We can extend this idea to define third, fourth, and higher derivatives. For instance, the third derivative of y with respect to x is written $d^3 y/dx^3$ and so on.

In our functional notation the second derivative of a function $f(x)$ is written as $f''(x)$ and the third as $f'''(x)$ and so on. Each 'dash' implies one differentiation.

6.5 Further worked examples

Example 1. The thermal decomposition of a chemical is such that its concentration after t min is

$$c(t) = (2.33)2^{-kt}$$

where the initial concentration is $2.33 \, \text{mol} \, l^{-1}$ and k is a constant. Show that the rate of change of concentration is proportional to $c(t)$.

Solution. We can differentiate 2^t to give $2^t \ln(2)$ but in this example we have 2^{-kt}. This is a composite function. If we let $u = -kt$, then $c(u) = (2.33)2^u$. Noting that $du/dt = -k$ and using the chain rule we see that

$$\frac{dc}{dt} = \frac{dc}{du} \cdot \frac{du}{dt}$$
$$= (2.33)2^u \ln(2) \cdot (-k)$$
$$= -k \ln(2) \cdot (2.33)2^{-kt}$$

This result is just $-k \ln(2)$ times $c(t)$. Thus the rate of change of concentration is proportional to $c(t)$.

Example 2. The Eckart potential V is given by the formula

$$V = \frac{Au}{(1+u)^2} + \frac{Bu}{(1+u)} \quad \text{where } u = e^{cx}$$

Find the rate of change of V with x when $x = 0$.

Solution. As written V is a function of u. So anticipating using the chain rule, we first obtain dV/du. This can be obtained by using the quotient rule as follows:

$$\frac{dV}{du} = A\left(\frac{(1+u)^2 \cdot 1 - u \cdot 2(1+u)}{(1+u)^4}\right) + B\left(\frac{(1+u) \cdot 1 - u \cdot 1}{(1+u)^2}\right)$$

$$= \frac{A(1-u^2)}{(1+u)^4} + \frac{B}{(1+u)^2}$$

The expression for dV/dx is $(dV/du) \cdot (du/dx)$. Since $u = e^{cx}$ it follows that $du/dx = c\, e^{cx}$. Thus

$$\frac{dV}{dx} = \frac{A(1-u^2)}{(1+u)^4} c\, e^{cx} + \frac{Bce^{cx}}{(1+u)^2} \quad \text{where } u = e^{cx}$$

When $x = 0$, the value of u is $e^0 = 1$. Thus $V'(0)$ is given by

$$V'(0) = A(0) + Bc/(1+1)^2 = Bc/4$$

The value of $V'(x)$ when $x = 0$ is therefore $0.25\, Bc$.

Example 3. The x and y coordinates of a point moving on the circumference of a circle of radius a, with constant angular velocity ω are

$$x = a \cos(\omega t) \text{ and } y = a \sin(\omega t)$$

where t is the time variable. Obtain expressions for dy/dx and d^2y/dx^2.

Solution. The quantities x and y can be regarded as being defined in terms of a parameter, which is t. Using parametric differentiation we see that

$$\frac{dy}{dx} = \frac{dy}{dt} \Big/ \frac{dx}{dt}$$

$$= (a \cos(\omega t) \cdot \omega)/(-a \sin(\omega t) \cdot \omega)$$

$$= -\cot(\omega t)$$

To obtain d^2y/dx^2 we note that this is $(d/dx)(dy/dx)$. Using the chain rule we see that

$$\frac{d}{dx}\left(\frac{dy}{dx}\right) = \frac{d}{dt}\left(\frac{dy}{dx}\right) \Big/ \frac{dx}{dt}$$

Thus we obtain d^2y/dx^2 by differentiating $-\cot(\omega t)$ with respect to t and then we divide the result by dx/dt. The derivative of $-\cot(\omega t)$ is $\operatorname{cosec}^2(\omega t) \cdot \omega$, so

$$\frac{d^2y}{dx^2} = \omega \operatorname{cosec}^2(\omega t)/(-a\omega \sin(\omega t))$$

$$= -\frac{1}{a}\operatorname{cosec}^3(\omega t)$$

This example serves to indicate that care needs to be exercised in obtaining the second derivative of a quantity when the variables involved are defined parametrically – as is often the case in science.

6.6 Exercises

1. If $y = 0.3x + 7$ find the average rate of change in y when x changes from 0.4 to 2.4.

2. If $y = 0.3x + 7$ find the limit of the ratio $(y_2 - y_1)/(x_2 - x_1)$ as x_1 and x_2 tend to the value 3.5.

3. If $s = 2 - 3t^2$ find the limit of the ratio $(s_2 - s_1)/(t_2 - t_1)$ as t_1 and t_2 tend to the value -1.

4. If $y = t^2 - 25t + 1$ find the value of dy/dx using its definition as a limit:
   (i) at the point $t = 1$
   (ii) at a general point $t = t_0$

5. Find an expression for dy/dx when:
   (i) $y = 7x$ (ii) $y = 8 + 4x$
   (iii) $y = 16x^2$ (iv) $y = 3x^3 + 2x^2 - x + 1$
   (v) $y = -x$ (vi) $y = 4x^7$
   (vii) $y = 2x^{19}$ (viii) $y = x^5$
   (ix) $y = 5$ (x) $y = 7x^9 - 3x^4$

6. Differentiate the following functions with respect to x:

   (i) (a) $5x^4$,   (b) $\dfrac{1}{x^2}$ (c) 15,

   (d) $(x+2)^{2.4}$,   (e) $\sqrt{(x+1)}$ (f) $3 \sin x + x^3$,

   (g) $\cos x + \ln(x)$,   (h) $\dfrac{1}{x^{1/3}}$ (i) $3e^x$,

   (j) e^{x+17},   (k) $\dfrac{x^3 + x^2}{x}$,   (l) $\dfrac{\cos x - \sin x}{2}$,

   (m) $x^{\ln(2)}$ (n) $2\ln(x+1)$,   (o) $\dfrac{1}{\sqrt{x}}$

   (ii) (a) $x^3 \sin x$,   (b) $e^x \cos x$,   (c) $x \ln(x)$
   (d) $\sin x \cos x$,   (e) $x^{\frac{1}{4}}(1+x)^{3/2}$,   (f) $x \cos x \tan x$,
   (g) $x^2 \cdot e^x \ln(x)$ (h) $x^{-1} \sin x$

   (iii) (a) $\dfrac{\sin x}{x}$,   (b) $\dfrac{x^2}{1+x}$,   (c) $\dfrac{\sin x}{e^x}$,

   (d) $\dfrac{1+x^2}{1-x^2}$,   (e) $\dfrac{\ln(x)}{x^3}$,   (f) $\dfrac{x + \sin x}{1 - \cos x}$,

   (g) $\dfrac{x^2 \sin x}{\cos x}$,   (h) $\dfrac{1 + x + x^2}{e^x}$,

   (iv) (a) e^{3x},   (b) $2\exp(x^2)$,   (c) $\sin(3x + \pi/2)$,
   (d) $\ln(1 + x^2)$,   (e) $\sin(e^x)$,   (f) $e^{\sin x}$,

   (g) $\ln\left(\dfrac{1+x}{1-x}\right)$ (h) $\sqrt{(1+x^2)}$

(v) (a) $\arctan(x)$,   (b) $\arcsin(x^2)$,   (c) $x\ln(1 + \frac{1}{2}\sin x)$,

(d) $\dfrac{e^x \sin x}{1+x}$,   (e) $\exp(x^2)\ln(x)$,   (f) $\sqrt{\left(\dfrac{x + \sin x}{1 - \cos x}\right)}$.

7. The height of a ball above the ground when thrown vertically upwards is given by the equation

$$s = 30t - 5t^2$$

where t is measured in s. Find the speed of the ball when (i) $t = 0\,$s and (ii) $t = 2\,$s.

8. The solubility S of a certain chemical in water at $T\,°$C is given by

$$S = 0.0003T^3 + 0.72T + 16$$

Find the rate of change of S with temperature at $100\,°$C.

9. Coughing causes the diameter of the windpipe to decrease due to the pressure required to empty the lungs. If Q is the volume of air passing through the windpipe and r is the radius of the windpipe then

$$Q = b\,(ar^4 - r^5)/k$$

where a, b and k are constants. Find the rate of change of Q with r when $r = a/2$.

10. A simple pendulum of length l has a period T given by $T = 2\pi\sqrt{(l/g)}$ where g is the acceleration due to gravity. Find the rate of change of period with length when l is 1 m.

11. The luminous intensity I at a depth of x m in clean sea water is given by

$$I = I_0 e^{-2x}$$

where I_0 is the intensity at the surface of the sea. Show that the rate of change of intensity with distance when x is 1m is $1/e^2$ that when at the surface.

12. The total energy of an electron is given by

$$E(r) = \frac{n^2 h^2}{8\pi^2 mr^2} - \frac{Z_e^2}{r}$$

Find the value of r for which $E'(r)$ is zero.

13. The velocity of blood in an artery of circular cross-section is given by

$$v(r) = 1000\,(R^2 - r^2)$$

in cm s^{-1} where R is the radius, in cm, of the artery and r is the distance of a point in the artery from the central axis of the artery. Find the rate of change of velocity with respect to r at a point when $r = 0.1$ cm. What does the negative sign in your answer mean?

14. Find df/dt for:

(i) $f = (3t^2 + 2t + 7)\,e^t$ (ii) $f = \sin(t)\cos(t)$
(iii) $f = (2t + 7)(t^2 + 4t + 1)\sin(t)$

(iv) $f = \dfrac{1}{t^6 + 3t^2 + 2t}$ (v) $\dfrac{1}{\sin(t)}$

(vi) $f = \dfrac{t^2}{t-1}$

15. Find df/dx for:

   (i) $f = \exp(\sin(x))$ (ii) $f = (3x+4)^7$

   (iii) $f = \sin(2x^2 + 4x - 1)$ (iv) $f = \sinh(x/3) + 3\cosh(5x)$

   (v) $f = \ln(\cosh(3x))$ (vi) $f = \sin(\arccos(x))$

16. Find df/dx and d^2f/dx^2 for:

   (i) $f = \sqrt{x}$ (ii) $f = (x^3 + 2x^2 + 7x)^{-7/2}$

   (iii) $f = \ln(2x^2 + x - 5)$

17. Find d^2s/dt^2 and d^3s/dt^3 for:

   (i) $s = 3t^4 + 5\cos(t)$

   (ii) $s = 4t^3 + 2t^2 - t$

   (iii) $s = 4\sin(t) - \exp(t)$

18. The von Bertalonffy growth curve is given by the expression

$$N = (a - be^{-kt})^3$$

where a, b and k are constants with $a > b$. Find the growth rate dN/dt and hence show that

$$\frac{dN}{dt} = 3k(a - N^{1/3})N^{-2/3}$$

19. For a thermistor used as a detector in microwave power measurements

$$R = R_0 \exp(a/(T + cp))$$

where R is the resistance, p is the applied power, T is the temperature and R_0, a and c are constants.

If the detector sensitivity s is defined as the rate of change of resistance with respect to power, show that

$$s = -\frac{cR}{a}(\ln(R/T))^2$$

20. Find an expression for the nth derivative with respect to x of each of the following functions:

   (i) $f = \cos(x)$

  (ii) $f = \ln[(1-x)/(1+x)]$

21. If $y = \arctan(x)$ prove that

$$(1 + x^2)\frac{d^2y}{dx^2} + 2x\frac{dy}{dx} = 0$$

and deduce that

$$(1 + x^2)\frac{d^{n+2}y}{dx^{n+2}} + 2(n+1)x\frac{d^{n+1}y}{dx^{n+1}} + n(n+1)\frac{d^ny}{dx^n} = 0$$

7

Differentiation 2: stationary points

Contents: 7.1 scientific context – 7.1.1 zero speed; 7.1.2 stability of a bead on a bent wire; 7.2 mathematical developments – 7.2.1 geometrical interpretation of the derivative; 7.2.2 stationary points; 7.3 worked examples; 7.4 further developments – 7.4.1 curve sketching 7.5 further worked examples; 7.6 exercises.

In Chapter 6 we introduced the idea of a derivative and saw how quantities of interest in science could often be related through differentiation. For example speed is the rate of change of distance, current is the rate of change of charge, and so on. We continue our illustration of 'differentiation in action' by using the technique of differentiation to investigate the properties of the functions themselves.

A knowledge of the maximum or minimum values of a function, known as the stationary values, helps us to sketch out the shape of the function. This sketch, and a knowledge of where these stationary values occur, is useful when the function concerned represents some variable of interest in a scientific system. Then we find we can often predict the behaviour of the system and hence discover its properties from an investigation of the function. Differentiation is of crucial importance in locating and identifying these stationary values and in curve sketching.

7.1 Scientific context

7.1.1 Example 1: zero speed. Consider once again experiment 2 from Chapter 1. A ball is thrown vertically upwards and its height h above the ground is measured at different times t. Fig. 7.1 again shows the graph of h against t. This graph should not be confused with the path of the ball which is actually a vertical line. The slope of the graph at any point P is the speed with which the ball is travelling.

Between 0 and A the ball is going up and between A and B it is coming down. At the point A, the highest point reached by the ball, the ball is instantaneously at rest. It is neither going up nor coming down. So its speed is zero, and therefore

$$\frac{dh}{dt} = 0$$

at A as shown in Fig. 7.2. Geometrically we know that dh/dt is the slope of the tangent to the curve. At A the tangent to the curve is parallel to the x axis so the slope is zero because θ, the angle between the tangent and the x direction, is 0.

An important feature of the graph in Fig. 7.2 is the point A which is an example of a *stationary point*. At this stationary point the derivative of h is zero, so finding

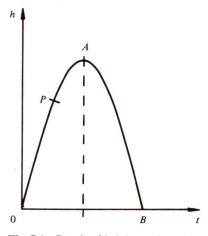

Fig. 7.1. Graph of height against time.

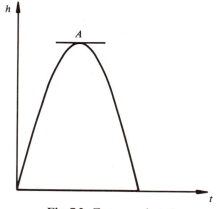

Fig. 7.2. Zero speed at A.

the position of the stationary point will help us to find the point where the ball is instantaneously at rest. It also allows us to obtain the maximum height of the ball as well.

We know from Chapter 1 that the formula for $h(t)$ is $h(t) = -\frac{1}{2}gt^2 + u_0t + 0.087$, where $u_0 = 4.48 \, \mathrm{m \, s^{-1}}$ is the initial speed of the ball. Hence

$$\frac{dh}{dt} = u_0 - gt$$

and this equals zero when $t = u_0/g$. Putting in the values for u_0 and g we have that $t = 0.46 \, \mathrm{s}$, which looks about right on the graph of h against t. The maximum height is thus $h(0.46) = 1.11 \, \mathrm{m}$. Differentiation, therefore, has enabled us to pinpoint when the ball is stationary and at what height this occurs.

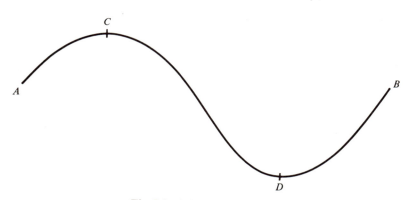

Fig. 7.3. A bent vertical wire.

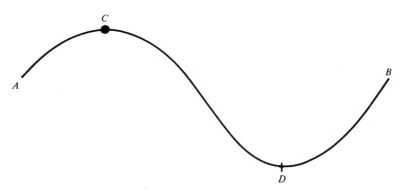

Fig. 7.4. A bead placed at *C*.

7.1.2 Example 2: the stability of a bead on a bent wire. A piece of wire *AB* is bent to form the shape shown in Fig. 7.3 and is held fixed in a vertical plane. Suppose we now thread a bead onto the wire and investigate the motion of the bead at various points on the wire.

First, suppose that we put the bead at the point *C* which is the highest point of the wire. If we are careful enough the bead will stay at rest at *C* as shown in Fig. 7.4.

Question: What will happen to the bead if it is given a small push?
Answer: Common sense tells us that the bead will slide down the wire, towards *A* if pushed to the left and towards *D* if pushed to the right. The bead will not return to the point *C* by itself. We would have to put it there.

Now let us place the bead at the lowest point of the wire, which is labelled *D*, as shown in Fig. 7.5.

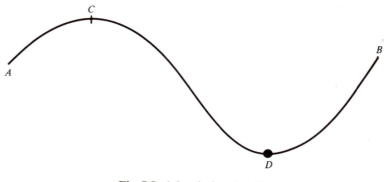

Fig. 7.5. A bead placed at *D*.

Question: What will happen to the bead if it is now given a small push?
  Answer: This time the bead will return to *D*. Of course it will not stop at *D* but will oscillate
      back and forth about the point *D*.

The points *C* and *D* are special points. When the bead is placed at these points
it will remain at rest. These points are therefore *positions of equilibrium* of the
bead. However when the bead is slightly displaced its behaviour at *C* is different
from that at *D*. The point *C* is called a position of *unstable* equilibrium and *D* a
position of *stable* equilibrium. If the equation of the curve representing the wire
is $y = f(x)$ then the potential energy of the bead is $mgf(x)$ and the function has
the same shape as the wire. At the points corresponding to *C* and *D* on this
potential energy curve the slope is also zero, so these points, too, are examples of
stationary points just like *C* and *D*. Note that the bead can remain at rest (or be
stationary) at *C* and *D*.
    At a stationary point $f'(x) = 0$, so we can express the condition for the bead to
be in equilibrium as

$$\frac{d}{dx}(\text{potential energy}) = 0$$

This condition is quite general in mechanics, where the potential energy is a
function of just one variable, and does not apply only to beads on wires. We will
see that conditions on the second derivative will determine whether or not the
equilibrium is stable.

7.2 Mathematical developments

7.2.1 Geometrical interpretation of the derivative. In Chapter 6 we defined the
derivative of a function as

$$\lim_{\delta x \to 0} \left[\frac{f(x + \delta x) - f(x)}{\delta x} \right]$$

and developed rules for differentiating the basic functions and various combinations of them.

Geometrically the ratio $(f(x + \delta x) - f(x))/\delta x$ can be viewed as the slope of the chord joining the two points A and B on the graph of $y = f(x)$ as shown in Fig. 7.6. This slope is $\tan(\phi)$ where ϕ is the angle shown. In the limit as $\delta x \to 0$ the chord becomes parallel to the tangent to the curve at A so that (Fig. 7.7)

$$f'(x) \text{ is the slope of the tangent at } A \qquad (1)$$

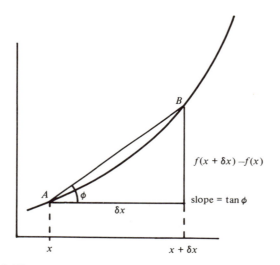

Fig. 7.6. The average rate of change of f is the slope of the chord.

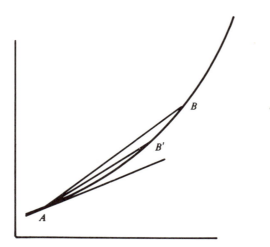

Fig. 7.7. In the limit, chords become the tangent at A.

This idea of the derivative being equal to the gradient of the curve is of course not new and we have already made use of this concept many times before. However, keeping in mind this geometrical interpretation of the derivative is all important when discussing and describing features of the graph of a function as we shall now see.

7.2.2 Stationary points. Consider the curve shown in Fig. 7.8 which is the graph of the function $f(x)$. There are some features of the graph that can be described in terms of its slope. To the left of the point A the value of y increases as the value of x increases; the tangent of the angle is positive and we say that the slope of the graph is positive. Between the points A and B the value of y decreases as the value of x increases, the tangent of the angle β is negative (because β lies between 90 and 180°) and we say that the slope of the graph is negative. Beyond B the slope is again positive. At both A and B the tangents to the graph are parallel to the x axis and the slope of the graph is therefore zero at these points.

Point A is called a *local maximum* of the function and point B is called a *local minimum*.

We attach the word 'local' to remind us that the points A and B are not necessarily overall or global maximum and minimum (i.e. the largest and smallest) values of the function. Any function may have more than one local maximum (or local minimum) but it will only have one global maximum value (or one global minimum value). Also, whereas the gradient is zero at a local maximum or minimum, the slope does not have to be zero at a global maximum or minimum. This can be appreciated in Fig. 7.9 for example. What we mean by a *local maximum* is a point where the y value is greater than any other in the immediate vicinity or neighbourhood. A *local minimum*, then, is a point that is *less* than any other in the immediate neighbourhood.

From this discussion it is clear that the important features of a graph can be

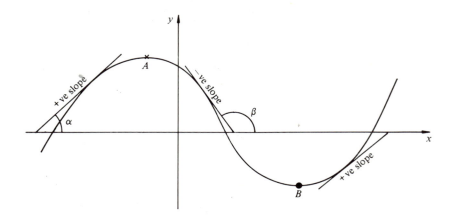

Fig. 7.8. The change in slope of a typical function f.

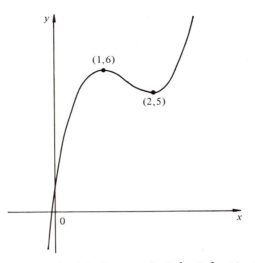

Fig. 7.9. Graph of the function $f = 2x^3 - 9x^2 + 12x + 1$.

described geometrically in terms of the slope of the graph, in other words in terms of the derivative. As we have said at a local maximum and a local minimum the slope of the graph is zero. We express this fact as:

$$f'(x) = 0 \tag{2}$$

and define those values of x for which $f'(x) = 0$ as the *stationary points* of $f(x)$.

Example: Where are the stationary points of the function $f(x) = 2x^3 - 9x^2 + 12x + 1$?
Solution: The stationary points occur where $f'(x) = 0$. If $f(x) = 2x^3 - 9x^2 + 12x + 1$ we
require values of x such that $f'(x) = 6x^2 - 18x + 12 = 0$. This equation has solutions
$x = 2$ and $x = 1$. The corresponding values of the function at these points are $y = 5$
and $y = 6$. Thus the points $(1, 6)$ and $(2, 5)$ are the stationary points.

A graph of the function $f(x) = 2x^3 - 9x^2 + 12x + 1$ is shown in Fig. 7.9 together with the stationary points. The graph shows us that the point $(1, 6)$ is a local maximum value, whilst the point $(2, 5)$ is a local minimum. However, to identify the nature of the stationary points that occur it is not always necessary to sketch out the graph of the function. There is an alternative way, which we now describe. Fig. 7.10 shows the plot of a function which possesses a local maximum at the point $x = c$. The slope of the curve before, at, and after this point is indicated on the graph. For this local maximum we can note that

(i) $f'(x) > 0$ for $x < c$ (positive slope);
(ii) $f'(x) = 0$ for $x = c$ (zero slope);
(iii) $f'(x) < 0$ for $x > c$ (negative slope).

The derivative $f'(x)$ as a function of x is always *decreasing* for x increasing near the point $x = c$ and so the slope of the function $f'(x)$ is negative at $x = c$. But the

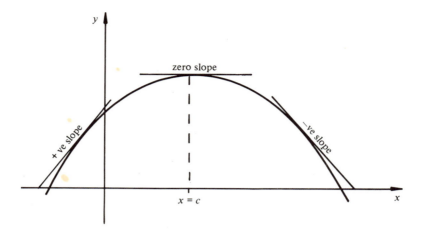

Fig. 7.10. A local maximum.

slope of $f'(x)$ is the *second derivative* of $f(x)$ namely $f''(x)$ and so

$$\text{if } f'(x) = 0 \text{ and } f''(x) < 0 \text{ when } x = c \text{ then } f(x)$$
$$\text{has a local maximum at } x = c \tag{3}$$

For the function that we considered in our last example, namely $f(x) = 2x^3 - 9x^2 + 12x + 1$, we can show that $x = 1$ is a local maximum by using the second derivative. We saw that $f'(x) = 6x^2 - 18x + 12$, so that $f''(x) = 12x - 18$. When $x = 1$ this value is negative and the point $(1, 6)$ is therefore a local maximum.

Let us now consider the plot of a function possessing a local minimum at $x = c$ as shown in Fig. 7.11. For this local minimum we can now note that

(i) $f'(x) < 0$ for $x < c$ (negative slope);
(ii) $f'(x) = 0$ for $x = c$ (zero slope);
(iii) $f'(x) > 0$ for $x > c$ (positive slope);

so that $f''(x) > 0$ at $x = c$. Thus

$$\text{if } f'(x) = 0 \text{ and } f''(x) > 0 \text{ when } x = c \text{ then } f(x)$$
$$\text{has a local minimum at } x = c \tag{4}$$

It is now easy to verify that $x = 2$, in the above example, corresponds to a local minimum. We see that $f''(2) = 6$, which is positive, showing that the point $(2, 5)$ is indeed a local minimum.

Example: What are the values of $f'(x)$ and $f''(x)$ for the function $f(x) = x^3 - 1$ when $x = 0$?
Solution: The values of $f'(0)$ and $f''(0)$ are both zero because $f'(x) = 3x^2$ and $f''(x) = 6x$.

This last example introduces another type of stationary point called a *point of inflexion*. A graph of $f(x) = x^3 - 1$ is shown in Fig. 7.12. At the point $(0, -1)$, $f'(x) = 0$ so the slope of the graph is zero and the tangent to the graph is parallel to the x axis. On either side of the point $(0, -1)$ the slope of the graph is positive.

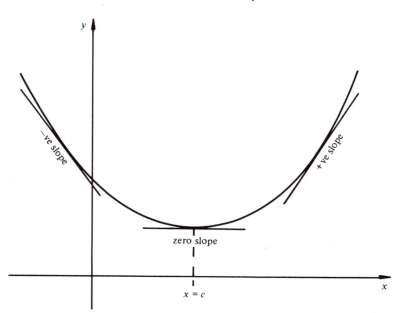

Fig. 7.11. A local minimum.

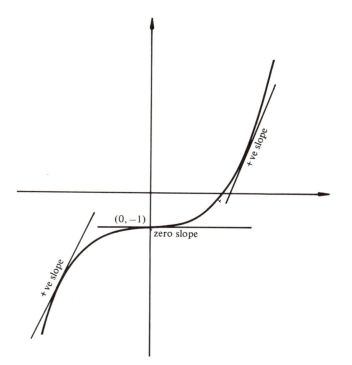

Fig. 7.12. A point of inflexion.

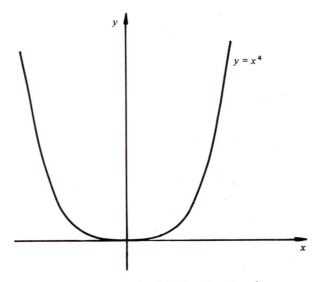

Fig. 7.13. Graph of the function $f = x^4$.

So this point is neither a local maximum nor a local minimum. We call this stationary point a *point of inflexion*.

One question we might ask in connection with identifying the nature of stationary points is if $f''(x) = 0$ *and* $f'(x) = 0$ when $x = c$, does the function always have a point of inflexion at this point? The answer to this question is NO. Consider the function $f(x) = x^4$ that is shown in Fig. 7.13. When $x = 0$, $f'(x) = 0$ and $f''(x) = 0$, but at $x = 0$ there is a local minimum value as the graph shows. So we have to be very careful. All we can say is that if $f''(x) = 0$ *and* $f'(x) = 0$ when $x = c$, then there is either a local maximum or a local minimum or a point of inflexion! To decide which type of stationary point it is, we investigate the value of the derivative either side of the stationary point. We note that for a point of inflexion the gradient both sides of the stationary point is either positive or negative. It does not change sign as in the case of a local maximum or local minimum.

Example: Where are the stationary points of the function $f(x) = x^5 - 5x^4 + 5x^3$? Classify them as local maxima, local minima or points of inflexion.

Solution: The stationary points are given by $f'(x) = 0$. The quantity $f'(x)$ is $5x^4 - 20x^3 + 15x^2$. This is zero when $5x^2(x^2 - 4x + 3) = 0$, i.e. when $5x^2(x - 1)(x - 3) = 0$. The solutions are $x = 0$, $x = 1$, and $x = 3$ and the stationary points are therefore $(0, 0)$, $(1, 1)$ and $(3, -27)$ respectively.

To classify the stationary points we need to find the values of $f''(x)$ at the points. We have that

$$f''(x) = 20x^3 - 60x^2 + 30x$$

When $x = 0$, $f''(x) = 0$, so we need to investigate $f'(x)$ either side of $x = 0$. When $x < 0$, $f'(x) > 0$ and when $x > 0$, $f'(x) > 0$. Thus the slope of the graph is positive each side of the stationary point and $(0, 0)$ is therefore a point of inflexion. When

$x = 1, f''(x) = -10$, so the point $(1, 1)$ is a local maximum value. Finally when $x = 3, f''(x) = 90$, so the point $(3, -27)$ is a local minimum value.

We can summarise the steps in finding and classifying the stationary points of a function with the following procedure:

step 1: find $f'(x)$;

step 2: solve the equation $f'(x) = 0$; suppose $x = c$ is a solution.

step 3: EITHER (a)

Investigate the sign of $f'(x)$ on either side of $x = c$

(i) if $f'(x) < 0$ for $x < c$ and $f'(x) > 0$ for $x > c$ then $x = c$ is a local minimum;

(ii) if $f'(x) > 0$ for $x < c$ and $f'(x) < 0$ for $x > c$ then $x = c$ is a local maximum;

(iii) if $f'(x) > 0$ for $x < c$ and $f'(x) > 0$ for $x > c$ or if $f'(x) < 0$ for $x < c$ and $f'(x) < 0$ for $x > c$ then $x = c$ is a point of inflexion.

OR (b)

Find $f''(x)$ and check the value of $f''(x)$ when $x = c$

(i) if $f''(c) > 0$ then $x = c$ is a local minimum;

(ii) if $f''(c) < 0$ then $x = c$ is a local maximum;

(iii) if $f''(c) = 0$ then investigate the sign of $f'(x)$ on either side of $x = c$ as above in step 3(a).

7.3 Worked examples

Example 1. The energy of an electron travelling round the nucleus of an atom, according to the Bohr model, is given by

$$E(r) = \frac{n^2 h^2}{8\pi^2 m r^2} - \frac{Z_e^2}{r}$$

The Bohr radius r for a *stable* orbit is given by a minimum of this function. Find the radius of stable orbits.

Solution. Possible minimum values of the function are given by

$$\frac{dE}{dr} = 0$$

and solutions of this equation for r are minimum values if

$$\frac{d^2 E}{dr^2} > 0$$

Differentiating $E(r)$ we obtain

$$E'(r) = \frac{n^2 h^2}{8\pi^2 m}\left(\frac{-2}{r^3}\right) + \frac{Z_e^2}{r^2}$$

The expression $E'(r)$ is zero when

$$r = \frac{n^2 h^2}{4\pi^2 m Z_4^2} = r_0, \quad \text{say}$$

Checking on the value of $E''(r)$ we see that

$$E''(r) = \frac{n^2 h^2}{8\pi^2 m}\left(\frac{6}{r^4}\right) - \frac{2Z_e^2}{r^3}$$

$$= \frac{n^2 h^2}{8\pi^2 m}\cdot\frac{2}{r^3}\left(\frac{3}{r} - \frac{Z_e^2 8\pi^2 m}{n^2 h^2}\right)$$

When $r = r_0$ we have

$$E''(r_0) = \frac{n^2 h^2}{8\pi^2 m}\cdot\frac{2}{r_0^3}\left(\frac{3}{r_0} - \frac{2}{r_0}\right) > 0$$

Thus the orbits given by

$$r = r_0 = \frac{n^2 h^2}{4\pi^2 m Z_e^2}$$

(for different values of the principal quantum number n) are the stable orbits of the electron.

Example 2. The power P delivered into the load X of a simple class A amplifier of output resistance R is given by

$$P = \frac{V^2 X}{(X + R)^2}$$

where V is the output voltage. Deduce an expression for X such that the maximum power will be delivered into the load given that V and R are constants for the amplifier.

Solution. For a maximum value for $P(X)$ we require $P'(X) = 0$ and $P''(X) < 0$.

$$P(X) = \frac{V^2 X}{(X + R)^2}$$

so

$$P'(X) = V^2\left(\frac{(X + R)^2\cdot 1 - X\cdot 2(X + R)}{(X + R)^4}\right)$$

$$= V^2\left(\frac{(X + R) - 2X}{(X + R)^3}\right)$$

$$= \frac{V^2(R - X)}{(X + R)^3}$$

This is zero when $X = R$. Looking now at $P''(X)$ we have

$$P''(X) = V^2\left(\frac{(X + R)^3\cdot(-1) - (R - X)\cdot 3(X + R)^2}{(X + R)^6}\right)$$

When $X = R$, this second derivative is $-V^2/8R^3$ which is negative. Thus when $X = R$ the power delivered is a maximum value.

Example 3. In mechanics the positions of equilibrium can be obtained by investigating the stationary points of the potential energy function, V. A position of equilibrium is stable if V has a local minimum value and is unstable if V has a local maximum value.

A bead of mass m can slide freely on a vertical circular wire of diameter a. A spring of natural length $3a/4$ and modulus of elasticity λ attaches the bead to the highest point of the wire. Investigate the stability of the equilibrium positions in the two cases (i) $\lambda = 12mg$ and (ii) $\lambda = mg$, given that the potential energy function

$$V(\theta) = \frac{\lambda a}{24}(4\cos(\theta) - 3)^2 + mg\,a\sin^2(\theta)$$

where θ is the angle shown in Fig. 7.14.

Solution.

(i) With $\lambda = 12mg$ the potential energy function is

$$V(\theta) = \frac{mg\,a}{2}(4\cos(\theta) - 3)^2 + mg\,a\sin^2(\theta)$$

The equilibrium positions occur where $V'(\theta) = 0$.

Differentiating $V(\theta)$ we obtain

$$V'(\theta) = mg\,a(4\cos(\theta) - 3)(-4\sin(\theta)) + 2\,mg\,a\sin(\theta)\cos(\theta)$$

$$= 2mg\,a\sin(\theta)(6 - 7\cos(\theta))$$

$V'(\theta) = 0$ when $\sin(\theta) = 0$ or when $\cos(\theta) = \frac{6}{7}$. Equilibrium occurs therefore when $\theta = 0°$ or $\pm 31°$.

To investigate the stability of these equilibrium positions we need to look at $V''(\theta)$.

$$V''(\theta) = 2\,mg\,a\sin(\theta)(7\sin(\theta)) + 2\,mg\,a\cos(\theta)(6 - 7\cos(\theta))$$

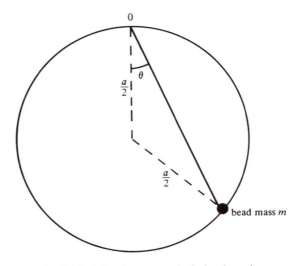

Fig. 7.14. A bead on a vertical circular wire.

When $\theta = 0$, $V''(0) = 2mg\,a(6-7)$, which is negative. The position is one of maximum potential energy and therefore unstable.

When $\cos(\theta) = \frac{6}{7}$ i.e. when $\sin^2(\theta) = \frac{1}{7}$ we see that

$$V''(0) = 2mg\,a \cdot 7 \cdot \tfrac{1}{7} + 2mg\,a \cdot \tfrac{6}{7}(0) = 2mg\,a,$$

which is positive. The two positions corresponding to $\theta = \pm 31°$ are therefore positions of minimum potential energy and the equilibrium is stable.

(ii) If $\lambda = mg$, the potential energy function is now

$$V(\theta) = \frac{mg\,a}{24}(4\cos(\theta) - 3)^2 + mg\,a\sin^2(\theta)$$

Differentiation gives

$$V'(\theta) = \frac{mg\,a}{12}(4\cos(\theta) - 3)(-4\sin(\theta)) + 2\,mg\,a\sin(\theta)\cos(\theta)$$

$$= mg\,a\sin(\theta)(1 + \tfrac{2}{3}\cos(\theta))$$

$V'(\theta) = 0$ when $\sin(\theta) = 0$ or when $\cos(\theta) = -\frac{3}{2}$. Since $\cos(\theta)$ cannot be $-\frac{3}{2}$, the only equilibrium position is when $\sin(\theta) = 0$, i.e. when $\theta = 0$.

Investigating the stability of this equilibrium position we see that

$$V''(\theta) = mg\,a\sin(\theta)(-\tfrac{2}{3}\sin(\theta)) + (1 + \tfrac{2}{3}\cos(\theta))\,mg\,a\cos(\theta))$$

When $\theta = 0$ we have

$$V''(0) = 0 + \tfrac{5}{3}mg\,a$$

Since $V''(0)$ is positive, the equilibrium position is one of minimum potential energy and is therefore stable.

7.4 Further developments

7.4.1 Curve sketching. The behaviour of a physical system can on many occasions be deduced from a graph representing the functional relationship between the variables used to model the system. The graph that we need is very often only a *sketch* of the function showing the important features and not necessarily an accurate plot.

Consider the graph shown in Fig. 7.15. The sort of information we would need to sketch the curve can be listed as follows. We would need a knowledge of

(1) where the stationary points are, in other words B and E, and what their nature is;
(2) where the graph cuts the axes. We would need to establish where the roots A, D and F are and the value of the intercept at C;
(3) the behaviour of the function for large values of x, both positive and negative, and what happens to the graph to the left of A and to the right of F;
(4) where any singularities occur on the graph and how the graph behaves near

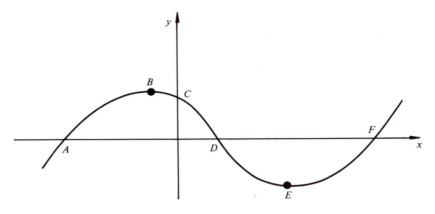

Fig. 7.15. The important points in curve sketching.

them. As it happens the plot in Fig. 7.15 does not have any singular points but this might not always be the case.

To illustrate the steps that we need to go through to sketch the graph of a function, consider the following two functions

    (i) $f(x) = x^2(x - 2)$;

and

    (ii) $g(x) = (2 - x)/(x - 1)$.

Let us begin by seeing how we might sketch the function $f(x) = x^2(x - 2)$. According to the list we could start by finding and identifying the stationary points.

The stationary points are given by $f'(x) = 0$. So we require

$$f'(x) = 3x^2 - 4x = 0$$

The values of x satisfying $f'(x) = 0$ are thus $x = 0$ and $x = \frac{4}{3}$, so the points $(0, 0)$ and $(\frac{4}{3}, \frac{-22}{27})$ are stationary points. To classify these stationary points we consider the second derivative which is

$$f''(x) = 6x - 4$$

When $x = 0$, $f''(x) = -4$, so the point $(0, 0)$ is a local maximum value. When $x = \frac{4}{3}$, $f''(x) = 4$, so that the point $(\frac{4}{3}, \frac{-22}{27})$ is a local minimum value. A graph of the function is now beginning to take shape as shown in Fig. 7.16. We can now see where the graph cuts the axes. The graph cuts (or touches) the y axis when $x = 0$ and the x axis when $y = f(x) = 0$. For our function, when $x = 0$, $x^2(x - 2) = 0$. The graph therefore cuts the y axis at the point $(0, 0)$. Also $f(x) = 0$ when $x = 2$ so that the graph cuts the x axis at the point $(2, 0)$. When $x = 0$ we know already that there is a local maximum value for f. Further, when $x = 2$, $f'(x) = 4$, so at the root $x = 2$, the slope is positive. The behaviour of the graph near the point $(2, 0)$ can now be included as shown in Fig. 7.17.

To complete the picture we need some idea of what happens for large, positive

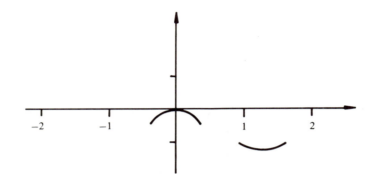

Fig. 7.16. Step 1: find the stationary points.

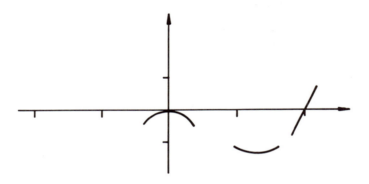

Fig. 7.17. Step 2: find where the graph cuts the *x*-axis.

and negative, values of *x*. If *x* is large and positive then $f(x) = x^2(x - 2)$ is large and positive. When *x* is large and negative $f(x)$ will be large and negative. We can now include this information on the sketch as shown in Fig. 7.18. Putting a smooth curve through the dotted lines will give us the desired sketch of our function.

There is, however, one further question that we should have asked, namely whether there are any singular points of the function for finite values of *x*. Clearly, for this function there are not. But let us now consider our second function $g(x) = (2 - x)/(x - 1)$.

Suppose we start by specifically looking for any singular points. When $x = 1$ the denominator is zero so clearly $x = 1$ is a *singular point* of the function. We need to investigate the behaviour of the graph near to $x = 1$. If *x* is just to the right of 1, say $x = 1.001$, then $g(x) = 999$. If $x = 1.0001$, then $g(x) = 9999$, so that as *x* tends to 1 'from above', $g(x)$ tends to $+\infty$. If now *x* is just to the left of 1, say, $x = 0.999$, then $g(x) = -1001$ and if $x = 0.9999$ then $g(x) = -10001$. So as *x* tends to 1 'from below', $g(x)$ tends to $-\infty$. The line $x = 1$ is an *asymptote* and the graph

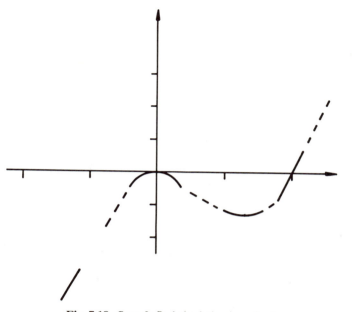

Fig. 7.18. Step 3: find the behaviour for large x.

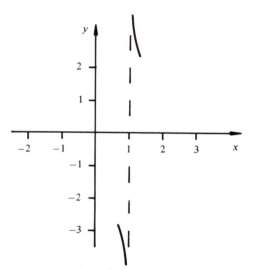

Fig. 7.19. Step 4: find the singular points.

of $g(x)$ near $x = 1$ is shown in Fig. 7.19. To complete the graph we need to find the stationary points, where the graph cuts the axes and consider what happens when x is large and positive, and when x is large and negative.

For $g(x) = (2 - x)/(x - 1)$ we have that $g'(x) = -1/(x - 1)^2$ and since $g'(x) \neq 0$ for any finite value of x there are *no stationary points*.

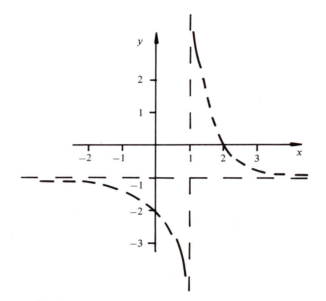

Fig. 7.20. Graph of the function $f = (2 - x)/(x - 1)$.

Also $g(x) = 0$ when $x = 2$, and $g(x) = -2$ when $x = 0$. The graph cuts the axes at the points $(2, 0)$ and $(0, -2)$. When $x = 0$, $g'(x) = -1$ so the slope of the graph is negative near the point $(0, -2)$. When $x = 2$, $g'(x) = -1$, so that near the root $x = 2$ the slope is negative.

Finally, $\lim_{x \to \infty}(g(x)) = -1$ 'from above' and $\lim_{x \to -\infty}(g(x)) = -1$ 'from below'.

Putting all these results together we can sketch the graph of the function. The sketch is shown in Fig. 7.20. Note that the line $y = -1$ is also an asymptote of our function along with $x = 1$.

The two examples we have considered are typical of many of the graphs obtained when a function is plotted out in that we have met roots, intercepts, stationary points, singular points, aymptotes and asymptotic behaviour. We can summarise the procedure for sketching a graph representing a function $f(x)$ in the following four steps:

step 1: find the position of any stationary points and classify them;
step 2: find where the graph cuts the axes and the slope of the function there;
step 3: find $\lim_{x \to \pm \infty}(f(x))$;
step 4: investigate the behaviour of $f(x)$ for values of x for which $f(x)$ is singular.

7.5 Further worked examples

Example 1. In worked example 2 we showed that a local maximum value of the function $P = V^2 X/(X + R)^2$ occurred when $X = R$, i.e. when the load connected to a class A amplifier equals the output resistance. By sketching a graph of the function $P(X)$ show that when $X = R$, P attains its maximum value possible.

Solution. From worked example 2 we know that the only stationary point is a local maximum occurring when $X = R$. At $X = R$ the value of P is $V^2/4R$. The graph cuts the P axis with an intercept equal to $P(0)$, i.e. 0. The point $(0, 0)$ is also the only root of P. The gradient of the curve is given by $P'(X) = V^2(R - X)/(X + R)^3$ and at $X = 0$ this is positive. For large values of positive X, P is always positive, but tends to zero as X tends to ∞. For large and negative values of X, P is always negative, but again tends to zero (from below) as X tends to $-\infty$.

Finally, we note that at $X = -R$ the function has a singular point as the denominator has the value zero. The line $X = -R$ is thus an asymptote. Just to the left of $X = R$ the value of P is large and negative. Just to the right the value is also large and negative. Thus we can sketch out the overall shape of P as shown in Fig. 7.21. It is obvious from the diagram that the function achieves its global maximum when $X = R$. We can usefully note, too, that in practice only positive values of X can occur. The analysis of the function for negative X values was strictly speaking not needed in this particular real example!

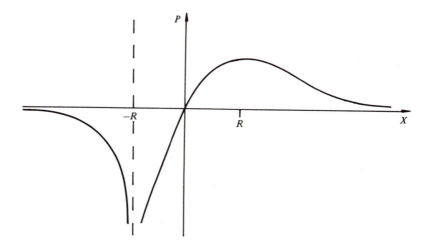

Fig. 7.21. Graph of the function $P = V^2 X/(X + R)^2$

Example 2. The electric current at time t flowing in a circuit consisting of a resistance R, an inductance L and a capacitance C is given by

$$i = Ae^{-Rt/2L}\cos(\sqrt{(4L/C) - R^2}t/2L),$$

where A is a constant.

In a particular experiment the values of R, L and C are given such that the current is

$$i = Ae^{-t/2}\cos(3t)$$

Draw a sketch of the graph of i against t for positive values of t.

Solution. The stationary points of i are given by $i'(t) = 0$. We have that

$$i'(t) = A(-\tfrac{1}{2})e^{-t/2}\cos(3t) - A(-3)\sin(3t) \, e^{-\frac{t}{2}}$$

This expression is zero when $\tan(3t) = -\tfrac{1}{6}$.

From a graph of the tangent function we see that there are many solutions to this equation, each solution differing by π. So the stationary points are

$$t = \arctan(-\tfrac{1}{6})/3 + n\pi, \quad n = 0, 1, 2, \ldots$$

To investigate the nature of these stationary points we evaluate $i''(t)$ in each case. Now

$$i''(t) = A(\tfrac{1}{4})e^{-t/2}\cos(3t) + A(\tfrac{3}{2})e^{-t/2}\sin(3t) - A(9)\cos(3t)\,e^{-\frac{t}{2}}$$
$$= Ae^{-t/2}\cos(3t)(-9 + \tfrac{1}{4} + \tfrac{3}{2}\tan(3t))$$
$$= Ae^{-t/2}\cos(3t)(\tfrac{-35}{4} + \tfrac{3}{2}\tan(3t))$$

At a stationary point $\tan(3t) = -\tfrac{1}{6}$, so the value of $i''(t)$ will be $Ae^{-t/2}\cos(3t)(-9)$, i.e. $-9Ae^{-t/2}\cos(3t)$, see Fig. 7.22.

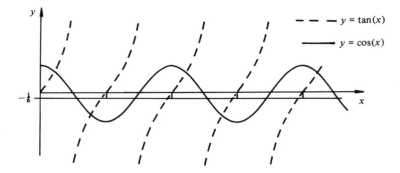

Fig. 7.22. Solutions of $\tan(x) = -\tfrac{1}{6}$.

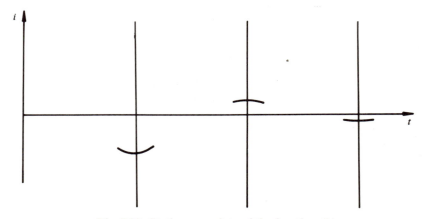

Fig. 7.23. Stationary points of the function $i(t)$.

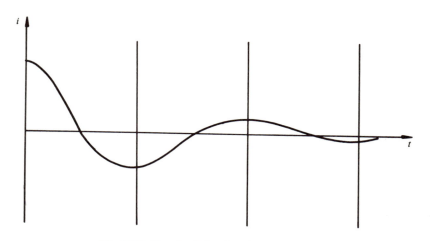

Fig. 7.24. Graph of the electric current $i(t)$.

From the graph of $\cos(3t)$ and $\tan(3t)$ we see that when $\tan(3t) = \frac{1}{6}$ the sign of $\cos(3t)$ changes and hence the sign of $i''(t)$ changes. When $n = 0$ the sign of $i''(t)$ is positive. When $n = 1$ it is negative, when $n = 2$ it is positive and so on. Hence the graph of $i(t)$ has alternately a minimum, a maximum, a minimum etc. The distance of these stationary points from the t axis decreases as t increases because $\cos(3t)$ retains the same value at each point, but $e^{-t/2}$ gets smaller. The stationary points are shown in Fig. 7.23. The graph of $i(t)$ cuts the t axis when $i(t) = 0$. These points are given by the zeros of $\cos(3t)$, which are $t = (2n + 1)(\pi/6)$ for $n = 0, 1, 2, \ldots$. The graph cuts the i axis when $t = 0$, i.e. when $i = A$. As t increases indefinitely $\cos(3t)$ always lies between -1 and $+1$, but $e^{-t/2}$ tends to zero. Hence as $t \to \infty$, $i(t) \to 0$. The function has no singular points and its graph is therefore that shown in Fig. 7.24.

7.6 Exercises

1. Find the equation of the tangent to the curve $y = x^3 + x^2$ at the point $(-1, 0)$.
   Find the coordinates of the local maximum and minimum points.

2. By considering the derivative of the function
   $$f = \sin(x)\tan(x) - 2\ln[\sec(x)]$$
   prove that f steadily increases as x increases from 0 to $\pi/2$.
   Show that the graph has no points of inflexion between these limits.
   Show that the function $2\sin(x)\tan(x) - 5\ln[\sec(x)]$ has one minimum value in the range $0 < x < \pi/2$.

3. Obtain the values of x for which $f = x^3(x - 1)^2$ is stationary, determining which are local maxima and which are local minima.

4. Find the stationary points of the following functions, and determine their nature:

(i) $x^2 - 6x + 8$ (ii) $16 - 6x - 3x^2$
(iii) $x^3 - 12x + 5$ (iv) $x^3 + 3x^2 + 20x - 10$
(v) $x^3 - 3x^2 + 3x - 1$ (vi) $x^4 - 8x^3 + 10x^2 + 40$
(vii) $(x-1)^2(x-2)$ (viii) $(x-1)^3(x-2)^2$

(ix) $\dfrac{(3x-3)^2}{(x+1)^3}$ (x) $\dfrac{x^2 - 2x + 4}{x^2 + 2x + 4}$

(xi) $\dfrac{x}{(x+8)(x+2)}$ (xii) $\sin(x) - \cos(x)$

(xiii) $e^x \cos(x)$ (xiv) $\cosh(x) - \cos(x)$
(xv) $x + \sin(x)$ (xvi) $\sin^3(x)\cos(x)$

5. Frequency stability in the cathode-coupled oscillator can be studied with the aid of the correction factor

$$f(\alpha) = 1 - \frac{L}{16C}(\alpha - 1/R)^2$$

where L, R and C are constants. Show that the maximum value of f is obtained when $\alpha = 1/R$.

6. An electric circuit with inductance L, capacitance C and resistance R in series, has a current with amplitude

$$I = V[R^2 + (wL - 1/wC)^2]^{-\frac{1}{2}}$$

where V is the amplitude and $w/2\pi$ is the frequency of the voltage. Show that I is a maximum at a frequency $1/(2\pi\sqrt{(LC)})$ cycles s^{-1}.

7. Sketch the curves representing the following functions:

(i) $y = x$ (ii) $y = x^2$
(iii) $y = x^3$ (iv) $y = x^4$
(v) $y = x^5$ (vi) $y = \sqrt{x}$
(vii) $y = x^{\frac{3}{2}}$ (viii) $y^2 = x$
(ix) $y^2 = x^3$ (x) $y = x^{\frac{2}{3}}$
(xi) $y^3 = x^2$ (xii) $y = x^2/(1 + x^2)$
(xiii) $y = x^2/(1 - x^2)$ (xiv) $y = x + 1/x$

(xv) $y = \dfrac{(2x-1)(x-8)}{(x-1)(x-4)}$ (xvi) $y = \dfrac{(x+2)(x-3)}{(x+1)(x-6)}$

(xvii) $y = \dfrac{1}{(x-2)}$ (xviii) $y = \dfrac{(x-1)(x-3)}{5x^2 + 4}$

(xix) $y = \dfrac{xe^x}{(x+1)}$ (xx) $y = \dfrac{e^x}{2 + 3e^x}$

8. Sketch the following curves which have the loops:

(i) $y^2 = x(x-2)^2$ (ii) $y^2 = x^2(3-x)$
(iii) $y^2 = x^2(4-x^2)$ (iv) $y^2 = x^2(x^2-9)$
(v) $y^2 = x(3-x)^2$ (vi) $y^2 = x^3(1-x)^2$
(vii) $y^2 = x(x-4)(x-3)$ (viii) $y^2 = (x-1)(x-2)(x-3)$

9. The intensity of illumination from a light source S at a point on a plane area varies directly as the cosine of the angle between the normal to the plane and the line joining the source to the point, and varies inversely as the square of the distance of the point from the source. Consider a circular disc of radius R. Find the height above centre of the disc of a source so that the intensity of illumination at the circumference is a maximum.

10. Mathematical models of traffic flow can help to show how to avoid hold-ups and to ensure optimum flow conditions in congested situations such as road tunnels. A simple model for the flow of cars along a straight level road is

$$f(v) = \frac{v}{L + vT + v^2/2a}$$

where v is the speed of the cars, L is a car length (all cars are assumed to have the same length), a is the maximum deceleration of a car and T is the thinking time of a driver. Find the speed of the cars v_M which gives a maximum flow rate. Sketch a graph of the function f.

If typical values of L, T and a are $4\,\text{m}$, $0.8\,\text{s}$ and $7\,\text{m s}^{-2}$, respectively, find a value for v_M in mil hr^{-1}.

11. In population dynamics an important mathematical model which describes the size of many populations as a function of time is the logistic equation

$$P(t) = \frac{a}{b + (a/P_0 - b)}$$

where P_0, a and b are constants. Sketch the graph of this function.

12. For damped oscillatory motion the displacement from the equilibrium position of a body x as a function of time is given by

$$x(t) = A\mathrm{e}^{-2t}\sin(3t).$$

Find, to two decimal places, the smallest value of t for x to have a maximum value.

13. For a belt drive the formula relating the power transmitted P to the speed of the belt v is given by

$$P(v) = Tv - av^3$$

where T is the tension in the driving side of the belt and a is a constant. Find the speed of v such that the belt delivers maximum power. Sketch a graph of the function P.

14. The velocity of a certain chemical reaction is given by the equation

$$v = k(c_1 - x)(c_2 + x),$$

where c_1, c_2 and k are positive constants and x is the amount of a product. Show that the maximum value of v is

$$k\left(\frac{c_1 + c_2}{2}\right)^2 \quad \text{if } c_1 \geqslant c_2,$$

and kc_1c_2 otherwise.

15. The pH factor is related to temperature, t, in °C according to the formula

$$pH = 4 + \frac{1}{2}\frac{(t-15)^2}{100}$$

Show that $dpH/dt = 0$ when $t = 15\,°C$, and show by sketching the curve of pH against t that $t = 15$ is a minimum.
Verify that d^2pH/dt^2 is positive at $t = 15$.

16. In an aqueous solution the product xy of the concentrations x and y of OH^- and H^+ ions respectively is a constant at constant temperature. Deduce under what conditions $x + y$ will be a minimum. Find the concentration $[H^+]$ under minimum conditions at 293 K when $[H^+]$ $[OH^-] = 10^{-14}\,mol^2\,dm^{-6}$.

17. The total annual inventory cost C in pounds sterling, as a function of the number of items q ordered each time stocks become depleted, for a particular item, is given by the expression

$$C = \frac{18\,000}{q} + 0.2q$$

Obtain the economic batch order quantity for this item that minimises C. Verify that C is a minimum and determine its value.

18. The free energy f per atom in a simple alloy system is given by

$$f = ZVc(1-c) + kT[c\ln(c) + (1-c)\ln(1-c)]$$

where c represents the concentration and k, T, Z and V are constants. Obtain an equation for df/dc and show that stationary values of f occur where

$$(2c-1) = \frac{kT}{ZV}\ln(c/(1-c))$$

19. In a tubular chemical reactor the concentration $[c]$, on 'dumping', of one of the products is given by

$$[c] = \frac{[A]k_1}{(k_2-k_1)}(\exp(-Vk_1/v) - \exp(-Vk_2/v))$$

where $[A]$, k_1, k_2 and V are all positive constants and v is the adjustable flow rate in the reactor. Assuming that the stationary value of $[c]$ is a maximum, show that the flow rate that maximises $[c]$ is given by

$$v = \frac{V(k_2-k_1)}{\ln(k_2) - \ln(k_1)}$$

8

Differentiation 3: approximation of functions

Contents: 8.1 scientific context – 8.1.1 errors in measurement; 8.1.2 simplifying models; 8.2 mathematical developments – 8.2.1 Taylor polynomials for small x; 8.2.2 Taylor polynomials about $x = \alpha(\alpha \neq 0)$; 8.3 worked examples; 8.4 further developments – 8.4.1 the Newton–Raphson method; 8.5 further worked examples; 8.6 exercises.

8.1 Scientific context

We have seen that it is very convenient to be able to approximate functions by polynomials. For example, in Chapter 2, we showed that values of the exponential function could be approximated by values of the following polynomial of degree 2:

$$f(x) = 1 + x + \frac{x^2}{2} \tag{1}$$

Using this polynomial we could find the value of $\exp(x)$ to within 2% of its correct value for values of x up to approximately 0.55.

In this chapter we are going to consider a rather special polynomial approximation to functions using what are known as *Taylor polynomials*. In 8.2 we formulate the general theory of Taylor polynomials of nth degree and then in 8.4.1 we see that the first-degree Taylor polynomial provides us with a powerful iterative technique for solving equations known as the *Newton–Raphson method*. The polynomial approximation above for $\exp(x)$ is an example of a Taylor polynomial of degree 2 and in fact the polynomial approximations to $\sin(x)$ and $\cos(x)$ that we wrote down in Chapter 3 are also Taylor polynomials. In 8.2 we will be able to derive these results using calculus.

8.1.1 Example 1: errors in measurement. There are many sources of error in obtaining numerical solutions to scientific problems. For example, when a mathematical model is used, this usually involves making simplifying assumptions and therefore omissions, and when calculations are carried out errors may creep in due to rounding or not keeping a sufficient number of significant figures and so on. However, one source of error familiar to all scientists is introduced in measuring or estimating the values of physical quantities. For example, in measuring a length with an ordinary ruler, graded in millimetres, we usually find that we can measure accurately to within about 0.5 mm so that for an exact length of 4.362 mm our measurement could be 4 mm, 4.5 mm (or even 5 mm on a bad day!). Whereas an accuracy of 0.5 mm might not be too bad if the length being measured is 1000 mm (i.e. 1 m) it is certainly more serious when the length is only about 4 mm. Such experimental errors, when introduced, will lead to errors in function values that depend on these readings. For example, in an experiment to determine the focal length L of a convex lens two lengths are measured; these are the distances u and v from the lens of an object and of its image respectively (Fig. 8.1). The well-known formula relating L to u and v is

$$\frac{1}{L} = \frac{1}{u} + \frac{1}{v} \quad \text{or} \quad L = \frac{uv}{u+v}$$

The distance u can often be measured much more accurately than the distance v. An experimental error in measuring v will lead to an error in L. Suppose that the exact value of v is v_0 and the measurement of v, v_m say, is in error by a small amount e, then we calculate L as

$$L = \frac{uv_m}{u + v_m}$$

But the correct value of L, L_0, say, is

$$L_0 = \frac{uv_0}{u + v_0}$$

so that the error in the focal length due to an error in measuring v is $L - L_0$. Replacing v_m by $v_0 + e$ we get

$$L_0 = \frac{u(v_m - e)}{u + v_m - e} = \frac{uv_m}{u + v_m}\left(1 - \frac{e}{v_m}\right)\left(1 - \frac{e}{u + v_m}\right)^{-1}$$

Thus if we could write $(1 - (e/\{u + v_m\}))^{-1}$ as a polynomial in terms of e then we would have a useful formula for the error $L - L_0$ in terms of powers of e and the measured distances.

The expression $(1 - \{e/u + v_m\})^{-1}$ is really just a special case of the function $f(x) = (1 + x)^n$ so we could expand it using the binomial expansion that we met before However, using Taylor polynomials we shall be able to obtain the identical result as well as showing where the binomial expansion formula comes from.

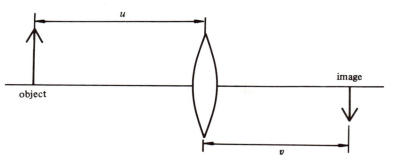

Fig. 8.1. Image of an object in a convex lens.

8.1.2 Example 2: simplifying models. One source of error in obtaining solutions to scientific problems is in the simplification of the mathematical model that describes the physical situation. For example, the equation of motion of a simple pendulum (see Fig. 8.2) neglecting air resistance and friction at the pivot is actually

$$ml\frac{d^2\theta}{dt^2} = -mg\sin(\theta)$$

Now this equation cannot be solved analytically to give a solution $\theta = \theta(t)$ for a known function θ. It is an example of a non-linear second order differential equation. (Differential equations are the subject of Chapters 12 and 13.) For a small initial displacement (for θ values less than about 0.175 radians (or 10°)) this equation can be simplified to the more familiar form

$$ml\frac{d^2\theta}{dt^2} = -mg\theta$$

In which case the techniques of Chapter 13 can be used to give an analytic solution

$$\theta = A\sin(wt) + B\cos(wt)$$

where $w^2 = g/l$. But for larger initial values, this solution will not describe the motion very accurately.

An alternative approach is to write $\theta(t)$ as a polynomial in t

$$\theta(t) = a_0 + a_1 t + a_2 t^2 + a_3 t^3 + \cdots + a_n t^n \tag{2}$$

and turn the problem into one of finding expressions for the coefficients $a_0, a_1, a_2, \ldots$ We might guess that the first coefficient a_0 is just the value of θ when $t = 0$ and that the coefficient of t, a_1, is the value of $d\theta/dt$ when $t = 0$. So as a start we have a linear approximation to θ as

$$\theta(t) \simeq \theta(0) + \frac{d\theta}{dt}(0)t$$

Fig. 8.2. A simple pendulum.

However, this approximation takes no account of the mathematical model describing the motion of the pendulum.

The coefficient of t^2 in (2), which is a_2, turns out to be

$$a_2 = \frac{1}{2!} \frac{\mathrm{d}^2\theta}{\mathrm{d}t^2}(0)$$

so that the value of a_2 can be found using the model and the polynomial approximation of degree 2 is then

$$\theta(t) \simeq \theta(0) + \frac{\mathrm{d}\theta}{\mathrm{d}t}(0)t + \frac{1}{2}\left(-\frac{g}{l}\sin\theta(0)\right)t^2$$

For example if initially $\theta(0) = \pi/8$ and $\mathrm{d}\theta/\mathrm{d}t = 0$, then

$$\theta(t) \simeq \pi/8 - \frac{g}{2l}\sin\frac{\pi}{8}t^2 = 0.3927 - 0.1913gt^2/l$$

This polynomial expansion is again an example of a Taylor polynomial, as is the original nth-degree polynomial we had to start with.

8.2 Mathematical developments

8.2.1 Taylor polynomials for small x. In Chapter 1 we introduced the idea of a polynomial function as a function whose rule can be expressed in the form

$$p(x) = a_0 + a_1x + a_2x^2 + a_3x^3 + \cdots + a_nx^n$$

where n is a positive integer or zero and the coefficients $a_0, a_1, \ldots, a_n$ are real numbers. In Chapter 1 we used $a, b, c, \ldots$ for the coefficients. Here it will be more convenient to work with the subscript notation. As before, the highest power of x is called the degree of the polynomial. For example, the following polynomials have degree 1, 3 and 7 respectively

$$
\begin{aligned}
f(x) &= 3 - 7x && \text{degree 1 – a 'linear' polynomial} \\
g(x) &= 1 + 5x^2 + 2x^3 && \text{degree 3 – a 'cubic' polynomial} \\
h(x) &= 2x + 3x^2 - 7x^4 + x^7 && \text{degree 7}
\end{aligned}
$$

In Chapters 2 and 3 we have seen that the functions e^x, $\sin(x)$ and $\cos(x)$ can be expressed as power series in the following way:

$$\exp(x) = \lim_{n\to\infty}\left(1 + \frac{x}{n}\right)^n = 1 + x + \frac{x^2}{2!} + \frac{x^3}{3!} + \cdots + \frac{x^j}{j!} + \cdots$$

$$\sin(x) = x - \frac{x^3}{3!} + \frac{x^5}{5!} - \cdots + \frac{(-1)^jx^{2j+1}}{(2j+1)!} + \cdots$$

$$\cos(x) = 1 - \frac{x^2}{2!} + \frac{x^4}{4!} - \cdots + \frac{(-1)^jx^{2j}}{(2j)!} + \cdots$$

Although the first of these series was derived from the definition of e^x, the expressions for $\sin(x)$ and $\cos(x)$ were assumed. We now develop a procedure for finding both power series and polynomial representations for any function. There are many advantages in having an approximate polynomial representation for a standard function, such as $\sin(x)$. One obvious application is in calculating the value of a standard function at a given value of x. Before the availability of calculators or microcomputers, the traditional method of finding the value of $\sin(30°)$, say, was to look up the value in a sine table. It is not really feasible for a calculator to store the sine of every possible angle. Instead the calculator stores a program for calculating values of $\sin(x)$ that effectively uses a polynomial representation of $\sin(x)$.

Let us begin then by finding polynomial approximations to functions which give good agreement with the function value at points near to the value $x = 0$. Suppose that we are given a function $f(x)$ and we want to determine a polynomial of degree n, call it $p(x)$, which is a good approximation to $f(x)$ near $x = 0$. So let

$$p(x) = a_0 + a_1 x + a_2 x^2 + \cdots + a_n x^n \tag{3}$$

What we require are values of $a_0, a_1, \ldots a_n$ such that

$$f(x) \simeq p(x)$$

We can make the value of the function $f(x)$ and its derivatives *agree* with those of $p(x)$ *at the point* $x = 0$ by choosing the coefficients of $p(x)$ in the following way:

$$p(0) = a_0 = f(0)$$

$$\frac{dp}{dx}(0) = a_1 = \frac{df}{dx}(0)$$

$$\frac{d^2 p}{dx^2}(0) = 2a_2 = \frac{d^2 f}{dx^2}(0)$$

$$\frac{d^3 p}{dx^3}(0) = 6a_3 = \frac{d^3 f}{dx^3}(0)$$

$$\vdots$$

$$\frac{d^j p}{dx^j}(0) = j! a_j = \frac{d^j f}{dx^j}(0)$$

This choice of coefficients follows from successively differentiating (3) and substituting $x = 0$ in the resulting identities. Hence the coefficients of $p(x)$ are given in terms of the derivatives of the function $f(x)$ evaluated at $x = 0$. The polynomial

$$p(x) = f(0) + \frac{x\,df}{dx}(0) + \frac{x^2}{2!}\frac{d^2 f}{dx^2}(0) + \cdots + \frac{x^n}{n!}\frac{d^n f}{dx^n}(0) \tag{4}$$

is called the *nth Taylor polynomial about* $x = 0$ for the function $f(x)$.

Example: Determine the 5th Taylor polynomial about $x = 0$ for the function

$$f(x) = \sin(x)$$

Solution: Differentiating $\sin(x)$ is easy and the values of $\sin(0)$ and $\cos(0)$ are 0 and 1 respectively. We get

$$f(x) = \sin(x) \qquad f(0) = 0$$

$$\frac{df}{dx} = \cos(x) \qquad \frac{df}{dx}(0) = 1$$

$$\frac{d^2 f}{dx^2} = -\sin(x) \qquad \frac{d^2 f}{dx^2}(0) = 0$$

$$\frac{d^3 f}{dx^3} = -\cos(x) \qquad \frac{d^3 f}{dx^3}(0) = -1$$

$$\frac{d^4 f}{dx^4} = \sin(x) \qquad \frac{d^4 f}{dx^4}(0) = 0$$

$$\frac{d^5 f}{dx^5} = \cos(x) \qquad \frac{d^5 f}{dx^5}(0) = 1$$

Hence the 5th Taylor polynomial about $x = 0$ approximating $\sin(x)$ is

$$p_5(x) = 0 + x + \frac{x^2}{2!}(0) + \frac{x^3}{3!}(-1) + \frac{x^4}{4!}(0) + \frac{x^5}{5!}(1)$$

$$p_5(x) = x - \frac{x^3}{3!} + \frac{x^5}{5!}$$

Note that in this example the polynomial we have found is just the first three terms of the power series expression for $\sin(x)$ that was used in Chapter 3.

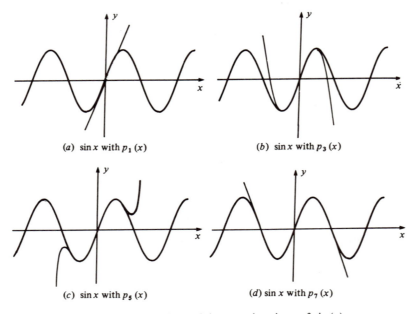

(a) $\sin x$ with $p_1(x)$ (b) $\sin x$ with $p_3(x)$

(c) $\sin x$ with $p_5(x)$ (d) $\sin x$ with $p_7(x)$

Fig. 8.3. Taylor polynomial approximations of $\sin(x)$.

Clearly, the coefficient of x^k is just

$$\frac{1}{k!}\frac{d^k f}{dx^k}(0) = \begin{cases} (-1)^{k/2}\sin(0) & \text{if } k \text{ is even} \\ (-1)^{(k-1)/2}\cos(0) & \text{if } k \text{ is odd} \end{cases}$$

so that we can now justify the formula for $\sin(x)$. There are no even powers of x since $\sin(0) = 0$ and the odd powers take the form (letting $k = 2j + 1$)

$$\frac{(-1)^j}{(2j+1)!}x^{2j+1}$$

Fig. 8.3 shows the graph of the function $\sin(x)$ and the Taylor polynomials $p_1(x)$, $p_3(x)$, $p_5(x)$ and $p_7(x)$. We see that the Taylor polynomials provide good approximations to $\sin(x)$ near $x = 0$. For example, in the interval $-\pi/2 \leqslant x \leqslant \pi/2$ the 5th Taylor polynomial $p_5(x)$ and $\sin(x)$ agree to three significant figures at the edge of the interval $x = \pi/2$. The agreement is better nearer to $x = 0$, for instance at $x = 0.4$ ($\simeq \pi/8$ or $22.5°$) $p_5(x)$ and $\sin(x)$ agree to six significant figures.

Example: Find the general formula for the nth Taylor polynomial about $x = 0$ for each of the following functions:
  (i) $f(x) = e^x$;
  (ii) $g(x) = 1/(1 + x)$;
  (iii) $h(x) = \ln(x)$.

Solution: (i) For $f(x) = e^x$ we have

$$\frac{df}{dx} = \frac{d^2 f}{dx^2} = \cdots = \frac{d^n f}{dx^n} = e^x$$

Hence

$$f(0) = \frac{df}{dx}(0) = \cdots = \frac{d^n f}{dx^n}(0) = 1$$

and the nth Taylor polynomial is

$$p(x) = 1 + x + \frac{x^2}{2!} + \cdots + \frac{x^n}{n!}.$$

(ii) Consider $g(x) = 1/(1 + x)$, then $g(0) = 1$.

$$\frac{dg}{dx} = -\frac{1}{(1+x)^2} \quad \text{and} \quad \frac{dg}{dx}(0) = -1$$

$$\frac{d^2 g}{dx^2} = \frac{2}{(1+x)^3} \quad \text{and} \quad \frac{d^2 g}{dx^2}(0) = 2$$

$$\vdots$$

$$\frac{d^n g}{dx^n} = \frac{(-1)^n n!}{(1+x)^{n+1}} \quad \text{and} \quad \frac{d^n g}{dx^n}(0) = (-1)^n n!$$

The nth Taylor polynomial of $1/(1 + x)$ near $x = 0$ is

$$p(x) = 1 - x + x^2 - x^3 + \cdots + (-1)^n x^n.$$

(iii) The function $h(x)$ and its derivatives are not defined at $x = 0$ so that a Taylor polynomial approximation to $\ln(x)$ at $x = 0$ does not exist!

The last part of this example provides a note of caution. We can only find Taylor polynomials about $x = 0$ for functions that are continuously differentiable at $x = 0$.

8.2.2 Taylor polynomials about $x = \alpha (\alpha \neq 0)$. If we cannot obtain $\ln(x)$ as a polynomial about $x = 0$, then perhaps we can obtain $\ln(x)$ as a polynomial about some other value and, of course, this is the case. Very frequently we find that we need approximations to functions about points other than $x = 0$. The same method of approach can be used to find a polynomial expansion about some point $x = \alpha$, by assuming the form

$$p(x) = a_0 + a_1(x - \alpha) + a_2(x - \alpha)^2 + \cdots + a_n(x - \alpha)^n$$

The constants $a_0, a_1, a_2, \ldots$ can again be specified by requiring that the value of $f(x)$ and its derivatives agree with those of $p(x)$ at the point $x = \alpha$. The polynomial is therefore given as follows:

$$p(x) = f(\alpha) + (x - \alpha)\frac{\mathrm{d}f}{\mathrm{d}x}(\alpha) + \frac{(x - \alpha)^2}{2!}\frac{\mathrm{d}^2 f}{\mathrm{d}x^2}(\alpha) + \cdots + \frac{(x - \alpha)^n}{n!}\frac{\mathrm{d}^n f}{\mathrm{d}x^n}(\alpha) \qquad (5)$$

and is called the *nth Taylor polynomial about* $x = \alpha$ for the function $f(x)$. In choosing α we have obviously to be able to evaluate $f(x)$ and its derivatives at $x = \alpha$. Clearly, $\alpha = 0$ is a special case giving the previous form.

Example: Find the 5th Taylor polynomial about $x = 1$ for $f(x) = \ln(x)$ and hence find an
approximate value of $\ln(0.5)$.
Solution: The derivatives of f are

$$\frac{\mathrm{d}f}{\mathrm{d}x} = \frac{1}{x}, \quad \frac{\mathrm{d}^2 f}{\mathrm{d}x^2} = -\frac{1}{x^2}, \quad \frac{\mathrm{d}^3 f}{\mathrm{d}x^3} = \frac{2}{x^3}, \quad \frac{\mathrm{d}^4 f}{\mathrm{d}x^4} = -\frac{6}{x^4}, \quad \frac{\mathrm{d}^5 f}{\mathrm{d}x^5} = \frac{24}{x^5}$$

and hence the coefficients of the 5th Taylor polynomial about $x = 1$ are

$$f(1) = \ln(1) = 0, \quad \frac{\mathrm{d}f}{\mathrm{d}x}(1) = 1, \quad \frac{\mathrm{d}^2 f}{\mathrm{d}x^2}(1) = -1, \quad \frac{\mathrm{d}^3 f}{\mathrm{d}x^3}(1) = 2,$$

$$\frac{\mathrm{d}^4 f}{\mathrm{d}x^4}(1) = -6, \quad \frac{\mathrm{d}^5 f}{\mathrm{d}x^5}(1) = 24$$

The 5th Taylor polynomial of $\ln(x)$ about $x = 1$ is

$$p_5(x) = (x - 1) - \frac{(x - 1)^2}{2} + \frac{(x - 1)^3}{3} - \frac{(x - 1)^4}{4} + \frac{(x - 1)^5}{5}$$

For $x = 0.5$, $p_5(x) = -0.689$ (to three decimal places)

The actual value of $\ln(0.5)$ is -0.693 (to three decimal places) so that the 5th Taylor polynomial and $\ln(0.5)$ agree to two decimal places. For a more accurate approximation we would need to take a higher degree polynomial representation. For example, the 9th Taylor polynomial about $x = 1$ agrees with $\ln(0.5)$ to four decimal places.

We have seen how Taylor polynomials can be used to evaluate function values; however it is the polynomial *approximation* itself as a *formula* representing the function that is useful in science. After all, a calculator or microcomputer can give very accurate function values directly without us having to evaluate values of polynomials!

One important application in science of Taylor polynomials is in estimating errors in calculations due to errors in experimental readings. If we measure values of a physical quantity x, say, and there is an error δx in x, then if we use x to calculate the value of another physical quantity, y say, given in terms of x by $y = f(x)$, the error in x will lead to an error in y. The first Taylor polynomial of f about α will often provide a good approximation for this error in y. We have

$$p_1(x) = f(\alpha) + (x - \alpha)\frac{\mathrm{d}f}{\mathrm{d}x}(\alpha)$$

and $f(x) \simeq p_1(x)$ near $x = \alpha$. Hence the error in calculating y as $f(x)$, i.e. $\delta y = f(x) - f(\alpha)$ is given by

$$\delta y = f(x) - f(\alpha) \simeq (x - \alpha)\frac{\mathrm{d}f}{\mathrm{d}x}(\alpha) = \delta x \frac{\mathrm{d}f}{\mathrm{d}x}(\alpha)$$

If we can calculate $(\mathrm{d}f/\mathrm{d}x)(\alpha)$ and estimate δx (or at least put bounds on it) then we can calculate the approximate error in y. There are instances where $(\mathrm{d}f/\mathrm{d}x)(\alpha) = 0$ so that we then use a higher degree Taylor polynomial (see example 2 in Section 8.3).

Example: Find the nth Taylor polynomial about $x = 0$ for the function $f(x) = (1 + x)^a$ where a is a constant.

Solution: For $f(x) = (1 + x)^a$ we have

$$\frac{\mathrm{d}f}{\mathrm{d}x} = a(1 + x)^{a-1}, \quad \frac{\mathrm{d}^2 f}{\mathrm{d}x^2} = a(a - 1)(1 + x)^{a-2}, \ldots$$

$$\frac{\mathrm{d}^j f}{\mathrm{d}x^j} = a(a - 1)\cdots(a - j + 1)(1 + x)^{a-j}$$

Hence $f(0) = 1$,

$$\frac{\mathrm{d}f}{\mathrm{d}x}(0) = a, \quad \frac{\mathrm{d}^2 f}{\mathrm{d}x^2}(0) = a(a - 1),\ldots \quad \frac{\mathrm{d}^j f}{\mathrm{d}x^j}(0) = a(a - 1)\cdots(a - j + 1)$$

so that the nth Taylor polynomial is

$$p(x) = 1 + ax + \frac{a(a - 1)}{2!}x^2 + \frac{a(a - 1)(a - 2)}{3!}x^3 + \cdots + \frac{a(a - 1)\cdots(a - n + 1)}{n!}x^n$$

It is important to note that the resulting Taylor polynomial in this last example is exactly the same as the first $n + 1$ terms in the *binomial expansion* of $(1 + x)^a$.

If a is a positive integer, m say, and m happens to be less than n, then all the terms after $j = m$ are zero since

$$\frac{\mathrm{d}^{m+1} f}{\mathrm{d}x^{m+1}}(0) = \frac{a(a - 1)\cdots(a - (m + 1) + 1)}{m!}$$

$$= \frac{m(m - 1)\cdots(m - (m + 1) + 1)}{m!}$$

$$= 0$$

Thus the binomial expansion of $(1 + x)^m$ is given by

$$(1 + x)^m = 1 + mx + \frac{m(m-1)}{2!}x^2 + \cdots x^m$$

8.3 Worked examples

Example 1. The focal length L of a convex lens can be calculated using the formula

$$\frac{1}{L} = \frac{1}{u} + \frac{1}{v}$$

where u and v are the distances from the lens of an object and of its image, respectively. In an experiment the value of u can be measured very accurately. However, the value of v is measured less accurately. Find a general formula to estimate the error in L due to the error δv in v.

If u and v are measured as 5 m and 0.4 m, respectively, and the maximum error in v is 0.05 m, estimate the focal length of the lens and the maximum error in this approximation.

Solution. The first Taylor polynomial about the value, α say, provides an approximation for the error in L.

If $\qquad\qquad \dfrac{1}{L} = \dfrac{1}{u} + \dfrac{1}{v} \qquad$ then $\qquad L = f(v) = \dfrac{uv}{u+v}$

The first Taylor polynomial for $f(v)$ is

$$p_2(v) = f(\alpha) + (v - \alpha)\frac{\mathrm{d}f}{\mathrm{d}v}(\alpha) = f(\alpha) + (v - \alpha)\left.\left(\frac{u(u+v) - uv}{(u+v)^2}\right)\right|^{v=\alpha}$$

and the error in L, $L - f(\alpha)$ is then approximately given by

$$L - f(\alpha) = \left(\frac{u}{u+\alpha}\right)^2 (v - \alpha) = \left(\frac{u}{u+\alpha}\right)^2 \delta v$$

If $u = 5$ and $v = 0.4$ the formula for L gives

$$\frac{1}{L} = \frac{1}{5} + \frac{1}{0.4} = 2.7$$

and the focal length of the lens is $\frac{1}{2.7} = 0.370$ (to 3 d.p.'s). This value of L, calculated from the measured values of u and v, will be in error due to v being in error.

If the maximum error in v is 0.05 we have

$$|\delta v| \leqslant 0.05$$

The actual error in v, being either an overestimate or an underestimate, will propagate into an error in L. The maximum error in L can therefore be approximated by

$$\left| \left(\frac{u}{u+\alpha} \right)^2 \delta v \right| = \left(\frac{5}{5.4} \right)^2 0.05$$

$$= 0.043$$

We conclude that the focal length of the lens is approximately 0.37 m with a maximum error of about 0.043 m.

Example 2 (the shot-putter's dilemma). In the athletic event of shot putting, a simple model gives the formula relating the range of a shot R to the initial speed v and angle of launch θ as

$$R = v^2 \sin(2\theta)/g$$

where g is the acceleration due to gravity. In this simple model it is assumed that the shot is projected from ground level. The maximum range is achieved when $\theta_0 = 45°$. However, a shot putter discovers that in attempting to achieve this angle he cannot launch the shot at his maximum speed of 14.8 m s^{-1}. Should the shot putter concentrate on achieving this angle of projection or should he try to achieve the maximum speed?

Solution. To attempt to resolve this dilemma we shall find the error in the range for small changes in angle and speed of projection.

(i) Suppose that the athlete launches the shot at his maximum speed v_0 but with an error of $\delta\theta$ for the maximum range angle θ_0. (Note: in this example $dR/d\theta = 0$ when $\theta = \theta_0$ (i.e. 45°) so we need to take the second Taylor polynomial here.) The second Taylor polynomial for $R(\theta)$ about $\theta = \theta_0$ gives

$$p_2(\theta) = R(\theta_0) + \delta\theta \frac{dR}{d\theta}(\theta_0) + \frac{\delta\theta^2}{2} \frac{d^2R}{d\theta^2}(\theta_0)$$

and since $R(\theta) = v_0^2 \sin(2\theta)/g$ we have

$$p_2(\theta) = \frac{v_0^2 \sin(2\theta_0)}{g} + \frac{2v_0^2 \cos(2\theta_0)\delta\theta}{g} - \frac{4v_0^2 \sin(2\theta_0)\delta\theta^2}{2g}$$

Hence an estimate for the reduction from the maximum range (given for $\theta_0 = 45°$) is

$$E_1 = p_2(\theta) - \frac{v_0^2 \sin(2\theta_0)}{g} = \frac{-2v_0^2}{g}(\delta\theta)^2$$

(ii) Now if the athlete achieves the ideal angle θ_0 but a reduced launch speed $v_0 - \delta v$, the first Taylor polynomial (with $v - v_0 = -\delta v$) gives

$$p_1(v) = R(v_0) + (v - v_0)\frac{dR}{dv}(v_0) = R(v_0) - \delta v \frac{dR}{dv}(v_0)$$

$$= \frac{v_0^2 \sin(2\theta_0)}{g} - \frac{2v_0 \sin(2\theta_0)}{g}\delta v$$

Hence an estimate for the error in the maximum range (again for $\theta_0 = 45°$) is

$$E_2 = -2v_0\delta v/g$$

The ratio of these errors is

$$\frac{E_1}{E_2} = \frac{v_0(\delta\theta)^2}{\delta v}$$

which can be rewritten in terms of fractional errors as

$$\frac{E_1}{E_2} = \frac{\theta_0^2(\delta\theta/\theta_0)^2}{(\delta v/v_0)} = \frac{0.61(\delta\theta/\theta_0)^2}{(\delta v/v_0)}$$

(since $\theta_0 = \pi/4 \simeq 0.78$). This formula implies that the error in the range due to small $\delta\theta$ is somewhat smaller than the error due to small δv. For instance, if v and θ each differ from their 'best values' by 10% (so that $\delta\theta/\theta_0 = \delta v/v_0 = 0.1$) then

$$\frac{E_1}{E_2} = 0.061$$

i.e. the error E_2 is roughly 16 times the error E_1. The advice that we offer to the shot putter is to concentrate on increasing the speed of launch at the expense of an accurate value for θ of 45°!

Example 3 (A simple pendulum). The equation of motion of a simple pendulum is given by the equation

$$\frac{d^2\theta}{dt^2} = -k\sin(\theta)$$

where $k = 10\,\text{rad s}^{-2}$. Initially the pendulum is displaced from rest and the initial angular displacement is 1 rad. Find the 5th Taylor polynomial for $\theta(t)$ about $t = 0$. Hence estimate the value of θ when $t = 0.5\,\text{s}$. Use this polynomial approximation to estimate the period.

Solution. Note that the initial angle, $\theta(0) = 1$ rad, is sufficiently large that it is not sensible to simplify the model by letting

$$\sin(\theta) \simeq \theta; \ (\sin(1) = 0.8415)$$

The 5th Taylor polynomial is

$$p_5 = \theta(0) + \frac{d\theta}{dt}(0)t + \frac{d^2\theta}{dt^2}(0)\frac{t^2}{2} + \frac{d^3\theta}{dt^3}(0)\frac{t^3}{3!} + \frac{d^4\theta}{dt^4}(0)\frac{t^4}{4!} + \frac{d^5\theta}{dt^5}(0)\frac{t^5}{5!}$$

Now $\theta(0) = 1$ and $d\theta/dt(0) = 0$ so using the equation of motion given above we have that

$$\frac{d^2\theta}{dt^2}(0) = -k\sin(\theta) = -10\sin(1) = -8.415$$

To evaluate subsequent coefficients we differentiate the given equation of motion

$$\frac{d^3\theta}{dt^3} = -k\cos\theta\frac{d\theta}{dt} = 0 \quad \text{when } t = 0 \quad \text{since } \frac{d\theta}{dt}(0) = 0$$

$$\frac{d^4\theta}{dt^4} = k\sin\theta\left(\frac{d\theta}{dt}\right)^2 - k\cos\theta\frac{d^2\theta}{dt^2} = 0 - k\cos(1)(-8.415)$$

$$= 45.46 \quad \text{when } t = 0$$

$$\frac{d^5\theta}{dt^5} = k\cos\theta\left(\frac{d\theta}{dt}\right)^3 + 3k\sin\theta\frac{d\theta}{dt}\frac{d^2\theta}{dt^2} - k\cos\theta\frac{d^3\theta}{dt^3}$$

$$= 0 \quad \text{when } t = 0$$

Hence the 5th Taylor polynomial about $t = 0$ is

$$p_5(t) = 1 - 8.415\frac{t^2}{2} + 45.46\frac{t^4}{4!}$$

When $t = 0.5$ an estimate for θ is given by $p_5(0.5)$, i.e.

$$\theta(0.5) \simeq p_5(0.5) = 0.067$$

The Taylor polynomial is not a periodic function like $\sin(wt)$ or $\cos(wt)$. However, we can obtain an estimate for the period because when $\theta = 0$ the pendulum bob has described a quarter of its periodic cycle. So if the time for θ to go from 1 rad to 0 is T s, then an estimate for the period is $4T$ s. Using the Taylor polynomial above, T satisfies

$$p(T) = 0 = 1 - 4.207T^2 + 1.894T^4$$

which is a quadratic in T^2. Solving we have

$$T^2 = \frac{4.207 \pm \sqrt{(4.207^2 - 4 \times 1.894)}}{2 \times 1.894} = 1.111 \pm 0.840$$

and the smallest positive value of T is when $\theta = 0$ for the first time, i.e. $T = 0.520$ s giving an estimate for the period as 2.08 s. (Simplifying the equation of motion to $d^2\theta/dt^2 = -100$ gives a periodic solution with period 1.99 s – a difference of about 4.5%.) The 6th Taylor polynomial gives the following equation for T

$$1 - 4.207T^2 + 1.894T^4 + 2.142T^6 = 0$$

The solution for T is now 0.535 s, so that the period estimate becomes 2.14 s.

We can continue to improve the estimate of the period by taking higher-degree Taylor polynomials. As we take more and more powers of T, the equation in T becomes impossible to solve with a simple rule like the rule for the solution of a quadratic equation. We have to use an iterative method to solve such equations.

In solving the last equation in T, we used an iterative method called the *Newton–Raphson method*. This is the subject of the next section.

8.4 Further developments

8.4.1 The Newton–Raphson method. In the course of solving mathematical problems, it is often necessary to obtain the roots of a function $F(x)$, i.e. find those

x values such that

$$F(x) = 0$$

For example, in Chapter 7 we saw that we can find the stationary points of a function f (i.e. the local maxima and minima) by solving the equation

$$f'(x) = 0$$

One method of solving such an equation is the 'bisection method' that we introduced in Chapter 1. In this section we introduce an iterative method that converges more quickly in many cases.

Suppose we want to find the roots of the equation

$$F(x) = 0$$

and we can guess that a root is near the point $x = x_0$. (Such a guess is always possible by drawing a graph, for example.) The first Taylor polynomial for $F(x)$ about x_0 is given by

$$p_1(x) = F(x_0) + (x - x_0) \frac{dF}{dt}(x_0)$$

Fig. 8.4 shows graphs of the functions F and p_1 near the point x_0. The graph of p_1 is a straight line touching the graph of F at the point $(x_0, F(x_0))$. The first Taylor approximation is often called *the tangent approximation* to $F(x)$ at x_0. The point x_0 is not a particularly good approximation to the root; however Fig. 8.4 suggests that the Taylor polynomial p_1 could provide a better approximation. The value of x for which $p_1(x) = 0$, i.e. the point labelled x_1 in Fig. 8.4, is closer to the root than x_0. So x_1 is an improved approximation to the root $x = r$, say.

Solving $p_1(x_1) = 0$ we get

$$x_1 = x_0 - \frac{F(x_0)}{\dfrac{dF}{dx}(x_0)}$$

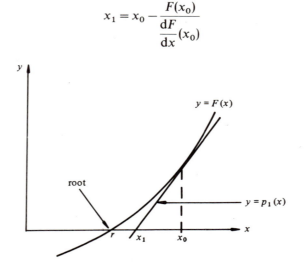

Fig. 8.4. The tangent approximation to $F(x)$.

The first Taylor approximation of F about x_1 is

$$q_1(x) = F(x_1) + (x - x_1)\frac{\mathrm{d}F}{\mathrm{d}x}(x_1)$$

Adding the graph of $q_1(x)$ to Fig. 8.4 we see that the solution of $q_1(x) = 0$, x_2 say, provides a better approximation to the root than both x_0 and x_1 (see Fig. 8.5).

Solving $q_1(x_2) = 0$ we get

$$x_2 = x_1 - \frac{F(x_1)}{\dfrac{\mathrm{d}F}{\mathrm{d}x}(x_1)}$$

Note the similarity between this equation for x_2 and the one for x_1.

We need not stop here. If we compute the first Taylor polynomial about x_2, we can obtain an even better approximation, x_3 say, and so on. Each successive value $x_1, x_2, x_3, \ldots$ is computed using the iterative formula

$$x_{n+1} = x_n - \frac{F(x_n)}{\dfrac{\mathrm{d}F}{\mathrm{d}x}(x_n)}$$

Example: Find the root close to $T = 30$ of the cubic equation

$$0.0003 T^3 + 0.72 T - 34 = 0.$$

(This equation was solved in Chapter 1, Subsection 1.5.2, by the bisection method to 2 d.p.'s in nine iteration steps.)

Solution: The iteration formula is

$$T_{n+1} = T_n - \frac{0.0003 T_n^3 + 0.72 T_n - 34}{0.0009 T_n^2 + 0.72}$$

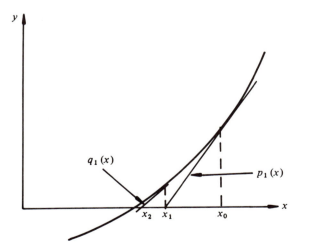

Fig. 8.5. Solving $F(x) = 0$ using tangent approximations.

and we start with $T_0 = 30$. The following table shows a systematic way of laying out the results (calculations are done to 4d.p.):

n	T_n	$0.0003T_n^3 + 0.72T_n - 34$	$0.0009T_n^2 + 0.72$	T_{n+1}
0	30	-4.3	1.53	32.8105
1	32.8105	0.2199	1.6889	32.6803
2	32.6803	5.2×10^{-4}	1.6812	32.6799
3	32.6799	3.6×10^{-6}	1.6812	32.6799

After only four iterations the T values agree to four decimal places and $F(T_3)$ is so small (3.6×10^{-6}) we can conclude that

$$T_3 = 32.6799$$

is a very close approximation (certainly to four decimal places) to the root.

This agrees with the value obtained in Chapter 1 using the bisection method. However, there it took nine iterations to find the answer correct to two decimal places, and here it took four iterations to obtain four decimal-place accuracy. There are two criteria for stopping the iterative method:

 (i) if $F(x_n)$ is very small then we can conclude that x_n is very close to the root; or

 (ii) if two successive approximations x_j and x_{j+1} differ by a very small amount then x_{j+1} is a good approximation to the root.

The method can be summarised in the following way:

The Newton–Raphson method. If x_0 is an initial approximation close to the root of $F(x) = 0$ then the iterative formula

$$x_{n+1} = x_n - \frac{F(x_n)}{\dfrac{\mathrm{d}F}{\mathrm{d}x}(x_n)} \tag{6}$$

provides successively better approximations $x_1, x_2, x_3 \cdots$.

Example: Give a pseudocode algorithm for solving the equation $F(x) = 0$ based on the Newton–Raphson method.

Solution: Let us assume that our initial guess is x_0 and that the relative difference between successive approximations has to be less than or equal to e. Suppose, too, that the number of iterations we shall carry out before abandoning computations is n. If the iterative formula is equivalent to $x_{n+1} = g(x_n)$, then the following pseudocode algorithm could be used for solving the equation $F(x) = 0$.

```
begin
input x₀, e, n
set x to x₀
set m to 0
repeat
        set x₀ to x
        set x to g(x₀)
        set m to m + 1
```

```
until (|(x − x₀)/x₀| ⩽ e) or (m > n)
if    (m ⩽ n) then
        set output to x
        print output
else
        set output to 'not converging'
        print output
endif
stop
```

Although the Newton–Raphson method works well in many problems, some caution is needed in using it. Particular care is required in the choice of x_0.

(i) If $dF/dx = 0$ near the root then the sequence of iterations converges for a suitable starting value x_0; however if $dF/dx = 0$ at the root then the method is definitely not suitable.

(ii) Fig. 8.6 illustrates another situation where caution is required.

Suppose we wish to find the root $x = r$ and there is a local maximum at the point $x = a$. If we choose x_0 between r and a then the Taylor polynomial does indeed give a closer approximation x_1' say. However, if we choose x_0 on the right of a, the iterations will converge to the root at s and not r.

So provided we choose x_0 carefully the Newton–Raphson method works very well. Any program written that incorporates the Newton–Raphson technique should make provision for the fact that convergence may be slow or may indeed not occur. The algorithm in the last example does this by halting computation if more than a stipulated number of iterations is needed.

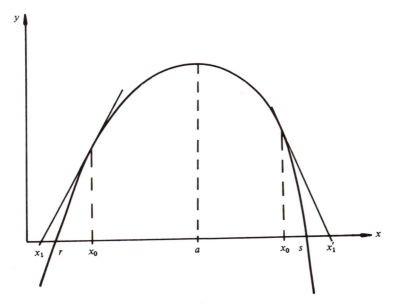

Fig. 8.6. Choose x_0 carefully.

8.5 Further worked examples

Example 1. A motor under load generates heat at a constant rate and radiates it at a rate proportional to the excess temperature θ, so that at some time t

$$\theta = \frac{10}{k} - \frac{10}{k} e^{-kt}$$

where k is a non-zero positive constant. It is found that after 10 min the temperature rise is $50\,^\circ\text{C}$ so that k is a root of the equation

$$e^{-10k} = 1 - 5k$$

Find the value for k correct to four decimal places, using the Newton–Raphson method.

Solution. In the Newton–Raphson method the first step is to make an initial guess at the value of the root. We usually do this by drawing a sketch of the function. It is easier in this example to draw the graphs of

$$f_1 = e^{-10k} \quad \text{and} \quad f_2 = 1 - 5k$$

and find where the two graphs cross each other. Fig. 8.7 shows these graphs for $k > 0$ (since we know that the physically meaningful root is positive). It appears that the root is in the interval $(0.1, 0.2)$. So as an initial guess we choose 0.15. (Note that this is quite an arbitrary choice provided we start the iterations away from $k = 0$. If we started near $k = 0$ then the iterations might converge to the root $k = 0$.) The iterations are shown in table:

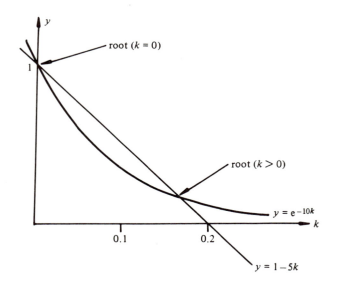

Fig. 8.7. Solutions of $1 - 5k = e^{-10k}$

n	k_n	$\exp(-10k_n) - 1 + 5k_n$	$-10\exp(-10k_n) + 5$	k_{n+1}
0	0.15	-2.687×10^{-2}	2.768 70	0.159 70
1	0.159 70	1.0176×10^{-3}	2.975 07	0.159 36
2	0.159 36	1.1859×10^{-6}	2.968 13	0.159 36

To four decimal places the value of k is 0.1594.

Example 2. The ratio of the voltage at the receiving and transmitting ends of a transmission line is a maximum when the length x (km) of the line is given by

$$\alpha \sinh(2\alpha x) = \beta \sin(2\beta x)$$

where $\alpha = 0.5 \times 10^{-3}\,\text{km}^{-1}$ and $\beta = 1.8 \times 10^{-3}\,\text{km}^{-1}$. Find the non-zero value of x correct to two decimal places using the Newton–Raphson method with $x = 1000\,\text{km}$ as an initial value.

Solution. The equation whose roots are required is

$$0.5 \times 10^{-3} \sinh(10^{-3}x) - 1.8 \times 10^{-3} \sin(3.6 \times 10^{-3}x) = 0$$

Since we are using an initial value of 1000 we change variables and write

$$X = x/1000$$

so that the equation to be solved in terms of X is

$$F(X) = 0.5 \sinh(X) - 1.8 \sin(3.6X) = 0$$

and

$$\frac{dF}{dx}(X) = 0.5 \cosh(X) - 1.8 \times 3.6 \cos(3.6X)$$

The following table shows the iterations:

n	X_n	$0.5\sinh(x_n)$ $-1.8\sin(3.6x_n)$	$0.5\cosh(x_n)$ $-6.48\cos(3.6x_n)$	X_{n+1}
0	1	1.384 14	6.582 53	0.789 726
1	0.789 726	-9.2288×10^{-2}	6.857 48	0.803 184
2	0.803 184	6.2857×10^{-4}	6.948 48	0.803 093
3	0.803 093	2.5029×10^{-8}	6.947 91	0.803 093

Hence the solution for X is 0.803 093 so that the solution for x to two decimal places is

$$x = 803.09\,\text{km}$$

8.6 Exercises

1. Find the Taylor polynomial for small x of each of the following functions writing down the terms up to x^3:

   (i) $f = e^x \sin(x)$
   (ii) $f = \exp(\sin(x))$
   (iii) $f = \ln(1 + e^x)$
   (iv) $f = \ln(1 + \sin(x))$

2. Write down the Taylor polynomial for small x of $\sec(x)$ in ascending powers of x as far as the term in x^4.
   Hence, show that if x is small so that the terms of higher order than x^4 can be neglected

   $$[3 + \sec^2(x)]^{\frac{1}{2}} = a + bx^2 + cx^4$$

   where $a = 2$, $b = \frac{1}{4}$ and $c = \frac{29}{192}$.

3. Write the power series for e^x and for $\sin(x)$ and use them to show that

   $$\exp(\sin(x)) = 1 + x + x^2/2 - x^4/8$$

4. Find the Taylor polynomial for small x of $\sin(x + \alpha)$ and $\tan(x + \alpha)$, where α is a constant.
   Use the answer to find the value of $\tan(46° \ 30')$ to four decimal places.

5. Find the Taylor polynomial about $x = 1$ of each of the following functions writing down the first four terms:

   (i) $f = e^x$
   (ii) $f = \sqrt{x}$
   (iii) $f = \ln(1 + x)$

6. Evaluate the following limits by using appropriate Taylor polynomials:

   (i) $\displaystyle\lim_{x \to 0} \frac{\sin(x) - x}{x^3}$

   (ii) $\displaystyle\lim_{x \to 0} \frac{\ln(1 + x) - x}{\sin^2(x)}$

   (iii) $\displaystyle\lim_{x \to 0} \frac{x^2(e^x - e^{-x})}{(1 + x^3)^4 - (1 - x^3)^4}$

   (iv) $\displaystyle\lim_{x \to 0} \frac{\ln[1 + \sin(x)] - \sinh(x)}{\sqrt{(1 + x)} - 1}$

7. Van der Waal's equation of state for a gas may be written as

   $$p = \frac{RT}{(V - b)} - \frac{a}{V^2}$$

   where p, V and T are the usual thermodynamical variables, and a and b are positive constants. The critical point is defined to be that point in the pVT space where dp/dV and d^2p/dV^2 are both zero. Show that this occurs on the isothermal for which $T = 8a/27Rb$ and where the volume is equal to $3b$.

Obtain the value of p at the critical point and obtain a third-order Taylor polynomial for the pressure about the point $V = 3b$, that is valid on the isothermal $T = 8a/27Rb$.

8. In a particular second-order chemical reaction, the amount of a product x is related to time t via the expression

$$x = \frac{0.01(1 - \exp(kt/125))}{(1 - 5\exp(kt/125))}$$

where k, the rate constant, is a positive number. Obtain x as a second-order Taylor polynomial about $t = 0$.

9. For a magnet, of magnetic moment M and length $2l$, the field strength H at a point on the axis a distance d from its centre is given by

$$H = \frac{M}{2l}\left\{\frac{1}{(d-1)^2} - \frac{1}{(d+1)^2}\right\}$$

If l is small compared to d expand the right-hand side using appropriate Taylor series to give H as a series in l/d. Hence show that $H = 2M/d^3$.

10. Solve the following equations using the Newton–Raphson method starting with the given values x_0 giving your answers correct to three decimal places:

   (i) $2x^3 + 3x - 3 = 0$ $x_0 = 0.4$
   (ii) $x - 2\sin(x) = 0$ $x_0 = -2$ and $x_0 = +2$
   (iii) $2x^3 - 5x^2 - 5x + 10 = 0$ $x_0 = -1.5$
   (iv) $xe^{-x} - 0.1 = 0$ $x_0 = 0.2$ and $x_0 = 3.5$

11. Solve the following equations using the Newton–Raphson method by first finding an approximate starting value:

   (i) $e^x - 3x = 0$ (ii) $x^3 + 13x^2 + 40x + 17 = 0$
   (iii) $\tan(x) = x$ (iv) $e^x - e^{-x} + 0.4x - 10 = 0$
   (v) $1 + \ln(x) = 0.5x$ (vi) $\sin(x) = 1/(2x)$

12. Use the Newton–Raphson method to solve the equation

$$x^3 - 5x + 3 = 0$$

starting with (i) $x_0 = 1.20$, (ii) $x_0 = 1.22$ and (iii) $x_0 = 1.23$
Explain with the aid of a diagram why they converge to three different roots.

13. Perform three iterations of the Newton–Raphson method to solve

$$x^3 - 4x - 7 = 0$$

starting with $x_0 = 1$.
Why does it not converge?

14. Derive an iterative solution for finding $\sqrt{A}$ using the Newton–Raphson method to solve $x^2 - A = 0$. Hence find $\sqrt{3}$ to four decimal places.

15. The stationary values of the free energy per atom in a simple alloy system

can be obtained by solving an equation of the type

$$A(2c - 1) = \ln(c/(1 - c))$$

for the concentration c where A is a constant depending on the type of alloy and its temperature. Derive an iterative formula for obtaining c, based on the Newton–Raphson technique.

16. Produce a pseudocode algorithm that will solve the equation given in Exercise 15, when $A = 0.3$, and give c accurate to 0.01%. An initial value of c equal to 0.5 may be assumed and a suitable message should be printed out if the number of iterations exceeds 30, at which point computation should cease.

17. The temperature T at which the dissociation pressure of strontium carbonate is 1 atmosphere satisfies the equation

$$\frac{12\,200}{T} - 0.77\ln(T) - 3.2 = 0$$

Produce a pseudocode algorithm, based on the Newton–Raphson method, that will obtain the value of T to within an accuracy of 0.1% and print out the result together with a recording of the number of iterations performed. The algorithm should be such that computation ceases if the number of iterations exceeds 50, and in this event a suitable message should be printed out. A first approximation to T can be taken as 1000 K.

9

Integration 1: introduction and standard forms

Contents: 9.1 scientific context – 9.1.1 speed–time graphs; 9.1.2 force–distance graphs; 9.2 mathematical developments – 9.2.1 definition of the integral as a summation; 9.2.2 evaluating integrals; 9.3 worked examples; 9.4 further developments – 9.4.1 numerical integration: Euler's method; 9.4.2 the trapezoidal method; 9.4.3 Simpson's method; 9.5 further worked examples; 9.6 Exercises.

We now begin three chapters about integration.

As the word implies, integration is about bringing things together. In mathematics the process of integration involves bringing together function values in a special way to form a summation. This sum of function values is called an integral and we shall see that there are many circumstances in which we calculate the value of some quantity by the process of integration. At first sight it will appear that integration is just the opposite of differentiation, i.e. in a sense what we appear to be doing is 'undoing' the differential of a function. This process is called anti-differentiation. For example, $\cos(x)$ is the derivative of $\sin(x)$, and $\sin(x)$ is the anti-derivative of $\cos(x)$.

However, integration and antidifferentiation are conceptually different. The fact that the two processes give the same answer is a consequence of an important theorem called *the Fundamental Theorem of Calculus*. In this first chapter we define what we mean by the integral of a function and investigate the results of integrating the basic functions we have met before.

9.1 Scientific context

9.1.1 Example 1: speed–time graphs. The speed of a car is an easy quantity to record from the speedometer. Suppose that on a car journey we recorded the speed of a car every five minutes and produced the following table of values:

time in min (t = 0 is start of journey)	speed in m.p.h.
5	30
10	30
15	40
20	45
25	50
30	50
35	50
40	40
45	35
50	30
55	25
60	0

If we were to assume that during each time interval the speed of the car is constant then the distance travelled during each time interval (in miles) is just the product of the speed times $\frac{1}{12}$ (5 min $= \frac{1}{12}$ hr). Adding all these products would give an estimate of the total distance travelled during the hour. Carrying out this calculation our estimate would be $\frac{425}{12} = 35.4$ mil. Now this estimate is only very approximate since it is based on the assumption that the speed of the car is constant during each time interval. Clearly, during the first interval the car is accelerating from 0 to 30 mil hr^{-1} and during the final interval the car slows down to 0 mil hr^{-1}. Furthermore, during the 15 min that the car appears to be travelling at 50 mil hr^{-1}, it could have stopped at roundabouts or traffic lights.

Question: How could we obtain a better estimate of the distance travelled?
Solution: If we shorten the time interval to one minute say then the speed is likely to vary less over this shorter time interval, so that the estimate of the distance travelled is likely to be more accurate. Shortening the time interval to $\frac{1}{2}$ min should lead to an even better estimate.

We can illustrate this idea more clearly if we draw a graph of the speed of the car against time. Fig. 9.1 shows such a graph for the car journey described above. We see from this graph that the speed does vary more that our simple model suggested.

The estimated distance travelled, 35.4 mil, is the sum of the areas of the rectangles shown in Fig. 9.2. Now if we reduce the time interval to 2 min, say, and hence increase the number of time intervals from 12 to 30, then the estimated distance travelled becomes 36.2 mil. This distance is the shaded area in Fig. 9.3.

Clearly, as we reduce the time interval still further the sum of the areas of the

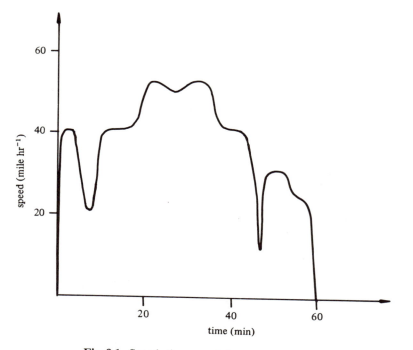

Fig. 9.1. Speed–time graph for a car journey.

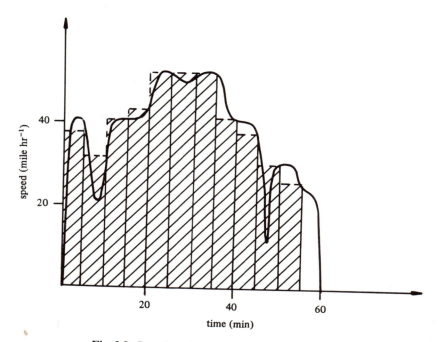

Fig. 9.2. Rough estimate of the distance travelled.

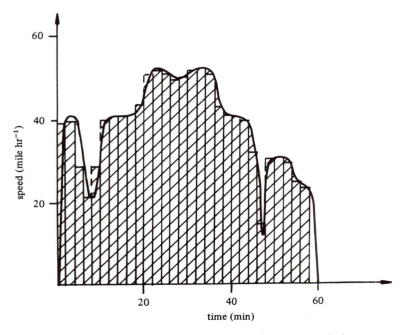

Fig. 9.3. An improved estimate of the distance travelled.

rectangles gradually approaches the value of the area between the curve, the t axis and the times $t = 0$ and $t = 60$. This area is the exact value of the distance travelled by the car during the journey of 1 hr. Symbolically we write this area in the following way: suppose that $F(t)$ is the function defining the speed as a function of time t, and we divide the t interval between $t = 0$ and $t = 60$ into N equal intervals of length dt ($dt = 60/N$); then if we write $F(t_1)$ as the value of F measured at time $t = t_i = i\,dt$ (see Fig. 9.3) then $F(t_i) \times dt$ is the approximate distance travelled during the ith interval. The area beneath the curve can then be *approximated* by the sum

$$(F(t_1) + F(t_2) + \cdots + F(t_n))\,dt = \sum_{i=1}^{n} F(t_i)\,dt$$

The limiting value of this sum as the number of intervals increases (and hence the width of the interval decreases) is clearly the *exact* area under the graph and represents the actual distance travelled during the 1-hr journey.

9.1.2 Example 2: force–distance graphs. In this second example we investigate the work done when a gas expands. As before we shall represent this work by the area under a graph. Fig. 9.4 shows a cylinder containing a gas with a movable piston. When the gas expands the piston moves to the right and the gas does work. Suppose that the pressure in the gas is P_g. If A is the area of the piston then the force exerted by the gas on the piston is $P_g A$. The work done by the gas

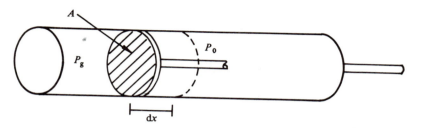

Fig. 9.4. A piston compressing a gas.

in moving the piston through a small distance dx is given by

$$dW = P_g A\, dx$$

Now $A dx$ is equal to the increase in the volume of the gas in the cylinder, so that we can write

$$dW = P_g\, dV$$

As the gas expands in the cylinder its pressure changes and P_g is a function of its volume V. If we are dealing with an ideal gas the pressure and volume are related via the ideal gas law namely $pV = RT$ so that $P_g = RT/V$ where R is the gas constant and T is the temperature. Under isothermal conditions P_g is just a constant divided by V as shown in Fig. 9.5. For each small movement of the piston work is done on it by the gas; the total work done as the gas changes volume from V_1 to V_2 is the summation of all the small elements of work and we have

$$W = \sum_i P_g(V_i)\, dV_i$$

The areas of the rectangles shown are the values of the products $P_g dV$ and the

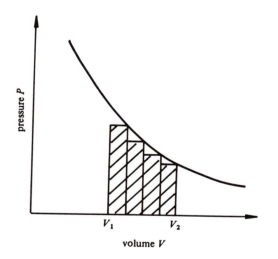

Fig. 9.5. $P = RT/V$: Boyle's law for an ideal gas.

summation representing the total work W is just an approximate value for the *area* under the graph in Fig. 9.5 between V_1 and V_2. As before, if we take smaller and smaller values of dV (and hence more and more rectangles) then the sum of the areas of the rectangles becomes a better approximation to the area under the graph and hence to the total work done by the gas.

Symbolically this limiting process is written as an integral; in the limit of dV tending to zero the work done on the system is written as $W = \int_{V_1}^{V_2} P_g dV$. This limiting process and the new notation will now be considered. We shall see that we can define rules for obtaining integrals without necessarily having to go through this limiting process explicitly – just as we can obtain derivatives without always having to appeal to the original definition of differentiation. However, the underlying interpretation of the integral as an area provides numerical methods for evaluating integrals.

9.2 Mathematical developments

9.2.1 Definition of the integral as a summation. In the last section we introduced the integral as the limit of a summation. This is our starting point for the definition of what we call *the definite integral*. Suppose that $f(x)$ is a non-negative function defined in the interval a to b. (You will soon see the significance of requiring f to be non-negative.) Fig. 9.6 shows a graph of such a function and suppose that the aim is to calculate the area of the shaded region below the graph of f, above the x axis and between the lines $x = a$ and $x = b$. To approximate the area we divide the interval (a, b) into n subintervals, and label them in the following way:

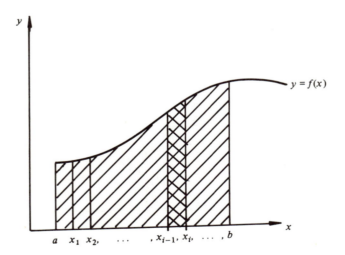

Fig. 9.6. The definite integral of f.

dx_1 is the length of the first subinterval ending at the point x_1,

dx_2 is the length of the second subinterval ending at the point x_2,

and so on.

The sum

$$f(x_1)\,dx_1 + f(x_2)\,dx_2 + \cdots + f(x_n)\,dx_n = \sum_{i=1}^{n} f(x_i)\,dx_i \qquad (1)$$

is then an approximate value of the required area. In principle the subintervals may have different lengths, though we usually choose their lengths to be the same. We define the value of this summation as the number of subintervals n increases indefinitely to be *the definite integral of f from a to b*. This is denoted by

$$\text{Area} = \int_a^b f(x)\,dx. \qquad (2)$$

The function f is called *the integrand* and the endpoints of the interval a and b are called *the limits of integration*; a is the lower limit and b is the upper limit. The limiting process is a complicated one. Clearly, there are many different choices for the subintervals dx_i. However, under suitable conditions it can be shown that the value of the limit of n becoming indefinitely large, and each dx_i therefore becoming vanishingly small, is independent of the method of subdividing the region.

Clearly, the definition of the integral given above is not very convenient for finding its value. We will now show that the integral of a function can be written in terms of the *derivative* of a different function.

Consider the integral of a non-negative function f between a and h then we can define a new function $F(h)$ as the area under the graph of f between a and h (see Fig. 9.7) which is therefore given by

$$F(h) = \int_a^h f(x)\,dx$$

Clearly, $F(b) - F(a)$ is the area under the graph of f between the lines $x = a$ and $x = b$.

Consider the derivative of the function F. From the definition of the derivative given in Chapter 6 we have

$$\frac{dF}{dh} = \lim_{dh \to 0} \frac{F(h + dh) - F(h)}{dh}$$

Now the numerator in the quotient on the right-hand side is the area beneath the curve between h and $h + dh$ of width dh shown in Fig. 9.7. We can approximate this area by the area of the rectangle of width dh and height $f(h)$ so we write

$$F(h + dh) - F(h) \simeq f(h)\,dh$$

and then the expression for the derivative of F becomes

$$\frac{dF}{dh} = \lim_{dh \to 0} f(h)\frac{dh}{dh}$$

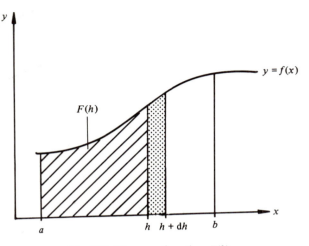

Fig. 9.7. The area function $F(h)$.

In the limit as dh approaches zero we have shown the derivative of F at the point h to be equal to the value of the integrand f evaluated at h. Now h was chosen as any arbitrary point so that instead of h we can adopt our usual notation of writing functions in terms of x and then we have the following method of evaluating a definite integral:

to compute the value of the definite integral of a function f between limits

a and b, i.e. $\displaystyle\int_a^b f(x)\,dx$, we find a new function $F(x)$ whose derivative

is equal to $f(x)$ at each point in (a, b) and then

$$\int_a^b f(x)\,dx = F(x)\Big|_a^b = F(b) - F(a) \tag{3}$$

This is called *the Fundamental Theorem of Calculus* and provides a link between the processes of differentiation and integration. Essentially to evaluate a definite integral we are 'undoing' the process of differentiation. Because of this integration is sometimes called anti-differentiation. The function F is called *a primitive* for the function f.

Before we illustrate the evaluation of definite integrals we return to the interpretation of the integral as an area. In the definition above we restricted the function f to be a non-negative function. Then the integral $\int_a^b f(x)\,dx$ is just the area beneath the graph of f, the x axis and the lines $x = a$ and $x = b$. Now suppose that the function f is non-positive (see Fig. 9.8), then summing the products $f(x_i)\,dx_i$ will give a negative number since $f(x)$ is negative in each subinterval. We can hardly interpret the value of this definite integral as an area because we usually consider areas to be positive. In this case we interpret the integral as minus the area of the shaded region.

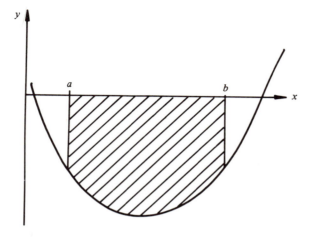

Fig. 9.8. The integral of a non-positive function.

Now consider a function which takes both positive and negative values in the interval (a, b) (see Fig. 9.9). Here the integral of f between a and c is a positive number A_1, say, and the integral between c and b is a negative number $- A_2$, say, where A_2 is the actual area of the shaded region. Here the value of the $\int_a^b f(x)\,dx$ is the area A_1 minus the area A_2 and not simply the sum of the areas. Clearly, we must be careful when interpreting the integral as an area. The Fundamental Theorem can be used to provide the following three properties of integration:

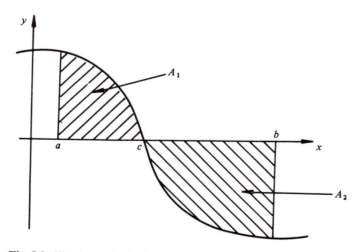

Fig. 9.9. The integral of a function which is positive and negative.

Property 1: Suppose that $F(x)$ is a primitive for the function $f(x)$, then

$$\int_a^c f(x)\,dx + \int_c^b f(x)\,dx = (F(c) - F(a)) + (F(b) - F(c))$$

$$= F(b) - F(a)$$

$$= \int_a^b f(x)\,dx \tag{4}$$

Property 2: Suppose that $F(x)$ is a primitive for the function $f(x)$ and C is a constant, then

$$\int_a^b C f(x)\,dx = CF(x)\Big|_a^b$$

$$= CF(b) - CF(a)$$

$$= C \int_a^b f(x)\,dx \tag{5}$$

Property 3: Suppose that $F(x)$ and $G(x)$ are primitives for the functions $f(x)$ and $g(x)$ respectively and c and e are constants, then $cF(x) + eG(x)$ is a primitive for the linear combination $cf(x) + eg(x)$; so that we can write

$$\int_a^b cf(x) + eg(x)\,dx = cF(x) + eG(x)\Big|_a^b$$

$$= cF(x)|_a^b + eG(x)|_a^b$$

$$= c \int_a^b f(x)\,dx + e \int_a^b g(x)\,dx \tag{6}$$

i.e. the integral of a sum of two functions equals the sum of the integrals of each function.

9.2.2 Evaluating integrals. According to the Fundamental Theorem of Calculus the process of integration is closely allied to the process of differentiation. To evaluate $\int_a^b f(x)\,dx$ we seek a function F such that

$$\frac{dF}{dx} = f$$

The rest of this section is devoted to the art of finding these functions F.

Example: If $f(x) = x$ evaluate $\int_0^2 f(x)\,dx$.
Solution: We must find a function F such that $dF/dx = f = x$. From our experience of differentiation we know that the derivative of $x^2/2$ is equal to x. In this case then, $F(x) = x^2/2$ and then

$$\int_0^2 x\,dx = F(2) - F(0) = 2^2/2 - 0^2/2 = 2$$

Example: Find the value of $\int_0^{\pi/2} \cos(x)\,dx$.
Solution: Here $f(x) = \cos(x)$.

Now the derivative of $\sin(x)$ is $\cos(x)$ so that $F(x) = \sin(x)$ since

$$\frac{dF}{dx} = \frac{d}{dx}(\sin(x)) = \cos(x)$$

Thus

$$\int_0^{\pi/2} \cos(x)\,dx = \sin(\pi/2) - \sin(0) = 1$$

In each of these examples we recalled the derivative of a function that was the same as the integrand in the problem. In each case we simply wrote down the function $F(x)$ and evaluated the difference between its value at each of the end points. However, the primitive of a function is not unique.

Consider the two functions

$$F(x) = \sin(x) \quad \text{and} \quad G(x) = \sin(x) + c$$

where c is a constant. Clearly, since the derivative of a constant is zero, we have

$$\frac{dF}{dx} = \cos(x) \quad \text{and} \quad \frac{dG}{dx} = \cos(x)$$

Clearly both the functions F and G can be used as primitives for f in the second example. Fortunately, however, this non-uniqueness in the primitive does not affect the value of the definite integral. For example, consider $\int_0^{\pi/2} \cos(x)\,dx$, then writing the primitive as $F(x) = \sin(x) + c$, we have

$$\int_0^{\pi/2} \cos(x)\,dx = (\sin(x) + c)\Big|_0^{\pi/2}$$

$$= (\sin(\pi/2) + c) - (\sin(0) + c)$$

$$= 1 \quad \text{as before}$$

To evaluate the integral of a function f the Fundamental Theorem tells us that we must first seek a function F such that $f = dF/dx$. We must do this irrespective of the limits of integration. Because of this we often write an integral without any limits to define a new function. For example, we may write

$$F(x) = \int \cos(x)\,dx = \sin(x) + c$$

where c is called the *constant of integration*. We call the function F *an indefinite integral*.

We now list a table of indefinite standard integrals:

Function $f(x)$	(Indefinite) integral $F(x)$
x^n $(n \neq 1)$	$x^{n+1}/(n+1) + c$
$1/x$	$\ln(x) + c$
e^{ax}	$e^{ax}/a + c$
$\sin(ax)$	$-(\cos(ax))/a + c$
$\cos(ax)$	$(\sin(ax))/a + c$

Note: In each case c is a constant.

Before proceeding to investigate the different approaches to evaluating an integral it should be noted that the value of the definite integral $\int_a^b f(x)\,dx$ depends on three things: the integrand f, the lower limit a and the upper limit b. The letter used to denote the independent variable, i.e. the 'x' in the integral, has no effect on the value of the integral. All of the following forms lead to exactly the same value of the integral:

$$\int_a^b f(x)\,dx = \int_a^b f(u)\,du = \int_a^b f(t)\,dt$$

For indefinite integrals, the table of standard integrals can be used equally well for symbols other than x. For example,

$$\int \cos(t)\,dt = \sin(t) + c$$

$$\int \cos(u)\,du = \sin(u) + c$$

where c is a constant of integration. The variable in the integrand is really a 'dummy' variable.

Example: Evaluate the following

(i) $\displaystyle\int_1^2 3x^6\,dx$;

(ii) $\displaystyle\int_0^2 2e^{3u}\,du$;

(iii) $\displaystyle\int 4\sin(3x) + 2x + \frac{1}{x}\,dx$;

(iv) $\displaystyle\int 3t^3 + 2\cos(5t)\,dt$.

Solution: In each case the integrand is a combination of basic functions so that we use the table of standard integrals to evaluate each part.

(i) $\displaystyle\int_1^2 3x^6\,dx = \frac{3x^7}{7}\bigg|_1^2 = \frac{3.2^7}{7} - \frac{3.1^7}{7} = \frac{381}{7}$;

(ii) $\displaystyle\int_0^2 2e^{3u}\,du = \frac{2e^{3u}}{3}\bigg|_0^2 = \frac{1e^6}{3} - \frac{2e^0}{3} = \tfrac{2}{3}(e^6 - 1)$;

(iii) $\displaystyle\int 4\sin(3x) + 2x + \frac{1}{x}\,dx = \tfrac{-4}{3}\cos(3x) + x^2 + \ln(x) + c$

(iv) $\displaystyle\int 3t^3 + 2\cos(5t)\,dt = \tfrac{3}{4}t^4 + \tfrac{2}{5}\sin(5t) + c$.

9.3 Worked examples

Example 1 (speed–time graphs). An object thrown vertically downwards from the top of a tall building falls with a speed given by the function $f = u_0 + gt$ where

u_0 is the initial speed, g is the acceleration due to gravity and t is the time. Find the distance travelled by the object in the first 6 s

(a) using a graphical method;

and

(b) using integration.

Solution. (a) Drawing a graph of the function representing the speed we obtain the distance travelled in the first 6 s as the area shown shaded in Fig. 9.10. We have

$$\text{distance} = \frac{(AD + BC)}{2} \times AB$$

$$= ((u_0) + (u_0 + 6g)) \times 6/2$$

$$= 6u_0 + 1\cancel{8}g$$

(b) As an integral the distance travelled is

$$\text{distance} = \int_0^6 f(t)\,dt = \int_0^6 (u_0 + gt)\,dt$$

$$= u_0 t + gt^2/2 \Big|_0^6$$

$$= 6u_0 + 1\cancel{8}g$$

As expected these are the same formula, illustrating that the definite integral may be interpreted geometrically as the area under the graph.

Example 2 (force–distance graphs). Find the work done by the gas in 9.1.2 when the volume of gas increases by 50%.

Solution. In 9.1.2 we derived a formula for the work done as the integral $p(V)\,dV$ where $p(V)$ is the pressure in the gas. For an isothermal expansion of a perfect

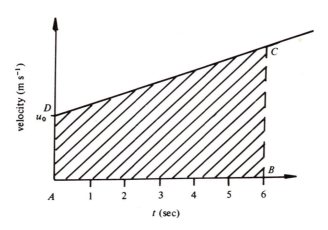

Fig. 9.10. Distance travelled in 6 seconds.

gas we have

$$p(V) = RT/V = k/V \quad \text{where } k \text{ is a constant}$$

so that

$$\text{work done} = \int_{V_1}^{V_2} k/V \, dV$$

Now if the volume increases by 50% then $V_2 = 1.5\,V_1$, and

$$\text{work done} = \int_{V_1}^{V_2} k/V \, dV$$

$$= k \ln(V) \Big|_{V_1}^{1.5V_1}$$

$$= k(\ln(1.5\,V_1) - \ln(V_1))$$

$$= k \ln(1.5)$$

Note that $\int 1/V \, dV$ is one of the standard forms:

$$\int 1/x \, dx = \ln(x) + c$$

Example 3. In the following example we show that it is often easier to find the limits of integration from the geometry of the problem.

    A straight wire of length $2L$ carries an electric current of i amps. At points near the wire a significant measurable magnetic field exists. Consider a point P positioned a distance h m from the wire as shown in Fig. 9.11. The contribution to the magnetic field at P due to the element of the wire dl is given by

$$dH = \frac{i \sin(\theta) \, dl}{4\pi r^2}$$

Adding all the contributions over the length of the wire, an expression for the

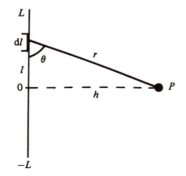

Fig. 9.11. The magnetic field near a current-carrying wire.

magnetic field at P is given by the integral

$$H = \int_{-L}^{L} \frac{i \sin(\theta) \, dl}{4\pi r^2}$$

Evaluate this integral.

Solution. Now r, θ and l are related so we must write the integrand in terms of a single variable. We have

$$l = \frac{h}{\tan \theta} \quad \text{and} \quad r = \frac{h}{\sin \theta}$$

then

$$\frac{dl}{d\theta} = \frac{-h \sec^2 \theta}{\tan^2 \theta}$$

Substituting for dl and r^2 in terms of θ gives

$$H = \int_{-L}^{L} \frac{i \sin \theta \, dl}{4\pi r^2} = \int_{\theta_1}^{\theta_2} \frac{i \sin \theta \sin^2 \theta}{4\pi h^2} \left(-\frac{h \sec^2 \theta}{\tan^2 \theta} \right) d\theta$$

$$= \int_{\theta_1}^{\theta_2} \left(-\frac{i}{4\pi h} \right) \sin \theta \, d\theta$$

$$= \frac{i}{4\pi h} \cos \theta \Big|_{\theta_1}^{\theta_2}$$

$$= \frac{i}{4\pi h} (\cos \theta_2 - \cos \theta_1)$$

where θ_1 and θ_2, the limits of the integration in θ, are the limiting values of θ at

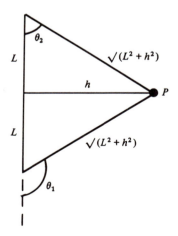

Fig. 9.12. Finding the limits of θ.

each end of the wire. Fig. 9.12 shows these values. Hence

$$\cos\theta_2 = \frac{L}{\sqrt{(L^2 + h^2)}} \quad \text{and} \quad \cos(\pi - \theta_1) = -\cos\theta_1 = \frac{L}{\sqrt{(L^2 + h^2)}}$$

and on substitution

$$H = \frac{iL}{2\pi h \sqrt{(L^2 + h^2)}}$$

9.4 Further developments

9.4.1 Numerical integration: Euler's method. The direct methods of integration described in 9.2 provide exact, usually called *analytical*, solutions of integrals. Further techniques of integration for supplying analytical results are discussed in the next two chapters. Often, however, in science we need to evaluate the integral of a function which is given in tabular form and we do not know the formula defining the function. For example, the data may have been obtained experimentally. We need some method of integrating just using these numerical values. Furthermore, it is often necessary to evaluate the definite integral $\int_a^b f(x)\,dx$ for a known function f for which the analytical methods of integration do not work. For example, $f(x) = e^{x^2}$ is a function that we cannot integrate analytically in terms of any of the functions we have defined so far. In such cases we can tabulate the function f over the range of the limits and again we then need a method of integration which uses these tabulated values.

The process of evaluating a definite integral from data given by a set of tabulated values is called *numerical integration* or *integration by quadrature*. Essentially, the method involves dividing the range of the limits (a, b) into subintervals, usually of the same width, and approximating the function within each interval by a polynomial function $p(x)$ which can then be integrated. This provides a formula for approximating the integral. For example, suppose we wish to evaluate the definite integral $\int_a^b f(x)\,dx$. We can divide the interval into N equal subintervals of width $h = (b - a)/N$ so that the subintervals are $(x_0, x_1), (x_1, x_2), \ldots$, as shown in Fig. 9.13. If we consider an arbitrary subinterval (x_i, x_{i+1}), we can approximate the function in this interval by its value at the beginning of the subinterval, i.e. let $p(x) = f(x_i)$. Then for this subinterval

$$\int_{x_i}^{x_{i+1}} f(x)\,dx \simeq \int_{x_i}^{x_{i+1}} p(x)\,dx = f(x_i)(x_{i+1} - x_i) = f(x_i)h$$

Adding the approximations for each of the subintervals we get

$$\int_a^b f(x)\,dx \simeq f(x_0)h + f(x_1)h + \cdots + f(x_i)h + \cdots + f(x_{n-1})h$$

$$= h(f(x_0) + f(x_1) + \cdots + f(x_i) + \cdots + f(x_{n-1})) \tag{7}$$

This method is called *Euler's method*, and is closely related to Euler's method for

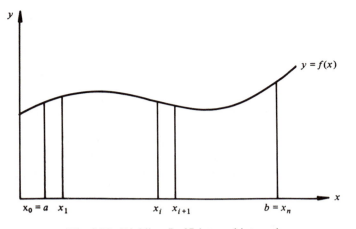

Fig. 9.13. Dividing $[a, b]$ into subintervals.

solving differential equations which we use in Chapter 12. Using the geometrical interpretation of the definite integral as the area under a curve, Fig. 9.14 shows a geometrical interpretation of this numerical method. Clearly, the method provides a very crude approximation to the value of an integral and this method is rarely used; however it does provide a simple illustration of the method of approach. In the next two subsections we introduce methods that provide better approximations where the approximating polynomials are linear and quadratic functions.

Example: Use Euler's method to approximate the integral $\int_0^1 e^x \, dx$ with $h = 0.2$.
Solution: Euler's method gives

$$\int_0^1 e^x \, dx = 0.2(e^0 + e^{0.2} + e^{0.4} + e^{0.6} + e^{0.8})$$

$$= 1.5522 \text{ to four decimal places}$$

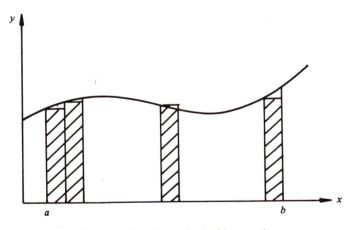

Fig. 9.14. Euler's method of integration.

We can compare this with the exact value because we can integrate e^x easily since it is one of our standard forms. We have

$$\int_0^1 e^x \, dx = e^x \Big|_0^1 = e^1 - e^0$$

$$= 1.7183 \text{ to four decimal places}$$

Thus we have an error of roughly 11% in the numerical solution.

9.4.2 The trapezoidal method. Suppose that we approximate the function in each subinterval by a linear polynomial $p(x) = cx + d$ such that $p(x) = f(x)$ at the beginning and end of each subinterval. Then for subinterval (x_i, x_{i+1}) we have

$$f(x_i) = cx_i + d$$

and

$$f(x_{i+1}) = cx_{i+1} + d$$

so that

$$c = (f(x_{i+1}) - f(x_1))/h \quad \text{and} \quad d = f(x_i) - cx_i$$

Integrating $p(x)$ and simplifying we obtain the formula

$$\int_{x_i}^{x_{i+1}} p(x) \, dx = h(f(x_{i+1}) + f(x_i))/2$$

If we approximate the integral in each of the subintervals and sum over the range (a, b) we obtain

$$\int_a^b f(x) \, dx \simeq h(f(x_0) + 2f(x_1) + \cdots + 2f(x_i) + \cdots + f(x_n))/2 \tag{8}$$

Geometrically this process approximates the area under the curve representing $f(x)$ by a set of trapeziums as shown in Fig. 9.15, and the method is called *the trapezoidal method.*

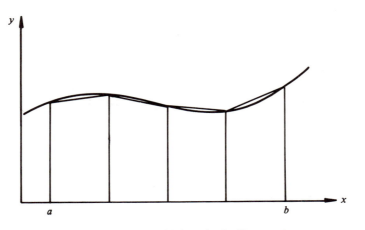

Fig. 9.15. The trapezoidal method of integration.

Example: Use the trapezoidal method to evaluate $\int_0^1 e^x \, dx$ with $h = 0.2$.

Solution: The formula (8) gives

$$\int_0^1 e^x \, dx = 0.2(e^0 + 2e^{0.2} + 2e^{0.4} + 2e^{0.6} + 2e^{0.8} + e^1)$$

$$= 1.7240 \text{ to four decimal places}$$

approximation to the integral. The analytical solution to four decimal places is 1.7183 so that the error is now 0.33%. The numerical solution is correct to three significant figures.

The approximation can be improved by taking smaller step lengths and hence more intervals. The trapezoidal rule is easy to apply and is more accurate than Euler's method.

9.4.3 Simpson's method. The third method of numerical integration is derived by using a quadratic polynomial to approximate the function. We divide the interval of integration into an *even* number of subintervals and find a quadratic polynomial $p(x)$ such that

$$p(x) = f(x) \text{ at the points } x_{i-1}, x_i \quad \text{and} \quad x_{i+1}$$

This is shown in Fig. 9.16. If we let $p(x) = ax^2 + bx + c$ then if

$$p(x_{i-1}) = f(x_{i-1})$$
$$p(x_i) = f(x_i)$$

and

$$p(x_{i+1}) = f(x_{i+1})$$

we can express a, b and c in terms of the known values $f(x_{i-1}), f(x_i)$ and $f(x_{i+1})$. Now we integrate $p(x)$ over the subinterval (x_{i-1}, x_{i+1}) and simplify the result.

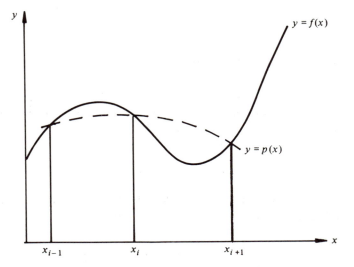

Fig. 9.16. Simpson's method: a quadratic polynomial approximation.

The algebra involved in this operation is somewhat tedious so we state the result.

$$\int_{x_i}^{x_{i+1}} f(x)\,dx = h(f(x_{i-1}) + 4f(x_i) + f(x_{i+1}))/3$$

Adding the contributions for each pair of subintervals gives

$$\int_a^b f(x)\,dx = h[f(x_0) + f(x_n) + 2(f(x_2) + f(x_4) + \cdots + f(x_{n-2}))$$
$$+ 4(f(x_1) + f(x_3) + \cdots + f(x_{n-1})]/3: \tag{9}$$

This simple formula is known as *Simpson's method*. This is a reasonably accurate and popular method of numerical integration.

Example: Use Simpson's method for evaluating the integral $\int_0^1 e^x\,dx$ with $h = 0.5$.
Solution: In this example we cannot use $h = 0.2$ as before because we must have an even number of subintervals. With $h = 0.2$ we have five subintervals. Using formula (9) with $h = 0.5$ gives

$$\int_0^1 e^x\,dx = 0.5(e^0 + 4e^{0.5} + e^1)/3$$
$$= 1.7189 \text{ to four decimal places}$$

Table 9.1. *The integral $\int_0^1 e^x\,dx$ using the three numerical methods introduced in this section given to four decimal places. Note how many steps are required for Euler's method*

Methods (step)	Number of evaluations of e^x	Answer
Euler (0.001)	1000	1.7174
trapezoidal (0.2)	5	1.7240
Simpson (0.5)	3	1.7189
analytical		1.7183

The Table 9.1 compares the number of steps required for two decimal-place accuracy for each numerical method. Although somewhat tedious to employ by hand these numerical methods can be incorporated into computer programs giving us efficient ways of evaluating integrals numerically. Several calculators now have the facility for evaluating integrals numerically at the press of a key.

9.5 Further worked examples

Example 1. During the launch of a rocket the speed was noted every second for 10 seconds and the following table of values obtained:

time (s)	0	1	2	3	4	5	6	7	8	9	10
speed (km hr^{-1})	0	32	80	128	176	224	272	320	368	400	448

Use (a) the Trapezoidal and (b) Simpson's methods to estimate the distance travelled by the rocket during the first 10 seconds.

Solution.

(a) Trapezoidal method:

From the formula for the trapezoidal method with $h = \frac{1}{3600}$ (i.e. 1 s in hr) we have

distance $= (1/3600) \times (0 + 448 + 2(32 + 80 + 128 + 176 + 224 + 272 + 320$
$+ 368 + 400))/2$
$= 0.618\,\text{km}$

(b) Simpson's method:

Since the number of intervals is even we can use Simpson's method with 1-s ($\frac{1}{3600}$-hr) subintervals. We have

distance $= (1/3600) \times (0 + 448 + 4(32 + 128 + 224 + 320 + 400)$
$+ 2(80 + 176 + 272 + 368))/3$
$= 0.616\,\text{km}$

Example 2. Write a program that will evaluate $\int_a^b f(x)\,dx$, for a given functional form $f(x)$, using Simpson's method. The user of the program should be able to choose the number of intervals that the program uses.

Solution. The following program assumes that the user will feed in an even number for n, the number of intervals. In the program the sums $f(x_2) + f(x_4) + \cdots + f(x_{n-2})$ and $f(x_1) + f(x_3) + \cdots + f(x_{n-1})$ are worked out separately and called sumofevens and sumofodds.

```
begin
input a, b, n
set h to (b − a)/n
set sumofodds to 0
set x to a + h
for i = 1 to n − 1 step 2 do
    set sumofodds to sumofodds + f(x)
    set x to x + h
endfor
set sumofevens to 0
set x to a + 2h
for i = 2 to n − 2 step 2 do
    set sumofevens to sumofevens + f(x)
    set x to x + h
endfor
set integral to h*(f(a) + f(b) + 2*sumofevens + 4*sumofodds)/3
print integeral
stop
```

9.6 Exercises

1. Integrate the following functions of x:

    (i) x^3 (ii) $4x^9$ (iii) $1/x^3$

    (iv) $16 + x^2$ (v) $\dfrac{x^2 + 2}{x}$ (vi) $\sin(3x)$

    (vii) $\cos(x/2)$ (viii) $\sec^2(x)$ (ix) $[\cos(x) - \sin(x)]/2$

    (x) e^{3x} (xi) $e^{(x-1)}$ (xii) $3x^{5/2} + 4x^2 + e^{2x}$

2. Integrate the following functions of x:

    (i) $\cos^2(x)$ (ii) $\sin^2(x)$

    (iii) $\cos^4(x)$ (iv) $\sin^4(x)$

    You may assume $\cos(2\theta) = 2\cos^2(\theta) - 1 = 1 - 2\sin^2(\theta)$

3. Evaluate the following definite integrals:

    (i) $\displaystyle\int_0^3 2x^2 + 4x - 1 \, dx$ (ii) $\displaystyle\int_{-1}^2 x^5 - x^2 \, dx$

    (iii) $\displaystyle\int_0^1 x(x^3 - 5x + 1) \, dx$ (iv) $\displaystyle\int_1^2 \dfrac{x^2 + 2x}{x} \, dx$

    (v) $\displaystyle\int_{\pi/4}^{\pi/2} 2\sin(\theta) + 3\cos(\theta) \, d\theta$ (vi) $\displaystyle\int_0^2 3e^{4x} - x^3 + 1 \, dx$

    (vii) $\displaystyle\int_0^{\pi/2} \sec^2(\theta/2) \, d\theta$ (viii) $\displaystyle\int_1^3 1/x \, dx$

    (ix) $\displaystyle\int_2^4 \dfrac{x^3 + 4x - 1}{x} \, dx$

4. Find the area enclosed between the following curves:

    (i) $y = \cos(x)$, $x = -\pi/2$, $x = \pi/2$, $y = 0$

    (ii) $y = 1/x$, $\quad x = 1$, $\quad x = 2$, $\quad y = 0$

    (iii) $y = x$, $\qquad y^2 = 9x$

    (iv) $y = -x$, $\qquad y = 2x + x^2$

    (v) $y^3 = 2x$, $\quad x = -4$, $\quad y = 2$

5. In a Klystron, the electric field strength in the space d between the two buncher grids is

$$E = \frac{V_1 \sin(wt)}{d}$$

where V_1 is the voltage between the grids. The acceleration of an electron between the bunchers at any time t is

$$\frac{dv}{dt} = eE/m$$

Find:

(i) the velocity of an electron at some time T if initially its velocity is v_0,

and

(ii) the distance travelled by the electron in the time interval T.

6. In the theory of an electronic flux meter, the following integral occurs

$$I = R \int_{t_1}^{t_2} i \, dt$$

If i is related to the charge by $i = dq/dt$, show that

$$I = R[q(t_2) - q(t_1)]$$

7. The current in an electric circuit $i(t)$ is given as a function of time by

$$i = 10 + 50 \sin(50 \pi t)$$

The root mean square (rms) value of a function y between $x = a$ and $x = b$ is defined by the integral $[\int_a^b y^2 \, dx/(b - a)]^{1/2}$

Find the rms value of the current over a time interval between $t = 0$ and $t = 0.04$

8. In an electric circuit containing a resistor R and an inductance L, the voltage V is related to the current I by

$$V = L\frac{dI}{dt} + RI$$

If $I = A \sin(wt)$ find the rms value and mean value of the product IV between $t = 0$ and $t = 2\pi/w$. (The mean value of a continuous function is $\int_a^b y \, dx/[b - a]$.)

9. A current $i(t)$ can be written as a sum of sine functions

$$i(t) = i_1 \sin(wt) + i_2 \sin(2wt) + i_3 \sin(3wt) + \cdots$$

Show that the mean value of $i(t)$ between 0 and $2\pi/w$ is zero.
Show that the rms value of $i(t)$ between 0 and $2\pi/w$ is

$$[(i_1^2 + i_2^2 + i_3^2 + \cdots)/2]^{1/2}$$

10. In an undamped resonant circuit the electric charge $q(t)$ is given by

$$q(t) = A \sin(wt + \phi) + \frac{I_0}{2Lw} t \sin(wt)$$

Find the mean value and rms value of the charge $q(t)$ between 0 and $2\pi/w$.

11. The thermodynamic equation of state of a certain substance takes the form

$$pV = RT\left(1 + \frac{a}{V} + \frac{b}{V^2}\right)$$

where R, a and b are constants.
Obtain, as an integral, the expression representing the work done by the gas in expanding from a volume V_1 to a volume V_2.
Evaluate this expression for an isothermal change (i.e. one in which the temperature T remains constant).

Show that if $V_2 = \alpha V_1$, and a and b are zero, the work done is independent of V_1 and V_2.

12. The shape of a thin flat piece of uniform material of length 2 m and thickness 1 cm, is that of a lamina bounded by the x-axis and the curve $y = x^2(2 - x)$ for $0 \leqslant x \leqslant 2$:

   (i) If the mass of the piece is 4 kg, obtain the density of the material from which it is made.

   (ii) Calculate the moment of inertia of the flat piece of material about the y-axis.

13. The mean life-time τ of an unstable particle is given by the expression

$$\tau = \lambda \int_0^T t e^{-\lambda t} \, dt$$

where λ and T are both positive constants:

   (i) Show that the derivative with respect to t of the expression

$$F(t) = -e^{-\lambda t}/\lambda - t e^{-\lambda t}$$

   is $\lambda t e^{-\lambda t}$

   (ii) Hence obtain the value of τ in terms of λ and T.

   (iii) If T is assumed to be very large, i.e. approaching infinity, show that τ is equal to $1/\lambda$.

14. In an adiabatic expansion of a gas

$$c_p \frac{dV}{V} + c_v \frac{dp}{p} = 0$$

where c_p and c_v are the two constant principal heat capacities. By dividing this equation by c_v, and subsequently integrating, show that

$$pV^\gamma = \text{constant}$$

where $\gamma = c_p/c_v$.

15. The Clausius–Clapeyron equation for the change of state from liquid to vapour is

$$\frac{1}{p} \frac{dp}{dT} = \frac{\Delta H^\theta}{RT^2}$$

where ΔH^θ is the mean molar enthalpy of vaporization. By integrating both sides with respect to T show that if p_2 and p_1 are vapour pressures at temperature T_2 and T_1, respectively, then

$$\ln(p_2/p_1) = \frac{\Delta H^\theta}{R} \frac{(T_2 - T_1)}{T_1 T_2}$$

16. Using the trapezoidal method calculate an approximate value for the integral $\int_0^{\pi/2} \sin(x) \, dx$ using a step size of $\pi/6$. Compare your answer with the true solution.

17. A function f is given in tabular form as

x	$f(x)$
1.8	6.050
2.0	7.389
2.2	9.025
2.4	11.023
2.6	13.464
2.8	16.445
3.0	20.086
3.2	24.533
3.4	29.964

Use (i) the trapezoidal method,
and
    (ii) Simpson's method,
to approximate the integral $\int_{1.8}^{3.4} f(x)\,dx$ with $h = 0.2$, 0.4 and 0.8.

18. Verify that Simpson's method is exact for a cubic polynomial
$$f(x) = a_0 + a_1 x + a_2 x^2 + a_3 x^3$$

19. The voltage in an electric circuit is recorded at regular intervals of $0.25\,s$ over a half-cycle of $2\,s$. The following list shows the results obtained:

    0,   12.07,   23.82,   35.04,   44.91,   52.17,   41.72,   19.62,   0

Use Simpson's rule to find the rms value of the voltage over the two second half-cycle.

20. An important integral that often occurs in science is
$$\int \sin(x)/x \, dx.$$

This integral has no analytical solution in terms of our basic functions. Evaluate $\int_{0.2}^{1} \sin(x)/x\,dx$ by the trapezoidal method using (i) two, (ii) four and (iii) eight subintervals.
Compare your answers.

21. Find an approximate value correct to three decimal places for the integral
$$I = \int_0^{\pi/2} \frac{x \sin(x)}{1 + \cos^2(x)} \, dx$$
using Simpson's method.

22. An important probability density function that occurs in statistics gives rise to the integral
$$I(x) = \int_0^x \exp(-t^2) \, dt$$

Using Simpson's method, find values of $I(0.1)$, $I(0.2), \ldots, I(1.0)$ correct to four decimal places.

23. The rate of flow of water into a reservoir, via a particular 'catch dyke' is monitored every hour. The readings for a particular 24-hr period are as follows (in units of 1000 gallons h^{-1}):

    time: 1.0 2.0 3.0 4.0 5.0 6.0 7.0 8.0 9.0 10.0

    flow: 213 219 227 247 266 278 291 293 295 288

    time: 11.0 12.0 13.0 14.0 15.0 16.0 17.0 18.0 19.0 20.0

    flow: 280 280 279 281 275 270 265 263 262 258

    time: 21.0 22.0 23.0 24.0 1.00

    flow: 255 253 250 250 245

    (i) Using Simpson's rule, with an interval of 1 hr, estimate the total volume of water flowing into the reservoir, via the catch dyke, in the 24-hr period 01.00–01.00.
    (ii) What difference is there in the calculation if only those readings collected every 2 hr are used to estimate this volume?

24. In an experiment it is necessary to estimate accurately the mean temperature $\bar{\theta}$ of a body over a period of time T. The quantity $\bar{\theta}$ is defined by the formula

$$\bar{\theta} = \frac{1}{T} \int_0^T \theta \, dt$$

    where θ is the temperature of the body at time t. Over a 5-min period the following results for θ, obtained every 30 s, were observed:

    time (min): 0.0 0.5 1.0 1.5 2.0 2.5 3.0 3.5 4.0 4.5 5.0

    $\theta(°C)$: 79 71 64 58 52 47 42 38 34 31 28

    Obtain the value of $\bar{\theta}$ over this 5-min period using
    (i) Simpson's rule;
    (ii) the trapezoidal rule.
    How do these results compare with simple average of the 11 temperature values?

25. The average level C ($\mu g\,g^{-1}$) of a certain pesticide, in the livers of herbivores, T yr after the pesticide was introduced into the ecosystem, is modelled by the expression

$$C = \frac{1}{e^{0.25T}} \int_0^T 0.32 e^{-0.64t} \, dt$$

    Obtain an explicit expression for C in terms of T, and show that C will reach its maximum value in a time T that is very nearly 2 yr after the pesticide was first introduced into the ecosystem.

10

Integration 2: techniques of integration

> **Contents:** 10.1 scientific context – 10.1.1 the centre of mass of a lamina; 10.1.2 chemical reactions; 10.2 mathematical developments – 10.2.1 direct substitutions; 10.2.2 indirect substitutions; 10.2.3 the substitution $t = \tan(x/2)$; 10.3 worked examples; 10.4 further developments – 10.4.1 use of partial fractions; 10.4.2 integration by parts; 10.4.3 reduction formulae; 10.5 further worked examples; 10.6 exercises.

In Chapter 9 we defined what we meant by a definite and an indefinite integral and used the Fundamental Theorem of Calculus to obtain the integrals of basic functions. We also introduced numerical methods of integration because there are many functions for which analytic integrals cannot be found.

In this chapter we concentrate on techniques of integration that can be used to reduce complicated looking integrands to a basic form that can be evaluated analytically. We should stress at the outset that there are no standard rules of integration that will *always* work, unlike the standard rules for differentiation. There is a certain amount of guesswork and experience involved with the evaluation of integrals. In 10.2 we introduce various substitutions that often help and in 10.4 we consider a method for tackling integrals of products of functions.

10.1 Scientific context

10.1.1 Example 1: the centre of mass of a lamina. One of the important properties of a body is the position of its centre of mass; this is the point at which all the mass can be thought to act when the body is moving in a straight line. For such translational motion the body can be modelled as a particle with its mass concentrated at this point. Finding the centre of mass is thus important and makes use of the technique of integration. Without any loss of generality let us consider how integration is used in finding the position of the centre of mass of a uniform lamina; this is often called the *centroid*.

Consider a plane lamina of uniform thickness t and density ρ, bounded by the two functions $y = f(x)$ and $y = g(x)$ as shown in Fig. 10.1. We can subdivide the lamina into rectangular elementary strips so that the ith element is at $x = x_i$ and has area $dA = (f(x_i) - g(x_i)) \, dx$. The mass of this elementary strip is then $dM = \rho t (f(x_i) - g(x_i)) \, dx$. The total mass of the lamina is the sum of the masses of the

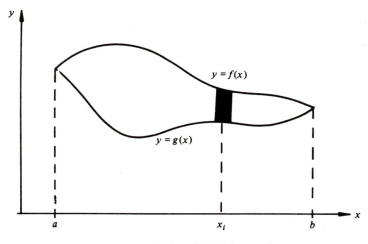

Fig. 10.1. A lamina divided into strips.

elementary strips,

$$M = \sum_{i=1}^{n} \rho t(f(x_i) - g(x_i)) \, dx$$

This is an approximation to the mass; as we take thinner strips (and more of them) then using the definition of the definite integral this summation becomes an integral equal to the actual mass of the lamina. We have

$$M = \int_{a}^{b} t\rho(f(x) - g(x)) \, dx$$

And since t and ρ are constants

$$M = t\rho \int_{a}^{b} f(x) - g(x) \, dx$$

Suppose that the coordinates of the centre of mass of this lamina are (X, Y). The centre of mass of the elementary strip is at the centre of the strip; i.e. at the point (x_i, y_i) where

$$y_i = (f(x_i) + g(x_i))/2$$

Taking moments about the x and y axes, the moment of the whole mass equals the sum of the moments of the elementary strips; thus

$$MX = \sum_{i=1}^{n} x_i \, dM_i$$

$$= \sum_{i=1}^{n} x_i t\rho(f(x_i) - g(x_i)) \, dx$$

$$MY = \sum_{i=1}^{n} y_i \, dM_i$$

$$= \sum_{i=1}^{n} y_i t\rho(f(x_i) - g(x_i)) \, dx$$

In the limit as the number of strips increases the summations become integrals; we obtain

$$MX = \int_a^b t\rho x(f(x) - g(x))\,dx$$

$$MY = \int_a^b t\rho(f(x) + g(x))(f(x) - g(x))/2\,dx$$

Substituting for M and dividing by t and ρ, the formulas for X and Y are

$$X = \frac{\int_a^b x(f(x) - g(x))\,dx}{\int_a^b (f(x) - g(x))\,dx}, \quad Y = (1/2)\frac{\int_a^b (f(x)^2 - g(x)^2)\,dx}{\int_a^b (f(x) - g(x))\,dx}$$

To find the position of the centre of mass from these formulae requires evaluating three integrals. The ease of these evaluations depends on the form of the functions $f(x)$ and $g(x)$. For example, if the lamina is bounded by the parabola $y = x^2$ and the lines $y = 0$, $x = 2$, then $f(x) = x^2$ and $g(x) = 0$, so that the integrals become

$$\int_0^2 x^2\,dx, \quad \int_0^2 x^3\,dx \quad \text{and} \quad \int_0^2 x^4\,dx$$

These integrals are easily evaluated using the basic formula for $\int x^n\,dx$. However, if the upper boundary is a portion of a circle with equation $y = \sqrt{(a^2 - x^2)}$, the integrals $\int x\sqrt{(a^2 - x^2)}\,dx$ and $\int \sqrt{(a^2 - x^2)}\,dx$ are not so straightforward. We shall see in Section 10.2 how the use of a substitution can change these integrals into standard forms.

However, the method of substitution does not work if the function f is given by $f(x) = e^x$. In Section 10.4 we introduce a method of integrating functions of the form $h(x)k(x)$ where $h(x)$ is a polynomial and $k(x)$ is a function for which a primitive can be found. Clearly, $\int xe^x\,dx$ is an integral of this type. The method of approach is called integration by parts.

10.1.2 Chemical reactions. Many integrals that arise in science can only be evaluated using partial fractions. We met partial fractions in Chapter 5 and to appreciate the need for partial fractions consider a chemical reaction involving two substances. In Chapter 12 we introduce a particular type of equation called *a differential equation*. In 12.1.1 we derive the equation that describes the chemical reaction of two substances. The equation is

$$\frac{dN}{dt} = k(a - N)(b - N)$$

where k, a and b are constants, N is the concentration of one of the substances at time t and dN/dt is the rate of change of this concentration. To find how long it takes for the concentration to reach a certain value, N_0 say, we need to evaluate the integral

$$T = \int_0^{N_0} \frac{1}{k(a - N)(b - N)}\,dN$$

Now this integral is not one of our standard forms. However if the right-hand side could be expressed as the sum of the two integrals, $\int 1/(a-N)\,\mathrm{d}N$ and $\int 1/(b-N)\,\mathrm{d}N$ the evaluation would be straightforward because

$$\int 1/(a+N)\,\mathrm{d}N = \ln(a+N) + C$$

The method of partial fractions allows the given integral for T to be written in terms of these two integrals, so this provides a method of evaluating integrals of the form $\int f(x)/g(x)\,\mathrm{d}x$ where $f(x)$ and $g(x)$ are polynomial functions. This and the integration of other products of functions is discussed in Section 10.4.

10.2 Mathematical developments

Integration by substitution. The need to find anti-derivatives is important in the process of integration. In this and the next chapter we shall develop certain techniques of integration. However, there are no general rules to suggest which particular technique can be used to find the primitive of a function. Furthermore, there are few set rules for evaluating integrals as there are for finding derivatives. You will find that methods of trial and error, or intelligent guesses based on a knowledge of differentiation, are very important in learning the techniques of integration. A table of standard derivatives and integrals, as introduced in Chapter 9, provides a valuable source of basic primitives. One technique that is important is performing a change of variable that will change a non-standard form into one that appears in the table of standard integrals. We now illustrate this technique.

10.2.1 Direct substitutions. The method of substitution depends on the chain rule for differentiating a composite function which we considered in Section 6.4.1.

Consider the function $f(x) = g(h(x))$, then the chain rule is the following formula:

$$\frac{\mathrm{d}f}{\mathrm{d}x} = \frac{\mathrm{d}g}{\mathrm{d}h}\frac{\mathrm{d}h}{\mathrm{d}x}$$

Many integrands have the form of the right-hand side in this formula and the method is to find a suitable substitution that reduces the integrand to a more familiar form that enables us to spot that the primitive we are looking for is $f(x)$. The following simple example illustrates the method:

Example: Evaluate $\int (x+7)^3\,\mathrm{d}x$.

Solution: Clearly, if the problem was that of obtaining $\int t^3\,\mathrm{d}t$ the answer would be $t^4/4 + c$. We might therefore try the substitution $t = (x+7)$ and check that $(x+7)^4/4 + c$ is the primitive we are looking for. When we differentiate $F(x) = (x+7)^4/4 + c$ using the chain rule we do indeed obtain

$$\frac{\mathrm{d}F}{\mathrm{d}x} = (x+7)^3$$

So we can write

$$\int (x+7)^3\,\mathrm{d}x = (x+7)^4/4 + c$$

This first example was rather easy because we easily spotted that $\int (x + 7)^3 \, dx$ was in principle the same as $\int t^3 \, dt$. All we had to do was identify $(x + 7)$ with t and note that the result had to be $t^4/4 + c$, i.e. $(x + 7)^4/4 + c$. What makes this example easy is the fact that $dt/dx = 1$. Not all substitutions that we have to make will be such that $dt/dx = 1$. This complicates matters slightly, but does not alter the overall approach.

Example: Obtain the following:
     (i) $\int \cos(3x^2)6x \, dx$;
     (ii) $\int 2(x + 1)(x^2 + 2x) \, dx$.

Solution: (i) We can integrate $\cos(t)$ by itself so let us try putting $t = 3x^2$. We then have that $dt/dx = 6x$, so that we can write

$$\int \cos(3x^2)6x \, dx = \int \cos(t) \, dt$$

Now $\int \cos(t) \, dt = \sin(t) + c$, hence substituting back for t we get

$$\int \cos(3x^2)6x \, dx = \sin(3x^2) + c$$

(ii) Consider

$$\int 2(1 + x)(x^2 + 2x) \, dx$$

If we let $t = x^2 + 2x$, then

$$\frac{dt}{dx} = 2x + 2 = 2(x + 1)$$

This latter expression appears in the integrand and we have

$$\int \underbrace{(x^2 + 2x)}_{t} \underbrace{2(1 + x) \, dx}_{dt} = \int t \, dt$$

This latter expression gives $t^2/2 + c$, so substituting back for t we have

$$\int 2(1 + x)(x^2 + 2x) \, dx = \frac{(x^2 + 2x)^2}{2} + c$$

This example illustrates the steps in the *method of integration by direct substitution*. In part (a), we have a function of 'something' (i.e. $3x^2$) multiplied by the derivative of the 'something' (i.e. $6x$). In part (b), we have 'something' (i.e. $(x^2 + 2x)$) times the 'derivative of the something', (i.e. $2(x + 1)$). In each case we put a new variable equal to the 'something'. And the trick is to be able to identify the 'something'.

Rule 1. When evaluating an integral which is in non-standard form, then ask the question:

Is the integrand in the form $f(g(x))$ times dg/dx or $g(x)$ times dg/dx? If the answer is yes then the substitution $t = g(x)$ will reduce the integral to a simplified form. If this is a standard form then integration is straightforward.

When integrating by substitution the following steps must be carried out:

Procedure for integration by direct substitution

step 1: choose a new variable $t = g(x)$;

step 2: differentiate g to find $dt/dx = dg/dx$;

step 3: substitute into the integral for $g(x)$ and $(dg/dx)dx$ to give

$$\int f(g(x)) \frac{dg}{dx} dx = \int f(t) dt;$$

step 4: if possible, find the primitive for $f(t)$ and substitute back for t.

Note that it is very important to substitute for both $g(x)$ *and* $(dg/dx)dx$ in terms of the new variable t.

Example: Obtain the following:

    (a) $\int 2xe^{x^2} dx$;

    (b) $\int (x^2 - x)^3 (4x - 2) dx$;

    (c) $\int 3x/(1 + x^2) dx$.

Solution: Adopting the procedure we have just obtained we have

    (a) *step 1:* put $t = x^2$;

        step 2: then $dt/dx = 2x$;

        step 3: on substitution with $2x dx = dt$ and $x^2 = t$ we have

$$\int e^{x^2} 2x \, dx = \int e^t \, dt;$$

        step 4: $\int e^t dt = e^t + c$ Replacing t by x^2 we have

$$\int e^{x^2} 2x \, dx = e^{x^2} + c;$$

    (b) *step 1:* put $t = (x^2 - x)$;

        step 2: then $dt/dx = 2x - 1$ and $(4x - 2)dx = 2dt$;

        step 3: on substitution we get

$$\int (x^2 - x)^3 (4x - 2) \, dx = \int t^3 (2dt);$$

        step 4: $\int t^3 2dt = 2t^4/4 + c$ Replacing t we have

$$\int (x^2 - x)^3 (4x - 2) \, dx = (x^2 - x)^4/2 + c;$$

    (c) *step 1:* put $t = 1 + x^2$;

        step 2: then $dt/dx = 2x$ and $x dx = dt/2$;

        step 3: on substituting we have

$$\int 3x/(1 + x^2) \, dx = \int (3/2t) \, dt;$$

        step 4: $\int (3/2t) dt = (\frac{3}{2}) \ln(t) + c$ Replacing t we have

$$\int 3x/(1 + x^2) \, dx = (3/2) \ln(1 + x^2) + c.$$

The third part of this example is different from the first two in that the integrand is a quotient and not a product of functions. However, the evaluation of the integral was quite straightforward by making a substitution for the function in the denominator. This worked here because the function in the numerator is the derivative of the denominator. This idea provides another type of rule.

Rule 2. When evaluating an integral which is in non-standard form, ask the following question:

Is the integrand in the form $c\,\mathrm{d}g/\mathrm{d}x$ divided by $g(x)$ (where c is a constant)? If the answer is yes then the substitution $t = g(x)$ will reduce the integral to a simplified form. If this is a standard form then integration is straightforward, and the integral is in terms of logs.

In the algorithm for integration by substitution it is step 3 that changes slightly. On substituting for $g(x)$ and $(\mathrm{d}g/\mathrm{d}x)\mathrm{d}x$ we get

$$\int \frac{(\mathrm{d}g/\mathrm{d}x)}{g(x)}\,\mathrm{d}x = \int \frac{1}{t}\,\mathrm{d}t$$

Example: Obtain the following:
    (a) $\int \sin(2x)/\cos(2x)\,\mathrm{d}x$;
    (b) $\int (12x + 8)/(3x^2 + 4x + 7)\,\mathrm{d}x$.

Solution:
  (a) *step 1*: put $t = \cos(2x)$;
     step 2: then $\mathrm{d}t/\mathrm{d}x = -2\sin(2x)$;
     step 3: on substitution with $\sin(2x)\,\mathrm{d}x = (-\frac{1}{2})\mathrm{d}t$ and $\cos(2x) = t$ we have

$$\int \frac{\sin(2x)}{\cos(2x)}\,\mathrm{d}x = \int \frac{(-1/2)}{t}\,\mathrm{d}t;$$

     step 4: $\int((-\frac{1}{2})/t)\mathrm{d}t = (-\frac{1}{2})\ln(t) + c$ Replacing t by $\cos(2x)$ we have

$$\int \frac{\sin(2x)}{\cos(2x)}\,\mathrm{d}x = -\ln(\cos(2x))/2 + c.$$

  (b) *step 1*: put $t = 3x^2 + 4x + 7$;
     step 2: then $\mathrm{d}t/\mathrm{d}x = 6x + 4$;
     step 3: on substitution with $(12x + 8)\mathrm{d}x = 2\mathrm{d}t$ and $3x^2 + 4x + 7 = t$ we have

$$\int \frac{12x + 8}{3x^2 + 4x + 7}\,\mathrm{d}x = \int \frac{2}{t}\,\mathrm{d}t;$$

     step 4: $\int(2/t)\mathrm{d}t = 2\ln(t) + c$. Replacing t by $3x^2 + 4x + 7$ we have

$$\int \frac{12x + 4}{3x^2 + 4x + 7}\,\mathrm{d}x = 2\ln(3x^2 + 4x + 7) + c.$$

10.2.2 Indirect substitutions. In the above examples it could be argued that the substitutions used were easily identified and the obvious ones to use anyway. But it is not so obvious which substitution to use in the following integral:

$$\int 1/(1 + x^2)\,\mathrm{d}x$$

The substitution $t = (1 + x^2)$ will not help in this case, because the integrand is not in the right form. We require $\mathrm{d}t/\mathrm{d}x$ to be part of the integrand and in this integral it is not. The aim now is to introduce other changes of variable which will simplify integrals of this type.

For $\int 1/(1 + x^2)dx$ consider the change of variable $x = \tan(t)$, then $dx/dt = \sec^2(t)$. Now on substitution into the integrand we get

$$1/(1 + x^2) = 1/(1 + \tan^2(t)) = 1/(\sec^2(t))$$

So

$$1/(1 + x^2)dx = (1/\sec^2(t))\sec^2(t)dt = dt$$

The integrand of our new integral is now 1 which is certainly easier to deal with than the one we started with. On using the substitution $x = \tan(t)$ we get

$$\int \frac{1}{1 + x^2} dx = \int dt = t + c$$

and substituting back for t we have

$$\int \frac{1}{1 + x^2} dx = \arctan(x) + c$$

This type of method is called *indirect substitution*. It is different from the direct substitutions of the earlier examples. There we substituted some function of x by t; for an indirect substitution we replace x by a suitable expression in t.

In either case the steps in the method of substitution are essentially the same, these are

Procedure for integration by indirect substitution.
step 1: choose a suitable change of variable $x = f(t)$;
step 2: differentiate to find a formula for dx in terms of dt,

$$dx = \frac{df}{dt} dt;$$

step 3: substitute into the integral for x and dx in terms of t and dt;
step 4: if possible, find the primitive of the new integral and substitute back to give a formula in x.

When using a method of substitution to evaluate a *definite* integral it is important that the limits of integration are changed from those given to appropriate limits in terms of this new variable. For example, consider the definite integral $\int_0^1 1/(1 + x^2)dx$; as we have seen above the substitution $x = \tan t$ will provide an easier integral, $\int_a^b dt$ where a and b are new limits. Now to find the new limits we have to solve the equations

$$0 = \tan a \quad \text{and} \quad 1 = \tan b$$

Solving we obtain $a = 0$ and $b = \pi/4$, and on substitution the definite integral becomes

$$\int_0^{\pi/4} dt = t \Big|_0^{\pi/4} = \pi/4$$

Thus when evaluating definite integrals it is not usually necessary to substitute back to give the solution in terms of x. There are other indirect substitutions that are useful in integration. The following table provides possible substitutions that should be tried when the integrand contains the function in the left column:

Function in integrand	Try the substitution
$a^2 + x^2$	$a\tan(t)$
$\sqrt{(a^2 - x^2)}$	$a\sin(t)$ or $a\cos(t)$
$\sqrt{(x^2 - a^2)}$	$a\cosh(t)$
$\sqrt{(a^2 + x^2)}$	$a\sinh(t)$

The following example illustrates the use of these substitutions.

Example: Evaluate the following integrals

(i) $\displaystyle\int_0^{\sqrt{3}} \frac{1}{\sqrt{(4 - x^2)}}\,dx;$

(ii) $\displaystyle\int_0^1 \frac{1}{(\sqrt{(1 + x^2)})^3}\,dx;$

(iii) $\displaystyle\int_{2.5}^3 \frac{1}{\sqrt{(x^2 - 4)}}\,dx.$

Solution: (i) Try the substitution $x = 2\sin(t)$ then

$$\frac{dx}{dt} = 2\cos(t) \quad \text{and} \quad \frac{1}{\sqrt{(4 - x^2)}} = \frac{1}{\sqrt{(4 - 4\sin^2(t))}} = \frac{1}{2\cos(t)}$$

Hence

$$\frac{1}{\sqrt{(4 - x^2)}}\,dx = \frac{1}{2\cos(t)}(2\cos(t)\,dt) = dt$$

The new limits of integration are solutions of the equations

$$0 = 2\sin(a) \quad \text{and} \quad \sqrt{3} = 2\sin(b)$$

Solving for a and b gives $a = 0$ and $b = \pi/3$. On substitution we get

$$\int_0^{\sqrt{3}} \frac{1}{\sqrt{(4 - x^2)}}\,dx = \int_0^{\pi/3} dt = \pi/3$$

(ii) Try the substitution $x = \sinh(t)$ so that

$$\frac{dx}{dt} = \cosh(t) \quad \text{and} \quad \frac{1}{(\sqrt{(1 + x^2)})^3} = \frac{1}{\cosh^3(t)}$$

Hence

$$\frac{1}{(\sqrt{(1 + x^2)})^3}\,dx = \frac{\cosh(t)}{\cosh^3(t)}\,dt = \text{sech}^2(t)\,dt$$

The new limits of integration are solutions of the equations

$$0 = \sinh(a) \quad \text{and} \quad 1 = \sinh(b)$$

Solving for a and b gives $a = 0$ and $b = 0.8814$. On substitution we get

$$\int_0^1 \frac{1}{(\sqrt{(1+x^2)})^3} \, dx = \int_0^{0.8814} \text{sech}^2(t) \, dt = \tanh(t) \Big|_0^{0.8814} = 0.7071$$

(iii) Try the substitution $x = 2\cosh(t)$ then

$$\frac{dx}{dt} = 2\sinh(t) \quad \text{and} \quad \frac{1}{\sqrt{(x^2-4)}} = \frac{1}{\sqrt{(4\cosh^2(t)-4)}} = \frac{1}{2\sinh(t)}$$

Hence

$$\frac{1}{\sqrt{(x^2-4)}} \, dx = \frac{1}{2\sinh(t)}(2\sinh(t)) \, dt = dt$$

The new limits of integration are solutions of the equations

$$2.5 = 2\cosh(a) \quad \text{and} \quad 3 = 2\cosh(b)$$

Solving for a and b gives $a = \cosh^{-1}(1.25) = 0.6931$ and $b = \cosh^{-1}(\frac{3}{2}) = 0.9624$. On substitution we get

$$\int_{2.5}^3 \frac{1}{\sqrt{(x^2-4)}} \, dx = \int_{0.6931}^{0.9624} dt = 0.9624 - 0.6931 = 0.2693$$

10.2.3 The substitution $t = \tan(x/2)$. There is one final substitution that may sometimes be useful when evaluating the integral of functions involving $\sin(x)$ and $\cos(x)$.

The substitution $t = \tan(x/2)$ for $-\pi < x < \pi$ is of particular use when the integrand contains the expressions $a\cos(x) + b\sin(x)$. However, the substitution should be thought of as a last resort since it is extremely cumbersome to use. We illustrate its use in the following example.

Example: Evaluate

$$\int_0^{\pi/2} \frac{1}{(\sin(x) + \cos(x) + 1)} \, dx$$

Solution: Consider the substitution $t = \tan(x/2)$. Using the expressions for $\sin(x)$ and $\cos(x)$ in terms of $(x/2)$ we have

$$\sin(x) = 2\sin(x/2)\cos(x/2)$$
$$= \frac{2\sin(x/2)\cos^2(x/2)}{\cos(x/2)}$$
$$= \frac{2\tan(x/2)}{(1 + \tan^2(x/2))}$$
$$= 2t/(1 + t^2).$$

$$\cos(x) = \cos^2(x/2) - \sin^2(x/2)$$
$$= \frac{1 - \tan^2(x/2)}{1 + \tan^2(x/2)}$$
$$= \frac{1 - t^2}{1 + t^2}.$$

Thus

$$\sin(x) + \cos(x) + 1 = \frac{2t}{1+t^2} + \frac{1-t^2}{1+t^2} + 1$$

$$= \frac{2t+2}{1+t^2}$$

Now $dt = \sec^2(x/2)/2\,dx = (1+t^2)/2\,dx$. On substitution and simplifying we get

$$\int_0^{\pi/2} \frac{1}{(\sin(x) + \cos(x) + 1)}\,dx = \int_0^1 \frac{1}{1+t}\,dt$$

$$= \ln(1+t)\Big|_0^1$$

$$= \ln(2)$$

$$= 0.6931$$

10.3 Worked examples

Example 1. A uniform lamina is defined by the region enclosed by part of the circle $x^2 + y^2 = 4$ and the lines $x = 0$, $x = 1$, $y = 0$. Find the position of the centroid of this region.

Solution. The shape of the lamina is shown in Fig. 10.2. Using the notation of 10.1.1 the functions $f(x)$ and $g(x)$ defining the boundaries of the region are

$$f(x) = \sqrt{(4 - x^2)}, \quad 0 < x < 1; \quad g(x) = 0, \quad 0 < x < 1$$

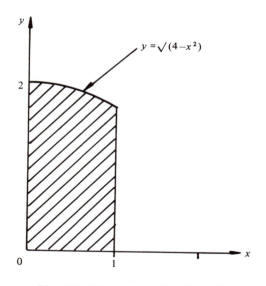

Fig. 10.2. The function $f = \sqrt{(4 - x^2)}$.

If the coordinates of the centroid are (X, Y) then the formulas for X and Y are

$$X = \frac{\int x f(x)\, dx}{\int f(x)\, dx} \quad \text{and} \quad Y = (1/2) \frac{\int f(x)^2\, dx}{\int f(x)\, dx}$$

Evaluating these integrals in turn we have

$$\int_0^1 f(x)\, dx = \int_0^1 \sqrt{(4 - x^2)}\, dx$$

To evaluate this integral we need to make the indirect substitution $x = 2 \sin(t)$ and then $dx = 2 \cos(t)\, dt$. The limits of integration become $t_1 = 0$ and $t_2 = \pi/6$. On substitution we obtain

$$\int_0^1 f(x)\, dx = \int_0^{\pi/6} 2 \cos(t) \cdot 2 \cos(t)\, dt$$

$$= 4 \int_0^{\pi/6} \cos^2(t)\, dt$$

$$= 4 \int_0^{\pi/6} (1 + \cos(2t))/2\, dt$$

$$= 4(t/2 + \sin(2t)/4) \Big|_0^{\pi/6}$$

$$= 4(\pi/12 + \sqrt{3}/8)$$

$$= \pi/3 + \sqrt{3}/2 = 2.547 \text{ (to three decimal places)}$$

Consider now the integral $I = \int x f(x)\, dx = \int_0^1 x \sqrt{(4 - x^2)}\, dx$. In this case we make the direct substitution $t = 4 - x^2$ and $dt = -2x\, dx$. The limits become

$$t_1 = 4 \quad \text{and} \quad t_2 = 3$$

On substitution we have

$$I = \int_4^3 \sqrt{t}\, dt/(-2)$$

$$= -(1/3)t^{3/2} \Big|_4^3$$

$$= -(1/3)(3\sqrt{3} - 4\sqrt{4})$$

$$= 0.935 \text{ (to three decimal places)}$$

Finally, consider the integral $\int f(x)^2\, dx$

$$J = \int_0^1 (4 - x^2)\, dx$$

No substitution is required, we can integrate 4 and x^2 directly. So

$$J = (4x - x^3/3) \Big|_0^1$$

$$= (4 - 1/3) = 3.667 \text{ (to three decimal places)}$$

Thus we can now give the centroid as the point (X, Y) where

$$X = 0.935/2.547 = 0.367$$

and

$$Y = (1/2)(3.667/2.547) = 0.720.$$

The solution to this example illustrates that although each integrand contains the same function $4 - x^2$, different substitutions are used.

Example 2. The fraction of internally radiated heat energy falling on a thermo-couple which is centrally placed on the axis of a hot tube of length $2L$ is given by

$$I = \frac{1}{2} \int_{-L}^{L} \frac{R^2 \, dl}{(R^2 + l^2)^{3/2}}$$

where R is the radius of the tube. Evaluate this for $L = 40 \text{ mm}$ and $R = 30 \text{ mm}$.

Solution. To evaluate this integral we try the indirect substitution $l = R \tan(t)$; then $dl = R \sec^2 t \, dt$ and the limits become $t_1 = \arctan(-L/R)$ and $t_2 = \arctan(L/R)$. The integrand becomes $R^2 + l^2 = (R \sec(t))^2$. On substitution we obtain

$$I = (1/2) \int_{u_1}^{u_2} \frac{\sec^2(t)}{\sec^3(t)} \, dt$$

$$= (1/2) \int_{u_1}^{u_2} \cos(t) \, dt$$

$$= (1/2)(\sin(t))|_{u_1}^{u_2}$$

$$= L/\sqrt{(L^2 + R^2)}$$

(since if $\tan(\theta) = L/R$ then $\sin(\theta) = L/\sqrt{(L^2 + R^2)}$). With $L = 40 \text{ mm}$ and $R = 30 \text{ mm}$ then $I = 0.8$.

10.4 Further developments

Integration of products and quotients of functions

10.4.1 Use of partial fractions. In 10.2 we noted a rule for differentiating quotients providing the integrands were in a particular form. A common integrand in science is of the form $h(x)/g(x)$ where $g(x)$ and $h(x)$ are polynomial functions. Now we can use Rule 2 in 10.2.1 providing $dg/dx = h$; however there are many integrands which are ratios of polynomials but which do not satisfy this condition. For example, $\int 1/(ax + bx^2) dx$ cannot to evaluated using Rule 2. As always, we try to transform the integral to a form that we recognise and in this case a *partial fraction expansion* solves the problem.

The following example illustrates the approach.

Example: Obtain

$$\int \frac{x+1\,dx}{x^2+x-2}$$

Solution: The factors of $x^2 + x - 2$ are $(x + 2)$ and $(x - 1)$, so we express the integrand as

$$\frac{x+1}{x^2+x-2} = \frac{x+1}{(x+2)(x-1)}$$

$$= \frac{A}{x+2} + \frac{B}{x-1}$$

Solving for A and B in the usual way (see Chapter 5) we obtain $A = \frac{1}{3}$ and $B = \frac{2}{3}$. Now the integral can be written as

$$\int \frac{x+1\,dx}{x^2+x-2} = \int \frac{(1/3)}{(x+2)}\,dx + \int \frac{(2/3)}{(x-1)}\,dx$$

$$= (1/3)\ln(x+2) + (2/3)\ln(x-1) + c$$

The method of approach in this example can be used for any integral where the integrand is the ratio of two polynomials. This provides a third rule for integration:

Rule 3. When evaluating an integral which is in non-standard form, ask the following question:

> Is the integrand a ratio of polynomials?
> If the answer is yes then if possible express the integrand in partial fraction form, and integrate each fraction in the sum.

The method of approach is the same as in our last example provided that the denominator has real linear factors $(x - x_i)$. However, there are many polynomials which have not and at best we can write such a polynomial in terms of quadratic functions. This then requires evaluation of integrals of the form

$$I = \int \frac{1}{ax^2+bx+c}\,dx$$

where a, b and c are constants. To evaluate integrals of this type we write the denominator as $A(x + B)^2 + C$ where A, B and C are constants in terms of a, b and c. Then we try one of the indirect substitutions suggested in 10.2.2.

We illustrate the method in the following example.

Example: Obtain $\int 4/(x^2 - 4x + 5)\,dx$.
Solution: Consider the quadratic function $x^2 - 4x + 5$. We complete the square and write this as

$$x^2 - 4x + 5 = (x - 2)^2 + 1$$

Let $t = (x - 2)$, so that $dt = dx$ and on substitution into the integral we obtain

$$\int \frac{4}{x^2-4x+5}\,dx = \int \frac{4}{t^2+1}\,dt$$

Now try the substitution $t = \tan(v)$ with $dt = \sec^2(v)\,dv$. Then

$$\int \frac{4}{t^2+1}\,dt = \int 4\,dv$$

$$= 4v + \text{constant}$$

Hence replacing v by $\arctan(t)$ and t by $(x-2)$ we have

$$\int \frac{4}{x^2-4x+5}\,dx = 4\arctan(x-2) + \text{constant}$$

10.4.2 Integration by parts. In this subsection we introduce a rule for integrating products of functions. As with all rules in the theory of integration the method is not guaranteed to be applicable in every case! This rule should be tried if Rule 1 of Section 10.2 does not apply, i.e. we have the integrand $f(x)g(x)$ but df/dx is not equal to $cg(x)$ for some constant c. For example, $\int x\sin(x)\,dx$ cannot be evaluated using the preceding three rules already introduced. The method of integration by parts exploits the product rule for differentiation.

Suppose we have to evaluate $\int f(x)g(x)\,dx$. Consider the product $H(x) = F(x)g(x)$, where $f(x) = dF/dx$. Then

$$\frac{dH}{dx} = \frac{dF}{dx}g + F\frac{dg}{dx}$$

$$= fg + F\frac{dg}{dx}$$

We rewrite this as

$$fg = \frac{dH}{dx} - F\frac{dg}{dx}$$

Now if we integrate each side between the limits a and b then the left-hand side contains the integral we require. We get

$$\int_a^b fg\,dx = \int_a^b \frac{dH}{dx}\,dx - \int_a^b F\frac{dg}{dx}\,dx$$

$$= H(x)\Big|_a^b - \int_a^b F\frac{dg}{dx}\,dx$$

$$= F(x)g(x)\Big|_a^b - \int_a^b F\frac{dg}{dx}\,dx \tag{1}$$

Thus we have written the integral of a product in terms of the primitive of one function in the product and the derivative of the other function. This may at first sight appear to make the integral more complicated; however, as the following example shows, in some problems the second integral is easier to evaluate than the original.

Example: Evaluate $\int_0^{\pi/2} x\sin(x)\,dx$.

Solution: The integrand is a product of two functions x and $\sin(x)$. We shall choose $g(x) = x$ to be the function to be differentiated and choose $f(x) = \sin(x)$ for which the primitive is $F(x) = -\cos(x)$. Then according to formula (1)

$$\int_0^{\pi/2} x\sin(x)\,dx = -\cos(x)\cdot x\,\Big|_0^{\pi/2} - \int_0^{\pi/2} -\cos(x)\cdot 1\,dx$$

$$= -\cos(x)\cdot x\,\Big|_0^{\pi/2} + \sin(x)\,\Big|_0^{\pi/2}$$

$$= 0 + 1 = 1$$

From this example we see that the first step is to find a primitive for one of the functions in the product so that this method of approach only integrates part of the integrand. The name of the method *integration by parts* comes from this step.

The procedure is summarised in the following five steps:

Procedure for the method of integration by parts.
step 1: identify that the integrand is a product of two functions,

$$f(x) \quad \text{and} \quad g(x) \text{ say}$$

choose to integrate one of them $f(x)$ and choose to differentiate the other;
step 2: find the primitive for $f(x)$ and let $F(x) = \int f(x)\,dx$;
step 3: find the derivative of $g(x)$, dg/dx;
step 4: substitute for $F(x)$, $g(x)$ and dg/dx in the formula (1)

$$\int_a^b f(x)g(x)\,dx = F(x)g(x)\,\Big|_a^b - \int_a^b F(x)\frac{dg}{dx}\,dx;$$

step 5: Evaluate the various expressions on the right-hand side.

The success of the method in many problems is that the new integral formed on the right-hand side in step 4 is easier to evaluate than the original one. Of course it is quite possible that at step 5 the task is more difficult than the original one. In such cases the remedy may be to reverse the roles of $f(x)$ and $g(x)$, choosing to integrate $g(x)$ and to differentiate $f(x)$.

The following example illustrates the steps in the procedure.

Example: Obtain the following:

(a) $\displaystyle\int_1^2 x^2 \ln(x)\,dx;$

(b) $\displaystyle\int_0^1 x^2 e^x\,dx;$

(c) $\displaystyle\int x\sec^2(x)\,dx.$

Solution:

(a) *step 1*: the integrand is the product of the two functions x^2 and $\ln(x)$. Choose to integrate x^2 and to differentiate $\ln(x)$;
step 2: $F(x) = \int x^2\,dx = x^3/3$;

step 3: $\mathrm{d}g/\mathrm{d}x = \mathrm{d}(\ln(x))/\mathrm{d}x = 1/x$;
step 4: on substitution into the formula we have

$$\int_1^2 x^2 \ln(x)\,\mathrm{d}x = (x^3/3)\ln(x)\Big|_1^2 - \int_1^2 (x^3/3)\cdot(1/x)\,\mathrm{d}x;$$

step 5: evaluating the various terms on the right-hand side gives

$$\int_1^2 x^2 \ln(x)\,\mathrm{d}x = (x^3/3)\ln(x)\Big|_1^2 - x^3/9\Big|_1^2$$
$$= 8\ln(2/3) - 7/9$$

(b) *step 1*: The integrand is the product of the two functions x^2 and e^x. Choose to differentiate x^2 and to integrate e^x.
step 2: $F(x) = \int e^x\,\mathrm{d}x = e^x$;
step 3: $\mathrm{d}g/\mathrm{d}x = 2x$;
step 4: on substitution into the formula we get

$$\int_0^1 x^2 e^x\,\mathrm{d}x = x^2 e^x\Big|_0^1 - \int_0^1 2x\,e^x\,\mathrm{d}x;$$

step 5: now we have an integral which is of similar form to the original but contains $2xe^x$ instead of $x^2 e^x$.

To evaluate this integral we must repeat the integration by parts procedure, choosing to integrate e^x and to differentiate $2x$. This second integral then becomes

$$\int_0^1 2xe^x\,\mathrm{d}x = 2xe^x\Big|_0^1 - \int_0^1 2\cdot e^x\,\mathrm{d}x$$

$$= 2xe^x\Big|_0^1 - 2e^x\Big|_0^1$$

Thus the original integral can now be evaluated:

$$\int_0^1 x^2 e^x\,\mathrm{d}x = (x^2 e^x - 2xe^x + 2e^x)\Big|_0^1$$
$$= e^1 - 2e^1 + 2e^1 - 2$$
$$= e - 2$$

(c) *step 1*: the integrand is a product of the two functions x and $\sec^2(x)$. Choose to integrate $\sec^2(x)$ and to differentiate x;
step 2: $F(x) = \int \sec^2(x)\,\mathrm{d}x = \tan(x)$;
step 3: $\mathrm{d}g/\mathrm{d}x = 1$;
step 4: on substitution into the formula we get

$$\int x \sec^2(x)\,\mathrm{d}x = x\cdot\tan(x) - \int 1\cdot\tan(x)\,\mathrm{d}x;$$

step 5: evaluating the various terms on the right-hand side

$$\int x \sec^2(x)\,\mathrm{d}x = x\tan(x) + \ln(\cos(x)) + c.$$

There are two important points worth making in connection with this example:

(1) Sometimes the integration by parts procedure has to be repeated more than once in a problem. This is illustrated by the integral in part (b). In the solution we had to integrate by parts twice.

(2) The method of integration by parts can be used to evaluate indefinite integrals. This is illustrated in part (c); we must remember to add on the unknown constant of integration in such problems.

With the method of integration by parts we can complete the list of integrals of our basic scientific functions. So far we have not integrated $\ln(x)$. The method of integration by parts help us to do so however, it may appear as rather a trick!

Consider $\int \ln(x)\,dx$. The integrand in this problem does not look like a product of two functions. However we choose 1 to be the second function, so that 1 and $\ln(x)$ are the two functions in the integrand. We choose to integrate 1 and to differentiate $\ln(x)$. We then have $F(x) = x$ and $dg/dx = 1/x$. Then

$$\int 1\cdot\ln(x)\,dx = x\cdot\ln(x) - \int x\cdot(1/x)\,dx$$

$$= x\cdot\ln(x) - \int dx + c$$

One more integration gives

$$\int \ln(x)\,dx = x\ln(x) - x + c$$

In each of the examples above we could eventually evaluate the integral explicitly, although we may have to repeat the procedure several times. There are integrals however where we cannot do this as the following example illustrates.

Example: Evaluate $\int_0^{\pi/2} e^x \cos(x)\,dx$.
Solution: The integrand is the product of the two functions e^x and $\cos(x)$.
Choose to integrate e^x and to differentiate $\cos(x)$. Then

$$F(x) = e^x \quad \text{and} \quad \frac{dg}{dx} = -\sin(x)$$

On substitution we get

$$\int_0^{\pi/2} e^x \cos(x)\,dx = e^x\cdot\cos(x)\Big|_0^{\pi/2} - \int_0^{\pi/2} e^x(-\sin(x))\,dx$$

Now the integral on the right-hand side cannot be evaluated explicitly. So we integrate by parts again, choosing to integrate e^x as before and to differentiate $(-\sin(x))$.
We then obtain

$$\int_0^{\pi/2} e^x(-\sin(x))\,dx = e^x(-\sin(x))\Big|_0^{\pi/2} - \int_0^{\pi/2} e^x(-\cos(x))\,dx$$

so that we appear to be back where we started! However, if we collect together the terms involving the original integral we have a formula for its value:

$$\int_0^{\pi/2} e^x \cos(x)\,dx = e^x\cos(x) - \left(-e^x\sin(x) + \int_0^{\pi/2} e^x\cos(x)\,dx\right)$$

$$2\int_0^{\pi/2} e^x \cos(x)\,dx = e^x\cos(x) + e^x\sin(x)\Big|_0^{\pi/2}$$

$$= -1 + e^{\pi/2}$$

$$\int_0^{\pi/2} e^x \cos(x)\,dx = (e^{\pi/2} - 1)/2 = 1.9052$$

10.4.3 Reduction formulae. In some of the problems in the previous subsection it was necessary to repeat the procedure of integrating by parts. This will always be the case if one of the functions in the integrand is a polynomial of degree greater than one.

Example: Obtain $\int x^n e^{-x}\,dx$.

Solution: The integrand is the product of the two functions x^n and e^{-x}. We choose to integrate e^{-x} and to differentiate x^n. Then $F(x) = -e^{-x}$ and

$$\frac{dg}{dx} = nx^{n-1}$$

Now on substitution we obtain

$$\int x^n e^{-x}\,dx = x^n \cdot (-e^{-x}) - \int nx^{n-1} \cdot (-e^{-x})\,dx$$

Now we have not evaluated the integral, but since the power of x has been reduced by one, we could continue to integrate by parts choosing to differentiate x^{n-1} and to integrate e^{-x}. After repeating the procedure n times we would arrive at the simple integral $\int e^{-x}\,dx$ for the last stage. Carrying out the integrations we would have

$$\int x^n e^{-x}\,dx = -x^n e^{-x} - nx^{n-1}e^{-x} - \cdots - n!\,e^{-x}$$

Now the work in such problems can be reduced considerably by using the similarity of the second integral, $\int nx^{n-1}e^{-x}\,dx$ to the original one. Suppose that we denote $\int x^n e^{-x}\,dx$ by I_n. Then after one integration by parts we obtain the formula

$$I_n = -x^n e^{-x} + nI_{n-1}$$

Such an expression is called *a reduction formula* and can be used to evaluate I_n for any n starting from I_0.

For example, consider the integral $\int x^5 e^{-x}\,dx$. From the reduction formula

$$I_5 = -x^5 e^{-x} + 5I_4$$
$$I_4 = -x^4 e^{-x} + 4I_3$$
$$I_3 = -x^3 e^{-x} + 3I_2$$
$$I_2 = -x^2 e^{-x} + 2I_1$$
$$I_1 = -x^1 e^{-x} + I_0$$

$$I_0 = \int e^{-x}\,dx = -e^{-x} + C$$

Thus

$$\int x^5 e^{-x}\,dx = -x^5 e^{-x} + 5x^4 e^{-x} - 5 \cdot 4x^3 e^{-x} - 5 \cdot 4 \cdot 3x^2 e^{-x} - 5 \cdot 4 \cdot 3 \cdot 2(xe^{-x} + e^{-x}) + C$$

This method of using an appropriate reduction is convenient for most integrals involving x raised to a power as one of the functions in the product.

Example: Obtain $\int x^4 \cos(x)\,dx$.

Solution: Let $I_n = \int x^n \cos(x)\,dx$. Then integrating by parts we have

$$I_n = x^n \sin(x) - \int n x^{n-1} \cdot \sin(x)\,dx$$

We are not quite back to the original integral, but if we integrate by parts again,

$$\int x^{n-1} \sin(x)\,dx = -x^{n-1}\cos(x) - \int (n-1)x^{n-2}\cdot(-\cos(x))\,dx$$

which can be written in terms of I_{n-2}. So we obtain

$$I_n = x^n \sin(x) - n[-x^{n-1}\cos(x) + (n-1)I_{n-2}]$$
$$I_n = x^n \sin(x) + n x^{n-1}\cos(x) - n(n-1)I_{n-2}$$

So setting $n = 4$ in this case we have

$$I_4 = x^4 \sin(x) + 4x^3 \cos(x) - 4.3 I_2$$
$$I_2 = x^2 \sin(x) + 2x \cos(x) - 2 I_0$$

$$I_0 = \int \cos(x)\,dx = \sin(x) + C$$

Thus

$$\int x^4 \cos(x)\,dx = x^4 \sin(x) + 4x^3 \cos(x) - 12x^2 \sin(x)$$
$$- 24x \cos(x) + 24 \sin(x) + \text{constant}$$

10.5 Further worked examples

Example 1. A solution of ethyl acetate $(0.01\,\text{mol}\,l^{-1})$ reacts with a solution of sodium hydroxide $(0.002\,\text{mol}\,l^{-1})$. The velocity constant of the reaction is $3\,\text{min}^{-1}$. If x is the concentration of ethyl acetate which has reacted in time t then

$$t = \int \frac{1}{3(0.01 - x)(0.002 - x)}\,dx$$

If initially the concentration x is zero, i.e. $x = 0$ when $t = 0$ find the time taken for the concentration to be $0.001\,\text{mol}\,l^{-1}$.

Solution. The time required is given by the definite integral

$$t = \int_0^{0.001} \frac{1}{3(0.01 - x)(0.002 - x)}\,dx$$

To evaluate this integral we use partial fractions:

$$\frac{1}{(0.01 - x)(0.002 - x)} = \frac{A}{(0.01 - x)} + \frac{B}{(0.002 - x)}$$

Solving for A and B we get $A = -\frac{1}{0.008}$ and $B = \frac{1}{0.008}$. The integral becomes

$$t = (1/0.024) \int_0^{0.001} 1/(0.002 - x) - 1/(0.01 - x)\,dx$$

$$= (1/0.024)(-\ln(0.002 - x) + \ln(0.01 - x))|_0^{0.001}$$

$$= (1/0.024)(-\ln(0.001/0.009) + \ln(0.002/0.01))$$

$$= (1/0.024)\ln(9/5)$$

$$= 24.5\,\text{mins (to one decimal place)}$$

Example 2. A semicircular wire is made from insulating material and is given an electric charge of density $e|\theta|$ per unit length. The radius of the wire is a (see Fig. 10.3). Find the electric field at the origin $(0,0)$.

Solution. The magnitude of the electric field at a point P due to a charge q a distance r from P is given by Coulomb's law, we have $E = q/r^2$. In this problem we consider the semicircle being made up of small arcs of length $ds = a\,d\theta$ each with charge $e|\theta|\,ds$ (Fig. 10.4). The electric field at the origin due to the charge on an elementary arc at Q, θ_i, is given by

$$dE = e|\theta_i|\,ds/a^2$$
$$= e|\theta_i|a\,d\theta/a^2$$
$$= e|\theta_i|\,d\theta/a$$

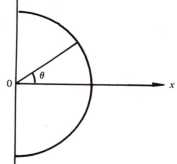

Fig. 10.3. A semicircular wire.

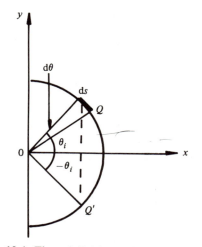

Fig. 10.4. The subdivisions of the wire are arcs.

The direction of the electric field vector is along Q to O. Now by symmetry the arc at Q' at an angle $-\theta_i$ will contribute an electric field of equal magnitude but direction Q' to O. The components of these quantities perpendicular to OP will cancel and the net electric field will act along the direction P to O. The magnitude of the component due to the arc at Q is $e|\theta_i|\cos|\theta_i|\,d\theta/a$. Thus the total field at O has magnitude

$$E = \sum_{i=1}^{n} e|\theta_i|\cos|\theta_i|\,d\theta/a$$

and direction P to O. In the limit this summation becomes the integral

$$E = \int_{-\pi/2}^{\pi/2} e|\theta|\cos(\theta)\,d\theta/a = (2e/a)\int_{0}^{\pi/2} \theta\cos(\theta)\,d\theta$$

This integral can be evaluated using the method of integration by parts, choosing to integrate $\cos(\theta)$ and to differentiate θ. We have

$$E = (2e/a)\left(\theta\sin(\theta)\Big|_0^{\pi/2} - \int_0^{\pi/2}\sin(\theta)\,d\theta \right)$$

$$= (2e/a)(\theta\sin(\theta)|_0^{\pi/2} - (-\cos(\theta))|_0^{\pi/2})$$

$$= (2e/a)(\pi/2 - 1).$$

Example 3. In the theory of the atomic structure of hydrogen, a consequence of Schrödinger's equation is that the probability of finding the electron between the distances r_1 and r_2 from the nucleus ($r_1 < r_2$) is given by the integral

$$P = \int_{r_1}^{r_2} \frac{4r^2}{a^3}e^{-2r/a}\,dr$$

where a is the Bohr radius, $a = 0.529 \times 10^{-10}$ m. Evaluate this definite integral and hence find the probability of finding the electron in a ball of radius $2a$ about the nucleus.

Solution. Although not absolutely necessary it is convenient in this problem to let $u = -2r/a$; this substitution removes the constants out of the integrand. We have $du = (-2/a)\,dr$ and on substitution the integral becomes

$$P = \int_{u_1}^{u_2} \frac{u^2}{a}e^u(-a/2)\,du$$

$$= (-1/2)\int_{u_1}^{u_2} u^2 e^u\,du$$

where $u_1 = -2r_1/a$ and $u_2 = -2r_2/a$. This integral is evaluated by integrating by parts twice. We choose to differentiate u^2 and to integrate e^u.

$$P = (-1/2)\left(u^2 e^u|_{u_1}^{u_2} - \int_{u_1}^{u_2} 2u e^u\,du \right)$$

$$= (-1/2)\left(u^2 e^u - 2u e^u|_{u_1}^{u_2} + \int_{u_1}^{u_2} 2e^u\,du \right)$$

$$= (-1/2)e^u(u^2 - 2u + 2)|_{u_1}^{u_2}$$
$$= -\tfrac{1}{2}e^{u_2}(u_2^2 - 2u_2 + 2) + \tfrac{1}{2}e^{u_1}(u_1^2 - 2u_1 + 2)$$

For a ball of radius $2a$ about the nucleus we have $r_1 = 0$ and $r_2 = 2a$ and hence $u_1 = 0$ and $u_2 = -4$. On substituting these values for u into the solution for the integral we obtain

$$P = (-1/2)e^{-4}(16 + 8 + 2) + (1/2)e^0(2)$$
$$= -13e^{-4} + 1$$
$$= 0.762 \text{ (to three decimal places)}$$

There is a probability of roughly 76% of finding the electron in a ball of radius $2a$. In fact, it can be shown that there is a 99% probability that the electron can be found within a ball of radius $4a$.

Example 4. A particle of mass m travels in a straight line in damped harmonic motion so that its speed at time t is given by $v = e^{-at}\sin(kt)$. Find the distance travelled by the particle in the time interval $t = 0$ to $t = \pi/k$.

Solution. Let $s(t)$ be the distance travelled by the particle in time t then $v = ds/dt$. The distance travelled in the time interval $t = 0$ to $t = \pi/k$ is the definite integral:

$$s = \int_0^{\pi/k} e^{-at}\sin(kt)\,dt$$

We can integrate this using the method of integration by parts twice. Choose to integrate e^{-at} and to differentiate $\sin(kt)$. We obtain

$$s = (e^{-at}/-a)\sin(kt)|_0^{\pi/k} - \int_0^{\pi/k} (e^{-at}/(-a))k\cos(kt)\,dt$$

$$= \frac{k}{a}\int_0^{\pi/k} e^{-at}\cos(kt)\,dt$$

$$= \frac{k}{a}\left((-e^{-at}/a)\cos(kt)|_0^{\pi/k} - \int_0^{\pi/k} (-e^{-at}/a)(-k\sin(kt))\,dt\right)$$

$$= \frac{k}{a^2}(1 - \cos(\pi)e^{-a\pi/k}) - \frac{k^2}{a^2}\int_0^{\pi/k} e^{-at}\sin(kt)\,dt$$

$$= \frac{2ke^{-a\pi/k}}{a^2} - \frac{k^2}{a^2}s$$

Taking the term involving s from the right-hand side to the left and simplifying, the distance travelled is given by

$$s = \frac{2e^{-a\pi/k}k}{(k^2 + a^2)}$$

10.6 Exercises

1. Evaluate the following integrals by making an appropriate direct substitution in each case:

    (i) $\displaystyle\int \sqrt{(x+1)}\,dx$ (ii) $\displaystyle\int (x+2)^2\,dx$

   (iii) $\displaystyle\int \frac{1.4}{(x+1.4)}\,dx$ (iv) $\displaystyle\int \frac{2x}{x^2+1}\,dx$

    (v) $\displaystyle\int \frac{x+1}{\sqrt{(x^2+2x+33)}}\,dx$ (vi) $\displaystyle\int x\exp(x^2)\,dx$

   (vii) $\displaystyle\int [\sin(x)-\cos(x)][(\cos(x)+\sin(x)]^3\,dx$

  (viii) $\displaystyle\int \frac{2\sin(x)-3\cos(x)}{2\cos(x)+3\sin(x)}\,dx$ (ix) $\displaystyle\int \sin(x)\cos(x)\,dx$

    (x) $\displaystyle\int \tan(x)\,dx$ (xi) $\displaystyle\int \frac{\sqrt{(x+1)}}{2+\sqrt{(x+1)}}\,dx$

2. Evaluate the following integrals by making an appropriate indirect substitution in each case:

    (i) $\displaystyle\int \frac{1}{(4+x^2)}\,dx$ (ii) $\displaystyle\int \sqrt{(1-x^2)}\,dx$

   (iii) $\displaystyle\int \frac{1}{x^2+4x+13}\,dx$ (iv) $\displaystyle\int \frac{1}{(4+x^2)^2}\,dx$

    (v) $\displaystyle\int \sqrt{(9x^2+1)}\,dx$ (vi) $\displaystyle\int \frac{1}{(1+x^2)^{3/2}}\,dx$

   (vii) $\displaystyle\int \frac{1}{[3+5\cos(x)]}\,dx$ (viii) $\displaystyle\int \frac{\sin(x)}{1-\sin(x)}\,dx$

3. Evaluate the following integrals, first expressing the integrand as partial fractions:

    (i) $\displaystyle\int \frac{1}{25-x^2}\,dx$ (ii) $\displaystyle\int \frac{1}{x^2-5x+6}\,dx$

   (iii) $\displaystyle\int \frac{1}{2x^2+7x-4}\,dx$ (iv) $\displaystyle\int \frac{1}{2x^3-5x^2+x+3}\,dx$

    (v) $\displaystyle\int \frac{1}{x^2(x+1)}\,dx$ (vi) $\displaystyle\int \frac{1}{x(x-1)^2(x+1)}\,dx$

4. Integrate the following by parts:

    (i) $\displaystyle\int x\ln(x)\,dx$ (ii) $\displaystyle\int x\sec^2(x)\,dx$

(iii) $\displaystyle\int x e^{-ax}\,dx$

(iv) $\displaystyle\int x^2 e^x\,dx$

(v) $\displaystyle\int \sin^{-1}(x)\,dx$

(vi) $\displaystyle\int \tan^{-1}(x)\,dx$

(vii) $\displaystyle\int x^2 \cos(x)\,dx$

(viii) $\displaystyle\int e^x \cosh(3x)\,dx$

(ix) $\displaystyle\int x^n \sinh(x)\,dx$

5. Find the values of the following definite integrals:

(i) $\displaystyle\int_0^1 \frac{1}{\sqrt{(1+x)}}\,dx$

(ii) $\displaystyle\int_1^2 \frac{1}{(x+2)}\,dx$

(iii) $\displaystyle\int_0^1 \frac{1}{x^2+2x+2}\,dx$

(iv) $\displaystyle\int_0^{\pi/2} \sin^3(\theta)\,d\theta$

(v) $\displaystyle\int_0^1 x^2 e^x\,dx$

(vi) $\displaystyle\int_0^a x\sqrt{(a^2-x^2)}\,dx$

(vii) $\displaystyle\int_0^1 \frac{x^3}{(x+2)}\,dx$

(viii) $\displaystyle\int_0^{\pi/2} \frac{1}{5+4\cos(\theta)}\,d\theta$

(ix) $\displaystyle\int_0^1 \frac{1}{(x+1)^2(x+2)}\,dx$

(x) $\displaystyle\int_0^{\pi/4} \sin^4(\theta)\cos^4(\theta)\,d\theta$

(xi) $\displaystyle\int_0^{\pi/2} \sin^5(\theta)\cos^2(\theta)\,d\theta$

(xii) $\displaystyle\int_{-2\pi}^{2\pi} \cos^4(\theta)\,d\theta$

6. By integrating the appropriate binomial expansions term by term, find the series expansions of $\tan^{-1}(x)$ and $\sin^{-1}(x)$ up to the term in x^7. Hence evaluate

$$\lim_{x\to 0} \frac{2\sin^{-1}(x) + \tan^{-1}(x) - 3x(1+x^4)^{\frac{1}{3}}}{x^5}$$

7. Show that if $I_n = \int \tan^n(x)\,dx$, then

$$I_n + I_{n-2} = \frac{\tan^{n-1}(x)}{(n-1)} \qquad n \neq 1$$

and hence evaluate $\int_0^{\pi/4} \tan^6(x)\,dx$.

8. Show that if $I_n = \int_0^\pi e^x \sin^n(x)\,dx, n > 1$, then

$$(n^2+1)I_n = n(n-1)I_{n-2}$$

Hence evaluate I_4 and I_5

9. (a) Using repeated integration by parts show that

$$\int e^{ax}\sin(bx)\,dx = \frac{a}{b^2}e^{ax}\sin(bx) - \frac{e^{ax}}{b}\cos(bx) - \int \frac{a^2 e^{ax}}{b^2}\sin(bx)\,dx$$

(b) The instantaneous current $i(t)$ in a particular electric circuit is given by the expression

$$i(t) = \frac{1}{e^{Rt/L}} \left\{ \frac{V_0}{L} \int e^{Rt/L} \sin(pt)\, dt + C \right\}$$

where R, L, V_0, C and p are all constants.
(i) Using the identity in (a) obtain the expression for the current i.
(ii) Obtain the value of the constant C, if $i = 0$ when $t = 0$.

10. By noting that $\int e^{ax} \sin(bx)\, dx$ is the imaginary part of the integral $\int e^{ax+ibx}\, dx$, work out this latter integral and, using the properties of complex numbers, give an alternative derivation of the identity expressed in Exercise 9(a) above.

11. The concentration, x, of one of the products in an nth order chemical reaction is related to the reaction time t by the formula

$$kt = \int_0^x \frac{dc}{(a - c)^n}$$

where a and k are constants. Show that:
(i) if $n = 1$, $kt = \ln(a/(a - x))$
(ii) if $n \neq 1$, $kt = [1/(a - x)^{n-1} - 1/a^{n-1}]/(n - 1)$
If T denotes the value of t when $x = a/2$, find the value of T in the two cases $n = 1$ and $n \neq 1$.

12. In a particular third-order chemical reaction the reaction time t is related to the amount of a product, x, via the equation

$$kt = \int_0^x \frac{dc}{(2 - c)(3 - c)(1 - c)}$$

where k is the velocity constant. Work out the integral and rearrange the equation connecting t and x in the form of a quadratic equation from which x could be calculated for a given t value.

13. The speed V, in m s^{-1}, of a body moving in a viscous medium at time t s, satisfies the equation

$$t = \int_V^{100} \frac{dv}{v(1 + kv)}$$

where $v = 100$ is the speed of the body at time $t = 0$:
(i) Obtain the expression that gives t explicitly in terms of V.
(ii) Find the time taken for V to become 50.
(iii) Obtain V explicitly in terms of t.
(iv) Find the value of V when t becomes very large (i.e. tends to infinity).

14. The mean magnetic moment $\bar{\mu}$ of atoms in a given solid, that is subject to a magnetic field of strength H, is given by the expression

$$\bar{\mu} = \frac{\displaystyle\int_0^\pi \exp(\mu H \cos(\theta)/kT)\mu \cos(\theta) \sin(\theta)\, d\theta}{\displaystyle\int_0^\pi \exp(\mu H \cos(\theta)/kT) \sin(\theta)\, d\theta}$$

where T is the temperature, and k is Boltzmann's constant. Show that $\bar{\mu}$ is given by

$$\bar{\mu} = \mu \coth(\mu H/kT) - kT/H$$

15. In crystallography it is necessary to be able to work out integrals of the form $\int x^n \sin(mx)\,dx$ and $\int x^n \cos(mx)\,dx$, for positive integer values of n. Obtain suitable reduction formulae for evaluating these integrals.

16. The electric potential V a distance r from the centre of a sphere of radius a, carrying a charge e, is given by

$$V(r) = \frac{e}{2a} \int_{-a}^{a} \frac{dx}{(r^2 - 2rx + a^2)^{\frac{1}{2}}}$$

Show that

(i) if $r > a$, then $V(r) = e/r$;

(ii) if $r < a$, then $V(r) = e/a$.

11

Integration 3: further techniques

Contents: 11.1 scientific context – 11.1.1 electric potential;
11.1.2 chemical reaction times; 11.2 mathematical developments – 11.2.1 infinite limits; 11.2.2 coping with singularities;
11.3 worked examples; 11.4 further developments – 11.4.1 the
length of a plane curve; 11.4.2 volumes of revolution; 11.4.3
the centre of mass and moment of inertia of a solid of
revolution; 11.5 further worked examples; 11.6 exercises.

11.1 Scientific context

11.1.1 Example 1: electric potential. Fig. 11.1 shows some of the field lines in the $x-y$ plane of an electric field E. Let us consider the work done in moving a unit point charge (a point charge of magnitude 1 coulomb) in this field. Suppose the point charge moves along the x axis from the point B, $(b, 0)$ to the point A, $(a, 0)$ at constant speed. Then the work done by the charge against the electric field in moving a small distance dx is

$$dW = E \cos(\alpha) \, dx$$

In moving from B to A the total work done is the integral

$$W = \int_b^a E \cos(\alpha) \, dx$$

Now E and $\cos(\alpha)$ in general depend on x as we move along the x axis. For many vector fields the integral depends only on the end-points A and B and not on the path joining them. In such cases we define the *electric potential* $V(x)$ in terms of the primitive of the function $E \cos(\alpha)$ as

$$V(x) = -\int E \cos(\alpha) \, dx$$

With this definition for V we have

$$\int_b^a E \cos(\alpha) \, dx = -V(x) \Big|_b^a$$
$$= -(V(a) - V(b))$$
$$= V(b) - V(a)$$

The quantity $V(b) - V(a)$ is a measure of the difference in potential between the two points A and B. It is minus the work done by the charge in moving from B to A.

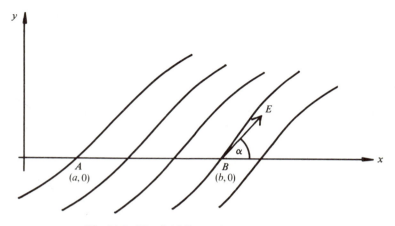

Fig. 11.1. The field lines of an electric field.

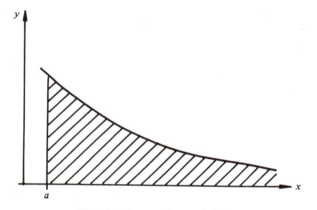

Fig. 11.2. Integrating to infinity.

Now it is often convenient in science to define the potential at a point and this requires a convention for a zero of potential. We usually choose the potential to be zero at infinity. The potential at a point A is then defined to be the work done in bringing a unit point charge from infinity to the point A in the field. In terms of the integral definition given above we then have

$$V(a) = -\int_{\infty}^{a} E\cos(\alpha)\,\mathrm{d}x$$

This integral for $V(a)$ introduces an infinite limit and is an example of *an improper integral*. In Section 11.2 we investigate how to evaluate such integrals.

Question: Could this integral be interpreted as an area?
  Answer: Geometrically this integral is the area shown in Fig. 11.2.

At first sight it might appear that since the distance between the limits is infinite then the value of the integral must be infinite too. However, we shall see that this is not always the case and improper integrals of this type do have finite values. This example illustrates a difficulty that may occur with the limits. The following example shows problems can occur in the integrand.

11.1.2 Example 2: chemical reaction times. In 10.1.2 we introduced an integral that occurs often in chemical kinetics. Consider a reaction between two substances, then the time taken for the concentration N to reach a certain value N_0 is given by the integral

$$T = \int_{N_1}^{N_0} \frac{1}{k(a-N)(b-N)} \, dN$$

where N_1 is the initial concentration. The function in the integrand in this example is singular at $N = a$ and $N = b$. Now if the limits N_0 or N_1 are equal to a or b or the range of the integration (N_0, N_1) contains either a or b then we are integrating up to or across the singularity. This may cause the integral to diverge and we have to evaluate such integrals with care. In the context of a chemical reaction the concentration cannot exceed the smaller of a and b, but nevertheless it can eventually reach this value. Thus, as we have said, we have to evaluate the corresponding integral with care.

11.2 Mathematical developments

The integrals that we have discussed in the last two chapters have all produced finite values when evaluated. But this is not always the case. Infinite integrals can occur when the limits of integration become infinite or the integrand itself becomes infinite in the region of integration. We begin this chapter by investigating these cases.

11.2.1 Infinite limits. Suppose that $F(x)$ is a primitive for the function $f(x)$ so that $F'(x) = f(x)$, then we define the integral with upper limit infinity in the following way:

$$\int_a^\infty f(x) \, dx = \lim_{s \to \infty} \int_a^s f(x) \, dx = \lim_{s \to \infty} (F(s) - F(a))$$

When this limit exists we say that the integral is *convergent*, otherwise it is *divergent*.

Example: Evaluate the following:

(i) $\displaystyle \int_2^\infty (1/x) \, dx;$

(ii) $\displaystyle \int_0^\infty e^{-2x} \, dx.$

Solution: (i) A primitive for the function $1/x$ is $\ln(x)$ so we write

$$\int_2^\infty (1/x)\,dx = \lim_{s\to\infty} \ln(x) \Big|_2^s$$

$$= \lim_{s\to\infty} \ln(s) - \ln(2)$$

$$= \infty \text{ as } s \to \infty$$

This integral is divergent.

(ii) A primitive for the function e^{-2x} is $-e^{-2x}/2$ so we write

$$\int_0^\infty e^{-2x}\,dx = \lim_{s\to\infty} -e^{-2x}/2 \Big|_0^s$$

$$= \lim_{s\to\infty} (-e^{-2s} + 1)/2$$

$$= 1/2 \text{ in the limit}$$

This integral is convergent and has the finite value $\frac{1}{2}$.

We adopt a similar approach if the lower limit is infinity. We define

$$\int_{-\infty}^b f(x)\,dx = \lim_{s\to\infty} F(x) \Big|_{-s}^b = \lim_{s\to\infty} (F(b) - F(-s))$$

The integral is convergent if the limit exists. Although these are formal definitions of integrals with infinite limits, we do not usually write the solution as a limit process. We treat ∞ as if it were a finite limit although we must be on our guard in case in the limit the integral is divergent.

Example: Evaluate $\int_{-\infty}^\infty 1/(a^2 + x^2)\,dx$.

Solution: A primitive in this case is $(1/a)\arctan(x/a)$, thus

$$\int_{-\infty}^\infty 1/(a^2 + x^2)\,dx = (1/a)\arctan(x/a) \Big|_{-\infty}^\infty$$

$$= (1/a)(\pi/2 - (-\pi/2))$$

$$= \pi/a$$

11.2.2 Coping with singularities. A second problem that can occur when evaluating integrals is the function in the integrand becoming infinite at the end-points or at other points within the range of integration. In other words, the function is singular at some point $x = c$. For example, in the integral $\int_0^1 1/\sqrt{(1 - x^2)}\,dx$ the integrand is undefined at the end-point $x = 1$ (see Fig.11.3(a)) and in the integral $\int_{-1}^1 1/x^2\,dx$ the integrand is undefined at the point $x = 0$ (see Fig. 11.3(b)).

To investigate what happens to the integral of such functions, we divide the range into intervals in which the integrand is continuous and investigate the limit of the integral as the end points are approached. The next example considers the case where the integrand is singular at one of the end-points.

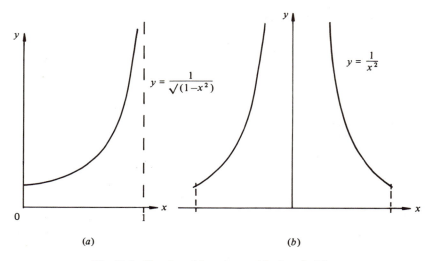

Fig. 11.3. Graphs of functions with singularities.

Example: Evaluate $\int_0^1 1/\sqrt{(1 - x^2)}\,dx$.

Solution: The integrand becomes infinite when $x = 1$ so consider the integral between $x = 0$ and $x = 1 - s$. Then

$$\int_0^{1-s} 1/\sqrt{(1 - x^2)}\,dx = \arcsin(x) \Big|_0^{1-s}$$

$$= \arcsin(1 - s) - 0$$

Now the original integral between the limits $x = 0$ and $x = 1$ is obtained if we take the limit as $s \to 0$. In this limit $\arcsin(1 - s)$ approaches the finite value $\pi/2$. So we write

$$\int_0^1 1/\sqrt{(1 - x^2)}\,dx = \lim_{s \to 0} \int_0^{1-s} 1/\sqrt{(1 - x^2)}\,dx = \pi/2$$

In general, then, if the integrand becomes infinite at either end-point of the range (or both) then we define the integral $\int_a^b f(x)\,dx$ in the following way:

if $F(x)$ is a primitive for $f(x)$ and both $F(a + s)$ and $F(b - s)$ tend to finite limits as $s \to 0$, then

$$\int_a^b f(x)\,dx = \lim_{s \to 0} F(b - s) - F(a + s)$$

Clearly if $F(a + s)$ or $F(b - s)$ do not tend to finite limits then the integral diverges. This definition may be extended to the integration of functions which are undefined at various points within the range of integration. We use Property 1 of 9.2.1.

Consider the integrand $f(x)$ which is singular at the point $x = c$ within the range $x = a$ and $x = b$. Property 1 gives

$$\int_a^b f(x)\,dx = \int_a^c f(x)\,dx + \int_c^b f(x)\,dx$$

The singular point is now a limit in each case so that we proceed as above. If $F(x)$ is a primitive for $f(x)$ then the integral exists providing both $F(c - s)$ and $F(c + s)$ approach a finite limit as s tends to 0.

The following example illustrates that care must be taken in such cases.

Example: Evaluate $\int_{-1}^{1} 1/x^2 \, dx$.

Solution: There is a singularity in the integrand at $x = 0$ so we write the integral as

$$\int_{-1}^{1} 1/x^2 \, dx = \int_{-1}^{0} 1/x^2 \, dx + \int_{0}^{1} 1/x^2 \, dx$$

Consider the first integral on the right-hand side. Writing the upper limit as $0 - s$ we have

$$\int_{-1}^{0-s} 1/x^2 \, dx = (-1/x) \Big|_{-1}^{0-s} = -(1/(0 - s) - 1/(-1))$$

and this has no finite limit as s tends to 0 so that the integral diverges. Similarly, the second integral on the right-hand side does not have a finite value either. Hence $\int_{-1}^{1} 1/x^2 \, dx$ diverges.

The care that must be taken is important when the integrand is a singular function. It would be *incorrect* to write the following:

$$\int_{-1}^{1} 1/x^2 \, dx = (-1/x) \Big|_{-1}^{1} = (-1/1) - (-1/-1) = -2$$

11.3 Worked examples

Example 1. The electric field produced by a positive point charge Q at the origin has magnitude Q/r^2 and direction radially outwards from the origin. Find the electric potential at any point in the region.

Solution. Consider a point P which is distance a from the origin (Fig. 11.4). If we bring a unit positive charge from a large distance to the point P along a radial

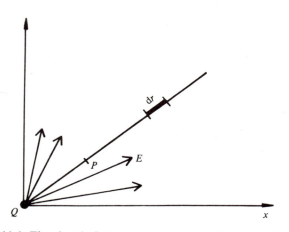

Fig. 11.4. The electric field produced by a positive point charge.

line, then the work done by the charge against the electric field is the integral

$$W = -\int_a^\infty \frac{Q}{r^2} \, dr$$

The electric potential of this electric field is given by the improper integral

$$V(a) = \int_a^\infty \frac{Q}{r^2} \, dr$$

(Note that this is the same as the integral in 11.1.1 with $\alpha = 0$ and $E = Q/r^2$.) A primitive for the function $1/r^2$ is $-1/r$, so

$$V(a) = -Q/r \Big|_a^\infty$$

$$= -0 + Q/a$$

This potential is usually written as Q/r and is called the potential field of a point charge Q placed at the origin.

Example 2. A motor under load generates heat at a constant rate and radiates it at a rate proportional to the difference in temperature between itself and the surroundings (T_0). The time taken for the temperature to rise from T_0 to T_1 is given by the integral

$$t = \int_{T_0}^{T_1} \frac{1}{10 - k(T - T_0)} \, dT$$

where T is the temperature and k is constant. Find the time taken for the motor to reach a temperature of $10/k + T_0$.

Solution. In this example the denominator of the integrand is zero at the upper limit $T_1 = 10/k + T_0$, i.e. there is a singularity in the integrand at the upper limit. We therefore write the upper limit as $T_1 - s$ and consider the limit as s tends to zero. We have

$$t = \int_{T_0}^{T_1 - s} \frac{1}{10 - k(T - T_0)} \, dT$$

$$= (-1/k) \ln(10 + kT_0 - kT) \Big|_{T_0}^{T_1 - s}$$

$$= (-1/k)(\ln(s) - \ln(10))$$

Now in the limit as s tends to 0 the integral diverges and does not have a finite limit.

However, there is a physical meaning that we can attach to the infinite value for the time t. It means that the temperature of the motor is only $10/k + T_0$ after a very long time. For example, a typical value of k is 0.16 so that the eventual temperature rise of the motor will be 62.5 °C. The following table showing the temperature rise at 5 min time intervals indicates that after 30 min the temperature rise is within 0.5 °C of 62.5 °C:

Time min	5	10	15	20	25	30
Temperature °C	34.4	49.9	56.8	60.0	61.4	62.0

11.4 Further developments

When integration was introduced in Chapter 9 we pointed out that integration was concerned with bringing things together. By summing together small areas we were able to interpret the definite integral as the area under a curve. In this section by adding together small lengths and small volumes, respectively, we illustrate how integration may be used to find lengths of curves and volumes of solids of revolution. We shall see that the integration skills developed in Chapters 9–11 can be put to use in such problems, in each case it is the definition of integration as the limit of a sum that provides the formula.

11.4.1 The length of a plane curve. Consider the curve drawn in the $(x-y)$ plane defined by the function $y = f(x)$ between $x = a$ and $x = b$. The length of the curve may be approximated by the perimeter of the polygon of line segments $P_0 P_1$, $P_1 P_2, \ldots, P_{n-1} P_n$ (see Fig. 11.5). We have

$$L = P_0 P_1 + P_1 P_2 + P_2 P_3 + \cdots + P_{i-1} P_i + \cdots + P_{n-1} P_n$$

Consider the ith line segment between P_{i-1} and P_i. If the x coordinates of these points differ by h and the y coordinates by k, then we may write

$$P_i P_{i-1} = \sqrt{(h^2 + k^2)}$$
$$= \sqrt{(1 + (k/h)^2)}h$$
$$= \sqrt{\left[1 + \left(\frac{f(x+h) - f(x)}{h}\right)^2\right]}h$$

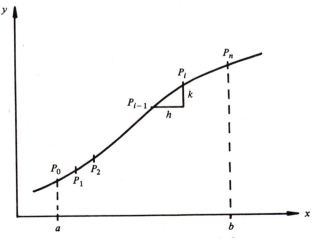

Fig. 11.5. The curve divided into straight segments.

Now if we choose another point between P_i and P_{i-1} it is clear that the length of the sum has increased although the sum of the lengths of the two lines is a closer approximation to the length of the curve. Now as we increase the number of points subdividing the curve, and hence increase the number of line segments, the sum L either increases to a finite limit (i.e. the length of the curve) or to an infinite limit. For a finite limit the curve is called *rectifiable*. For such a curve the length is approximated by the following summation:

$$L = \sum_i \sqrt{\left[1 + \left(\frac{f(x_i + h) - f(x_i)}{h}\right)^2\right]} h$$

Hence in the limit as the number of line segments increases we have

$$\text{curve length} = \int_a^b \sqrt{[1 + (f'(x))^2]} \, dx \qquad (1)$$

This formula is correct provided that $\lim_{h \to 0}$ (arc length $P_{i-1}P_i$/line segment $P_{i-1}P_i) \to 1$. (This condition can be shown to exist provided $f'(x)$ is continuous.)

Example: Find the length of the curve $y = (x + 2)^{\frac{3}{2}}$ between $x = 0$ and $x = 2$.
Solution: To apply the formula we must first find $f'(x)$. We have $f'(x) = 3(x + 2)^{\frac{1}{2}}/2$. Thus the required curve length is given by the integral

$$L = \int_0^2 \sqrt{[1 + 9(x + 2)/4]} \, dx$$

To integrate this we make the substitution $u = 1 + 9(x + 2)/4$, then $du = \left(\frac{9}{4}\right) dx$ and in terms of u the integral becomes

$$L = \int_{11/2}^{10} (4/9) u^{\frac{1}{2}} \, du$$

$$= (4/9)(2/3) u^{\frac{3}{2}} \Big|_{11/2}^{10}$$

$$= 5.55 \text{ (to two decimal places)}$$

In many cases the curve is defined parametrically by $x = x(t)$ and $y = y(t)$ and it is convenient to write the expression for the length in terms of $x'(t)$ and $y'(t)$ and integrate over the parameter t between appropriate limits. Consider Fig. 11.5 again; the length of the line segment $P_i P_{i-1}$ can be expressed as $ds = \sqrt{(dx^2 + dy^2)}$ where we have written $h = dx$ and $k = dy$. The length of the curve is then given as the integral $\int ds$ or using the chain rule $\int (ds/dt) \, dt$

Now

$$\frac{ds}{dt} = \sqrt{\left[\left(\frac{dx}{dt}\right)^2 + \left(\frac{dy}{dt}\right)^2\right]}$$

Hence the length of the curve given in parametric form is

$$L = \int \sqrt{[x'(t)^2 + y'(t)^2]} \, dt \qquad (2)$$

Example: Find the length of the curve given parametrically by $x = 1 + t^2/2$,
$y = t \cosh(t) - \sinh(t)$ between the values $t = 0$ and $t = 1$.

Solution: First we find the derivatives of $x(t)$ and $y(t)$. These are $x'(t) = t$ and $y'(t) = t \sinh(t)$.
Using the parametric form for the curve length (2) we have

$$L = \int_0^1 \sqrt{[t^2 + t^2 \sinh^2(t)]}\, dt$$

$$= \int_0^1 t \cosh(t)\, dt$$

(using the identity $\cosh^2(t) - \sinh^2(t) = 1$). Now integrating by parts

$$L = t \sinh(t) \Big|_0^1 - \int_0^1 \sinh(t)\, dt$$

$$= t \sinh(t) \Big|_0^1 - \cosh(t) \Big|_0^1$$

$$= \sinh(1) - \cosh(1) + 1$$

$$= 1.1752 - 1.5431 + 1$$

$$= 0.6321$$

11.4.2 Volumes of revolution. In a similar way that a plane area can be found by summing small area elements, the volumes of many solids can be found by dividing the solid into elementary volumes and summing. For example, consider the solid shown in Fig. 11.6. The volume of the solid can be found by summing the volumes of the slices of area $A(x)$ and thickness dx. We have

$$V = \sum_i A_i(x)\, dx$$

In the limit as the number of slices increases the right-hand side becomes an integral and then

$$V = \int A(x)\, dx$$

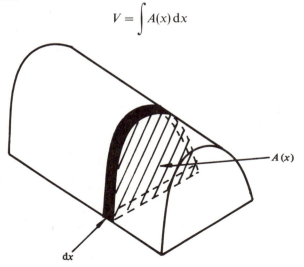

Fig. 11.6. Using integration to find volumes.

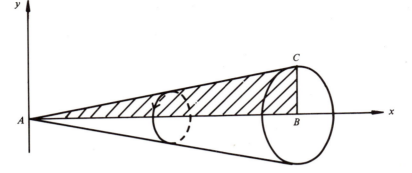

Fig. 11.7. A cone formed by rotating a triangle.

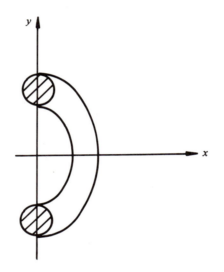

Fig. 11.8. A torus formed by rotating a circle.

This method of approach is particularly useful when the solid is a *solid of revolution*. Such a solid is formed by rotating a plane area about an axis (Fig. 11.7). For example, a cone is formed when the triangle ABC is rotated about the x axis and a 'doughnut' shape or torus, as it is called, is formed when the circle shown in Fig. 11.8 is rotated about the x axis.

To find the volume of such solids we use the symmetry of the solid about the axis of rotation. Consider the solid formed by rotating about the x axis the area between the curve $y = f(x)$, the x axis and the lines $x = a$, $x = b$ (see Fig. 11.9). The solid can be divided up into n elementary volume elements which are discs each with the same thickness dx. The ith disc has radius y_i. Its volume is

$$\mathrm{d}V = \pi y_i^2 \, \mathrm{d}x$$

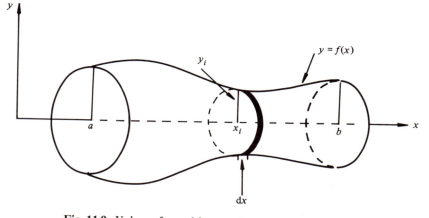

Fig. 11.9. Volume formed by rotating f about the x-axis.

Since the area is bounded by the curve $y = f(x)$ the volume element can be written in terms of f. We obtain

$$dV_i = \pi f(x_i)^2 \, dx$$

The volume of the solid of revolution is found by summing all the elementary volumes:

$$V = \sum_i dV_i = \sum_i \pi f(x_i)^2 \, dx$$

In the limit as the thickness of the volume elements tends to zero, and the number of volume elements increases indefinitely, the volume of the solid is given by the integral

$$V = \int_a^b \pi f(x)^2 \, dx$$

Example: Find the volumes of the following solids:

    (i) a cone of length h and base radius a;

    (ii) a sphere formed by rotating the semicircle of radius a, centre the origin about the x axis.

Solution: The first step in finding these volumes is to express each boundary curve in the x–y plane as function of x.

    (i) For the cone he curve is a straight line with equation $y = f(x) = ax/h$ (Fig. 11.10). The area between this line, $x = 0$ and $x = a$, is rotated about the x axis to form the cone. Hence the volume of the cone is given by

$$V = \int_0^h \pi(ax/h)^2 \, dx$$

$$= \pi(a/h)^2 x^3/3 \Big|_0^h$$

$$= \pi a^2 h/3$$

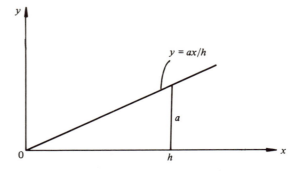

Fig. 11.10. The boundary curve of a cone.

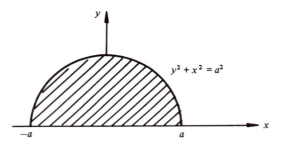

Fig. 11.11. The boundary curve of a sphere.

This is recognisable as the well-known formula 'a third the area of the base times the height'.

(ii) For the sphere the semicircle has equation $y = f(x) = \sqrt{(a^2 - x^2)}$ (Fig. 11.11). The volume of the sphere formed by rotating this curve around the x axis is then

$$V = \int_{-a}^{a} \pi(a^2 - x^2)\,dx$$

$$= \pi(a^2 x - x^3/3)\Big|_{-a}^{a}$$

$$= 2\pi(2a^3/3)$$

$$= 4\pi a^3/3$$

11.4.3 The centre of mass and moment of inertia of a solid of revolution. In the previous subsection we used the symmetry of a solid of revolution to find its volume. A similar approach can be used to find other properties of such a solid. Consider, for example, the problem of finding the centre of mass and moment of inertia of a cone of constant density ρ (Fig. 11.12). If we divide the cone into elementary discs as before, then the centre of mass of the disc is at its centre, i.e. the point $(x_i, 0)$. By symmetry the centre of mass of the cone lies on the x axis at the point $(\bar{x}, 0)$, say. If we take moments about the y axis summing the contributions

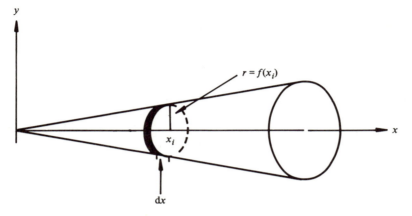

Fig. 11.12. Subdividing a cone into discs.

from each elementary disc we obtain

$$M\bar{x} = \sum_i dV_i x_i \rho$$

$$= \sum_i \pi \rho f(x_i)^2 x_i\, dx$$

where M is the mass of the cone.

Proceeding to the limit in the usual way gives the following integral

$$\bar{x} = \int_0^h \pi \rho x f(x)^2\, dx / M$$

For the cone $f(x) = ax/h$ and $M = \pi \rho a^2 h/3$. Thus

$$\bar{x} = \int_0^h \rho x \pi (ax/h)^2\, dx / (\pi \rho a^2 h/3)$$

$$= (3/h^3) x^4/4 \Big|_0^h$$

$$= 3h/4$$

Consider the moment of inertia of the cone about the x axis. We begin by considering the moment of inertia of the disc about the x axis. This is given by

$$dI = dM r^2/2$$
$$= \rho \pi\, dx\, f(x_i)^4/2$$

since $r = f(x_i)$ in this case. The moment of inertia of the cone about the x axis is the sum of the moments of inertia of the individual discs. Summing and proceeding to the limit in the usual way gives the integral

$$I = \int_0^h \pi \rho f(x)^4/2\, dx$$

For the cone $f(x) = ax/h$ so that

$$I = \pi \rho a^4 h / 10$$

11.5 Further worked examples

Example 1. A telegraph wire hanging freely between two posts has the shape of a catenary defined by the equation $y = c \cosh(x/c)$ where the parameter c can be found from the 'sag'. Find a formula for the length of wire between the posts. Consider a wire suspended between poles 100 m apart with a sag of 100 cm. Find the length of wire used.

Solution. Suppose that the posts are a distance $2a$ apart and placed at the points $x = a$ and $x = -a$ so that the wire is symmetrical about the y axis. The length of the curve defined by the function $f(x)$ is given by the integral

$$L = \int_{-a}^{a} \sqrt{[1 + f'(x)^2]} \, dx$$

In this case $f(x) = c \cosh(x/c)$ so that $f'(x) = \sinh(x/c)$. Now $1 + \sinh^2(x/c) = \cosh^2(x/c)$ so the integral becomes

$$L = \int_{-a}^{a} \cosh(x/c) \, dx$$

$$= c \sinh(x/c) \Big|_{-a}^{a}$$

$$= 2c \sinh(a/c)$$

In this problem we have $a = 50$ m and the sag $= 100$ cm, i.e. sag $= 1$ m. Now the sag is given by

$$f(50) - f(0) = c \cosh(50/c) - c$$
$$= 1$$

Dividing throughout by c and letting $u = 50/c$ the equation to solve is

$$0.02u = \cosh(u) - 1$$

To solve this equation we can use the method of interval bisection (introduced in 1.4.2) or the Newton–Raphson method (of 8.4.1). To two decimal places the solution is $u = 0.04$ so that $c = \frac{50}{0.04} = 1250$. The length of the telegraph wire is

$$L = 1250 \times 2 \times \sinh(50/1250) = 100.03 \text{ m}$$

Example 2. A gun shell has the shape of a solid of revolution defined by rotating about the x-axis the area bounded by the curve $y^2 = x/16$ and the lines $y = 0$, $x = 1$. The shell is made from a constant density material. When in motion the shell spins about the axis of symmetry. Find the mass of the shell, the position of its centre of mass and the moment of inertia of the shell about the axis of symmetry.

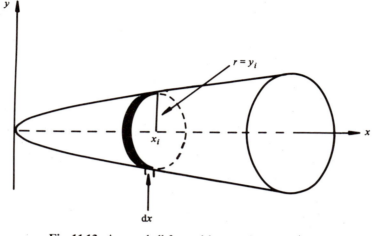

Fig. 11.13. A gun shell formed by rotating $f = \sqrt{x}/4$.

Solution. A diagram of the shell is shown in Fig. 11.13. Suppose we subdivide the shell into elementary discs like the one shown at the point x_i. The radius of the disc r is given by

$$r^2 = y_i^2$$
$$= x_i/16$$

(i) The volume of the elementary disc is $dV = \pi r^2 dx$, and its mass $dM = \rho dV$ where ρ is the (constant) density. The mass of the shell is given by the definite integral

$$M = \int_0^1 \pi\rho r^2 \, dx$$
$$= \int_0^1 \pi\rho x/16 \, dx$$
$$= \pi\rho x^2/32 \Big|_0^1 = \pi\rho/32$$

(ii) The centre of mass lies on the axis of symmetry, the x axis; let the x coordinate of the centre of mass be $\bar{X}$. Then summing the moments about the y axis and proceeding to the limit we have

$$M\bar{X} = \int x \, dM = \int_0^1 x\pi\rho r^2 \, dx$$
$$= \int_0^1 \pi\rho x^2/16 \, dx$$
$$= \pi\rho/48$$

Now dividing by the mass from part (i) we obtain

$$\bar{X} = 2/3$$

(iii) For the moment of inertia of the shell we sum the moments of inertia of the elementary discs. The moment of inertia of a disc of radius r about a perpendicular axis through its centre is $dMr^2/2$. The moment of inertia of the shell is then

$$I = \int (1/2)r^2 \, dM$$

$$= (1/2) \int_0^1 \rho(\pi r^2)r^2 \, dx$$

$$= (1/2) \int_0^1 (\pi\rho x^2/256) \, dx$$

$$= \pi\rho/1536$$

$$= M/48$$

where M is the mass of the shell.

11.6 Exercises

---

1. Find the values of the following integrals, when they exist:

   (i) $\displaystyle\int_1^\infty \frac{1}{x^3} dx$ (ii) $\displaystyle\int_0^\infty e^{-x} dx$

   (iii) $\displaystyle\int_1^\infty \frac{1}{\sqrt{x}} dx$ (iv) $\displaystyle\int_0^\infty \frac{x}{1+x^2} dx$

   (v) $\displaystyle\int_1^\infty \frac{1}{x^2 + 4x + 13} dx$ (vi) $\displaystyle\int_{-\infty}^0 xe^{2x} dx$

   (vii) $\displaystyle\int_0^\infty e^{-x} \sin(x) dx$

2. Find the values of the following integrals, when they exist:

   (i) $\displaystyle\int_0^1 \frac{1}{x^{2/3}} dx$ (ii) $\displaystyle\int_0^a \frac{x}{\sqrt{(a^2 - x^2)}} dx$

   (iii) $\displaystyle\int_0^1 1/x \, dx$ (iv) $\displaystyle\int_0^4 1/(x-1)^2 \, dx$

   (v) $\displaystyle\int_0^1 (1+x)/(1-x) \, dx$ (vi) $\displaystyle\int_1^2 1/(x^2 - 1) dx$

3. In working out the mean speed of molecules in a gas it is necessary to work out an integral of the form $I = \int_0^\infty x \exp(-\beta x^2) dx$. By first evaluating

$\int_0^L x \exp(-\beta x^2) \, dx$ and letting L tend to infinity, show that the value of I is $\frac{1}{2}\beta$.

4. The mean life time τ of a particular unstable particle is given by

$$\tau = \frac{1}{a} \int_0^\infty t e^{-t/a} \, dt$$

where $a = 2.1 \times 10^{12} \, \mathrm{s}^{-1}$. By first considering $\int_0^L t e^{-t/a} \, dt$, and letting $L \to \infty$, obtain the value of τ.

5. Van der Waal's equation of state gives the pressure p of a gas in terms of its volume V and temperature T as

$$p = \frac{RT}{(V-b)} - \frac{a}{V^2}$$

In an isothermal change the work done in compressing a gas from a volume V_1 to a volume V_2 is $\int_{V_1}^{V_2} p \, dV$. Obtain this work done using the expression for p given, and investigate whether it is possible to compress a gas to the volume given by $V_2 = b$.

6. The phase change suffered by a light wave in a particular scattering experiment is estimated to be given by an expression involving the following integral

$$\int_{-\infty}^\infty \frac{b \, dx}{[R^2 + (b+x)^2]}$$

where b and R are constants, having dimensions of length. Show that the value of the integral exists and obtain its value.

7. In electrostatics, the electric potential V around a charged oblate conducting spheroid is conveniently expressed in terms of one of the oblate spheroidal coordinates u as

$$V = \int_u^\infty \frac{A \, dc}{\cosh(c)}$$

where A is a constant, related to the charge. By making the substitution $x = e^c$, show that

$$V = 2A \cot^{-1}(e^c)$$

8. (a) A parabola can be regarded as the curve whose equation is the relation $y^2 = 4ax$. Any point (x, y) on the parabola can be parametrised by

$$x = at^2, \quad y = 2at$$

Obtain the length of the curve between the origin, given by $t = 0$, and the point P given by

$$x = aq^2, \quad y = 2aq.$$

(b) A projectile fired with speed u at an angle α to the horizontal moves under gravity in a parabolic trajectory. Find the actual distance travelled by the projectile in achieving its maximum height. N.B. The distance travelled will be the same as the arc length on a parabola, with $a = u^2 \cos^2(\alpha)/2g$, between points where $t = 0$ and $t = \tan(\alpha)$.

9. The trajectory of any planet about the sun is an ellipse. An ellipse can be regarded as the set of points (x, y) satisfying

$$\frac{x^2}{a^2} + \frac{y^2}{b^2} = 1$$

where a and b are the lengths of the major and minor axes of the ellipse.

(i) Show that a parameterization of the ellipse is given by

$$x = a \sin(\theta), \quad y = b \cos(\theta), \quad 0 \leqslant \theta < 2\pi$$

(ii) Show that the circumference, S, of an ellipse is given by the expression

$$S = 4 \int_0^{\pi/2} [a^2 \sin^2(\theta) + b^2 \cos^2(\theta)]^{1/2} \, d\theta$$

(iii) If $a = 1$, $b = 0.8$, use Simpson's rule with ten intervals to obtain the value of S.

10. In many applications in mechanics and engineering it is necessary to know the moment of inertia of a solid sphere. If the sphere has radius R and density ρ obtain the moment of inertia of the sphere about any axis through its centre, and express the result in terms of R and the mass M of the sphere.

12

First-order ordinary differential equations

Contents: 12.1 scientific context – 12.1.1 chemical reactions; 12.1.2 Newton's law of cooling; 12.2 mathematical developments – 12.2.1 classification of differential equations; 12.2.2 variables separable; 12.2.3 linear; 12.2.4 finding particular solutions; 12.3 worked examples; 12.4 further developments – 12.4.1 a numerical solution; 12.4.2 improving the accuracy; 12.5 further worked examples; 12.6 exercises.

Many physical situations are represented mathematically by equations that involve a variable and its rate of change. For example, a mathematical model of the natural cooling of a mug of coffee relates the temperature T of the coffee to its rate of change. In symbolic form we would write this as

$$\frac{dT}{dt} = \alpha T + \beta$$

Such an equation cannot be integrated directly with respect to the time variable t using the techniques of the previous three chapters, because the right-hand side involves the function T. We need a new set of techniques to solve equations of this type. An equation which involves the dependent variable (in this example T) and its derivatives is called an *ordinary differential equation* and in the next two chapters we investigate methods of solving them.

We begin with differential equations involving only the first derivative of the dependent variable.

12.1 Scientific context

12.1.1 Example 1: chemical reactions. Mathematical models in science can often be expressed in terms of an equation of balance. For example, the balance of heat in a system implies that the heat lost from one part of the system is gained by another part of the system. If the property that we are balancing changes continuously then the equation of balance will contain derivatives and so lead to a differential equation. The equation of balance may often be expressed in the simple form of an input–output process in the following way:

$$\text{quantity input} - \text{quantity output} = \text{accumulation}$$

This simple equation is often one of the basic steps in formulating a mathematical model of a physical situation. Take, for example, a model of the reactions between chemical materials.

305

Chemical reactions are, in general, complicated processes so that a mathematical description is difficult. However, the following simple reaction illustrates the use of the input–output equation and the occurrence of a differential equation in chemistry. Consider a simple irreversible reaction in which a chemical substance A is transformed into another chemical substance B. Suppose that at some time t the number of molecules of A and B are n_a and n_b respectively, and in a short time δt these increase by δn_a and δn_b. First we formulate a model that describes the change in the amount of substance A. Since there is no input of the substance A into the reaction once it has started we have

$$\text{quantity input} = 0$$

To model the amount of the substance A that is transformed into a substance B during the time interval δt we assume that it is proportional to (i) the amount of substance A that is present at the start of the time interval, and (ii) the time interval δt. We then have

$$\text{quantity output} = K n_a \delta t$$

This seems a reasonable model. If there is no substance A, i.e. $n_a = 0$, then the reaction will stop because there is no quantity to output. Also the longer the time interval then the more of substance A will be converted into substance B.

The accumulation of substance A is just

$$(n_a + \delta n_a) - n_a$$

i.e. the amount at the end of the time interval minus the amount at the beginning. Putting these three terms into the input–output equation we have

$$0 - K n_a \delta t = (n_a + \delta n_a) - n_a = \delta n_a$$

and K is called the reaction constant. This equation can be written as

$$\frac{\delta n_a}{\delta t} = - K n_a$$

In the same way as we express rates of change at instants of time in Chapter 6, we usually model chemical reactions as occurring continuously and not over discrete time intervals δt. So that by taking the limit as $\delta t \to 0$ in the usual way, the reaction equation becomes a differential equation

$$\frac{\mathrm{d} n_a}{\mathrm{d} t} = - K n_a$$

This is an example of a *linear* differential equation. The equation for the rate of change of substance B can be found in the same way, we have

$$\text{quantity input} = K n_a \delta t$$
$$\text{quantity output} = 0$$
$$\text{accumulation} = (n_b + \delta n_b) - n_b$$

so that the input–output equation becomes

$$K n_a \delta t = \delta n_b$$

and in the limit

$$\frac{dn_b}{dt} = Kn_a$$

The input–output process applied to this simple reaction leads to a *pair* of differential equations. A more complicated situation is that of a bimolecular reaction.

In a bimolecular reaction there are two substances present, A and B. The reaction between them proceeds as the result of many encounters between the molecules, in each of which one molecule of A combines with one molecule of B to form a new substance C. A simple model would be to *assume* that the rate of formation of molecules of C is proportional to the concentrations of A and B.

Suppose that, at time t, N is the number of molecules of C, and that n_a and n_b are the numbers of molecules of A and B respectively.

The assumption above can be expressed mathematically using the input–output principle as

$$Kn_a n_b \delta t - 0 = (N + \delta N) - N$$

which in terms of the derivatives of N becomes in the limit

$$\frac{dN}{dt} = Kn_a n_b$$

where K is again a reaction constant.

Now if the original numbers of molecules of substances A and B were a and b say, and if one molecule of A combines with one molecule of B to produce two molecules of C we can express n_a and n_b each in terms of N: we have

$$\text{number of molecules of A 'lost'} = a - n_a = N/2$$

and

$$\text{number of molecules of B 'lost'} = b - n_b = N/2$$

so that the differential equation becomes

$$\frac{dN}{dt} = K\left(a - \frac{N}{2}\right)\left(b - \frac{N}{2}\right)$$

Question: Is this differential equation linear?
Answer: No. The right-hand side is a quadratic function of N.

This is an example of a *non-linear* differential equation.

12.1.2 Example 2: Newton's law of cooling. In this second example we consider the problem of modelling the cooling of a quantity of liquid. For example, we may be interested in the cooling of the water in a domestic hot water tank heated by a central heating system, or the cooling of a liquid in an industrial process or even the cooling of a cup of tea or coffee.

Hot substances cool 'naturally' by transferring heat energy to the surroundings. There are two main modes of heat transfer in such problems, conduction and convection. In *conduction*, energy is transferred from one portion of a material to another because of the contact between the molecules or atoms making up the substance. The molecules or atoms in the hotter part of the material move faster than in the cooler part, and this excites the molecules or atoms in the cooler part to move faster. Energy is passed through the material in this way, but the material itself shows no apparent bodily movement.

Convection is associated with liquids and gases (i.e. fluids). Energy is transferred by virtue of the movement of the fluid. The fluid in the hotter region becomes less dense and rises to mix with the cooler fluid. When the hotter fluid mixes with the cooler fluid, heat energy is transferred mainly by conduction. This movement of the fluid can be seen in the form of convection currents, i.e. a flow in the fluid due to the temperature differences. Experiments show that the rate of heat transfer is proportional to the temperature difference between the liquid and the surroundings. We may write this as

$$Q \propto (T - T_0)$$

where Q is the rate of heat transfer.

This is a mathematical model that describes the cooling process and includes both the processes of conduction and convection. The heat energy is transferred through the container by conduction and to the surroundings by convection. We can model each process separately by a linear equation, so that a linear model for the more complicated combination for 'natural cooling' is a reasonable assumption.

The heat content of a material can be expressed in terms of its temperature T so that Q is also proportional to the rate of change of the temperature of the substance, i.e.

$$Q \propto \frac{dT}{dt}$$

Putting together the two expressions for Q we can write

$$\frac{dT}{dt} = K(T - T_0)$$

K is a constant for the substance and is negative if the substance is cooling. This differential equation is often called 'Newton's law of cooling'.

If we put $T - T_0 = \theta$ then the differential equation takes on the same form as the equation for n_a in 12.1.1. We have

$$\frac{d\theta}{dt} = K\theta$$

This linear, first-order differential equation can be used to model many physical situations. The cooling (or heating) of a substance is one example and the concentration of a substance in a simple chemical reaction is another. Further examples are:

(i) A radioactive substance gives off harmful radiation and by doing so changes into another substance. The rate at which the radioactive material decays at some

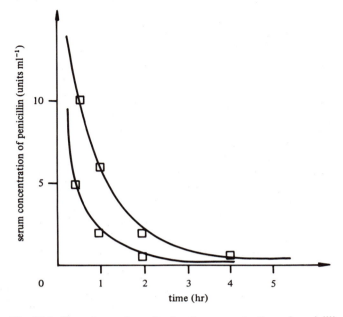

Fig. 12.1. Experimental results for the concentration of penicillin.

time t is proportional to the mass of the material present at time t. This can be modelled by the equation

$$\frac{dm}{dt} = -\alpha m$$

for some constant α which is related to the half-life of the material.

(ii) The study of the way in which a drug loses its concentration in the blood of a patient is fundamental to pharmacology. Fig. 12.1 shows the results of an experiment that measures the concentration of penicillin in the bloodstream. Clearly, the concentration is not linear with time. The simplest model is to assume that the rate of change in the concentration is proportional to the concentration. Again in symbolic form we have the first-order linear differential equation,

$$\frac{dc}{dt} = -\alpha c$$

Experiments show that this model gives a good representation of reality.

Question: Can the differential equations formulated in this section be written in a standard form?

Answer: Yes they can. If we choose y as the dependent variable and x as independent variable, then all the differential equations can be written as $dy/dx = f(x, y)$ where f is a function of x and y only.

We now begin the search for solutions of equations of this type.

12.2 Mathematical developments

12.2.1 Classification of differential equations. We begin by defining a first-order ordinary differential equation (or o.d.e. in our shorthand notation) as an equation which contains an independent variable x, a dependent variable y and the first-order derivative dy/dx. We have seen that such equations occur as models of physical reality. Two examples of first-order ordinary differential equations are:

(i) radioactive decay

$$\frac{dm}{dt} = -\alpha m;$$

in this example the independent variable is time (t) and the dependent variable is mass (m);

(ii) bimolecular chemical reaction

$$\frac{dN}{dt} = -K\left(a - \frac{N}{2}\right)\left(b - \frac{N}{2}\right);$$

in this example the independent variable is time (t) and the dependent variable is the number of molecules (N).

The two examples illustrate an important way in which ordinary differential equations may be classified. In the first example the derivative varies linearly with the dependent variable, whereas in the second example the variation is non-linear (in fact it is quadratic $ab - \frac{1}{2}(a + b)N + \frac{1}{4}N^2$). In general we may write a first order o.d.e. as

$$\frac{dy}{dx} = f(x, y) \tag{1}$$

and we say that it is *linear* if $f(x, y)$ is a function of x or a linear function of y; otherwise (1) is *non-linear*.

Example: Which of the following o.d.e.s are linear?

(i) $\dfrac{dy}{dx} = \alpha y + \beta;$

(ii) $\dfrac{dy}{dx} = e^x;$

(iii) $\dfrac{dy}{dx} = e^y;$

(iv) $y\dfrac{dy}{dx} = x^2 + 4.$

Solution: Clearly, according to our definition (i) and (ii) are linear and (iii) and (iv) are non-linear. Example (iv) is posed in non-standard form and at first sight it may appear linear because $x^2 + 4$ is a function of x only. However, the function $f(x, y)$ is $(x^2 + 4)/y$ which is clearly non-linear.

A *solution* to an o.d.e. is any function $y(x)$ whose derived function dy/dx equals the function $f(x, y(x))$. Finding the solution of an o.d.e. involves integration so that we would expect unknown constants of integration to appear as they did in the chapters involving integration. For example, the function $y = Ae^{3x} - \frac{2}{3}$ is a solution to the first-order o.d.e. $dy/dx = 3y + 2$, because if

$$y = Ae^{3x} - \tfrac{2}{3}$$

then

$$\frac{dy}{dx} = 3Ae^{3x}$$

and

$$3y + 2 = 3(Ae^{3x} - \tfrac{2}{3}) + 2 = 3Ae^{3x}$$

Hence we say that $y = Ae^{3x} - \frac{2}{3}$ satisfies the o.d.e. Notice that the solution also contains *one* unknown constant A. To find the value of A in a particular problem we would require one further piece of information.

But how do we find the form of y when all we know is that

$$\frac{dy}{dx} = 3y + 2?$$

Direct integration of the right-hand side, i.e.

$$y = \int 3y + 2 \, dx$$

is *not possible* until we know the functional form for y, and that is the problem!

12.2.2 Variables separable. Many first-order differential equations, linear and non-linear, are such that the function $f(x, y)$ in (1) can be written as the product of a function of x and a function of y, i.e. we may write

$$\frac{dy}{dx} = f(x, y) = h(x) \cdot g(y) \tag{2}$$

We can now 'separate' the variables writing the y's on one side and the x's on the other in the following way:

$$\frac{1}{g(y)} \, dy = h(x) \, dx$$

and each side can be integrated using familiar techniques.

For example, consider the first-order o.d.e.

$$\frac{dy}{dx} = x(1 + y)$$

This may be written as

$$\frac{1}{1 + y} \, dy = x \, dx$$

Integrating the left-hand side gives $\int [1/(1+y)]\,dy = \ln(1+y) + A$ and integrating the right-hand side gives $\int x\,dx = \frac{1}{2}x^2 + B$. Equating these we have

$$\ln(1+y) + A = \frac{1}{2}x^2 + B$$

and since A and B are arbitrary constants we put $C = B - A$ and then

$$\ln(1+y) = C + \frac{1}{2}x^2$$

is the solution of the o.d.e. $dy/dx = x(1+y)$.

Now for each value of C we have a solution to the o.d.e. and each solution can be represented by a curve in the x–y plane. Fig. 12.2 shows four possible solution curves. There is a whole family of curves that represent the set of solutions to this o.d.e. However if we are seeking a solution to an o.d.e. which satisfies a particular condition, for instance we may be told that $y = 0$ when $x = 0$, then the solution is the particular curve that passes through the origin $(0,0)$. This has equation

$$\ln(1+y) = \frac{1}{2}x^2$$

or

$$y = \exp\left(\tfrac{1}{2}x^2\right) - 1$$

The set of solutions, $\ln(1+y) = C + \frac{1}{2}x^2$, is called *the general solution* to the o.d.e. $dy/dx = x(1+y)$. The function $y = \exp(\frac{1}{2}x^2) - 1$ which satisfies the extra condition

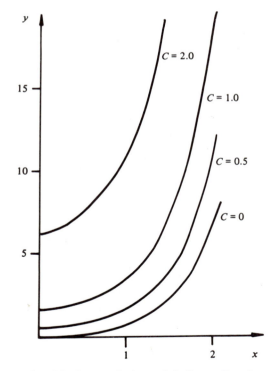

Fig. 12.2. Some solutions of $dy/dx = x(1+y)$.

that $y = 0$ when $x = 0$ is called a *particular solution*. In general we find particular solutions algebraically as the following example illustrates:

Example: Find the particular solution of the first-order o.d.e.

$$y\frac{dy}{dx} - x^2 = x^2 y$$

that satisfies the condition $y = 1$ when $x = 0$.

Solution: The first step is to find the general solution of the o.d.e. We can rewrite it so that it is a variables separable standard form. We have

$$\frac{dy}{dx} = x^2 \frac{(1 + y)}{y}$$

and then

$$\left(\frac{y}{1 + y}\right) dy = x^2 \, dx$$

Taking each side separately we have

$$\int \frac{y}{1 + y} \, dy = \int 1 - \frac{1}{1 + y} \, dy = y - \ln(1 + y) + A$$

and

$$\int x^2 \, dx = \tfrac{1}{3} x^3 + B$$

so that

$$y - \ln(1 + y) = \tfrac{1}{3} x^3 + C$$

(where $C = B - A$). This is the form of the general solution. For the particular solution that passes through the point $(0, 1)$, we put $x = 0$ and $y = 1$ into the general solution. Then

$$C = 1 - \ln(2) - 0$$

The particular solution is then

$$y - 1 - \ln\left(\frac{1 + y}{2}\right) = \tfrac{1}{3} x^3$$

12.2.3 Linear.

A first-order differential equation is said to be *linear* if it can be written in the form

$$\frac{dy}{dx} = l(x) + m(x)y \tag{3}$$

i.e. if the right-hand side of equation (1) is linear in y. We begin by observing that if we could write equation (3) in the form

$$\frac{1}{R(x)} \frac{d}{dx} [R(x)y] = f(x) \tag{4}$$

then the solution would follow quite easily. To see this multiply both sides of (4) by $R(x)$ and integrate; we have

$$R(x)y = \int R(x) f(x) \, dx + A$$

and hence we can find y. The aim is to find a function $R(x)$ which makes (3) (i.e. the equation we are given) identical with (4) (i.e. the equation we can solve).

If we expand equation (4) we get

$$\frac{1}{R}\left(\frac{\mathrm{d}R}{\mathrm{d}x}y + R\frac{\mathrm{d}y}{\mathrm{d}x}\right) = f$$

which can be written in standard form as

$$\frac{\mathrm{d}y}{\mathrm{d}x} = f - \frac{1}{R}\frac{\mathrm{d}R}{\mathrm{d}x}y$$

To make this equation the same as (3) we *must choose* $f(x) = l(x)$ and

$$\frac{1}{R}\frac{\mathrm{d}R}{\mathrm{d}x} = -m(x)$$

i.e.

$$\frac{\mathrm{d}}{\mathrm{d}x}(\ln R) = -m(x)$$

and on integration we have

$$\ln R = \int -m(x)\,\mathrm{d}x$$

giving $R(x)$ as an integral in the form

$$R = \mathrm{e}^{\int -m(x)\mathrm{d}x} \qquad (5)$$

$R(x)$ is a special function that allows us to transform a linear first-order differential equation into a form that we can solve. $R(x)$ is called an *integrating factor*. No arbitrary constant is necessary in the integration to find R because we are looking for one way of changing the original differential equation into a form that can be integrated. The steps involved in solving a linear differential equation are:

step 1: bring the differential equation into the form $\mathrm{d}y/\mathrm{d}x = l(x) + m(x)y$;

step 2: find $\int -m(x)\,\mathrm{d}x$ and hence the integrating factor $R(x)$;

step 3: substitute for $R(x)$ and $l(x)$ into the equation

$$y = \frac{1}{R}\left[\int R(x)l(x)\,\mathrm{d}x + A\right]$$

which is the general solution for y.

The following example illustrates the procedure.

Example: Find the general solution of the following differential equations:

(i) $\dfrac{\mathrm{d}y}{\mathrm{d}x} + \dfrac{1}{x}y = x^2$;

(ii) $\dfrac{\mathrm{d}y}{\mathrm{d}x} = \dfrac{x^2 + x\sin 2x}{\cos x} - \dfrac{y\sin x}{\cos x}$.

Solution: Both differential equations in this example are linear.

(i) *step 1*: writing the equation in standard form, we have

$$\frac{dy}{dx} = x^2 - \frac{1}{x}y$$

then $l(x) = x^2$ and $m(x) = -1/x$.

step 2: integrating $m(x)$ we have $\int -m(x)dx = \int (1/x)dx = \ln x$ and the integrating factor is

$$R(x) = e^{\ln x} = x$$

step 3: the general solution for y is then given by

$$y = \frac{1}{x}\left[\int x \cdot x^2 \, dx + A\right]$$

i.e.

$$y = \frac{A}{x} + \frac{x^3}{4};$$

(ii) *step 1*: the differential equation is given in standard form so that we can identify $l(x)$ and $m(x)$ immediately. We have $l(x) = (x^2 + x\sin 2x)/\cos x$ and $m(x) = \sin x/\cos x$

step 2: integrating $-m(x)$ we have

$$\int -m(x)dx = \int \frac{\sin x}{\cos x}dx = -\ln(\cos x)$$

and the integrating factor is

$$R(x) = e^{-\ln(\cos x)} = \frac{1}{\cos x}$$

step 3: the general solution for y is then given by

$$y = \frac{1}{(\cos x)}\left[\int \frac{1}{\cos x}\left(\frac{x^2 + x\sin 2x}{\cos x}\right)dx + A\right]$$

$$y = \cos x(x^2\tan x + A)$$

$$y = x^2\sin x + A\cos x$$

12.2.4 Finding particular solutions.

The general solution of a first-order differential equation contains one arbitrary (unknown) constant. To find the value for this constant in a particular problem we need one extra piece of information, i.e. we require a value of y at a given value of x. For example, the function $y = A\exp(x^2)$ is a solution of the differential equation $dy/dx = 2xy$ for any value of the constant A. However, if we are told that when $x = 0$, $y = 1$ then this information is sufficient for us to deduce that the value of A is 1. The solution $y = \exp(x^2)$ is then called a *particular solution* of the differential equation.

Finding particular solutions is not difficult.

Example: Find the particular solution of the differential equation

$$\frac{dy}{dx} = x^2 + y$$

that satisfies the condition $y = 2$ when $x = 1$.

Solution: First we must find the general solution of the differential equation. In this example the equation is linear so we solve it by finding an integrating factor. We have $l(x) = x^2$ and $m(x) = 1$ so that

$$R(x) = \exp\left(\int -m(x)\,dx\right) = \exp\left(\int -1\,dx\right) = \exp(-x)$$

The general solution for y is then

$$y = \frac{1}{e^{-x}}\left[\int e^{-x}x^2\,dx + A\right]$$

Integrating by parts twice we get

$$y = \frac{-1}{e^{-x}}(x^2 e^{-x} + 2xe^{-x} + 2e^{-x} + A)$$

$$= -x^2 - 2x - 2 + Ae^x$$

This is the general solution.

When $x = 1$, the general solution gives a value for y,

$$y = Ae^1 - 5$$

The statement of the problem tells us that when $x = 1$, $y = 2$, so we have the following equation for A,

$$2 = Ae^1 - 5$$

and then

$$A = 7e^{-1}$$

giving the particular solution

$$y = -x^2 - 2x - 2 + 7e^{x-1}$$

12.3 Worked examples

Example 1. In a room held at temperature $20\,°C$ a heated body cools from $100\,°C$ to $60\,°C$ in 20 min. How long will it take the body to cool to $30\,°C$?

Solution. If a body has a temperature T and if the surrounding environment temperature is T_0 then the rate at which the body loses heat is proportional to the temperature difference $(T - T_0)$. This leads to the following first-order differential equation for T,

$$\frac{dT}{dt} = -K(T - T_0)$$

where K is a positive constant.

We can solve this differential equation either by using separation of variables or by finding an integrating factor. Usually when we have such a choice the method of separation of variables is more straightforward.

Writing the differential equation as

$$\frac{1}{T - T_0}\,dT = -K\,dt$$

we can integrate directly to give

$$\ln(T - T_0) = -Kt + C$$

and solving algebraically for T we get

$$T = T_0 + Ae^{-Kt}$$

In the solution $T_0 = 20$; A and K can be found using the conditions

$$T = 100 \text{ when } t = 0$$

and

$$T = 60 \text{ when } t = 20$$

We have

$$100 = 20 + Ae^0 = 20 + A$$

and

$$60 = 20 + Ae^{-20K}$$

Solving for A and K we get $A = 80$, and then

$$K = -\frac{1}{20} \ln\left(\frac{40}{80}\right) = \frac{1}{20} \ln(2)(\simeq 0.015)$$

The problem statement asks us to find t when the temperature of the body is $30°C$. If $T = 30$, then $30 = 20 + 80 \exp(-\frac{1}{20}\ln(2)t)$. Solving algebraically for t we have

$$t = -\frac{20}{\ln(2)} \ln(\tfrac{1}{8}) = 20 \frac{\ln(8)}{\ln(2)} = 60$$

It will take the body 60 min to cool to $30°C$.

Example 2. In a reflecting telescope the light rays that are coming from a distant star are reflected from a highly polished surface in such a way that the rays are focussed onto one point. The shape of the reflecting surface can be defined by the differential equation

$$\frac{dy}{dx} = \frac{y + \sqrt{(x^2 + y^2)}}{x}$$

By introducing a new variable $u = y/x$ and eliminating y find the general solution of this equation. Draw the solution curves that pass through the points $(0, 1)$, $(0, 2)$ and $(0, 3)$.

Solution. The differential equation

$$\frac{dy}{dx} = \frac{y + \sqrt{(x^2 + y^2)}}{x}$$

is neither linear nor of the variables separable form. However, it is an example of an equation that can be reduced to a variable separable type by letting $y = ux$. We then have

$$\frac{dy}{dx} = \frac{du}{dx} x + u = u + \sqrt{(1 + u^2)}$$

giving the differential equation for u,

$$\frac{du}{dx} = \frac{\sqrt{(1 + u^2)}}{x}$$

This technique always works when the right-hand side of the differential equation can be written as a function of y/x. Separating the variables in the equation for u we have

$$\frac{1}{\sqrt{(1+u^2)}}\,du = \frac{1}{x}\,dx$$

and integrating we have

$$\sinh^{-1}u = \ln x + A$$

Solving algebraically for u,

$$u = \sinh(A + \ln x) = \frac{\exp(A + \ln x) - \exp(-(A + \ln x))}{2}$$

$$= \frac{\exp(A)x}{2} - \frac{\exp(-A)}{2x}$$

and the solution for y is

$$y = x^2 \frac{c}{2} - \frac{1}{2c}$$

where, for convenience, we have written $c = \exp(A)$.

If we draw a graph of y against x then for each value of c we obtain a different solution curve.

For the solution curve to pass through the point $(0, 1)$ we must have $1 = 0 - (1/2c)$, i.e. $c = -\frac{1}{2}$.

Similarly, for the two solution curves through the points $(0, 2)$ and $(0, 3)$ we have $c = -\frac{1}{4}$ and $c = -\frac{1}{6}$ respectively. The three particular solution curves are shown in Fig. 12.3. Each of these solution curves could be used as a geometrical model

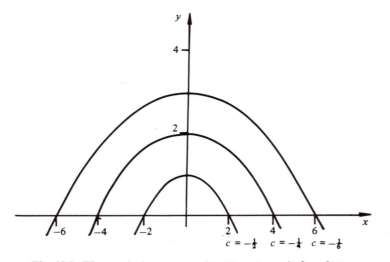

Fig. 12.3. Three solution curves of $dy/dx = [y + \sqrt{(x^2 + y^2)}]/x$.

for a mirror in a telescope. Light rays arriving in a parallel beam from a source at $y = -\infty$ will all be focussed on the origin.

Example 3. A solution of ethyl acetate $(0.01 \, \text{mol} \, l^{-1})$ reacts with a solution of sodium hydroxide $(0.002 \, \text{mol} \, l^{-1})$. If x is the concentration of ethyl acetate which has reacted in time t and K is the velocity constant of the reaction, then

$$\frac{dx}{dt} = K(0.01 - x)(0.002 - x)$$

If, originally, $x = 0$ when $t = 0$, obtain the particular solution to this differential equation.

Solution. The differential equation is an example of one that occurs quite often in the physical sciences. The right-hand side is a quadratic in the dependent variable. In this example it is used to model a chemical reaction but it also occurs as a model for the growth of a population of species. The differential equation $dy/dx = f(y)$ where $f(y)$ is a quadratic function of y is called the logistic equation.

In this example we have

$$\frac{dx}{dt} = K(0.01 - x)(0.002 - x)$$

and this equation can be solved by the method of separation of variables. Separating variables we have

$$\frac{1}{(0.01 - x)(0.002 - x)} \, dx = K \, dt$$

and integrating each side we have

$$\int \frac{1}{(0.01 - x)(0.002 - x)} \, dx = \int K \, dt$$

The right-hand side becomes

$$\int K \, dt = Kt + A$$

For the left-hand side we need to use partial fractions. We get

$$\frac{1}{(0.01 - x)(0.002 - x)} = \frac{1}{(0.008)} \left[\frac{1}{(0.002 - x)} - \frac{1}{(0.01 - x)} \right]$$

Then

$$\int \frac{1}{(0.01 - x)(0.002 - x)} \, dx = \int \frac{1}{0.008} \left(\frac{1}{0.002 - x} - \frac{1}{0.01 - x} \right) dx$$

$$= \frac{1}{0.008} \left[\ln(0.01 - x) - \ln(0.002 - x) \right] + B$$

Finally, combining A and B we have

$$\ln\left(\frac{0.01 - x}{0.002 - x}\right) = 0.008\, Kt + c$$

The constant c can be found by putting $x = 0$ when $t = 0$. We then have

$$\ln\left(\frac{0.01}{0.002}\right) = 0 + c$$

giving

$$c = \ln(5)$$

The particular solution is

$$\ln\left(\frac{0.01 - x}{0.002 - x}\right) = 0.008\, Kt + \ln(5)$$

and solving algebraically for x we have

$$x = \frac{0.01 - 0.01e^{0.008Kt}}{1 - 5e^{0.008Kt}}$$

12.4 Further developments

12.4.1 A numerical solution. The models for solving first-order differential equations that we have introduced in this chapter have given the solutions as formulas involving x and y. In many cases we can manipulate the solution and write y as a function of x such as $y = (A/x) + (x^3/4)$. These are *exact methods* of solution with no approximations, and have the advantage that being general solutions we can fit any conditions that may apply to find particular solutions. However there are many differential equations for which exact methods do not work and an approximate method must be used. Such methods lead to a solution of a differential equation which is not a formula but is given by a table of values, accordingly we often call such a method a *numerical method* and the corresponding solution, a *numerical solution*. In this section we will introduce one numerical method known as *Euler's method*.

Consider the first-order differential equation

$$\frac{dy}{dx} = f(x, y) \tag{6}$$

and the condition

$$y = y_0 \text{ when } x = x_0 \tag{7}$$

One disadvantage of numerical methods of solution is that, in general, we cannot solve the differential equation to find a general solution. We must use any given conditions at the outset and thus find a particular solution.

Now by the definition of a derivative the left-hand side of (6) can be written as

$$\frac{dy}{dx} = \lim_{h \to 0} \frac{y(x + h) - y(x)}{h}$$

and for sufficiently small h, the ratio

$$\frac{y(x + h) - y(x)}{h}$$

is a good approximation to the derivative dy/dx. Hence we can write (6) in the approximate form

$$\frac{y(x + h) - y(x)}{h} \simeq f(x, y) \tag{8}$$

This approximation holds for all values of x.

Suppose that at some point $x = x_n$ we know the corresponding value of y then at a neighbouring point $(x_n + h)$ we can find an approximate value of the solution of the differential equation (6) by using (8). We have

$$y(x_n + h) \simeq y(x_n) + hf(x_n, y(x_n)) \tag{9}$$

This is a recurrence relation in which if we know an approximation $(x_n, y(x_n))$ then we can find the next approximation $x_{n+1}, y(x_{n+1})$ where $x_n + h$ has been written as x_{n+1}. Since the initial conditions x_0, y_0 are known we *can* use this recurrence relation to find subsequent approximate solutions.

To illustrate the method consider the differential equation

$$\frac{dy}{dx} = \frac{x}{y}$$

and the condition $y = 1$ when $x = 0$. Equation (9) becomes

$$y(x_{n+1}) \simeq y(x_n) + \frac{hx_n}{y(x_n)}$$

Suppose that we choose $h = 0.1$. We shall see that the choice of h is important for obtaining accurate solutions, but that although a smaller h improves accuracy, it results in many more calculations. So we have to find a balance between the amount of work needed and the accuracy that we require. With $h = 0.1$:

$$x_1 = x_0 + h = 0 + 0.1 = 0.1.$$

$$y(x_1) = y(x_0) + \frac{hx_0}{y(x_0)} = 1 + \frac{0.1 \times 0}{1} = 1$$

$$x_2 = x_1 + h = 0.2$$

$$y(x_2) \simeq y(x_1) + \frac{hx_1}{y(x_1)} = \frac{1 + 0.1 \times 0.1}{1} = 1.01$$

$$x_3 = x_2 + h = 0.3$$

$$y(x_3) \simeq y(x_2) + \frac{hx_2}{y(x_2)} = \frac{1.01 + 0.1 \times 0.2}{1.01} = 1.029\,801\,98\ldots.$$

and so on.

In numerical mathematics when the solution to a problem is a string of numbers it is a good idea to lay the results out in the form of a table; for example:

n	x_n	$y(x_n)$	$f_n = x_n/y(x_n)$	hf_n	$y(x_{n+1})$
0	0	1	0	0	1
1	0.1	1	0.1	0.01	1.01
2	0.2	1.01	0.198 019 8	0.019 801 98	1.029 801 98
3	0.3	1.029 801 98	0.291 318 1	0.029 131 81	1.050 933 7
4	0.4	...			

For this differential equation we can find an exact solution by the method of separation of variables and the solution is $y = \sqrt{(x^2 + 1)}$. The actual value of $y(0.4)$ is $\sqrt{(1.16)}$ or 1.077 03 (to five decimal places), and the numerical solution is 1.058 93 (to five decimal places). We can see that there is some deviation from the exact solution.

It is more usual to write $y(x_n)$ as y_n and $y(x_{n+1})$ as y_{n+1}. In solving the recurrence relation (8) it is important to remember that we are calculating the approximate solution of the differential equation (5). The points (x_1, y_1), (x_2, y_2), $(x_3, y_3) \cdots$ do not lie on the solution curve of the differential equation. Fig. 12.4 shows this very clearly. The solid line is the exact solution and the dotted line represents the approximate solution for a value of h equal to 0.1.

We can explain the deviation of the approximate solution from the exact solution

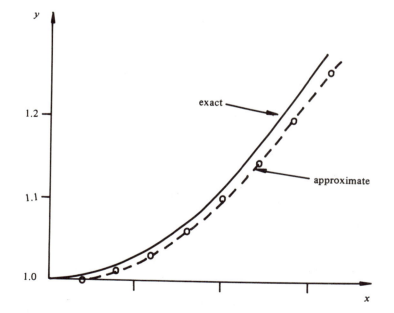

Fig. 12.4. Comparison between the numerical and the exact solution.

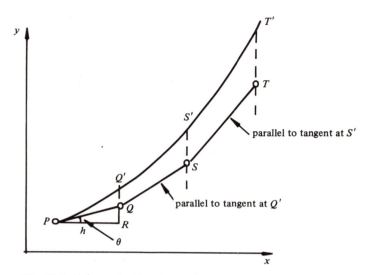

Fig. 12.5. Geometrical interpretation of the numerical solution.

geometrically. In Fig. 12.5 the solid curve represents the exact solution of the o.d.e. and the circles are the approximate solutions.

Suppose that we know the exact solution at the point P. Then PQ' represents the exact solution over a step length h. Consider the tangent to the solution curve at P. This is the line PQ and is such that

$$\tan \theta = \text{gradient at } P = f(x_P, y_P)$$

using the geometrical interpretation of the derivative and the equation

$$\frac{dy}{dx} = \text{gradient} = f(x, y)$$

From the triangle PQR we have

$$\tan \theta = \frac{QR}{PR} = \frac{y_Q - y_R}{x_Q - x_R} = \frac{y_Q - y_R}{h}$$

and since $\tan \theta = f(x_P, y_P)$ we have

$$y_Q = y_P + h f(x_P, y_P)$$

The equation for y_Q is the same as the recurrence relation (8) from which we have found approximate solutions to the differential equation. Thus the point we have calculated is a point on the tangent to the solution curve through P.

The whole process is now repeated starting from the point Q to compute the next approximation by going along a line starting at Q and parallel to the tangent at Q'. Because we are successively using lines which are parallel to tangents to the exact solution curve, Euler's method is often called the *tangent method of solution*.

12.4.2 Improving the accuracy. Whenever we use Euler's method to solve a differential equation we begin by choosing a value of h, which is called the *step length*. We introduced h in approximating the derivative dy/dx by $[y(x+h)-y(x)]/h$ and we know from the definition of the derivative that the smaller we take h, then the closer to dy/dx does the approximation become. The choice of a suitable step length when using Euler's method is very important. As an example consider again the differential equation $dy/dx = x/y$ and the condition $y = 1$ when $x = 0$. The following table shows the approximations to $y(1)$ and their deviations from the exact value for different values of h. (We have given the numbers to an accuracy of five decimal places.)

h	Approximation to $y(1)$	Deviation	Number of steps between 0 and 1
0.2	2.488 32	0.229 96	5
0.1	2.593 74	0.124 54	10
0.05	2.653 30	0.064 81	20
0.01	2.704 81	0.013 47	100
0.001	2.716 92	0.001 36	1000

The deviations are shown plotted against the step size h on Fig. 12.6. We see that

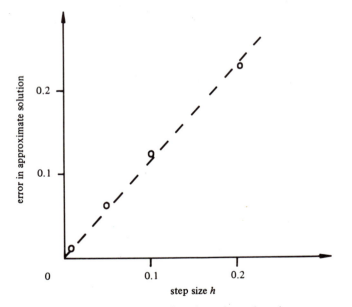

Fig. 12.6. Error as a function of step length.

the deviation is roughly linearly proportional to h, so that the smaller is h then the more accurate is our approximate solution.

In fact, it can be shown that by making h small enough we can make the error as small as we please, and the error approaches zero as the step size approaches zero. However, as the table shows, the increased accuracy obtained by reducing h leads to many more calculations. So we must balance accuracy against number of calculations. In practice what we would do is to repeat the calculations by halving the step length until the difference between two solutions is less than some predetermined amount. For example, suppose that we wished to find a value for $y(1)$ in the solution of the differential equation $dy/dx = x + y^2$ with $y = 0$ when $x = 0$, to an accuracy of two decimal places. Euler's method gives the recurrence relation

$$y_{n+1} = y_n + h(x_n + y_n^2)$$

for the approximate values of y, and the initial conditions are $x_0 = 0$ and $y_0 = 0$.

We begin by choosing a value of h for which there are not too many steps between 0 and 1. Suppose we choose $h = 0.2$. (This is a fairly arbitrary choice.) Then solving the recurrence equation five times we predict an approximate solution to $y(1)$ is 0.415 05 (to five decimal places).

Now suppose that we halve the step length to $h = 0.1$. The approximate solution for $y(1)$ is then 0.480 83 (to five decimal places).

We can see that these two approximations do not agree to one decimal place and we require the solution correct to two decimal places. The following table shows the results as we continually halve the value of h:

h	Approximation to $y(1)$
0.05	0.517 35
0.025	0.536 79
0.012 5	0.546 85
0.006 25	0.551 97
0.003 125	0.554 56
0.001 562 5	0.555 86
0.000 781 25	0.556 51
0.000 390 625	0.556 84

Now at last we have two approximations that agree to two decimal places, so we could say that the solution of the problem $dy/dx = x + y^2$, $y(0) = 0$ is $y(1) = 0.56$ correct to two decimal places. Clearly, this sort of numerical work is an ideal task for a computer.

Finally, there is one word of caution here. These remarks about the deviation of the approximate solution from the exact solution are only valid if the calculations are done to enough decimal places. When working with a calculator or a computer, the number of decimal places is restricted and errors are introduced because the numbers are rounded off. These are called *rounding errors* and after a certain point, any increase in accuracy obtained by reducing the size h will be swamped by these rounding errors.

12.5 Further worked examples

Example 1. Uranium 238 decays in the following way:

$$^{238}_{92}U \xrightarrow{K_1} {}^{234}_{90}Th \xrightarrow{K_2} {}^{234}_{91}Pa$$

The amount y of thorium 234 present at time t satisfies the differential equation

$$\frac{dy}{dt} + K_2 y = K_1 a \exp(-K_1 t)$$

If a is the initial amount of uranium 238, and $y = 0$ when $t = 0$, show that

$$y = \frac{K_1 a}{(K_2 - K_1)} [\exp(-K_1 t) - \exp(-K_2 t)]$$

Solution. The differential equation in this example is linear in y, because we can write it as

$$\frac{dy}{dt} = K_1 a \exp(-K_1 t) - K_2 y = l(t) + m(t)y$$

The integrating factor is given by

$$R(t) = \exp\left(\int -m(t)\,dt\right) = \exp\left(\int K_2\,dt\right) = \exp(K_2 t)$$

since $m(t) = -K_2$. The solution is then

$$y = \frac{1}{\exp(K_2 t)} \left[\int \exp(K_2 t) K_1 a \exp(-K_1 t)\,dt + A\right]$$

$$= \frac{1}{\exp(K_2 t)} \left[\frac{aK_1}{K_2 - K_1} \exp((K_2 - K_1)t) + A\right]$$

$$= \frac{aK_1}{K_2 - K_1} (\exp(-K_1 t) + A \exp(-K_2 t))$$

and this is the general solution for y. If $y = 0$ when $t = 0$ then $A = -aK_1/K_2 - K_1$; hence the particular solution for y is given by

$$y = \frac{aK_1}{K_2 - K_1} (\exp(-K_1 t) - \exp(-K_2 t))$$

as required.

Example 2. A chemical reaction proceeds at a rate given by the equation

$$\frac{dx}{dt} = K(2 - x)(3 - x)(5 - x)$$

where K is a constant and x is the amount of a product of the reaction. Find an expression for the time t in terms of x if originally $x = 0$ when $t = 0$.

Solution. The differential equation can be solved using the method of separation of variables; dividing both sides by $(2 - x)(3 - x)(5 - x)$ we can see that the variables have separated out. We have

$$\int \frac{1}{(2 - x)(3 - x)(5 - x)} dx = \int K \, dt = Kt + A$$

For the left-hand side, using partial fractions

$$\int \frac{1}{(2 - x)(3 - x)(5 - x)} dx = \int \frac{1/3}{2 - x} + \frac{-1/2}{3 - x} + \frac{1/6}{5 - x} dx$$

$$= -\tfrac{1}{3} \ln (2 - x) + \tfrac{1}{2} \ln (3 - x) - \tfrac{1}{6} \ln (5 - x)$$

$$= \ln \left[\frac{(3 - x)^{\frac{1}{2}}}{(2 - x)^{\frac{1}{3}}(5 - x)^{\frac{1}{6}}} \right]$$

Hence

$$\ln \left[\frac{(3 - x)^{\frac{1}{2}}}{(2 - x)^{\frac{1}{3}}(5 - x)^{\frac{1}{6}}} \right] = Kt + A$$

is the general solution of the differential equation. Using the condition that $x = 0$ when $t = 0$ we have

$$A = \ln \left[\frac{3^{\frac{1}{2}}}{2^{\frac{1}{3}}5^{\frac{1}{6}}} \right]$$

and then

$$t = \frac{1}{K} \ln \left[\frac{(1 - x/3)^{\frac{1}{2}}}{(1 - x/2)^{\frac{1}{3}}(1 - x/5)^{\frac{1}{6}}} \right]$$

Example 3. The current i in an electrical circuit consisting of a 4.3 ohm resistor, non-linear inductor 10.2 H, and subject to a linear excitation may be described by the first-order differential equation

$$(10.2 - 0.048i^2) \frac{di}{dt} + 4.3i = 2.6t$$

If the initial current is 0.0023 amps, use Euler's method with a step length of 0.1 s to find the value of the current i after 2 s.

Solution. This differential equation is non-linear so that the linear and variables separable methods cannot be used here. Writing the derivative as $(i_{n+1} - i_n)/h$ we have the following recurrence relation to solve:

$$i_{n+1} = i_n + h \left(\frac{2.6t_n - 4.3i_n}{10.2 - 0.048i_n^2} \right)$$

Choosing $h = 0.1$ and solving until $t = 2$ (i.e. 20 iterations) we obtain the

First-order differential equations

following table:

n	t_n	$i_n = i(t_n)$	i_{n+1}
0	0	0.0023	0.0022
1	0.1	0.0022	0.0047
2	0.2	0.0047	0.0096
3	0.3	0.0096	0.0168
4	0.4	0.0168	0.0263
5	0.5	0.0263	0.0379
6	0.6	0.0379	0.0516
7	0.7	0.0516	0.0673
8	0.8	0.0673	0.0848
9	0.9	0.0848	0.1042
10	1.0	0.1042	0.1253
11	1.1	0.1253	0.1481
12	1.2	0.1481	0.1724
13	1.3	0.1724	0.1983
14	1.4	0.1983	0.2256
15	1.5	0.2256	0.2544
16	1.6	0.2544	0.2844
17	1.7	0.2844	0.3158
18	1.8	0.3158	0.3484
19	1.9	0.3484	0.3821
20	2.0	0.3821	

The approximate solution for $i(2)$ using Euler's method is 0.382 amps.

12.6 Exercises

1. *Separable variables*: Solve the following differential equations:

(i) $\dfrac{dy}{dx} = \dfrac{x^2}{y}$

(ii) $\dfrac{dy}{dx} = e^x \tan(y)$

(iii) $\dfrac{dy}{dx} = \dfrac{y}{(1+x)}$

(iv) $\dfrac{dy}{dx} = (y+1)\sin(x)$

(v) $(x^2+1)\dfrac{dy}{dx} = y^2 + 4$

(vi) $y\dfrac{dy}{dx} = e^{x+2y}\cos(x)$

(vii) $2x^3 \dfrac{dy}{dx} = y^2 + 3xy^2$ given that $y = 1$ when $x = 1$,

(viii) $\dfrac{dy}{dx} = \dfrac{1}{e^y(1+x^2)}$ given that $y = 0$ when $x = 1$.

2. *Integrating factor*: Solve the following differential equations:

   (i) $\sin(x)\dfrac{dy}{dx} - 2y\cos(x) = e^x \sin^3(x)$

   (ii) $x^2\dfrac{dy}{dx} + xy = \ln(x)$

   (iii) $x(1+x)\dfrac{dy}{dx} - y = 3x^4$

   (iv) $\tan(x)\dfrac{dy}{dx} + 2y = x\csc(x)$

   (v) $\dfrac{dy}{dx} = y + x^2$ given that $y = 5$ when $x = 1$

   (vi) $\dfrac{dy}{dx} = \sin(x) + y\tan(x)$ given that $y = \frac{1}{2}$ when $x = 0$

3. Use the appropriate method to evaluate the following differential equations:

   (i) $x^3\dfrac{dy}{dx} = 2x + 3$

   (ii) $\dfrac{dy}{dx} = xy + x^3$

   (iii) $\dfrac{dy}{dx} = \dfrac{1}{(ax+b)}$

   (iv) $y\sec^2(x) + \tan(x)\dfrac{dy}{dx} = 0$

   (v) $\dfrac{dy}{dx} = \sqrt{(1+y)}$

   (vi) $(2x+1)\dfrac{dy}{dx} = x - 2y$

4. Classify the following differential equations giving the order and linearity of each equation:

   (i) $\dfrac{dy}{dx} = x + y^2$

   (ii) $2x\dfrac{d^2y}{dx^2} + 4x^2\dfrac{dy}{dx} - 3y = 0$

   (iii) $\tan(x)\dfrac{dy}{dx} - 3y = e^x$

(iv) $y\dfrac{d^2y}{dx^2} - 2x^2y^2 = 0$

(v) $\dfrac{d^4y}{dx^4} - \dfrac{3d^2y}{dx^2} + y = x^2$

(vi) $\dfrac{dy}{dx} = \dfrac{x}{\sqrt{(1+y)}}$

5. Solve completely using the given conditions:

(i) $L\dfrac{di}{dt} + Ri = 0$ given that $i = i_0$ when $t = 0$

(ii) $\dfrac{dy}{dx} - xy - x^3 = 0$ where $y = 0$ when $x = 0$

(iii) $\dfrac{dp}{dz} = \dfrac{gp}{R(T_0 - 0.0061z)}$ where $p(0) = p_0$ and g, R, T_0 and p_0 are constants.

6. Find the value of 'a' given that

$$x^3\dfrac{dy}{dx} = a - x$$

and that $y = 0$ when $x = 2$ and when $x = 6$.

7. Living tissue contains 10^{-22} g of carbon 14 in every gram of carbon. A sample of dead wood is estimated to contain 3.5×10^{-23} g of carbon 14 in every g of carbon. We assume that the proportion of carbon 14 in carbon in living tissue was the same when the wood died as it is in living tissue today. Using a model for the rate of decay of the radioactive material given by

$$\dfrac{dm}{dt} = -km$$

estimate how long ago the wood died.

8. A rabbit population can be modelled by the differential equation

$$\dfrac{dP}{dt} = bP - cP$$

where b and c are the birth and death rates respectively. Female rabbits, on average, rear one off-spring to maturity per month and if we assume that half the rabbits are female and all female rabbits reproduce then we may take b to be 0.5. The rabbit population is hit by myxomatosis resulting in a death rate of about 60% of the population per month giving a value for c of 0.6. Find the solution of the differential equation and predict the time it takes for the population to drop to half its initial value.

9. A motor under load generates heat at constant rate and radiates it at a rate proportional to the excess temperature T, so that

$$\dfrac{dT}{dt} = 10 - kT$$

where t denotes time and k is a constant:
(i) Solve this differential equation for T given that $T = 0\,°C$ initially and show that the ultimate rise in temperature is $(10/k)\,°C$.
(ii) If after 10 min the rise is $50\,°C$, show that k is a root of the equation

$$e^{-10k} = 1 - 5k$$

Use the Newton–Raphson numerical method to find the non-zero positive root of this equation correct to four decimal places.

10. Use Euler's method with step size $h = 0.1$ to approximate the solution to the initial value problem

$$\frac{dy}{dx} = x\sqrt{y}$$

$$y(1) = 4$$

at the points $x = 1.1,\ 1.2,\ 1.3,\ 1.4,$ and 1.5
Find the analytic solution of the problem and hence work out the percentage error in the value of $y(1.5)$

11. For each of the following initial value problems, use Euler's method to find the solution correct to three significant figures:

(i) $\dfrac{dy}{dx} = x - y^2$ $\qquad y(1) = 0$

find $y(2)$.

(ii) $\dfrac{dy}{dx} = 1 + x\sin(xy),\quad y(0) = 0$

find $y(1)$.

(iii) $\dfrac{dy}{dx} = x^2 - y^2$ $\qquad y(0) = \tfrac{1}{2}$

find $y(1)$.

12. Under certain conditions the rate of reaction between bromic and hydrobromic acids can be modelled by a differential equation of the form

$$\frac{dx}{dt} = k(na + x)(a - x)$$

where k, a and n are constants, and x is a measure of the extent of reaction at time t. Solve the differential equation when $n = 1$ to give x in terms of t if originally $x = 0$ when $t = 0$.
Give a rough sketch of x against t.

13. A chemical reaction of the form

$$A \underset{k_1}{\overset{k}{\rightleftharpoons}} B$$

can be modelled to a reasonable degree of accuracy by a differential equation of the form

$$\frac{dx}{dt} = k(a - x) - k_1 x$$

where k and k_1 are constants, a is the initial concentration of A and x is the concentration of B after time t. Show that if $x = 0$ initially, then, according to the model

$$x = \frac{ka[1 - \exp - (k + k_1)t]}{(k + k_1)}$$

Calculate the ultimate (equilibrium) value of x and the time taken for x to become equal to half this value.

14. In a particular gas reaction of the type

$$2A + B \rightarrow products$$

the rate equation may be written as:

$$\frac{dx}{dt} = k(2 - x)^2(3 - x)$$

where x is the amount of a product after time t, and k is the reaction rate constant. Solve the differential equation, giving t in terms of x, if originally $x = 0$ when $t = 0$. Calculate the time taken for x to become equal to half its equilibrium value.

15. In a certain autocatalytic reaction, the rate equation is given by

$$\frac{dx}{dt} = kx(M - x)$$

where k and M are constants, and the initial value of x, when $t = 0$, is $M/2$. Solve the differential equation giving x in terms of t.

16. Salt, in solution, flows into a tank of volume V. Salt solution also flows out of the tank. If the flow of liquid in and out of the tank is the same, and equal to $v\,\mathrm{m}^3\mathrm{s}^{-1}$, and the tank is full, the differential equation governing the concentration c of salt in the tank (and output stream) is

$$\frac{dc}{dt} + \frac{vc}{V} = \frac{vc_{in}}{V}$$

where c_{in} is the concentration of salt $(\mathrm{kg\,m}^{-3})$ in the input stream. Solve the differential equation to give c as a function of t for $0 < t < \infty$ if
(i) $c = 0$ at time $t = 0$, and $c_{in} = c_0$ for all t,
(ii) $c = 0$ at time $t = 0$, and $c_{in} = c_0$ for $0 < t \leqslant 2$, and $c_{in} = 0$ for $2 < t < \infty$.

17. An equation for the velocity of a body moving in a viscous medium at time $t\,\mathrm{s}$ with velocity $v\,\mathrm{ms}^{-1}$ is

$$\frac{dv}{dt} + v + kv^2 = 0$$

where k is a positive constant. Obtain v in terms of t given that $v = 100$ when $t = 0$. Find the terminal velocity of the body.

18. In the reaction $H_2O + CO = H_2 + CO_2$ the equilibrium constant k changes with temperature T according to the equation

$$\frac{1}{k}\frac{dk}{dT} = \frac{Q}{RT^2}$$

where Q is the heat of reaction and R is the gas constant. Assuming that $Q = -42\,000$ joules and $R = 8.314$ joules K^{-1} mol^{-1}, find k when $T = 1200\,K$ if it is known that $k = 3.26$ when $T = 1000\,K$.

19. A simple heat exchanger consists of a steam jacket enclosing a pipe of circular cross-section through which fluid, to be heated, flows. The temperature T of the fluid, a distance z from the beginning of the steam jacket, satisfies the following differential equation

$$\frac{dT}{dz} = \frac{2\pi Rk}{vs}(T_s - T)$$

where R is the radius of the pipe, k the heat transfer coefficient between the pipe and the fluid, T_s is the steam temperature, v is the rate of flow ($kg\,s^{-1}$) of the fluid in the pipe, and s is the specific heat of the fluid. Solve the differential equation if the value of T when $z = 0$ is T_A.
If the steam jacket is of length L, obtain an expression for the rate of flow if a temperature of T_0 is required in the fluid as it leaves the steam jacket. It may be assumed that T_0 is less than T_s, but greater than T_A.

20. In population dynamics a more general model for the population of a species at time t is given by the differential equation

$$\frac{dP}{dt} = aP - bP^r \quad P(0) = P_0$$

where a, b and $r > 1$ are constants. ($r = 2$ gives the logistic model.) If $a = 6$, $b = 2$ and $P_0 = 1$, use Euler's method, with a step length $h = 0.25$, to find a numerical solution of this equation in the interval $0 \leqslant t \leqslant 5$ for each of the cases $r = 1.5$, $r = 3$ and $r = 4$.
Sketch a graph of $P(t)$ against t in each case and compare your results with the logistic model.

21. The velocity of a body of unit mass falling freely under gravity is given by the differential equation

$$\frac{dv}{dt} = g - kv^r \quad v(0) = v_0$$

where k and r are constants. In this equation kv^r is the force due to air resistance. If $g = 9.8$, $k = 2$ and $v_0 = 0$ use Euler's method to find an approximate value for $v(5)$ when $r = 1$ and $r = 2$.

13

Second-order ordinary differential equations

Contents: 13.1 scientific context – 13.1.1 cooling fins; 13.1.2 vibrations; 13.2 mathematical developments – 13.2.1 homogeneous equations; 13.2.2 the method of solution; 13.2.3 interpreting the solution; 13.3 worked examples; 13.4 further developments – 13.4.1 non-homogeneous equations; 13.4.2 the method of solution; 13.4.3 trial functions; 13.5 further worked examples; 13.6 exercises.

13.1 Scientific context

In Chapter 12 we introduced the idea of an ordinary differential equation as a model to describe physical systems. The type of equation discussed there was first-order and we saw the need to develop different methods depending on the form of the differential equation. We continue our discussion of ordinary differential equations by investigating the solution of second-order equations. Such equations occur commonly in modelling. For example, in dynamics, Newton's second law equates the force to the product of mass and acceleration, which is the second time-derivative of position. Thus the mathematical statement, $F = m\,\mathrm{d}^2x/\mathrm{d}t^2$ is an example of a second-order ordinary differential equation.

13.1 Example 1: cooling fins. There are many situations where it is desirable to transfer heat away from one region to another. For example, Fig. 13.1 is a photograph of a motor-cycle engine showing the metal fins which provide better cooling of the engine and help to avoid problems of over-heating. The same idea is used in refrigerators and the output transistors of a hi-fi set. In designing such fins it is important to know the relationship between the heat input from the source and the temperature distribution along the fin in the steady state situation when the fin has finished heating up.

To formulate a model that describes the temperature distribution we shall take a single fin which is heated at one end and this heat energy is transferred to the atmosphere (see Fig. 13.2). Assume that the atmosphere is at constant temperature, T_0 say, and neglect the heat transfer from the edges of the fin.

If we consider a thin strip of the fin of width δx then the input–output principle, where the accumulation term is zero, for the rate of heat energy gives

$$\text{(input)} \quad \text{(output)}$$
$$q_1 \quad = q_2 + q_3 \tag{1}$$

where q_3 is the rate of heat energy transfer out from the bottom and top

Fig. 13.1. Cooling fins on an engine.

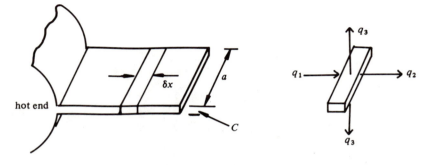

Fig. 13.2. Modelling the heat flow for a single fin.

faces of the strip by the process of convection. We have a simple model for convection,

$$q_3 = h \times \text{area} \times \text{temperature difference}$$

where h is a constant called the convective heat transfer coefficient. Thus

$$q_3 = h \times 2a\delta x \times (T - T_0). \tag{2}$$

The process of transferring heat along the fin is by conduction and Fourier's law provides a relation between the rate of heat transfer and the temperature gradient, we have

$$q = -K \times \text{area} \times dT/dx \tag{3}$$

where K is a constant.

Equation (1) can now be written as

$$q_1 - q_2 = q_3 = 2ha(T - T_0)\delta x$$

writing q_1 as $q(x)$ and q_2 as $q(x + \delta x)$ and dividing by δx we have

$$\frac{q(x) - q(x + \delta x)}{\delta x} = 2ha(T - T_0)$$

Taking the limit as $\delta x \to 0$ in the usual way this equation can be written in terms of the derivative of q. We have

$$\lim_{\delta x \to 0} \frac{q(x + \delta x) - q(x)}{\delta x} = \frac{dq}{dx}$$

hence

$$-\frac{dq}{dx} = 2ha(T - T_0).$$

Replacing q in terms of temperature according to (3) we have

$$K(aC)\frac{d^2 T}{dx^2} = 2ha(T - T_0) \tag{4}$$

If we define a new variable $\theta = T - T_0$ then since T_0 is a constant we can write

(4) as the following *second-order ordinary differential equation* in θ,

$$\frac{d^2\theta}{dx^2} - \left(\frac{2h}{KC}\right)\theta = 0$$

It is second order because the highest-order derivative is the term $d^2\theta/dx^2$.

Question: Can you guess a possible solution of this equation?
   Answer: There is one function which when differentiated twice takes on the same form and
      that is the exponential function. A solution is $e^{\alpha x}$ where $\alpha^2 = 2h/KC$. We shall see
      that all possible solutions can be written in terms of the exponential function.

13.1.2 Example 2: vibrations. Many mechanical systems exhibit oscillations or vibrations. For example, the stylus on a record player is forced to vibrate due to the variations in the grooves; the seismograph, an instrument that detects earthquake waves, is forced to vibrate by the displacements of the earth's surface; when air flows past a body, vortices are generated and detach from the body and this causes the body to vibrate from side to side.

Being able to model vibrations of a system is very important to the scientist and engineer. At certain frequencies the induced vibrations of a structure can become so violent as to cause catastrophic results. Two incidents, at the Ferrybridge power station in 1963 and in Washington in 1940, were caused by the wind. At Ferrybridge one of the large concrete cooling towers collapsed due to the shedding of vortices caused by the wind blowing past a row of towers in front of it. But perhaps more famous was the collapse of the Takoma Narrows road bridge[†] in Washington (USA) in 1940. A wind of moderate speed of 42 m.p.h. caused large oscillations of the bridge. Eventually the bridge crashed into the river below.

The rather idealised motion of a simple pendulum can be used to investigate the essential features of more complicated physical systems such as damping, forcing and resonance. The simple pendulum is just a heavy mass suspended from a fixed point by a light inextensible string (see Fig. 13.3). If we displace the mass by a small angle, θ_0 say, from its equilibrium position then experience tells us that the mass will swing back and forth with ever decreasing angles θ until (after a very long time) it stops. The forces on the mass are those due to gravity, air resistance and the tension in the string.

Using Newton's second law in the transverse direction we have

$$-ml\frac{d^2\theta}{dt^2} = mg\sin\theta + R \tag{5}$$

The resistance R can be modelled by a linear function of the speed of the body so that

$$R = mk\dot\theta$$

[†] For a short film showing this collapse see the Open University video tape for the course MST322.

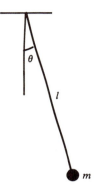

Fig. 13.3. A simple pendulum.

If the initial displacement θ_0 is small then θ remains small and we can write

$$\sin \theta \simeq \theta.$$

(This simplification is within $\frac{1}{2}\%$ for angles up to $10°$.) Equation (5) becomes

$$- ml \frac{d^2\theta}{dt^2} = mg\theta + mk\dot{\theta}$$

Thus the angle θ satisfies the second-order ordinary differential equation

$$\frac{d^2\theta}{dt^2} + \frac{k}{l}\frac{d\theta}{dt} + \frac{g}{l}\theta = 0$$

This equation differs from the one developed in 13.1.1 in that it has a term involving the first-order derivative $d\theta/dt$. We will see that it is the relative size of the coefficient of this term that allows us to predict how heavily damped the system is (i.e. how quickly the pendulum stops).

   The two differential equations that we have formulated to model the temperature distribution in a cooling fin and the vibration of a pendulum have two features in common, they are both second-order and have constant coefficients. It is the solution of equations of this type that we investigate in this chapter.

13.2 Mathematical developments

13.2.1 Homogeneous equations. We begin by introducing some more terminology.
   A general second-order *linear* ordinary differential equation can be written in the form

$$a(x)\frac{d^2y}{dx^2} + b(x)\frac{dy}{dx} + c(x)y = f(x) \tag{6}$$

where a, b, c and f are known continuous functions of the independent variable x. The coefficient $a(x)$ is not allowed to be zero because then we would have a first-order equation.

If $f(x) = 0$ the equation is said to be *homogeneous*, otherwise we have a non-homogeneous or inhomogeneous equation.

Example: Which of the following differential equations is second-order, linear and homogeneous?

(i) $4x \dfrac{d^2 y}{dx^2} + 3x^2 \dfrac{dy}{dx} - x^3 = 0$;

(ii) $\dfrac{d^2 y}{dx^2} + 6\left(\dfrac{dy}{dx}\right)^2 - 15y = 0$;

(iii) $3 \dfrac{d^2 y}{dx^2} + 5 \dfrac{dy}{dx} - 2y = 0$;

(iv) $y \dfrac{d^2 y}{dx^2} + x \dfrac{dy}{dx} - 4y = 0$.

Solution: We compare each o.d.e. with standard form (6).

(i) This o.d.e. is linear because the coefficients are $4x$, $3x^2$ and 0. It is non-homogeneous because $f = x^3$.

(ii) This o.d.e. is non-linear because it contains the term $(dy/dx)^2$.

(iii) This is linear (the coefficients are constant) and is homogeneous.

(iv) This o.d.e. is non-linear because of the term $y(d^2 y/dx^2)$.

In this section we shall be particularly concerned with homogeneous equations which have *constant coefficients*. An example of such an equation is

$$3 \frac{d^2 y}{dx^2} + 5 \frac{dy}{dx} - 2y = 0$$

13.2.2 The method of solution. Consider the equation

$$a \frac{d^2 y}{dx^2} + b \frac{dy}{dx} + cy = 0 \tag{7}$$

in which a, b and c are constants. To indicate the idea of the method for solving a general constant-coefficient second-order equation consider the equation for which $c = 0$; we then have

$$a \frac{d^2 y}{dx^2} + b \frac{dy}{dx} = 0$$

and if we put $z = dy/dx$, this becomes the following first-order linear equation in z,

$$a \frac{dz}{dx} + bz = 0$$

Using the variables separable method or the integrating factor method to solve this equation we have the solution

$$z = C \exp\left(-\frac{b}{a} x\right)$$

where C is an arbitrary constant. The equation for y becomes

$$\frac{dy}{dx} = C \exp\left(-\frac{b}{a}x \right)$$

which can be integrated to give

$$y = -\frac{a}{b} C \exp\left(-\frac{b}{a}x \right) + D$$

This solution contains two essential features of the solution of the more general equation: it contains an exponential function and it contains two arbitrary constants C and D. This suggests that we attempt to find a solution of the more general equation in the form $y = e^{\alpha x}$ for some constant parameter α. Let us assume therefore that

$$y = e^{\alpha x}$$

then

$$\frac{dy}{dx} = \alpha e^{\alpha x}$$

and

$$\frac{d^2 y}{dx^2} = \alpha^2 e^{\alpha x}$$

so that on substitution into (7) we get

$$a(\alpha^2 e^{\alpha x}) + b(\alpha e^{\alpha x}) + c e^{\alpha x} = (a\alpha^2 + b\alpha + c)e^{\alpha x} = 0$$

Since $e^{\alpha x}$ is in general non-zero, $y = e^{\alpha x}$ is a solution of (7) provided

$$a\alpha^2 + b\alpha + c = 0 \qquad\qquad (8)$$

This equation is called *the auxiliary equation* (or the indicial equation) of differential equation (7).

Example: Find the auxiliary equation of the differential equation

$$3\frac{d^2 y}{dx^2} + 5\frac{dy}{dx} - 2y = 0$$

Solution: Put $y = e^{\alpha x}$, $dy/dx = \alpha e^{\alpha x}$ and $d^2 y/dx^2 = \alpha^2 e^{\alpha x}$, then on substitution α satisfies the equation $3\alpha^2 + 5\alpha - 2 = 0$. This is the auxiliary equation.

Equation (8) is a quadratic equation in α. Such equations can be divided into three cases:
(1) if $b^2 - 4ac > 0$ the equation has two distinct, real, solutions;
(2) if $b^2 - 4ac = 0$ the equation has one repeated solution;
(3) if $b^2 - 4ac < 0$ the equation has two, conjugate, complex solutions.
For instance, in the example, the quadratic equation is of type 1 since $5^2 - 4(3)(-2) = 49 > 0$ and the two values of α are $\alpha = -2$ and $\alpha = \frac{1}{3}$. The functions $\exp(-2x)$ and $\exp(\frac{1}{3}x)$ are both solutions of the o.d.e.

$$3\frac{d^2 y}{dx^2} + 5\frac{dy}{dx} - 2y = 0$$

In this example we have two different solutions to the o.d.e. Now we can prove a very important theorem for the solution of second-order linear differential equations. This states that if $u_1(x)$ and $u_2(x)$ are two solutions of the linear o.d.e. (7) then so is the sum $Au_1(x) + Bu_2(x)$ where A and B are constants.

The proof is fairly straightforward. If $u_1(x)$ and $u_2(x)$ are solutions of the o.d.e. then

$$a\frac{d^2u_1}{dx^2} + b\frac{du_1}{dx} + cu_1 = 0$$

and

$$a\frac{d^2u_2}{dx^2} + b\frac{du_2}{dx} + cu_2 = 0$$

That $u = Au_1(x) + Bu_2(x)$ is also a solution follows by substitution.

$$a\frac{d^2u}{dx^2} + b\frac{du}{dx} + cu = a\frac{d^2}{dx^2}(Au_1 + Bu_2) + b\frac{d}{dx}(Au_1 + Bu_2) + c(Au_1 + Bu_2)$$

$$= A\left(a\frac{d^2u_1}{dx^2} + b\frac{du_1}{dx} + cu_1\right) + B\left(a\frac{d^2u_2}{dx^2} + b\frac{du_2}{dx} + cu_2\right)$$

since A and B are constants. Thus

$$a\frac{d^2u}{dx^2} + b\frac{du}{dx} + cu = A(0) + B(0) = 0$$

and $u = Au_1(x) + Bu_2(x)$ is also a solution. We call $u(x)$ the *general solution* of the second-order o.d.e. (7). Our aim is to find two different solutions $u_1(x)$ and $u_2(x)$ and the auxiliary equation is our starting point.

The three cases for the solution of the auxiliary equation lead to three cases for the solution of the original differential equation (7). We consider each case separately.

Case 1: two real solutions. If $b^2 - 4ac > 0$ then the quadratic equation (8) has the two real solutions

$$\alpha_1 = \frac{-b + \sqrt{(b^2 - 4ac)}}{2a} \quad \text{and} \quad \alpha_2 = \frac{-b - \sqrt{(b^2 - 4ac)}}{2a}$$

Thus there are two possible exponential solutions of the differential equation (7). These are $\exp(\alpha_1 x)$ and $\exp(\alpha_2 x)$. The general solution of (7) is then

$$y = A\exp(\alpha_1 x) + B\exp(\alpha_2 x)$$

where A and B are arbitrary constants.

Example: Solve the differential equation

$$\frac{d^2y}{dx^2} + 4\frac{dy}{dx} + 3y = 0.$$

Solution: The auxiliary equation is

$$\alpha^2 + 4\alpha + 3 = 0$$

Its solutions are $\alpha = -3$ and $\alpha = -1$ so that $\exp(-3x)$ and $\exp(-x)$ are both solutions of the differential equation. The general solution of the differential equation is

$$y = A\exp(-3x) + B\exp(-x)$$

where A and B are arbitrary constants.

Case 2: one solution. If $b^2 - 4ac = 0$ then the quadratic equation (8) has only one real solution

$$\alpha = -\frac{b}{2a}$$

A second-order differential equation has two 'genuinely different' functions that satisfy it because to find a solution we must, in effect, 'integrate twice'. The method of solution adopted here has provided one function $\exp[-(b/2a)x]$.

A second function $x\exp[-(b/2a)x]$ is also a solution in this case. We can check this by substituting for $y = x\exp[-(b/2a)x]$ into the differential equation (7). We have

$$\frac{dy}{dx} = \exp\left(-\frac{b}{2a}x\right) - \frac{b}{2a}x\exp\left(-\frac{b}{2a}x\right)$$

$$\frac{d^2y}{dx^2} = -\frac{b}{2a}\exp\left(-\frac{b}{2a}x\right) - \frac{b}{2a}\left(\exp\left(-\frac{b}{2a}x\right) - \frac{bx}{2a}\exp\left(-\frac{b}{2a}x\right)\right)$$

$$= -\frac{b}{a}\exp\left(-\frac{b}{2a}x\right) + \frac{b^2x}{4a^2}\exp\left(-\frac{b}{2a}x\right)$$

On substitution into the left-hand side of (7) we have

$$a\left(-\frac{b}{a}\exp\left(-\frac{b}{2a}x\right) + \frac{b^2}{4a^2}x\exp\left(-\frac{b}{2a}x\right)\right)$$

$$+ b\left(\exp\left(-\frac{b}{2a}x\right) - \frac{b}{2a}x\exp\left(-\frac{b}{2a}x\right)\right) + cx\exp\left(-\frac{b}{2a}x\right)$$

Collecting together terms

$$\exp\left(-\frac{b}{2a}x\right)\left(-b + \frac{b^2}{4a}x + b - \frac{b^2}{2a}x + cx\right) = x\exp\left(-\frac{b}{2a}x\right)\left(-\frac{b^2}{4a} + c\right) = 0$$

since $b^2 - 4ac = 0$ in this case. Hence the function

$$y = x\exp\left(-\frac{b}{2a}x\right)$$

is also a solution of the differential equation.

The general solution is then

$$y = A\exp\left(-\frac{b}{2a}x\right) + Bx\exp\left(-\frac{b}{2a}x\right)$$

where A and B are arbitrary constants.

Example: Solve the differential equation

$$\frac{d^2y}{dx^2} + 2\frac{dy}{dx} + y = 0$$

Solution: The auxiliary equation is

$$\alpha^2 + 2\alpha + 1 = 0$$

It has one solution $\alpha = -1$ so that e^{-x} and xe^{-x} are both solutions of the differential equation. The general solution is

$$y = Ae^{-x} + Bxe^{-x}$$

where A and B are arbitrary constants.

Case 3: two complex solutions. If $b^2 - 4ac < 0$ then the quadratic equation (8) has two conjugate, complex solutions

$$\alpha_1 = -\frac{b}{2a} + i\omega \quad \text{and} \quad \alpha_2 = -\frac{b}{2a} - i\omega$$

where $\omega = \sqrt{(4ac - b^2)}/2a$. We can write the two solutions as exponential functions $\exp(\alpha_1 x)$ and $\exp(\alpha_2 x)$ but it is often more convenient to replace the complex exponential by sines and cosines using Euler's formula $e^{i\theta} = \cos\theta + i\sin\theta$ (see 3.4.4). With $\theta = \omega x$ we have

$$e^{i\omega x} = \cos\omega x + i\sin\omega x$$

and with $\theta = -\omega x$ we have $e^{-i\omega x} = \cos\omega x - i\sin\omega x$

The general solution becomes

$$y = A\exp(\alpha_1 x) + B\exp(\alpha_2 x)$$

$$= \exp\left(-\frac{b}{2a}x\right)[A(\cos\omega x + i\sin\omega x) + B(\cos\omega x - i\sin\omega x)]$$

$$= \exp\left(-\frac{b}{2a}x\right)[\cos\omega x(A + B) + \sin\omega x(iA - iB)]$$

Since A and B are arbitrary constants we can write $A + B = C$ and $iA - iB = D$, where C and D are now arbitrary constants. The general solution of the differential equation is then

$$y = \exp\left(-\frac{b}{2a}x\right)(C\cos\omega x + D\sin\omega x)$$

Example: Solve the differential equation

$$\frac{d^2y}{dx^2} + 4\frac{dy}{dx} + 8y = 0$$

Solution: The auxiliary equation is

$$\alpha^2 + 4\alpha + 8 = 0$$

It has two complex, conjugate solutions

$$\alpha_1 = -2 + 2i \quad \text{and} \quad \alpha_2 = -2 - 2i$$

The general solution of the differential equation is

$$y = e^{-2x}(C \cos 2x + D \sin 2x)$$

The physical interpretation of each of the three cases outlined above is given in 13.2.3. We summarise the method in the following procedure:

Procedure for solving second-order differential equations. To solve the homogeneous second-order differential equation

$$a\frac{d^2y}{dx^2} + b\frac{dy}{dx} + cy = 0$$

where a, b and c are constants, and $a \neq 0$.

    (1) Use the trial function $y = e^{\alpha x}$ to obtain the auxiliary equation $a\alpha^2 + b\alpha + c = 0$.

    (2) Solve the auxiliary equation.

    (3)   (i) If the auxiliary equation has two real, distinct solutions α_1 and α_2 then the general solution is

$$y = A\exp(\alpha_1 x) + B\exp(\alpha_2 x)$$

        (ii) If the auxiliary equation has one solution α then the general solution is

$$y = Ae^{\alpha x} + Bxe^{\alpha x}$$

        (iii) If the auxiliary equation has two, complex, conjugate solutions $\alpha_1 = \alpha + i\omega$ and $\alpha_2 = \alpha - i\omega$ then the general solution is

$$y = e^{\alpha x}(C \cos \omega x + D \sin \omega x)$$

13.2.3 Interpreting the solution. Ordinary differential equations of the type introduced in this section occur frequently in science. One of the most common applications is in modelling vibrations and we can give a physical interpretation of the solutions of 13.2.2 by investigating the motion of a pendulum.

Simple harmonic motion. The motion of a simple pendulum, for which the resistance to the motion of the body is neglected, can be described by differential equation (5) with $R = 0$.

$$\frac{d^2\theta}{dt^2} + \frac{g}{l}\theta = 0$$

This is a second-order ordinary differential equation. Using the procedure of 13.2.2, the auxiliary equation is

$$\alpha^2 + \frac{g}{l} = 0$$

This quadratic equation has two complex roots

$$\alpha_1 = i\sqrt{\left(\frac{g}{l}\right)} \quad \text{and} \quad \alpha_2 = -i\sqrt{\left(\frac{g}{l}\right)}$$

The general solution of the differential equation is

$$\theta = A \cos \sqrt{\left(\frac{g}{l}\right)} t + B \sin \sqrt{\left(\frac{g}{l}\right)} t$$

If the pendulum is initially at rest at the angle $\theta = \theta_0$, then mathematically this can expressed as

$$\frac{d\theta}{dt} = 0 \quad \text{and} \quad \theta = \theta_0 \text{ when } t = 0$$

Using these conditions in the solution for θ we have

$$\theta = A = \theta_0 \qquad \text{when } t = 0$$

and

$$\frac{d\theta}{dt} = B \sqrt{\left(\frac{g}{l}\right)} = 0 \qquad \text{when } t = 0$$

Hence the particular solution satisfying the required initial conditions is

$$\theta = \theta_0 \cos \sqrt{\left(\frac{g}{l}\right)} t$$

We can interpret this solution by drawing a graph of θ against t, Fig. 13.4 shows this graph. The graph shows that θ always lies between the values θ_0 and $-\theta_0$, and the basic cosine curve repeats itself every $2\pi/\sqrt{(g/l)}$ s. This quantity is called the *period* of the pendulum and θ_0 is the *amplitude*. The solution implies that the body on the end of the string swings back and forth returning to its starting point every $2\pi/\sqrt{(g/l)}$ s.

Such a motion is called *simple harmonic* and according to the solution of the differential equation it 'goes on for ever'. Vibrating systems do not, in general, behave in simple harmonic motion. They tend to slow down and stop, although it may take a long time. More realistically we should include air resistance in the model.

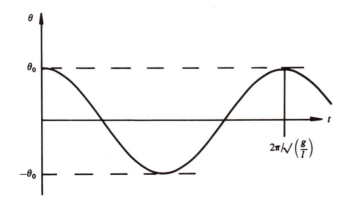

Fig. 13.4. A graph of θ against t for a simple pendulum.

The damping of a pendulum. Consider the motion of a simple pendulum for which the body is opposed by a resistive force which is proportional to its speed. According to the model developed in 13.1.2 the angle θ satisfies the differential equation

$$\frac{d^2\theta}{dt^2} + \frac{k\,d\theta}{l\,dt} + \frac{g}{l}\theta = 0$$

and we assume that the pendulum starts from rest with $\theta = \theta_0$. The solution of this second-order differential equation has three forms depending on the value of $(k/l)^2 - 4(g/l)$. This physical situation of a pendulum in a resistive medium provides a means of interpreting these three forms of solution.

Case 1: $(k/l)^2 - 4(g/l) > 0$. In this case the auxiliary equation

$$\alpha^2 + \frac{K}{l}\alpha + \frac{g}{l} = 0$$

has two real roots: so that the solution of the differential equation is

$$\theta = A\exp(\alpha_1 t) + B\exp(\alpha_2 t)$$

where

$$\alpha_1 = \frac{-k + \sqrt{(k^2 - 4gl)}}{2l} \quad \text{and} \quad \alpha_2 = \frac{-k - \sqrt{(k^2 - 4gl)}}{2l}$$

Using the initial conditions $\theta = \theta_0$ and $d\theta/dt = 0$ when $t = 0$ we have

$$A + B = \theta_0$$

and

$$\alpha_1 A + \alpha_2 B = 0$$

The solution of these two equations is

$$A = \frac{-2\alpha_2\theta_0 l}{\sqrt{(k^2 - 4gl)}}, \quad B = \frac{2\alpha_1\theta_0 l}{\sqrt{(k^2 - 4gl)}}$$

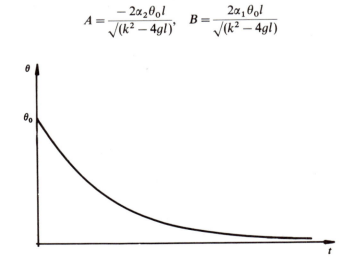

Fig. 13.5. Angular displacement for an overdamped pendulum.

so that

$$\theta = \frac{2\theta_0 l}{\sqrt{(K^2 - 4gl)}} \exp\left(-\frac{kt}{2l}\right) [\alpha_1 \exp[-\sqrt{(k^2 - 4gl)}t/2l]$$
$$- \alpha_2 \exp[+\sqrt{(k^2 - 4gl)}t/2l]]$$

Fig. 13.5 shows a graph of this function. There are no oscillations of the pendulum at all. Physically this situation occurs because the resistive force dominates the system. Such a system is said to be *overdamped* and in such a case we would not expect the pendulum to oscillate.

Case 2: $(k/l)^2 - 4(g/l) = 0$. In this case the auxiliary equation has one root

$$\alpha = -k/2l$$

and the solution of the differential equation is

$$\theta = (A + Bt)\exp\left(-\frac{k}{2l}t\right)$$

The initial conditions give

$$\theta_0 = A \quad \text{and} \quad B = \frac{k}{2l}\theta_0$$

so that

$$\theta = \theta_0\left(1 + \frac{k}{2l}t\right)\exp\left(-\frac{k}{2l}t\right)$$

Fig. 13.6 shows a graph of this function. Again the pendulum does not oscillate. We say that the pendulum is critically damped. This is very similar to case 1.

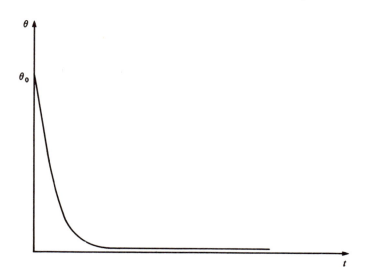

Fig. 13.6. Angular displacement for a critically damped pendulum.

Measuring instruments are constructed so that the pointer is critically damped for then the pointer returns to the equilibrium position as quickly as possible, a galvanometer is one such instrument.

Case 3: $(k/l)^2 - 4(g/l) < 0$. In this case the auxiliary equation has two complex roots

$$\alpha_1 = -\frac{k}{2l} + \frac{i\omega}{2} \quad \text{and} \quad \alpha_2 = -\frac{k}{2l} - \frac{i\omega}{2}$$

where $\omega = \sqrt{[4(g/l) - (k/l)^2]}$, and the general solution of the differential equation is

$$\theta = \exp\left(-\frac{kt}{2l}\right)\left(A\cos\frac{\omega}{2}t + B\sin\frac{\omega}{2}t\right)$$

so that

$$\theta = \theta_0 \exp\left(-\frac{kt}{2l}\right)\left(\cos\frac{\omega}{2}t + \frac{k}{2l\omega}\sin\frac{\omega}{2}t\right)$$

This solution, which includes the trigonometric functions sine and cosine, has a similar form to the solution for simple harmonic motion but the amplitude slowly decays (see Fig. 13.7). The pendulum oscillates with a fixed period $T = 2\pi/\sqrt{[(g/l) - (k/2l)^2]}$ but the amplitude $\theta_0\exp(-kt/2l)$ is an exponentially decaying function of time. Notice that if we put $k = 0$ we recover simple harmonic motion with constant amplitude θ_0 and period $2\pi/\sqrt{(g/l)}$. This solution is more in keeping with our experience of a pendulum. During each cycle the string does not quite get back to its starting point for the previous cycle and eventually the oscillations are very small. This is called an *underdamped harmonic motion* and occurs when the resistive force is small compared with other forces, in this case the force due to gravity.

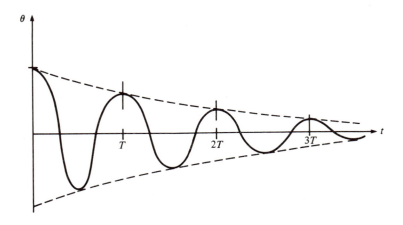

Fig. 13.7. Angular displacement for an underdamped pendulum.

13.3 Worked examples

Example 1: the cooling fin. The steady state temperature distribution in a rectangular metal sheet, one end of which is heated, is modelled by the equation

$$\frac{d^2\theta}{dx^2} - 10\theta = 0$$

where θ is the temperature difference between the plate and the air. A 5 kW electric motor is cooled by eight symmetrically situated fins, each of length 0.25 m, thickness 0.01 m and width 1 m, and thermal conductivity 220 W m^{-1} °C^{-1}. The motor stands in a room at temperature 16 °C. Estimate the temperature of the motor when running at full power if it loses 20% of its power as heat.

Solution. The general solution of the differential equation is obtained in the following way: putting $\theta = e^{\alpha x}$ we get the auxiliary equation

$$\alpha^2 - 10 = 0$$

with solutions $\alpha = \sqrt{(10)}$ and $\alpha = -\sqrt{(10)}$. The general solution is

$$\theta = A \exp[\sqrt{(10)}x] + B \exp[-\sqrt{(10)}x]$$

To find A and B we must obtain two boundary conditions.

At the end of the fin, not attached to the motor, the flow of heat is zero because there is no metal to conduct heat.

Hence when $x = 0.25$,

$$q = -KA\frac{d\theta}{dx} = 0$$

The heat generated by the motor (at $x = 0$) is dissipated by conduction along each of the eight fins. Since the motor loses 20% of the 5 kW of power as heat, $\frac{1}{8}$ kW is conducted along each fin.

Hence when $x = 0$,

$$q = -KA\frac{d\theta}{dx} = \tfrac{1}{8} \times 10^3 \quad \text{(in watts)}$$

where the thermal conductivity $K = 220$ W m^{-1} °C^{-1} and $A = 0.01$ m^2. These boundary conditions can be used to find A and B.

We have

$$\frac{d\theta}{dx} = \sqrt{(10)}[A \exp(\sqrt{(10)}x) - B \exp(-\sqrt{(10)}x)]$$

when $x = 0$,

$$\frac{d\theta}{dx} = \sqrt{(10)}(A - B) = -\frac{1/8 \times 10^3}{220 \times 0.01} = 56.8$$

when $x = 0.25$,

$$\frac{d\theta}{dx} = \sqrt{(10}[A \exp(\sqrt{(10)}/4) - B \exp(-\sqrt{(10/4)})] = 0$$

Thus

$$A - B = -\frac{56.8}{\sqrt{(10)}} = -17.96$$

and

$$Ae^{0.79} - Be^{-0.79} = 0$$

Solving these two equations gives

$$A = 4.66 \quad \text{and} \quad B = 22.62$$

The solution for θ is then

$$\theta = 4.66 \exp(\sqrt{(10}x) + 22.62 \exp(-\sqrt{(10)}x)$$

When $x = 0$, $\theta = 27.28\,°\text{C}$ so that the temperature of the motor is estimated to be $27.28\,°\text{C} + 16\,°\text{C} = 43.28\,°\text{C}$.

Example 2: an electric circuit. Vibrational problems exhibiting damping also occur in non-mechanical applications and one example is the oscillating electrical network. Fig. 13.8 shows a simple electrical circuit consisting of a condenser C, an inductance L and a resistance R. If j is the current and if q is the charge on the plate of the condenser; then providing there is no external e.m.f. the sum of the voltage drops across each component is zero and this leads to the following two equations relating j and q

$$L\frac{dj}{dt} + Rj + \frac{q}{c} = 0 \quad \text{and} \quad j = \frac{dq}{dt}$$

Show that q satisfies a second-order homogeneous ordinary differential equation and comment on the different solutions for q.

Solution. By substituting for j, the charge q satisfies the second-order o.d.e.

$$\frac{d^2q}{dt^2} + \frac{R}{L}\frac{dq}{dt} + \frac{1}{LC}q = 0$$

This is homogeneous since the right-hand side is zero. The charge is not oscillatory if $R^2/L^2 \geqslant 4/LC$ and its behaviour is similar to the damped pendulum shown in

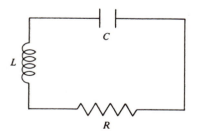

Fig. 13.8. A simple electric circuit.

Figs. 13.5 and 13.6. However, if $R^2/L^2 < 4/LC$ then the charge oscillates with an exponentially decaying amplitude in a way similar to that shown in Fig. 13.7.

13.4 Further developments

13.4.1 Non-homogeneous equations. The vibrating systems discussed in 13.2.3 may oscillate naturally or may be 'forced' to oscillate in some way. For example, the electrical circuit of Example 2 in Section 13.3 could include an applied voltage $E(t)$, say, and then the charge q will satisfy a non-homogeneous differential equation whose general form is

$$a\frac{d^2y}{dx^2} + b\frac{dy}{dx} + cy = f(x) \qquad (9)$$

in which a, b and c are constants, x is the independent variable and y is the dependent variable. Our aim is to find y as a function of x.

13.4.2 The method of solution. The key to solving the non-homogeneous differential equation (9) is to find one *particular solution*, $y_p(x)$, say, and the general solution of (9) is then the sum of the particular solution y_p and the general solution to the homogeneous equation

$$a\frac{d^2y}{dx^2} + b\frac{dy}{dx} + cy = 0 \qquad (10)$$

The following example shows how the method works.

Example: Find the general solution of the equation

$$\frac{d^2y}{dx^2} - y = 3$$

Solution: Consider first the homogeneous equation obtained by replacing the right-hand side of the equation by zero, i.e.

$$\frac{d^2y}{dx^2} - y = 0$$

The general solution of this equation is then

$$y_c(x) = Ae^x + Be^{-x}$$

(using the procedure of 13.2.2).

Now we find any solution of the non-homogeneous equation. For this example this is quite easy. If we choose $y_p = -3$ then since $(d^2/dx^2)(-3) = 0$, $y_p = -3$ satisfies the differential equation

$$\frac{d^2y}{dx^2} - y = 3$$

Finally, we assert that the general solution of this non-homogeneous equation is the sum of y_c and y_p; i.e.

$$y = Ae^x + Be^{-x} - 3$$

The method appears straightforward but it is not always so easy to 'spot' the particular solution $y_p(x)$. However, for some specific forms of the right-hand side of (9), we can use certain 'trial functions' that will provide particular solutions.

The method works for all linear differential equations. The solution of the homogeneous equation is called the *complementary function* so that the general solution of the non-homogeneous equation can be written as

general solution = complementary function + particular solution

The complementary function is found using the procedure of Section 13.2 and now we develop a method for finding the particular solution in a systematic way.

13.4.3 Trial functions. The form of the particular solution for the non-homogeneous differential equation (9) depends on the form of the function $f(x)$. The particular solution is *any* function that satisfies the full equations so that the technique is to use *trial functions* containing unknown constants and then, by substituting into the differential equation find particular values for these constants. We shall illustrate the method by examples.

> If $f(x)$ is a *polynomial*, the trial function is a polynomial of the same degree

Example: Find a particular solution of the equation

$$3\frac{d^2y}{dx^2} + 2\frac{dy}{dx} + y = x^2 - 2x + 1$$

Solution: Consider the trial function

$$y_p(x) = mx^2 + nx + p$$

On substituting this into the differential equation we have

$$3(2m) + 2(2mx + n) + (mx^2 + nx + p) = x^2 - 2x + 1$$

i.e.

$$mx^2 + (4m + n)x + (6m + 2n + p) = x^2 - 2x + 1$$

This solution can only be true for all values of x if

$$m = 1, \quad 4m + n = -2 \quad \text{and} \quad 6m + 2n + p = 1$$

We can solve these simultaneous equations to give

$$m = 1, \quad n = -6 \quad \text{and} \quad p = 7$$

Thus $y_p(x) = x^2 - 6x + 7$ is a particular solution of the given differential equation. The complementary function is $y_c(x) = \exp(-x/3)(A\cos\frac{\sqrt{2}}{3}x + B\sin\frac{\sqrt{2}}{3}x)$ and the general solution of the given differential equation is

$$y = \exp\left(-\frac{x}{3}\right)\left(A\cos\frac{\sqrt{2}}{3}x + B\sin\frac{\sqrt{2}}{3}x\right) + (x^2 - 6x + 7).$$

> If $f(x)$ is an *exponential function*, the trial function is an exponential with the same exponent as f

Example: Find a particular solution of the equation

$$3\frac{d^2y}{dx^2} + 2\frac{dy}{dx} + y = 2e^{3x}$$

Solution: Consider the trial function

$$y_p(x) = me^{3x}$$

Then

$$\frac{dy_p}{dx} = 3me^{3x} \quad \text{and} \quad \frac{d^2y_p}{dx^2} = 9me^{3x}$$

so that on substituting we have

$$27me^{3x} + 6me^{3x} + me^{3x} = 2e^{3x}$$

Thus me^{3x} is a particular solution of the differential equation if

$$34me^{3x} = 2e^{3x} \quad \text{i.e. if } m = \tfrac{2}{34}$$

Hence $\frac{1}{17}e^{3x}$ is a particular solution. The general solution of the given differential equation is

$$y = \exp\left(-\frac{x}{3}\right)\left(A\cos\frac{\sqrt{2}}{3}x + B\sin\frac{\sqrt{2}}{3}x\right) + \frac{1}{17}\exp(3x)$$

If $f(x)$ is a *trigonometric function*, the trial function is a linear combination of sine and cosine

Example: Find a particular solution of the equation

$$3\frac{d^2y}{dx^2} + 2\frac{dy}{dx} + y = 3\sin 2x$$

Solution: Consider the trial function

$$y_p(x) = m\cos 2x + n\sin 2x$$

Then

$$\frac{dy_p}{dx} = -2m\sin 2x + 2n\cos 2x$$

and

$$\frac{d^2y_p}{dx^2} = -4m\cos 2x - 4n\sin 2x$$

so that on substituting we have

$$3(-4m\cos 2x - 4n\sin 2x) + 2(-2m\sin 2x + 2n\cos 2x)$$
$$+ (m\cos 2x + n\sin 2x) = 3\sin 2x$$

i.e.

$$(-11m + 4n)\cos 2x + (-11n - 4m)\sin 2x = 3\sin 2x$$

Hence $m\cos 2x + n\sin 2x$ is a particular solution of the differential equation if

$$-11m + 4n = 0 \quad \text{and} \quad -11n - 4m = 3$$

Solving for m and n gives

$$m = -\frac{12}{137} \quad \text{and} \quad n = -\frac{33}{137}$$

The particular solution is

$$-\frac{12}{137}\cos 2x - \frac{33}{137}\sin 2x$$

This last type of equation, where the right-hand side is a sine or cosine function is very important in science and engineering. There are many situations in which a system is forced to undergo vibrations; and a mathematical model to describe the system is often a second-order non-homogeneous differential equation with sines or cosines on the right-hand side. The general solution is then a combination of the *natural oscillations* described by the complementary function and *forced oscillations* described by the particular solution.

There are differential equations for which our trial functions do not provide a particular solution. For example, for the equation

$$\frac{d^2y}{dx^2} + \omega^2 y = 2 \sin \omega x$$

the trial function $y_p = m \cos \omega x + n \sin \omega x$ gives on substitution

$$\frac{d^2y}{dx^2} + \omega^2 y = 0$$

So there are no values of m and n which make $m \cos \omega x + n \sin \omega x$ a particular solution. This difficulty always occurs when the function on the right-hand side of a non-homogeneous differential equation is a solution of the associated homogeneous equation and thus forms part of the complementary function.

For example, the complementary function of the given differential equation is

$$y_c = A \cos \omega x + B \sin \omega x$$

and $\sin \omega x$ appears on the right-hand side of the differential equation. When the trial function fails in this way we usually try multiplying the suggested trial function by x.

For example, we attempt to find a particular solution in the form

$$y_p = x(m \cos \omega x + n \sin \omega x)$$

Then

$$\frac{dy_p}{dx} = (m \cos \omega x + n \sin \omega x) + x\omega(-m \sin \omega x + n \cos \omega x)$$

and

$$\frac{d^2y_p}{dx^2} = 2\omega(-m \sin \omega x + n \cos \omega x) + x\omega^2(-m \cos \omega x - n \sin \omega x)$$

so that on substituting we have

$$2\omega(-m \sin \omega x + n \cos \omega x) + x\omega^2(-m \cos \omega x - n \sin \omega x)$$
$$+ \omega^2[x(m \cos \omega x + n \sin \omega x)]$$
$$= 2\omega(-m \sin \omega x + n \cos \omega x) = 2 \sin \omega x$$

This equation can only be true for all x if $n = 0$ and $m = -1/\omega$. Thus the given non-homogeneous differential equation has particular solution

$$y_p = -\frac{1}{\omega} x \cos \omega x$$

13.5 Further worked examples

Example 1. An electrical network consists of an inductance L, a resistance R and a condenser of capacity C connected in series with an applied voltage $E(t) = a \sin \omega t$. The charge on a plate of the condenser satisfies the equation

$$L\frac{d^2q}{dt^2} + R\frac{dq}{dt} + \frac{1}{C}q = a \sin \omega t$$

If $R < 2\sqrt{(L/C)}$, find the general solution for q at any time t. For the case when $L = 1$ henry, $R = 100$ ohms, $C = 10^{-4}$ farads, $\omega = 100$ and $a = 1000$ volts show the response of the circuit graphically.

Solution. The solution consists of the sum of

(i) the complementary function which satisfies the non-homogeneous equation

$$L\frac{d^2q}{dt^2} + R\frac{dq}{dt} + \frac{1}{C}q = 0$$

and

(ii) a particular solution using the trial function

$$q_p = m\cos \omega t + n \sin \omega t$$

(i) For the homogeneous equation, let $q = e^{\alpha t}$ giving the auxiliary equation

$$L\alpha^2 + R\alpha + \frac{1}{C} = 0$$

with solution

$$\alpha = [-R \pm \sqrt{(R^2 - 4L/C)}]/2L$$

Since $R < 2\sqrt{(L/C)}$ we have an oscillating solution, i.e. the roots for α are complex conjugate solutions. We have

$$q_c = e^{-Rt/2L}\left(A\cos\left[\frac{\sqrt{((4L/C) - R^2)}t}{2L}\right] + B\sin\left[\frac{\sqrt{((4L/C) - R^2)}t}{2L}\right]\right)$$

(ii) With the trial function $q_p = m\cos \omega t + n \sin \omega t$ we have

$$L\frac{d^2q}{dt^2} + R\frac{dq}{dt} + \frac{1}{C}q = -L\omega^2(m\cos \omega t + n\sin \omega t)$$

$$+ R\omega(-m\sin \omega t + n\cos \omega t) + \frac{1}{C}(m\cos \omega t + n\sin \omega t)$$

$$= \left[\left(-L\omega^2 + \frac{1}{C}\right)m + Rn\omega\right]\cos \omega t$$

$$+ \left[\left(-L\omega^2 + \frac{1}{C}\right)n - R\omega m\right]\sin \omega t$$

$$= a\sin \omega t$$

Hence m and n are given as solutions to the simultaneous equations

$$\left(-L\omega^2 + \frac{1}{C}\right)m + Rn\omega = 0$$

and

$$\left(-L\omega^2 + \frac{1}{C}\right)n - Rm\omega = a$$

Solving we have

$$m = \frac{R\omega a}{R^2\omega^2 - \left(-L\omega^2 + \frac{1}{C}\right)^2} \quad \text{and} \quad n = \frac{-\left(-L\omega^2 + \frac{1}{C}\right)a}{R\omega^2 - \left(-L\omega^2 + \frac{1}{C}\right)^2}$$

The general solution for q is then $q = q_c + q_p$. If $L = 1$, $R = 100$, $C = 10^{-4}$, $\omega = 100$ and $a = 1000$ then

$$q_c = e^{-50t}(A\cos(50\sqrt{3}t) + B\sin(50\sqrt{3}t))$$

and

$$q_p = 0.1\cos 100t$$

The complementary function q_c describes the natural oscillations that can occur in the circuit and the particular solution q_p is the oscillation forced in the circuit by the applied voltage. Fig. 13.9 shows the graph of the charge against time. We see that the amplitude of the response q_p (0.1) is much less than the amplitude of the input voltage (1000). The 'natural oscillations' will eventually die out and q_c is called *the transient*. Eventually the solution is given by $q = q_p$ which is called the *steady state solution*.

Example 2: Resonance. In the last example the steady state response from the electrical circuit is much smaller than the input voltage. Now, in the first section

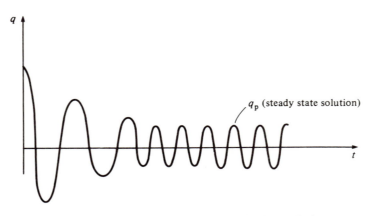

Fig. 13.9. The charge approaches a steady state solution.

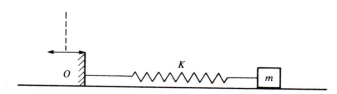

Fig. 13.10. A forced harmonic oscillator.

of this chapter we described a physical situation, the Ferrybridge cooling tower collapse, for which a forced vibration had a catastrophic effect. In this example we investigate a simple vibrating system which exhibits a large response to an input, similar to that which occurred at Ferrybridge.

Fig. 13.10 shows a body of mass m on a horizontal smooth surface joined to a point O by an elastic spring of stiffness K. The point O is forced to vibrate so that its displacement at any time t is $a \sin \omega t$, where $\omega^2 = K/m$. The extension of the spring x is governed by the equation

$$\frac{d^2x}{dt^2} + \frac{K}{m}x = a\omega^2 \sin \omega t$$

Solve this differential equation and investigate the physical significance of the solution.

Solution. The general solution of this equation is the sum of the complementary function and a particular solution. The complementary function x_c is a solution of the homogeneous equation

$$\frac{d^2x}{dt^2} + \frac{K}{m}x = 0$$

We have

$$x_c = A \cos \sqrt{\left(\frac{K}{m}\right)}t + B \sin \sqrt{\left(\frac{K}{m}\right)}t$$

For the particular solution we use a trial function

$$x_p = m \cos \omega t + n \sin \omega t$$

then

$$\frac{dx_p}{dt} = -m\omega \sin \omega t + n\omega \cos \omega t \quad \text{and} \quad \frac{d^2x_p}{dt^2} = -m\omega^2 \cos \omega t - n\omega^2 \sin \omega t$$

On substitution we have

$$\frac{d^2x}{dt^2} + \frac{K}{m}x = -\omega^2(m \cos \omega t + n \sin \omega t) + \frac{K}{m}(m \cos \omega t + n \sin \omega t)$$

and since $\omega^2 = K/m$, the terms involving cosine and sine cancel out so that our trial function will not give $a\omega^2 \sin \omega t$.

This differential equation is one for which the function on the right-hand side appears in the complementary function. In this case we try

$$x_p = t(m \cos \omega t + n \sin \omega t)$$

Then

$$\frac{d^2 x_p}{dt^2} + \frac{K}{m} x_p = -\omega^2 t(m \cos \omega t + n \sin \omega t) + 2\omega(-m \sin \omega t + n \cos \omega t)$$

$$+ \frac{K}{m} t(m \cos \omega t + n \sin \omega t)$$

$$= 2\omega(-m \sin \omega t + n \cos \omega t)\left(\text{since } \omega^2 = \frac{K}{m}\right)$$

$$= a\omega^2 \sin \omega t$$

Hence for this to be true for all t, we have $m = -a\omega/2$ and $n = 0$. The particular solution is $x_p = -(a\omega/2)t \cos \omega t$. The complementary function x_c describes the motion of the body if the point O was fixed. Given a small displacement the body would describe simple harmonic motion and

$$x_c = A \cos \sqrt{\left(\frac{K}{m}\right)} t + B \sin \sqrt{\left(\frac{K}{m}\right)} t$$

The forcing term $a\omega^2 \sin \omega t$, produced by making O vibrate, has a response described by

$$x_p = -\frac{a\omega}{2} t \cos \omega t$$

A graph of this function is shown in Fig. 13.11. We see that the motion is still oscillatory but the amplitude $a\omega t/2$ now increases so that the body will oscillate with ever increasing magnitude. This is a phenomenon called *resonance*. It occurs

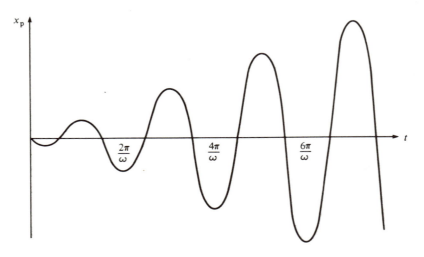

Fig. 13.11. The solution shows the amplitude growing with time.

when the frequency of the forcing terms equals the natural frequency of the system. This simple example shows that in designing an engineering structure, it is important to know what the natural frequency of the system is and to ensure that any forced vibrations have a frequency which is different from this natural frequency.

An occasion when resonance proved disastrous was in 1831 when a column of soldiers marching over Broughton suspension bridge (near Manchester) set up a forced vibration, which had a frequency approximately equal to one of the natural frequencies of the bridge. The bridge collapsed. Ever since this disaster, troops have been ordered to break step when crossing bridges!

13.6 Exercises

1. Solve the following differential equations:

   (i) $\dfrac{d^2 s}{dt^2} = t^2(t+7)$

   given that $ds/dt = 2$ when $t = 1$ and $s = 0$ when $t = 0$

   (ii) $\dfrac{d^2 y}{dx^2} = x \sin(x)$

   given that $dy/dx = 0$ when $x = \pi/2$ and $y = 1$ when $x = 0$

2. Find the general solution of the following differential equations:

   (i) $\dfrac{d^2 y}{dt^2} + 2\dfrac{dy}{dt} - 15y = 0$ (ii) $\dfrac{d^2 y}{dx^2} - 6\dfrac{dy}{dx} + 9y = 0$

   (iii) $\dfrac{d^2 y}{dx^2} - 4\dfrac{dy}{dx} + 13y = 0$ (iv) $2\dfrac{d^2 y}{dx^2} + \dfrac{dy}{dx} + y = 0$

3. Find the general solution of the following differential equations

   (i) $\dfrac{d^2 y}{dx^2} + 3\dfrac{dy}{dx} + 2y = 6e^x$

   (ii) $\dfrac{d^2 y}{dx^2} - 5\dfrac{dy}{dx} + 6y = 5x^2$

   (iii) $\dfrac{d^2 y}{dx^2} - 2\sin(a)\dfrac{dy}{dx} + y = 2\cos(a)\sin(x)$ where $a \neq n\pi/2$

   (iv) $2\dfrac{d^2 y}{dx^2} - 9\dfrac{dy}{dx} - 35y = \sin(2x)$

   (v) $\dfrac{d^2 y}{dx^2} - 5\dfrac{dy}{dx} + 4y = 4e^x$

4. In appropriate units, the time-independent Schrödinger equation of a

fixed-axis rigid rotor reduces to

$$-\frac{d^2\psi}{d\theta^2} = m^2\psi$$

where θ is the angle turned through and m is the angular momentum of the rotor. Show, by solving the differential equation, that the condition $\psi(\theta) = \psi(\theta + 2\pi)$ restricts m to integer values.

5. The damped oscillations of a body are described by the solution of the differential equation

$$\frac{d^2x}{dt^2} + 2n\frac{dx}{dt} + w^2x = 0$$

where n and w are positive constants. Show that the oscillations are heavily damped for $n^2 > w^2$ and lightly damped for $n^2 < w^2$. Obtain the solution to this differential equation when $n = w$, subject to the initial conditions $x = d$ and $dx/dt = 0$.

6. In a heavily shunted galvanometer, the deflection θ satisfies the differential equation

$$\frac{d^2\theta}{dt^2} + 6\frac{d\theta}{dt} + 9\theta = 2$$

Solve this equation assuming that $\theta = 0$, $d\theta/dt = 0$ when $t = 0$ and hence find the final value of the deflection θ.

7. A sphere of mass m and radius a is falling vertically through a light liquid of viscosity η. At time t it has fallen a distance x and acquired a velocity v downwards where x and t are related by the differential equation

$$m\frac{d^2x}{dt^2} + 6\pi\eta a\frac{dx}{dt} = mg$$

If at $t = 0$ $v = 0$, obtain v in terms of t, m, η, a and g.

8. Find the particular solutions of the following problems:

(i) $5\dfrac{d^2y}{dt^2} + 2\dfrac{dy}{dt} + y = 2t + 3$

given that $y = -1$ and $dy/dt = 0$ when $t = 0$;

(ii) $\dfrac{d^2y}{dt^2} - 10\dfrac{dy}{dt} + 24y = 24t$

given that $y = 0$ and $dy/dt = 0$ when $t = 0$;

(iii) $\dfrac{d^2y}{dx^2} + 2\dfrac{dy}{dx} + 5y = 4e^{-x}$

given that $y = 0$ and $dy/dx = 0$ when $x = 0$;

(iv) $\dfrac{d^2y}{dx^2} + 4\dfrac{dy}{dx} + 4y = x^2 + e^{-2x}$

given that $y = \frac{1}{2}$ and $dy/dx = 0$ when $x = 0$;

(v) $\dfrac{d^2x}{dt^2} + 9x = e^{-t}$

given that $x = 0$ and $dx/dt = 1$ when $t = 0$.

9. When a particular measuring instrument is used, the pointer on the scale of the instrument oscillates before settling down to a final value. If x is the distance at time t of the pointer from its final position, then the oscillation of the pointer is described by the differential equation

$$\frac{d^2x}{dt^2} + 3\frac{dx}{dt} + 10x = 0$$

   (i) Find the general solution of the above equation for x.
   (ii) If $x = 0.1$ and $dx/dt = 0$ when $t = 0$, find the arbitrary constants in the general solution for x.
   (iii) With the conditions in part (ii), what length of time will elapse before the amplitude of the oscillations of the pointer has decreased to 10% of its original value?

10. The time-independent Schrödinger equation, representing a charged particle of mass m in a potential well, of width L, with perfectly reflecting walls, may be written as

$$\frac{-h^2}{8\pi^2 m}\frac{d^2\psi}{dx^2} = E\psi, \quad 0 \leqslant x \leqslant L$$

where h is Planck's constant, E is the energy of the particle, and $\psi(x)$ is the wave function of the particle:
   (i) Solve the differential equation to obtain the general expression for $\psi(x)$, assuming E is always positive.
   (ii) Given that $\psi(0) = \psi(L) = 0$, show that the energy of the particle is quantized, and find the expression for $\psi(x)$ in the state of lowest energy. You may assume that $\psi(x) = 0$ is not an acceptable wave function.
   (iii) Obtain a unique expression for $\psi(x)$ in this 'ground state', by noting that $\int_0^L \psi(x)^2\, dx = 1$ is a condition that must be satisfied by all the positive wave functions.

11. The displacement y of a body in a damped mechanical system, with no external forces, satisfies the following differential equation

$$2\frac{d^2y}{dt^2} + 6\frac{dy}{dt} + 4.5y = 0$$

where t represents time. If initially, when $t = 0$, $y = 0$ and $dy/dt = 4$, solve the differential equation for y in terms of t.

12. In the consecutive first-order chemical reaction

$$A \xrightarrow{k_1} B \xrightarrow{k_2} C$$

the various rate equations for the concentrations [A], [B] and [C] are

$$\frac{d[A]}{dt} = -k_1[A]$$

$$\frac{d[B]}{dt} = k_1[A] - k_2[B]$$

$$\frac{d[C]}{dt} = k_2[B]$$

Show that [B] satisfies the second-order differential equation given by

$$\frac{d^2[B]}{dt^2} + (k_1 + k_2)\frac{d[B]}{dt} + k_1 k_2[B] = 0$$

Solve this differential equation using the fact that [B] = 0 and [A] = a_0 when $t = 0$.

13. An electrical circuit consists of an inductor L, a resistor R, and a capacitor C, in series. The circuit is driven by an emf $E(t)$. The charge $q(t)$ on the plates of the capacitor satisfies the following differential equation:

$$L\frac{d^2q}{dt^2} + R\frac{dq}{dt} + \frac{1}{C}q = E$$

The charge arising as a result of an alternating emf of angular frequency w can be obtained by substituting the following expressions for q and E in the above equation:

$$q(t) = Q_0 e^{iwt}, \quad E(t) = E_0 e^{iwt}$$

(i) Carry out the substitution and hence obtain an expression for the complex number Q_0.

(ii) The current in the circuit is given by $i = dq/dt$, show that i can be represented by the complex expression

$$i = \frac{E_0 e^{iwt}}{R + i(wL - 1/wC)}$$

14. An oscillatory system, of negligible damping, is perturbed by a periodic force of amplitude F in such a way that the displacement y of the system from its equilibrium position satisfies the inhomogeneous equation

$$\frac{d^2y}{dt^2} + w^2 y = F\cos(pt)$$

(i) Solve this equation and show that

$$y = A\cos(wt) + B\sin(wt) + \frac{F\cos(pt)}{(w^2 - p^2)}$$

where A and B are constants.

(ii) Obtain the particular solution for which $y = 1$ and $dy/dt = 0$ when $t = 0$.

(iii) Show that if $w \simeq p$ then, to a good approximation, the particular solution becomes

$$y = \cos(wt) + \frac{Ft}{2w} \sin(wt)$$

15. In a strong electrolyte solution, the electrical potential V, a distance r from the anode, satisfies the following differential equation

$$\frac{d^2V}{dr^2} + \frac{2}{r}\frac{dV}{dr} = k^2V$$

where k is a constant:

(i) Show that the substitution $z = rV$ transforms the equation to the form

$$\frac{d^2z}{dr^2} = k^2z$$

(ii) Solve the equation in the normal way to show that in general

$$V = \frac{Ae^{-kr}}{r} + B\frac{e^{kr}}{r}$$

where A and B are constants.

(iii) If $V \to 0$ as $r \to \infty$, show that $V = Ae^{-kr}/r$.

16. A steam pipe, at temperature T_s, is supported by a metal hanger from the ceiling. The hanger is of length L and has a square cross-section of side R. The temperature $T(z)$ a distance z from the steam pipe, in the hanger, satisfies the following differential equation

$$\frac{d^2T}{dz^2} - \frac{4hT}{kR} = -4\frac{hT_A}{kR}$$

where T_A is the temperature of the ceiling and surrounding air, k is the thermal conductivity of the metal hanger, and h is the convective heat transfer coefficient between the metal hanger and the air. Solve this differential equation to give T as a function of z, if $T = T_s$ when $z = 0$ and $T = T_A$ when $z = L$.

14

Statistics 1: frequency distributions and associated measures

Contents: 14.1 scientific context – 14.1.1 radiation decay data; 14.1.2 velocity of light measurements; 14.2 mathematical developments – 14.2.1 Organisation of data: frequency distributions; 14.2.2 presentation of data; 14.2.3 frequency curves and relative frequency curves; 14.3 worked examples; 14.4 further developments – 14.4.1 measures of centre; 14.4.2 measures of scatter; 14.4.3 some numerical considerations; 14.5 further worked examples; 14.6 exercises.

We now begin a study of statistics. As well as being familiar with the mathematics that underpins the theoretical side of science we must also know how to analyse experimental data; for carrying out experiments, making measurements and collecting data are just as important in science as mathematical modelling.

Analysing experimental data is often fraught with difficulties, because one very important property, possessed by virtually any set of data values, is that different measurements of a given quantity will almost certainly show variation. This variation usually exists for one of two reasons. Firstly, variation could be present in what is being measured. For example, heights of people vary, daily hours of sunshine vary, the number of radioactive particles entering a geiger counter in a given time interval varies, and so on. Or, secondly, when measuring some fixed quantity, variation could arise due to errors of one kind or another creeping into the measuring process. We might, for example, take a sample of blood and analyse the blood for alcohol content. Even though we might use the same analytical technique, repeated determinations of the fixed concentration of alcohol would vary due to experimental error.

This variability not only affects the sort of conclusions we can draw from experimental data, but inevitably it means that in the first place we have to make repeated measurements of a quantity before we can say anything sensible about its value. For example, only by repeating measurements can we find out about the nature and extent of any real variation that might be present in what we are measuring. Or, in the case of some fixed quantity, like the blood alcohol level, only by repeated measurements can we specify limits within which we are reasonably confident its true value might lie.

Representing measurements is sometimes an easy task and occasionally we have the luxury of having masses of data to analyse. On other occasions, due to lack

364

of time or because of expense, we have to draw conclusions from only a few readings. Being able to analyse data within the context of this variability, whether there are masses of data or only a few readings, is of crucial importance in science and techniques for doing this are developed in the next three chapters on statistics. These three chapters provide only a modest introduction to the subject of statistics, but nevertheless the techniques we consider are at the heart of most computer packages that carry out statistical data analysis. We start off, in this chapter, by considering how to organise, present and summarise large amounts of data.

14.1 Scientific context

14.1.1 Example 1: radiation decay data. Suppose we use a geiger counter to measure the number of particles being emitted from a radioactive source in consecutive 10-s periods. The first 500 readings are shown in Table 14.1. We have here a lot of data! These data obviously contain valuable information about the radioactive source, but at the moment it is hidden from us by the large number of readings that we see. We note that there is variation present in what is being measured, but simple details of this variation, such as what proportion of readings there are below 5 counts, or between 6 and 8 counts and so on, are impossible to deduce from a cursory glance at the data. It is clear that the data need to be organised and summarised so that this information can be obtained more easily.

There are several ways we can organise the data to provide a summary. What is often done is to draw up a table listing the possible readings in some convenient

Table 14.1. *Radiation decay rate*

4	5	6	4	2	3	7	5	4	9	3	7	1	2	4	5	3	8	5	6	6	5	7	5	3
4	2	3	6	0	4	6	3	8	6	2	4	6	7	5	4	6	2	4	7	1	7	9	3	2
3	4	3	8	10	7	4	5	3	6	1	7	6	6	3	4	6	9	2	4	8	5	3	7	4
2	6	5	3	4	4	5	4	2	6	8	7	11	5	4	3	6	1	4	3	4	7	3	2	6
7	2	3	6	8	2	3	7	9	8	3	4	5	5	3	7	1	6	4	7	5	6	4	7	5
4	2	1	6	7	4	3	6	2	7	4	10	5	3	6	2	6	3	0	5	6	4	6	3	8
7	3	4	1	2	9	6	5	3	4	6	4	5	5	4	7	6	4	6	3	2	8	4	3	5
4	1	6	3	2	4	3	5	2	6	3	7	1	4	2	4	6	7	3	6	8	5	4	2	9
2	3	4	1	3	6	5	3	7	10	7	2	5	8	0	3	5	7	5	7	4	5	4	5	4
5	3	5	4	3	1	11	4	2	2	8	4	5	2	3	3	4	7	3	5	4	6	2	5	4
9	3	4	5	7	6	5	7	4	3	5	6	9	1	7	3	5	6	8	2	4	6	3	7	8
2	3	5	4	3	5	4	6	3	4	2	5	3	4	8	7	4	12	1	6	5	2	6	7	5
7	10	8	7	9	4	2	7	5	3	4	6	3	6	3	6	7	4	2	3	2	4	8	5	3
6	6	4	6	1	3	2	9	4	5	3	1	5	4	6	5	6	5	3	5	7	4	5	7	0
3	4	8	2	6	4	6	5	4	3	7	4	2	6	3	1	5	6	5	2	4	7	3	5	4
5	2	8	9	5	5	2	5	3	5	4	3	3	4	1	2	11	6	4	3	8	5	7	4	3
4	7	5	3	5	4	5	7	5	3	4	8	9	4	5	2	5	3	2	6	6	4	10	5	6
2	4	5	8	1	4	2	7	3	7	6	4	6	3	10	7	4	2	6	3	1	4	3	5	4
7	9	3	2	4	7	6	3	1	4	5	8	1	5	2	4	3	5	8	3	2	4	3	5	3
1	5	2	0	4	3	4	5	7	5	2	3	6	8	7	2	4	7	3	1	6	3	4	6	7

order, together with the number of times each reading appears in the data. This is called a *frequency table*.

Question: What are the possible readings in our example?
  Answer: Even though we have 500 readings it will be seen that only the numbers 0–12 appear. These 13 numbers are therefore the possible readings.

Having drawn up a frequency table we then usually seek to summarise the data further by providing concise measures that adequately describe the distribution of data values. In many cases we find that we can describe a whole mass of data by just two readings: a measure of centre, telling us where the bulk of readings occurs, and a measure of spread, telling us what the typical amount of scatter is about this centre. We shall consider such measures later, but for now let us have a look at a second example.

14.1.2 Example 2: velocity of light measurements. Suppose 50 students each measure the value of the velocity of light using the same experimental technique and obtain the results given in Table 14.2. The readings are given in $km\,s^{-1}$ and have been rounded to the nearest whole number. Once more we see that there is variation in the data, but since the value of the velocity of light is a constant, the variation now has to be due to experimental error.

Question: What is the best way of drawing up a frequency table for these data?
  Answer: We note that the number of different readings is now much larger than in the radiation counts data. The highest reading is 299 817 and the smallest 299 745, giving a spread of some 73 possible readings. A frequency table that assigns the 50 measurements that we have here to these 73 different possibilities is not going to prove all that useful. It will be far better if we assign the individual values to *groups* of possible readings. It would seem sensible, therefore, to draw up a frequency table (Table 14.4) by assigning the data, for example, to eight equal-sized groups say: 299 741–299 750, 299 751–299 760,..., 299 801–299 810, 299 811–299 820.

Such groups are commonly referred to as *classes* and we are at liberty to choose the most appropriate number of classes for a given set of data. In choosing the

Table 14.2. *Velocity of light data*

299 760	299 771	299 788	299 773	299 769
299 755	299 780	299 817	299 763	299 776
299 800	299 790	299 772	299 783	299 794
299 782	299 749	299 805	299 758	299 782
299 775	299 798	299 775	299 795	299 776
299 764	299 761	299 787	299 766	299 803
299 792	299 798	299 810	299 781	299 786
299 787	299 768	299 775	299 745	299 774
299 757	299 781	299 784	299 778	299 765
299 806	299 777	299 766	299 783	299 779

number of classes we have to bear in mind that we do require a summary of the data, so we do not want too many classes (73, for example, is far too many here). But, on the other hand, we do not want too few classes, as valuable information about the variability in the data will be lost. The number of classes we choose should be a sensible choice that takes into account the range of data values, the detail required, and the number of readings that we have. Most statistical packages, these days, allow you to specify the number of classes and, once the data have been input, these classes are automatically worked out and the resulting frequency table drawn up.

As we have said, the variation in the data is due to experimental error because the quantity we are trying to measure – the velocity of light – is a constant.

Question: So what value should we take for the velocity of light?

Answer: Strictly speaking, there is no hard and fast rule to guide us here. But one natural thing to do is to average the results by adding them all together and dividing by the number of measurements we have, namely 50. The result ($29\,979.18\,\mathrm{km\,s^{-1}}$) is the *arithmetic mean*, which is just one of the possible measures of centre that we mentioned earlier.

Having said that there is really no hard and fast rule to guide us, we do find that the arithmetic mean is often used as an estimate of the quantity we are trying to measure in those situations where the only source of variation is due to experimental error. The reasons for this will be explained more fully in later chapters; let us now start to develop some basic statistical ideas.

14.2 Mathematical developments

14.2.1 Organisation of data: frequency distributions

Frequency tables. One concept of fundamental importance in statistics is the idea of a *frequency distribution*. To focus ideas, consider the radioactive decay data in Table 14.1. These 500 readings are what we call *raw data*. As we have already said, the possible readings are the numbers 0–12 and these are effectively the *classes* that we spoke about earlier. Note, in this example, that a class is now one reading as opposed to a group of readings. The number of data values in each class, in the raw data list, is called the *frequency* of that class. To find the frequencies we simply scan through the raw data and assign the readings to their respective classes. The frequency distribution of data values is illustrated in the *frequency table* shown in Table 14.3. It is important to note that in organising the data into a frequency distribution, as in Table 14.3, we have in fact made more explicit the information that was there in the data.

Example: In the radiation counts data what proportion of readings are there below 5 counts, and what proportion lie between 6 and 8 counts?

Table 14.3. *Frequency table for radiation decay rate*

Class	Tally	Frequency
0	卌	5
1	卌 卌 卌 卌 |||	23
2	卌 卌 卌 卌 卌 卌 卌 卌 卌 卌 |||	53
3	卌 卌 卌 卌 卌 卌 卌 卌 卌 卌 卌 卌 卌 卌 卌 卌 ||	82
4	卌 卌 卌 卌 卌 卌 卌 卌 卌 卌 卌 卌 卌 卌 卌 卌 卌 卌 |||	93
5	卌 卌 卌 卌 卌 卌 卌 卌 卌 卌 卌 卌 卌 卌 卌 |||	78
6	卌 卌 卌 卌 卌 卌 卌 卌 卌 卌 卌 卌 卌	65
7	卌 卌 卌 卌 卌 卌 卌 卌 卌 卌 |||	53
8	卌 卌 卌 卌 卌	25
9	卌 卌 |||	13
10	卌		6
11					3
12			1

Solution: We can easily see from the summary in Table 14.3 that the number of readings below 5 counts is 5 + 23 + 53 + 82 + 93, which is 256. The proportion is thus 256 in 500, or 0.512. Between 6 and 8 counts there are 65 + 53 + 25 i.e. 143 readings. The proportion is thus 143 in 500 or 0.286.

The summary contains all the information that was present in the raw data list. The order in which the various counts appeared in the list is not now known, but this is of no consequence except in time-dependent situations. It is important to note, therefore, that storing just the class and corresponding frequencies will be equivalent, from an informational point of view, to storing the original 500 data values. Organising data can thus give us a considerable saving in data storage space.

If we look now at the velocity of light data in Table 14.2, and use the eight classes suggested earlier, we can again assign the readings to classes by scanning through the data. The frequency distribution is shown in Table 14.4. In summarising the data as we have done here, we have to note that some of the information present in the raw data list is now no longer present in the frequency table. All we know, for example, looking at Table 14.4, is that in the class of readings 29 781–790 there are 12 readings.

We do not know what the individual readings are. However, as we shall see, this loss of information is usually of little or no consequence. Of course, if we had fewer classes, by having larger groups of readings, the loss of information would be significant by the description being too aggregated. To repeat what we said before, *the number of classes should be a sensible choice that takes into account the range to be covered, the detail required, and the number of readings making up the raw data.* With such a sensible choice, storing just the classes and the corresponding frequencies will again be equivalent, from an informational point of view, to storing the original raw data. Once more, organising the data gives a saving in storage.

Table 14.4. *Frequency table for velocity of light data*

Class	Tally	Frequency
299 741–750	||	2
299 751–760	||||	4
299 761–770	||| |||	8
299 771–780	||| ||| |||	13
299 781–790	||| ||| ||	12
299 791–800	|||		6
299 801–810	||||	4
299 811–820	|	1

Some definitions. Further quantities can be defined when classes are groups of readings:

The class end marks are the two readings specifying the class – the upper and lower end marks.

The class mid-points are the averages of the respective class end marks. For example they are 299 745.5, 299 755.5, 299 765.5, ..., etc. for the velocity of light data.

The class interval, usually denoted by c, (and not to be confused with the value of the velocity of light!) is the difference between adjacent class mid-points. Thus $c = 10$ for the velocity of light data.

The class boundaries are defined to be equal to the class mid-points plus or minus $c/2$. The class boundaries for the velocity of light data are thus 299 740.5 and 299 750.5 for the first class, 299 750.5 and 299 760.5 for the second class, and so on.

Continuous data and contiguous classes. So far, in the examples we have been considering, we have been dealing with data that are *discrete* – whole numbers in this case. When counting the number of clicks on a geiger counter we cannot, for example, observe half a count or 4.7312··· counts. We can only observe whole numbers. Similarly, in the velocity of light measurements, all readings were rounded up or down to the nearest whole number. But in science not all data values will be whole numbers and we may well have to cater for *continuous* data. Continuous data arise when what we are measuring are the values of variables that are continuous, and no rounding up or down takes place. Measurements of mass, length, and time might fall into this category. When we are dealing with such variables, the classes we choose must first of all be groups of readings and, to cover a given continuous range, the class end marks of these groups must also be the same as the class boundaries. Such classes are known as *contiguous classes*. Thus a continuous variable whose range is between about 2 and 10 may be catered for using the contiguous classes 2–3, 3–4, ..., 9–10, for example. The problem then arises deciding in which class a reading of 3 would belong. By interpreting the classes 2–3, 3–4, ..., etc., as 2 up to but not including 3, 3 up to but not including 4 and so on, we see that 3 would, in fact, belong to the class 3–4. This convention is fairly standard in statistics and overcomes the problem of deciding within which class a reading belongs when it happens to fall exactly on a class boundary.

Strictly speaking, it could be argued that all data values, whether they are measurements of a continuous variable or not, will be discrete, because instruments can only record the value of quantities to a given accuracy. Any subsequent numerical representation of the value will therefore only be accurate to a certain number of decimal places, at which point rounding up or down may as well take place. However, for simplicity, it is often useful to work with contiguous classes of the type just described when dealing with data that are measurements of a continuous variable.

14.2.2 Presentation of data. Having drawn up a frequency table there is often the need to illustrate the distribution graphically. A frequency table helps us appreciate the nature of the distribution but a graphical presentation gives us a visual picture of exactly what is happening. Most data analysis packages have graphics output that can provide us with a picture of the distribution. There are several ways of illustrating frequency distributions:

The bar chart. A *bar chart* is simply a plot of frequency against class, where the class frequencies are represented by bars. A bar chart is used when the classes are single readings. It is appropriate therefore for us to illustrate the radiation counts data in this way and the bar chart is shown in Fig. 14.1.

The histogram. When we have a distribution whose classes are groups of readings we draw a slightly different plot called a *histogram*. A histogram is a plot

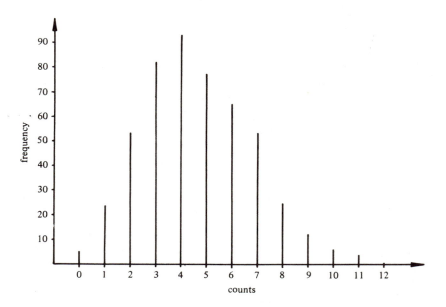

Fig. 14.1. Bar chart for the radiation counts data.

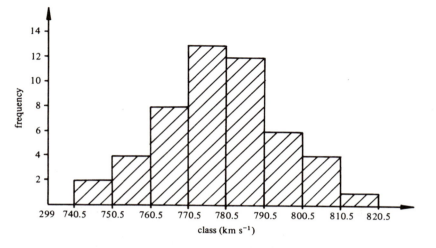

Fig. 14.2. Histogram for the velocity of light measurements.

of frequency against class where the class frequencies are represented by solid columns of width equal to the class interval. The histogram for the velocity of light readings is shown in Fig. 14.2.

The frequency polygon. This is a plot of frequency against class mid-point, where each point is joined to the next by a straight line, thus producing a polygon. These plots are useful for comparing, on one graph, several distributions that share the same classes. With colour graphics it is now possible, of course, to compare several histograms or bar charts on one graph. Even so, it is probably easier to make comparisons using frequency polygons. The frequency polygon for the velocity of light data is shown in Fig. 14.3.

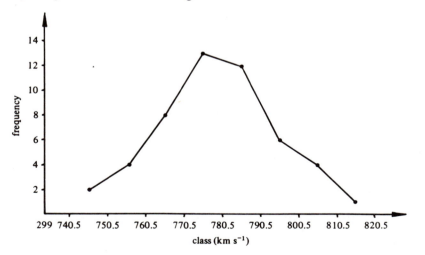

Fig. 14.3. A frequency polygon.

Table 14.5

Class	Frequency (f)	ucb	cf
299 741–750	2	299 750.5	2
299 751–760	4	299 760.5	6
299 761–770	8	299 770.5	14
299 771–780	13	299 780.5	27
299 781–790	12	299 790.5	39
299 791–800	6	299 800.5	45
299 801–810	4	299 810.5	49
299 811–820	1	299 820.5	50

The cumulative frequency polygon. This is a plot of cumulative frequency (*cf*) against upper class boundary (*ucb*). The cumulative frequency of a class is the total number of readings occuring below the upper class boundary of that class. Thus, for the velocity of light readings, the cumulative frequencies and upper class boundaries are as shown in Table 14.5. The cumulative frequency polygon is shown in Fig. 14.4. Again each point is joined to the next by a straight line giving a polygon as opposed to a curve. Such plots are useful for estimating the proportion of readings that occur between various given values in a distribution. Cumulative

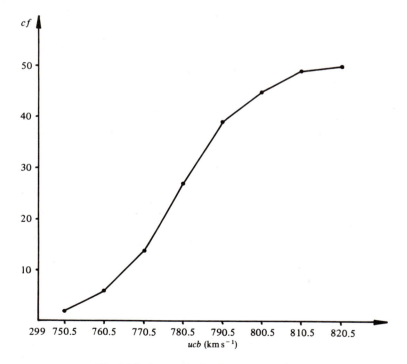

Fig. 14.4. A cumulative frequency polygon.

frequency polygons are often called *ogives* because their characteristic arch-like shape is very similar to that of an ogive in architecture.

14.2.3 Frequency curves and relative frequency curves

Frequency curves. If we were asked to indicate the shape of the distribution in Fig. 14.2 or Fig. 14.3 we would probably draw the shape shown in Fig. 14.5 – i.e. we would draw what is called a *frequency curve* as opposed to reproducing the frequency histogram or polygon. What we are doing when we draw a frequency

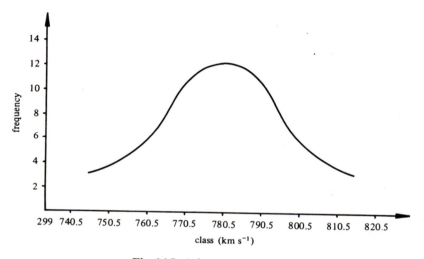

Fig. 14.5. A frequency curve.

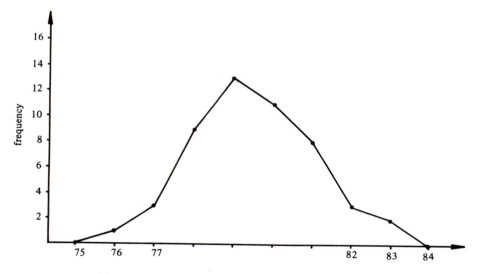

Fig. 14.6. The frequency polygon for 50 measurements.

curve is indicating, consciously or unconsciously, what we *expect* the shape of the distribution to be like. A frequency curve is therefore a model.

For data that are measurements of some continuous variable it is possible to see how frequency curves might be produced in practice. To do this let us recall the problem we mentioned right at the beginning of the chapter, that of measuring the alcohol content in a sample of blood. Suppose, instead of having a blood sample to work with, we have made up a very large synthetic sample of alcohol in water (79 mg/100 ml, say), so that we can go on obtaining measurements of the concentration almost indefinitely. To begin with, suppose we make 50 measurements of the concentration and suppose we draw up the data into a frequency distribution having a class interval of say 1 mg/100 ml. Let the frequency distribution, when illustrated as a frequency polygon, be as shown in Fig. 14.6.

If we now took 150 more measurements of the concentration, so that in total we had 200 readings, these could then be grouped into a frequency distribution having, say, half the class interval of 0.5 mg/100 ml. The frequency polygon might then be as shown in Fig. 14.7. What is happening is that the frequency polygon is beginning to look more like a curve. In the limit of a large number of readings and a very small class interval, the shape of the distribution will, to all intents and purposes, be a curve – a frequency curve.

Relative frequency curves. There is, however, one small problem associated with our example. The overall frequency plot will only look like a curve if changes in adjacent frequency values, which must always be whole numbers, are small in

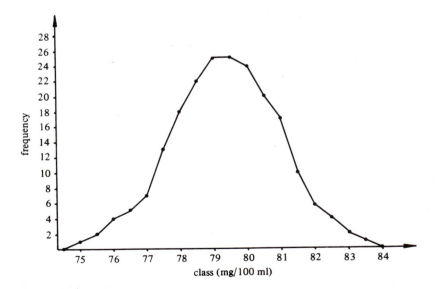

Fig. 14.7. The frequency polygon for 200 measurements.

comparison with typical frequency values – in other words if all frequency values are large. But large frequency values are not always what we might want. We did, after all, draw a frequency curve, in Fig. 14.5, with no frequency value apparently exceeding 14. So how, you might ask, could we draw the frequency curve that is illustrated in Fig. 14.5? Frequency curves indicating the shape of an expected distribution can be realised in practice through the use of *relative frequency plots*, as we shall now describe.

Returning, then, to the velocity of light data, to which Fig. 14.5 actually refers. Suppose we want to determine accurately what to expect when we analyse a typical sample of 50 estimates of the velocity of light. What we could do is to have the 50 students in our example repeatedly measure the value of the velocity of light so that in total we have a large number, N, of results – the larger the value of N the better. We could then draw up a frequency distribution using the same classes as before but with possibly extra classes at each end of the range to cater for results that did not occur previously. If we then divide the class frequencies by N and mutiply each result by 50 we will have excellent estimates of the class frequencies to expect in a typical sample of 50 readings. A smooth curve drawn through a plot of scaled frequency against class mid-point will give the appropriate *relative frequency* curve for this situation. We should note, of course, that a scaled frequency need not be a whole number.

Relative frequency distributions and relative frequency curves are common in statistics as models and we shall be meeting them again in the next two chapters.

14.3 Worked examples

Example 1. The following results are the next 50 readings in the radiation decay example:

```
2  4  3  5  4  3  6  5  4  3
5  6  2  4  6  7  2  4  3  5
6  4  7  3  2  3  5  8  7  5
4  2  5  8  3  4  4  6  5  5
5  3  4  5  6  7  2  4  6  3
```

(i) Draw up these data into a frequency table and illustrate the distribution by drawing a bar chart.

(ii) Using the first 500 readings in Table 14.1, construct the relative frequency curve that represents the distribution of results to expect in a typical sample of size 50 – such as the sample of results in this example.

Solution. (i) The possible readings in the sample are the numbers 2, 3, 4, 5, 6, 7 and 8. Scanning through the data and assigning the readings to these classes we obtain the following frequency table:

Class	Tally	Frequency
2	卌 |	6
3	卌 ||||	9
4	卌 卌 |	11
5	卌 卌 |	11
6	卌 ||	7
7	||||	4
8	||	2

Fig. 14.8 gives the bar chart for this distribution of results.

(ii) The first 500 readings given in Table 14.1 have the frequency distribution shown in Table 14.3. If we take the frequencies in this table and divide each one by 500 and then multiply by 50, the sum of the scaled frequencies will be 50. This scaled frequency distribution will be a good representation of the distribution of results to expect in a typical sample of size 50. Scaling the frequencies in Table 14.3, as described, we obtain the following relative frequency distribution:

Class	Frequency	Relative frequency
0	5	0.5
1	23	2.3
2	53	5.3
3	82	8.2
4	93	9.3
5	78	7.8
6	65	6.5
7	53	5.3
8	25	2.5
9	13	1.3
10	6	0.6
11	3	0.3
12	1	0.1
Total:	500	Total: 50.0

Plotting relative frequency v class and drawing a smooth curve through the points gives the appropriate relative frequency curve shown in Fig. 14.9.

Example 2. Draw up the velocity of light data, given in Table 14.2, into a frequency distribution having ten equal classes and represent the distribution as a histogram.

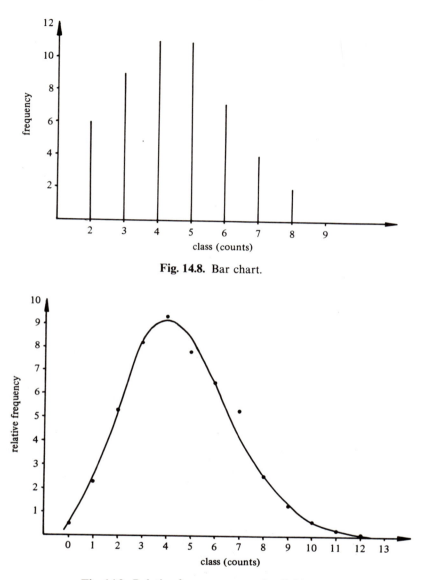

Fig. 14.8. Bar chart.

Fig. 14.9. Relative frequency curve for Table 14.1.

Solution. As we have already noted, the highest recorded reading in the data in Table 14.2 is 299 817 and the smallest is 299 745. The difference is 72 and, if we are to work in whole numbers, we therefore require a class interval of 8. We can thus conveniently choose the following classes: 299 741–299 748, 299 749–299 756,...,299 813–299 820. Obviously, we could have chosen slightly different classes, say 299 742–299 749, 299 750–299 757,..., 299 814–299 821 – the real consideration is that the ten classes we choose should sensibly cover the range of data values that we have. With our first choice of classes the frequency distribution produced by allotting the raw data to these classes is as follows:

Class (km s^{-1})	Tally	Frequency
299 741–299 748	│	1
299 749–299 756	‖	2
299 757–299 764	ЖҬ │	6
299 765–299 772	ЖҬ ‖	7
299 773–299 780	ЖҬ ЖҬ │	11
299 781–299 788	ЖҬ ЖҬ │	11
299 789–299 796	‖‖	4
299 797–299 804	‖‖	4
299 805–299 812	‖│	3
299 813–299 820	│	1
	Total:	50

The histogram representing this distribution is thus shown in Fig. 14.10.

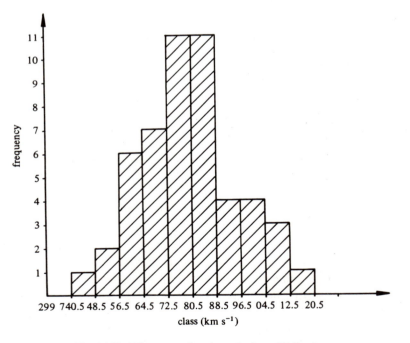

Fig. 14.10. Histogram for the velocity of light data.

Example 3.

    (i) From the frequency distribution in the last example, construct an ogive and use it to estimate the proportion of readings that will occur between $299\,765\,\mathrm{km\,s^{-1}}$ and $299\,796\,\mathrm{km\,s^{-1}}$ in future samples.

    (ii) Compare this proportion with that obtained working with the raw data given in Table 14.2.

Solution. (i) To draw up an ogive (i.e. a cumulative frequency polygon) we need to plot cumulative frequency against upper class boundary. The upper class boundaries (*ucb*) and cumulative frequencies (*cf*) for the distribution in example 2 are as follows:

Class ($\mathrm{km\,s^{-1}}$)	Frequency	*ucb* ($\mathrm{km\,s^{-1}}$)	*cf*
299 741–299 748	1	299 748.5	1
299 749–299 756	2	299 756.5	3
299 757–299 764	6	299 764.5	9
299 765–299 772	7	299 772.5	16
299 773–299 780	11	299 780.5	27
299 781–299 788	11	299 788.5	38
299 789–299 796	4	299 786.5	42
299 797–299 804	4	299 804.5	46
299 805–299 812	3	299 812.5	49
299 813–299 820	1	299 820.5	50

The ogive can now be drawn as in Fig. 14.11.

    To estimate the likely proportion of readings in future samples that will occur between $299\,765\,\mathrm{km\,s^{-1}}$ and $299\,795\,\mathrm{km\,s^{-1}}$ we use the ogive as follows. An estimate of the number of readings below $299\,795\,\mathrm{km\,s^{-1}}$ in our present sample is given by the cumulative frequency corresponding to the *ucb* value of 299 795, which, from the plot, is 41.3 readings. Similarly, an estimate of the number of readings below 299 765 in our sample is 9.5. The number of readings in our sample between $299\,765\,\mathrm{km\,s^{-1}}$ and $299\,795\,\mathrm{km\,s^{-1}}$ is therefore estimated to be $41.3 - 9.5$, which is 31.8 readings. The proportion of readings in our sample that occur between these limits is thus estimated as 31.8 in 50, i.e. as 0.64. This value is therefore our estimate of the proportion of readings in future samples that will occur in this range.

    (ii) Working directly with the raw data in Table 14.2 we see, in fact, that between the values $299\,765\,\mathrm{km\,s^{-1}}$ and $299\,795\,\mathrm{km\,s^{-1}}$ there are 33 readings. The real proportion in our sample is therefore 0.66. This value is thus our estimate of the proportion of readings to expect in future samples between $299\,765\,\mathrm{km\,s^{-1}}$ and $299\,795\,\mathrm{km\,s^{-1}}$. The two values of 0.64 and 0.66 do not differ by all that much,

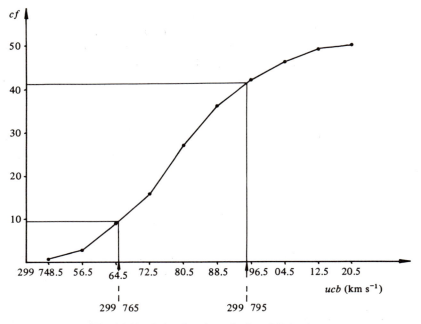

Fig. 14.11. Ogive for the velocity of light data.

confirming that little or no information is lost when we choose to work with a summary of a set of data values as opposed to the data values themselves.

14.4 Further developments

We have seen how we can organise data into a frequency distribution and illustrate this summary graphically. We are now going to consider how we can summarise the data further by providing concise measures of centre and scatter that adequately describe the distribution. We shall begin by looking at measures of centre which tell us where the bulk of the readings in a distribution are to be found.

14.4.1 Measures of centre. The centre of a distribution is indicated usually by one of three measures:

1 The arithmetic mean. The *arithmetic mean* of N numbers $x_1, x_2, \ldots, x_N$, denoted by m, is given by

$$m = (x_1 + x_2 + \cdots + x_N)/N.$$

Using the sigma notation this can be written as

$$m = \sum_i x_i/N \tag{1}$$

For data that have already been drawn up into a frequency distribution, with h

classes, the arithmetic mean is defined to be

$$m = (x_1 f_1 + x_2 f_2 + \cdots + x_h f_h)/(f_1 + f_2 + \cdots + f_h)$$

so that in terms of the sigma notation we have

$$m = \sum_i x_i f_i \bigg/ \sum_i f_i \qquad (2)$$

In this expression, x_i are now the class mid-points and f_i are the respective class frequencies.

In the definition of m, given in (2), the implicit assumption is that the average value of the readings in a class will be equal to the class mid-point. In a class some readings are larger than the class mid-point and some are smaller, so this assumption is not unreasonable. It does mean, however, that the value of m, worked out using (2), will actually be different to the mean of the raw data from which the frequency distribution was drawn up. But for large samples of say 30 or more readings, and a sensible number of classes, the difference is usually quite small and in many cases negligible. The next example serves to illustrate this fact.

Example: Obtain the mean of the summarised velocity of light data given in Table 14.4 and compare this value with that obtained directly from the raw data in Table 14.2.
Solution: The value of m worked out using the formula in (2) is easily obtained as follows:

Class	Frequency (f)	Class mid-point (x)	xf
299 741–299 750	2	299 745.5	599 491.0
299 751–299 760	4	299 755.5	1 199 022.0
299 761–299 770	8	299 765.5	2 398 124.0
299 771–299 780	13	299 775.5	3 897 081.5
299 781–299 790	12	299 785.5	3 597 426.0
299 791–299 800	6	299 795.5	1 798 773.0
299 801–299 810	4	299 805.5	1 199 222.0
299811–299 820	1	299 815.5	299 815.5
	$\sum_i f_i = 50$		$\sum_i x_i f_i = 14\,988\,955.0$

The mean of the summarised data is thus $14\,488\,955.0/50$ which is $299\,779.10\,\mathrm{km\,s^{-1}}$. To see how this compares with the mean of the raw data, recall that the average of the 50 results was $299\,779.18\,\mathrm{km\,s^{-1}}$. The difference is therefore $0.08\,\mathrm{km\,s^{-1}}$, which compared to the difference between the largest and smallest recorded values of the velocity of light (namely $72\,\mathrm{km\,s^{-1}}$) is very small. In fact, as a percentage, it is only 0.11%.

2 The median. The *median* of a set of N numbers arranged in ascending or descending order of magnitude is the middle number of the set, if N is odd, and the mean of the two middle numbers if N is even. The median of 6, 7, 9, 12, 16, 20, for example, is thus $(9 + 12)/2$, i.e. 10.5.

When we have a large number of results that have already been arranged into

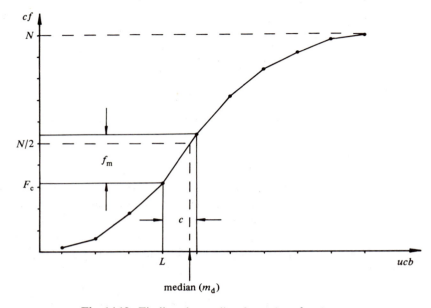

Fig. 14.12. Finding the median from the *c.f.* polygon.

a frequency distribution, the median can easily be read off from the cumulative frequency polygon as shown in Fig. 14.12. The median m_d is thus that value corresponding to a cumulative frequency of $N/2$.

A formula for the median can easily be obtained using linear interpolation, and involves quantities relating to the *median class*. The median class is the class containing the median and is therefore the lowest class whose cumulative frequency exceeds $N/2$. Thus if F_c is the cumulative frequency of the class immediately below the median class, f_m is the frequency of the median class, L is the upper class boundary of the class immediately below the median class, and c is the class interval, then, with reference to Fig. 14.12 we have:

$$\left(\frac{m_d - L}{c}\right) = \left(\frac{N/2 - F_c}{f_\lambda}\right)$$

which, after rearranging, gives

$$m_d = L + c(N/2 - F_c)/f_m \tag{3}$$

Example: What is the median of the velocity of light data as summarised in Table 14.4?
Solution: Using Table 14.5, that relates directly to Table 14.4, we see that the median class is the class of values 299 771–299 780 km s⁻¹. The quantities L, c, N, F_c and f_m in the formula for m_d are, respectively, 299 770.5, 10, 50, 14 and 13. The median is thus equal to 299 770.5 + (1050 − 14)/13, which is 299 779.0 km s⁻¹.

We should note that, strictly speaking, we need only use the formula in (3), or the cumulative frequency plot, to obtain the median when the classes in the distribution are groups of readings. When the classes are single readings the median can be read off directly from a table of cumulative frequencies. The median is the median class itself.

Example: What is the median of the radiation data summarised in Table 14.3?
Solution: Working from Table 14.3 we can easily deduce what the cumulative frequencies are and see that the class whose cumulative frequency first exceeds 250 is the class corresponding to 4 counts. The median is thus 4 counts.

3 The mode. The *mode* is defined to be that reading which occurs most often in a distribution. For data grouped into a frequency distribution, the class for which the frequency is greatest is called the *modal* class and the mid-point of this class is often taken as the mode (although some texts refer to this as the *crude mode*).

Example: What is the mode of the velocity of light data as summarised in Table 14.5?
Solution: The modal class is seen to be the class of readings 299 771–299 780. The mid-point of this class is the reading 299 775.5. The mode can thus be taken as 299 775.5 km s^{-1}.

The concept of the mode, as the most likely reading, is used extensively in model fitting, but, like the median, it tends not to be used as often as the mean in the analysis of experimental data. This is probably because the mean happens to be a parameter characterising many important theoretical distributions. As we shall see, when these distributions are used in the further analysis of data, usually one of the first quantities required is the value of the arithmetic mean. Few theoretical distributions require the mode or median.

14.4.2 Measures of scatter. As well as knowing where the bulk of data values occurs in a distribution, it is important to know how much scatter about the centre there is in the readings. Quantifying the scatter, for example, in a distribution of readings, all of which are estimates of some fixed quantity, gives us a measure of the precision to associate with the measuring process. There are several measures of scatter:

1 The standard deviation. The *standard deviation* is the square root of a quantity called the *variance* and the variance is the mean-square deviation of all the readings from the mean. Thus for N data values $x_1, x_2, \ldots, x_N$, whose mean is m, the variance, denoted by s^2, is given by

$$s^2 = ((x_1 - m)^2 + (x_2 - m)^2 + \cdots + (x_N - m)^2)/N$$

Thus we can write:

$$s^2 = \sum_i (x_i - m)^2 / N \tag{4}$$

For data drawn up into a frequency distribution with h classes the variance is defined as

$$s^2 = \sum_i (x_i - m)^2 f_i \Big/ \sum_i f_i \tag{5}$$

The standard deviation, denoted by s, in each case is simply the square root of s^2, and is thus the root-mean-square deviation of the readings from the mean.

If we wanted to work out the variance or the standard deviation of some data we would not normally use (4) and (5) as they stand because they are inefficient as formulae. They are basically inefficient because the mean must first be worked out before the individual deviations from the mean can be obtained. Used in a program to calculate the variance, (4) and (5) could reduce the program's efficiency by about 50%.

A more convenient and efficient form for calculational purposes can easily be obtained. If we consider the expression for s^2 in (4), for example, we see that

$$s^2 = \sum_i (x_i - m)^2/N$$

$$= \sum_i (x_i^2 - 2x_i m + m^2)/N$$

$$= \sum_i x_i^2/N - 2m \sum_i x_i/N + Nm^2/N$$

$$= \sum_i x_i^2/N - 2m^2 + m^2$$

i.e.

$$s^2 = \sum_i x_i^2/N - m^2 \tag{6}$$

This expression for the variance is called, appropriately, the *computing form* for the variance. For data organised into a frequency distribution the computing form is:

$$s^2 = \sum_i x_i^2 f_i \bigg/ \sum_i f_i - m^2 \tag{7}$$

To illustrate the advantages to be gained using the computing form, let us look at an algorithm that will obtain the variance of N readings and print out the result. Using x as the variable name for the various readings, and V for the variance, the algorithm, written in *pseudocode* might be:

```
begin
input N
set sum x to 0
set sum x² to 0
for i = 1 to N step 1 do
    input xᵢ
    set sum x to sum x + xᵢ
    set sum x² to sum x² + xᵢ²
endfor
set V to sum x²/N − (sum x/N)²
print V
stop
```

In this algorithm the variance is obtained and printed out with the computer having made only one pass through the data values. A program to calculate the variance using (4), however, would not only be longer, but would involve a double pass through the data values before the variance could be printed out. This is because the mean has to be worked out first (one pass through the data) before the values of $(x_i - m)^2$ can be obtained (second pass through the data). There is thus a saving in computing time when the computing form for the variance is used.

The standard deviation, like the arithmetic mean, is a natural parameter characterising several important theoretical distributions and, because of this fact, it is a commonly used measure of *scatter*.

2 The mean deviation. The *mean deviation MD* of N data values is defined to be

$$MD = \sum_i |x_i - m|/N \tag{8}$$

and measures how far, on average, the readings are from the mean. For data organised into a frequency distribution the expression is

$$MD = \sum_i |x_i - m| f_i / \sum_i f_i \tag{9}$$

At first sight the mean deviation might appear to be a more natural measure of *scatter* than the standard deviation. However, its value is relatively more time consuming to work out compared to the standard deviation because no computing form exists for it. Further, and more importantly, few theoretical distributions, if any, require a knowledge of the mean deviation for their use and this explains why the standard deviation tends to be used in favour of the mean deviation as a measure of scatter about the mean.

3 The range. The *range* is simply the difference between the largest and smallest readings in a distribution. It is easily and quickly worked out and for this reason has been used extensively in quality control work – although, with data-logging devices and computers now being commonplace, this property is no longer an overriding advantage. As a measure of scatter it tends to be rather crude, for it says little about the way in which the data points cluster about the centre of the distribution. Distributions having quite different standard deviations can have very similar ranges.

4 The semi-interquartile range. The *interquartile range* is the range within which the middle 50% of readings lie. It is easily read off from the cumulative frequency polygon as indicated in Fig. 14.13. The *semi-interquartile range* is $(Q_2 - Q_1)/2$. Using linear interpolation as we did when we obtained a formula for the median, we can obtain formulae for the *upper* and *lower* quartiles Q_2 and Q_1 respectively. They are:

$$Q_2 = L_2 + c(3N/4 - F_{c_2}/f_{Q_2}) \tag{10}$$

$$Q_1 = L_1 + c(N/4 - F_{c_1}/f_{Q_1}) \tag{11}$$

The quantities in the formulae are those indicated in Fig. 14.13. Unlike the range, the semi-interquartile range has all the sensitivity, as a measure of scatter, that is possessed by the standard deviation. However, it is little used in scientific

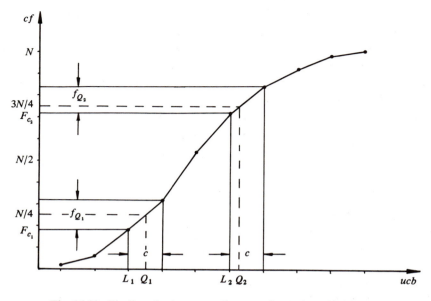

Fig. 14.13. Finding the interquartile range from the $c.f.$ polygon.

applications of statistics, and finds more use in the social sciences where the median is a more natural measure of centre than the mean.

14.4.3 Some numerical considerations. The calculations involved in working out the mean and the standard deviation can become very tedious, even using a calculator, whenever the classes or class mid-points are large numbers. Not only do calculations become tedious, but, what is more important, unless we are very careful errors can creep into calculations when we are using a calculator or computer.

Suppose we require to work out the standard deviation of the velocity of light data summarised in Table 14.4. The class mid-point of the first class, for example, is 299 745.5. Its square is 89 847 364 770.25, which has 13 figures in its decimal representation. The actual $x^2 f$ values that we would need are likely to have up to 15 significant figures in their representation. In many calculators and computers an accurate representation of such numbers is impossible and more than likely, when the quantity

$$\sum_i x_i^2 f_i \bigg/ \sum_i f_i - m^2$$

is worked out, subtractive cancellation will occur causing the result to be significantly in error.

Errors can be eliminated to a large extent if we use so-called *coding formulae* in our programs and calculations. Coding formulae exist for working out the mean and the standard deviation and are based on the two principles of working with an *assumed mean* and working in *units of the class interval*.

Suppose we want to calculate the mean of the five numbers: 93, 97, 92, 96, 87. The answer will be approximately 90, so we may take this as a first approximation and calculate the correction to it. The correction will be $(3 + 7 + 2 + 6 - 3)/5$, i.e. 3. The true mean is thus $90 + 3$, i.e. 93. The number 90 is the *assumed mean* here.

Now, suppose we want to obtain the mean of the numbers 600, 200, 300, 500. Working in hundreds we see that the mean is $(6 + 2 + 3 + 5)/4$, i.e. 4 (hundred). This short cut is based on the device of working in *units of the class interval* – here equal to 100. The exact nature of the role of the class interval will become clear as we work with summarised data as opposed to raw data.

Let us develop, then, a coding formula for the mean of a set of data that have been drawn up into a frequency distribution. The formula for the mean, given in (2), is

$$m = \sum_i x_i f_i \bigg/ \sum_i f_i$$

Let m' be an assumed mean and write x_i as $m' + x_i - m'$. Thus we see that

$$m = \sum_i (m' + x_i - m') f_i \bigg/ \sum_i f_i$$

$$= m' \sum_i f_i \bigg/ \sum_i f_i + \sum_i (x_i - m') f_i \bigg/ \sum_i f_i$$

$$= m' + \sum_i (x_i - m') f_i \bigg/ \sum_i f_i$$

If c is the class interval in our distribution, we can rewrite this last expression for m as

$$m = m' + c \sum_i \left(\frac{x_i - m'}{c} \right) f_i \bigg/ \sum_i f_i$$

So, with $(x_i - m')/c$ put equal to t_i, say, the *coding formula for the mean* is:

$$m = m' + c \sum_i t_i f_i \bigg/ \sum_i f_i \qquad (12)$$

We note that the t_i values give the correction to m' of each x_i value, worked out in units of the class interval. It is good practice to choose m' to be the crude mode of the distribution, or at least some value near the centre of the distribution. Choosing a class mid-point to be this value has the added advantage that all the t_i are then integers, making hand calculations very easy and straightforward.

The coding formulae for the variance and the standard deviation are obtained by starting from the definition of the variance given in (7). This can be rewritten as

$$s^2 = \sum_i x_i^2 f_i \bigg/ \sum_i f_i - \left(\sum_i x_i f_i \bigg/ \sum_i f_i \right)^2$$

Since effectively the variance relates only to the 'width' of the distribution, its value should not be changed if from every x_i value we take away the value of an assumed mean, m'. Thus, an equivalent expression for the variance is

$$s^2 = \sum_i (x_i - m')^2 f_i \Big/ \sum_i f_i - \left(\sum_i (x_i - m') f_i \Big/ \sum_i f_i \right)^2$$

Working in units of the class interval we have that

$$s^2 = c^2 \left(\sum_i \left(\frac{x_i - m'}{c} \right)^2 f_i \Big/ \sum_i f_i - \left(\sum_i \left(\frac{x_i - m'}{c} \right) f_i \Big/ \sum_i f_i \right)^2 \right)$$

$$= c^2 \left(\sum_i t_i^2 f_i \Big/ \sum_i f_i - \left(\sum_i t_i f_i \Big/ \sum_i f_i \right)^2 \right)$$

The *coding formula for the standard deviation* is therefore:

$$s = c \left(\sum_i t_i^2 f_i \Big/ \sum_i f_i - \left(\sum_i t_i f_i \Big/ \sum_i f_i \right)^2 \right)^{\frac{1}{2}} \tag{13}$$

Use of these coding formulae will now be considered in the further worked examples.

14.5 Further worked examples

Example 1. Work out the mean and the standard deviation of the summarised data in Table 14.4 using the appropriate coding formulae.

Solution. The crude mode of the distribution that is summarised in Table 14.4 is the class mid-point of the class 299 771–299 780 which is 299 775.5. This will be our m' value. The class interval is 10, so this is our c value. To use formulae (12) and (13) we thus require the following table:

Class	Frequency (f)	Class mid-point (x)	t	t^2	tf	t^2f
299 741–299 750	2	299 745.5	-3	9	-6	18
299 751–299 760	4	299 755.5	-2	4	-8	16
299 761–299 770	8	299 765.5	-1	1	-8	8
299 771–299 780	13	299 775.5	0	0	0	0
299 781–299 790	12	299 785.5	1	1	12	12
299 791–299 800	6	299 795.5	2	4	12	24
299 801–299 810	4	299 805.5	3	9	12	36
299 811–299 820	1	299 815.5	4	16	4	16

$$\sum_i f_i = 50 \qquad\qquad \sum_i x_i f_i = 18 \quad \sum_i x_i^2 f_i = 130$$

The mean is thus given by $m = 299\,775.5 + 10 \times \frac{18}{50}$, which is $299\,779.1\,\mathrm{km\,s^{-1}}$. The standard deviation is $s = 10(\frac{130}{50} - (\frac{18}{50})^2)^{\frac{1}{2}}$, which is $15.72\,\mathrm{km\,s^{-1}}$. It is worthwhile noting that the calculation of the standard deviation by hand would be extremely difficult without the use of the coding formula. Using a ten-digit display scientific calculator directly without coding leads to a result which is in error by approximately 1%.

Example 2. For the radiation data summarised in Table 14.3, estimate, using an ogive, the proportion of readings that occur within (i) one standard deviation of the mean, (ii) two standard deviations of the mean, and (iii) three standard deviations of the mean.

Solution. To begin with we require the mean and standard deviation of the distribution. Using a working mean of 4 counts and noting that the class interval is 1 count, we can draw up the following table:

Class	Frequency (f)	t	t^2	tf	$t^2 f$
0	5	-4	16	-20	80
1	23	-3	9	-69	207
2	53	-2	4	-106	212
3	82	-1	1	-82	82
4	93	0	0	0	0
5	78	1	1	78	78
6	65	2	4	130	260
7	53	3	9	159	477
8	25	4	16	200	400
9	13	5	25	65	325
10	6	6	36	36	216
11	3	7	49	21	147
12	1	8	64	8	64

$$\sum_i f_i = 500 \qquad \sum_i t_i f_i = 320 \quad \sum_i t_i^2 f_i = 2548$$

The mean m is therefore $4 + \frac{320}{500} = 4.640$ counts. The standard deviation s is $(\frac{2548}{500} - (\frac{320}{500})^2)^{\frac{1}{2}} = 2.165$ counts. To draw up the ogive we require the following cumulative frequency data:

Class	Frequency	ucb	cf
0	5	0.5	5
1	23	1.5	28
2	53	2.5	81
3	82	3.5	163
4	93	4.5	256
5	78	5.5	334
6	65	6.5	399
7	53	7.5	452
8	25	8.5	477
9	13	9.5	490
10	6	10.5	496
11	3	11.5	499
12	1	12.5	500

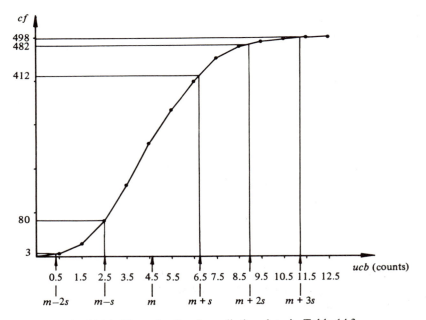

Fig. 14.14. The ogive for the radiation data in Table 14.3.

The ogive is thus as shown in Fig. 14.14. Entering the values m, $m \pm s$, $m \pm 2s$ and $m + 3s$ on the ogive, together with the corresponding cumulative frequencies, we can now work out the required proportions of readings:

(i) We see that the number of readings within one standard deviation of the mean can be estimated as the difference $412 - 80$, i.e. 332. The proportion of readings is thus $\frac{332}{500}$ or 0.664.

(ii) Between two standard deviations of the mean the proportion can be estimated as $482 - 3$ readings in 500, i.e. as $\frac{479}{500}$ or 0.958.

(iii) Within three standard deviations are 498 readings giving a proportion of $\frac{498}{500}$ or 0.996.

14.6 Exercises

1. The following data are the percentage losses in weights of 30 calcium carbonate samples as a result of storage:

6.7	9.9	6.2	8.0	9.4	10.8
9.1	6.9	11.1	9.2	11.9	7.7
6.4	8.7	8.9	10.1	10.3	10.0
8.4	12.3	10.1	7.6	7.1	8.6
7.3	11.0	9.1	12.2	9.2	9.3

(i) Using your calculator, obtain the mean and standard deviation of these data, giving your answer correct to two decimal places.

(ii) Group the data into classes 5.95–6.95, 6.95–7.95, ..., 11.95–12.95 and illustrate the distribution of readings using a histogram.

(iii) Using the grouped data recalculate the mean and standard deviation to two decimal places and compare these results with those in (i).

2. 37 mallard ducklings were ringed immediately after hatching. In the course of time their lifespans were noted and the results, in years, are as follows:

6.18	9.63	14.62	6.57	10.68	16.32	7.52	14.25
9.81	10.41	14.31	11.46	15.31	10.85	12.43	14.97
8.40	12.62	11.76	12.83	17.91	9.97	15.48	11.48
16.52	13.51	11.54	13.64	10.89	7.40	12.45	11.67
12.68	13.32	6.93	12.94	9.93			

(i) By constructing a tally chart, group the data into the classes 6.00–7.50, 7.50–9.00, 9.00–10.50, ..., etc.

(ii) Draw on the same plot a frequency histogram and a frequency polygon. Hence identify the modal class and determine the mode of the data.

(iii) Use your calculator to determine the mean and standard deviation of the raw data.

(iv) Use the grouped data to estimate the mean and standard deviation of the raw data. The coding method should be used. Compare your results with (iii).

3. Regroup the data given in Exercise 2 into the classes 6.00–9.00, 9.00–12.00, 12.00–15.00, etc., and recalculate the mean and standard deviation of these grouped data. Compare your results with those obtained in 2(iii) and 2(iv).

4. Given the following set of grouped data:

class	frequency
0–1	1
1–2	6
2–3	8
3–4	14
4–5	18
5–6	11
6–7	15
7–8	12
8–9	5
9–10	5
10–11	2
11–12	2
12–13	1

estimate the so-called skewness of the distribution using the expression

$$\text{skewness} = (\text{mean} - \text{mode})/\text{standard deviation}$$

5. The efficiency of a new computer operating system is being tested on a mainframe computer. A total of 40 runs is carried out and in each run the same number of jobs, each chosen to be representative of the particular computer environment, is submitted as a batch to the machine. For each run the throughput rate, measured in jobs per minute, is determined. The results of the 40 runs are as follows:

3.22	3.18	3.25	3.24	3.28	3.21	3.26	3.19	3.30	3.33
3.23	3.14	3.22	3.35	3.23	3.27	3.23	3.26	3.37	3.21
3.24	3.25	3.34	3.19	3.27	3.28	3.28	3.26	3.18	3.15
3.29	3.31	3.30	3.17	3.23	3.25	3.20	3.29	3.22	3.24

   (i) By constructing a tally chart, group the data into the classes 3.12–3.16, 3.16–3.20, 3.20–3.24, ..., etc.

   (ii) Calculate the mean and standard deviation of the grouped data, using the coding method.

   (iii) Draw a cumulative frequency polygon and use it to estimate the median and semi-interquartile range of the data.

   (iv) Estimate the % of runs with throughput rates which lie outside the interval which extends from one standard deviation below the mean to one standard deviation above the mean.

6. 100 metal castings were selected at random from a large batch produced

by a foundry. The masses of the rough castings were obtained and the results are summarised in the table below:

mass (kg)	44–46	46–48	48–50	50–52	52–54	54–56	56–58
frequency	2	8	24	32	23	9	2

(i) Estimate, using the coding formula, the mean and standard deviation of the 100 readings.

(ii) Using an ogive, estimate the percentage of castings in the batch that will have masses that are
   (a) less than 49 kg,
   (b) between 49 kg and 52 kg.

7. The following is a summary of the distribution of 57 measurements of hardness (Vickers pyramid hardness number) of the surface of a metal specimen after heat treatment

hardness	92–96	96–100	100–104	104–108	108–112	112–116
frequency	1	9	21	18	7	1

From this summary estimate the proportion of hardness readings likely to be found:

(i) within one standard deviation of the mean;
(ii) within two standard deviations of the mean;
(iii) within three standard deviations of the mean.

8. A new hybrid apple is developed with the aim of producing larger apples than a particular previous hybrid. In a sample of 1000 apples, the distribution of weights of the apples was as follows:

weight (g)	frequency
0–50	20
50–100	42
100–150	106
150–200	227
200–250	205
250–300	241
300–350	106
350–400	53

(i) Apples can only be sold to a particular retail outlet with a weight greater than 218 g. What proportion of the new hybrid would be rejected by this retail outlet?

(ii) How many grammes, above this weight of 218 g, is the mean weight of apples?

(iii) What is this difference in weights in units of the standard deviation of apple weights?

9. The table below shows the frequency distribution of the percentage of copper in 150 specimens of an alloy

% copper	frequency
0.070–0.075	2
0.075–0.080	7
0.080–0.085	21
0.085–0.090	24
0.090–0.095	36
0.095–0.100	30
0.100–0.105	18
0.105–0.110	9
0.110–0.115	2
0.115–0.120	1

(i) From the ogive obtain the median, and the range within which the middle 50% of the readings occur.

(ii) Estimate the skewness of this distribution using the formula:

skewness = (mean–median)/standard deviation

10. The pH level in a river is monitored five times a day. The following twenty sets of five readings were obtained on 20 consecutive days

```
6.7  6.3  6.2  6.1  7.0  7.0  7.1  6.9  6.8  6.2
6.3  6.4  6.3  6.2  7.1  7.0  7.0  6.8  6.9  6.3
6.6  6.2  6.1  6.4  7.1  6.8  7.3  6.8  7.0  6.4
6.9  6.1  6.0  6.4  6.8  6.5  7.1  6.3  7.1  6.4
6.6  6.6  6.5  6.8  6.7  6.9  7.0  6.4  6.3  6.4

6.3  6.4  6.5  6.4  6.3  6.1  5.9  5.9  5.8  6.1
6.3  6.3  6.5  6.5  6.2  7.0  5.9  5.9  5.9  6.1
6.2  6.6  6.7  6.6  6.2  6.9  6.0  6.1  5.9  6.0
6.1  6.6  6.8  6.3  6.6  6.6  6.0  6.3  5.8  6.2
6.4  6.3  6.3  6.1  7.0  6.4  5.9  6.1  6.1  6.3
```

(i) Using your calculator obtain the mean and standard deviation of these 100 readings.

(ii) Obtain the mean pH levels for each of the 20 consecutive days and plot them on a chart showing pH as a function of day.

(iii) A warning should be flagged if a mean pH level on any given day lies outside the range $m \pm 1.96s_m$, where m is the mean of the 100 readings and $s_m = s/\sqrt{5}$, where s is the standard deviation of the 100 readings. Identify those days on which a warning would be flagged.

11. Produce a pseudocode algorithm that will read in a set of data values, all of which can be assumed to be positive real numbers, that will calculate their mean and standard deviation, and print these values out together with a record of the number of data values entered. The algorithm should be designed in such a way that the reading in of data is terminated by entering a negative number.

15

Statistics 2: probability and probability distributions

Contents: 15.1 scientific context – 15.1.1 radiation decay data revisited; 15.1.2 the alcohol in water experiment; 15.2 mathematical developments – 15.2.1 the concept of probability; 15.2.2 the probability scale; 15.2.3 simple rules; 15.2.4 permutations and combinations; 15.2.5 general rules; 15.2.6 probability distributions; 15.2.7 the concept of expectation; 15.3 worked examples; 15.4 further developments – 15.4.1 the binomial distribution; 15.4.2 the Poisson distribution; 15.4.3 the Normal distribution; 15.5 further worked examples; 15.6 exercises.

In the last chapter we saw how experimental data could be organised and how measures of centre and scatter could be worked out to give a collective description of the data. This description was useful because it helped tell us what to expect from future experiments. However, when making predictions about future experiments, we have to bear in mind that the predictions we make are going to be based on only a sample of results and will necessarily have a measure of uncertainty associated with them. Being able to express numerically the inevitable uncertainties in the results of statistical analyses is very important in science, but it requires first of all an understanding of probability and of probability distributions. In this chapter we cover the basics of probability and consider some of the theoretical probability distributions that are important in handling scientific data.

15.1 Scientific context

15.1.1 Example 1: radiation decay data revisited. To provide us with some data for discussing the basic ideas of probability, let us return to the radiation decay data that we considered in Chapter 14. For convenience we present the 500 data values again in Table 15.1. Suppose we want to know what the chances are of observing fewer than 4 counts in another 10-s exposure of the geiger counter to the radioactive source. Since the words 'chance' and 'probability' mean the same thing, then what we are really wanting to work out here is nothing more than a *probability* – the probability of observing fewer than 4 counts. One way of working out this probability is simply to count how many zeros, ones, twos and threes there are in

Table 15.1. *Radiation decay data*

4	5	6	4	2	3	7	5	4	9	3	7	1	2	4	5	3	8	5	6	6	5	7	5	3
4	2	3	6	0	4	6	3	8	6	2	4	6	7	5	4	6	2	4	7	1	7	9	3	2
3	4	3	8	10	7	4	5	3	6	1	7	6	6	3	4	6	9	2	4	8	5	3	7	4
2	6	5	3	4	4	5	4	2	6	8	7	11	5	4	3	6	1	4	3	4	7	3	2	6
7	2	3	6	8	2	3	7	9	8	3	4	5	5	3	7	1	6	4	7	5	6	4	7	5
4	2	1	6	7	4	3	6	2	7	4	10	5	3	6	2	6	3	0	5	6	4	6	3	8
7	3	4	1	2	9	6	5	3	4	6	4	5	5	4	7	6	4	6	3	2	8	4	3	5
4	1	6	3	2	4	3	5	2	6	3	7	1	4	2	4	6	7	3	6	8	5	4	2	9
2	3	4	1	3	6	5	3	7	10	7	2	5	8	0	3	5	7	5	7	4	5	4	5	4
5	3	5	4	3	1	11	4	2	2	8	4	5	2	3	3	4	7	3	5	4	6	2	5	4
9	3	4	5	7	6	5	7	4	3	5	6	9	1	7	3	5	6	8	2	4	6	3	7	8
2	3	5	4	3	5	4	6	3	4	2	5	3	4	8	7	4	12	1	6	5	2	6	7	5
7	10	8	7	9	4	2	7	5	3	4	6	3	6	3	6	7	4	2	3	2	4	8	5	3
6	6	4	6	1	3	2	9	4	5	3	1	5	4	6	5	6	5	3	5	7	4	5	7	0
3	4	8	2	6	4	6	5	4	3	7	4	2	6	3	1	5	6	5	2	4	7	3	5	4
5	2	8	9	5	5	2	5	3	5	4	3	3	4	1	2	11	6	4	3	8	5	7	4	3
4	7	5	3	5	4	5	7	5	3	4	8	9	4	5	2	5	3	2	6	6	4	10	5	6
2	4	5	8	1	4	2	7	3	7	6	4	6	3	10	7	4	2	6	3	1	4	3	5	4
7	9	3	2	4	7	6	3	1	4	5	8	1	5	2	4	3	5	8	3	2	4	3	5	3
1	5	2	0	4	3	4	5	7	5	2	3	6	8	7	2	4	7	3	1	6	3	4	6	7

the data that we have – and there are 163 of them – and claim that this proportion of readings, namely 163 in 500 or 0.326, is the required probability. As we shall see later, this value is probably a good approximation to the truth and, indeed, one very good way of obtaining a probability is by counting the number of times an event of interest occurs and expressing this as a fraction of the total number of events considered.

But counting by itself cannot always be used to give us a probability. If we look at the raw data in Table 15.1 we will see that counts of 13 or more do not appear. We might be tempted to think that the probability of observing 13 or more counts is therefore zero. Instinct tells us however that this probability is not zero but something very small; for if we wait long enough we will surely observe a count of 13 or more.

Question: How, then, can we work out this probability in the absence of data values on 13 or more counts?

   Answer: We will see that this probability can be estimated by the help of a *theoretical probability distribution* called the *Poisson distribution*.

The Poisson distribution is just one of the three theoretical distributions that we will be considering later. Why the Poisson distribution happens to be appropriate will be explained, but let us now consider a second example.

15.1.2 Example 2: the alcohol in water experiment. The example just considered concerned a discrete variable – the number of counts in a 10-s period – which had to be a whole number. Many measurements in science, however, are values of

Table 15.2. *Frequency table for alcohol concentrations*

Level of alcohol (mg/100 ml)	No. of determinations
75.5–76.5	1
76.5–77.5	3
77.5–78.5	9
78.5–79.5	13
79.5–80.5	11
80.5–81.5	8
81.5–82.5	3
82.5–83.5	2

variables that are continuous, so let us now consider such a situation and see if we need to view probabilities differently or introduce any new concepts for continuous variables.

Again, let us return to a situation that we met previously in Chapter 14, that of determining the concentration of a synthetic sample of alcohol in water. Suppose we have made 50 determinations of the concentration, using the same technique of analysis, and suppose the distribution of results, which is summarised in Table 14.2, is that which was illustrated in the previous chapter in Fig. 14.6. Looking at the results we see, for example, that 13 out of the 50 readings are in the class 78.5–79.5 mg/100 ml. We may conclude, therefore, that if we made another determination of the alcohol concentration, the probability of the result being in the same class would be about $\frac{13}{50}$, i.e. 0.26. Similarly, by dividing each of the other class frequencies by 50 we can come up with the various probabilities of finding a new reading in any of the other classes. Thus by simply counting and dividing by the total number of readings we can work out probabilities as before.

But the interpretation we have to put on these probabilities is different to that for the probabilities we met in the first example on radiation decay. There it was meaningful to talk about working out probabilities of no counts, or one count, or two counts, etc., i.e. of working out probabilities of single results. Here we are not talking about the probability of observing a single result, but of observing a result in a particular range. The probabilities here, therefore (when divided by the range to which they apply), are essentially *probability densities* and not probabilities of single readings. In fact for a continuous variable it is meaningless to ask for the probability of obtaining a single reading because it will always turn out to be zero! Always we should ask for the probability of finding a result in a particular range. This, therefore, is one of the major differences in the way we handle discrete and continuous variables in probability theory. We shall be discussing probability density functions in more detail in the next section.

As well as serving as an introduction to probability density functions, this example also serves as an introduction to another theoretical probability distribution which is important in science – the *Normal distribution*. The *Normal* distribution has a characteristic, symmetrical, bell shape and has a definite mathematical form. Many distributions that occur in science have this symmetrical bell shape and many of them are in fact Normal distributions.

Question: Is it likely that these alcohol concentrations come from a Normal distribution?
  Answer: Distributions in which the only source of variation is due to random errors of
      one kind or another that creep into the measuring process are Normal distributions
      in many cases. Here in our particular example it is worth noting that in determining
      the alcohol concentration the only source of variation is experimental error, so
      there is every possibility that the sample of 50 readings does come from a Normal
      distribution.

Later, we shall in fact show that this is actually the case and we shall thus be able to look to the theory of the Normal distribution to provide us with an alternative way of working out relevant probabilities in this example. Let us now develop some theory.

15.2 Mathematical developments

15.2.1 The concept of probability. It is customary to talk about *events* and *trials* in connection with probability.

An event is a possible outcome of a trial. In science a trial could be anything from the simplest measuring process to carrying out a full-scale experiment, and an event might be the act of obtaining a single reading or of obtaining a result in a given range and so on. In the radiation decay example the act of subjecting the geiger counter to the source for a 10-s period constitutes the trial and obtaining, say, 4 counts in that period would be a possible event. In the second example the act of determining the concentration of alcohol would be the trial and obtaining a reading in the range 80.5–81.5 mg/100 ml, for example, would be a typical event.

The starting point in probability theory is to give a definition of what we mean by the probability of an event in a single trial. This can be given as follows:

Definition of probability. If the number of times a particular event occurs in N trials is n, and if the sequence of relative frequencies n/N, obtained for larger and larger values of N, approaches a limit, then this limit is the probability of the event occurring in a single trial.

As an illustration we can use this definition together with the raw data on radiation counts in Table 15.1 to obtain an estimate of the probability of observing a count rate of 3 or less in any given period. If we look, for example, at the first 30 readings in the list we see that there are ten readings of 3 counts or less. If we then look at the next 50 readings there are 15 such readings. In the next 80 there are 26; in the next 120 there are 40 and in the remaining 220 readings there are 72. Overall, in the 500 readings, there are thus 163 values equal to three or less. These findings can be represented as shown in Table 15.3.

Table 15.3. *Probabilities of number of readings*

No. of readings (N)	No. equal to 3 or less (n)	n/N
30	10	0.333
50	15	0.300
80	26	0.325
120	40	0.333
220	72	0.327
500	163	0.326

Working with successively larger samples of readings we see that the relative frequencies do look to be converging to some limit in the region of 0.33. We can conclude, therefore, that the probability of observing three or less counts in another exposure to the source is probably going to be in the region of 0.33. Ideally, of course, we do need far more than 500 readings to be certain of the value of the probability, but with the modest number of readings that we do have, we can be almost certain that the probability in question does exist and has a definite value that is not drifting or changing in any way over time. One pitfall of simply looking at all the available results at once, and not dealing with successively larger samples, is that any systematic drift in the values of the relative frequency would not be spotted. Since there is no discernable drift in our example, or so it seems, it is meaningful to regard $\frac{163}{500}$ as the best estimate of the required probability. *It is very important to work with larger and larger samples when setting out to evaluate a probability.*

The definition of probability that we have just been considering is a general definition that can be applied to all situations where it is meaningful to calculate a probability. There are, however, many cases where the probability may be stated or worked out without first collecting endless data. Without making trials we can say that the probability of obtaining a head when a coin is spun is $\frac{1}{2}$. Also we know that the probability of obtaining a 1 (or a 2, 3, 4, 5 or 6) when throwing an honest die is $\frac{1}{6}$ and so on. We can give a definition of probability where several *equally likely events* can occur as follows:

Alternative definition of probability. If an event can happen in S ways and fail to happen in F ways, and if each of these $S + F$ ways is equally likely to occur, the probability of the event happening in a single trial is

$$p = \frac{S}{S + F} \tag{1}$$

and the probability of its failing to happen is

$$q = \frac{F}{S + F} \tag{2}$$

These formulae, as we shall see, are most useful in helping us work out probabilities when we can actually enumerate all the possible outcomes of a trial.

15.2.2 The probability scale. It is easily seen from (1) and (2) and from the definition of probability as the limit of a sequence of relative frequencies, that probabilities lie between 0 and 1. The *probability scale* thus runs from 0 to 1. Any event with a probability of zero means that the event will never happen. A probability of 1 corresponds to a certainty. Negative probabilities have no meaning and are not allowed. It is always wise to check the value of a probability to see in fact that it does lies between 0 and 1!

15.2.3 Simple rules. There are two basic rules that enable us to develop most of the theory that will be needed. These two rules tell us essentially when to *add* and when to *multiply* probabilities together.

The first rule applies to *mutually exclusive events*. If two or more events are such that not more than one of them can occur in a single trial, they are said to be mutually exclusive. For mutually exclusive events we have the following rule:

Rule 1. If $p_1, p_2, \ldots, p_r$ are separate probabilities of r mutually exclusive events, then the probability P that some one of these will occur in a single trial is

$$P = p_1 + p_2 + \cdots + p_r \tag{3}$$

Thus the probability of drawing an ace, a king, or the queen of clubs when one card is drawn from a regular 52-card pack is

$$P = \tfrac{4}{52} + \tfrac{4}{52} + \tfrac{1}{52}, \text{ i.e. } \tfrac{9}{52}$$

Or, if we look at the radiation data again we see that the probability of no counts (working with the 500 readings available) is $\tfrac{4}{500}$, that the probability of a count of 1 is $\tfrac{23}{500}$, of a count of 2 is $\tfrac{54}{500}$, and of a count of 3 is $\tfrac{82}{500}$. It follows therefore that the probability of either a count of 0, 1, 2 or 3 is

$$P = \tfrac{4}{500} + \tfrac{23}{500} + \tfrac{54}{500} + \tfrac{82}{500}, \text{ i.e. } \tfrac{163}{500}$$

which is what we worked out previously. That we have mutually exclusive events in these two cases is easily appreciated. When a card is drawn it excludes all other cards from occurring in that particular trial. Similarly, when the number of counts in a 10-s period has been noted, its value automatically excludes all other possible values for that particular period. We therefore have mutually exclusive events.

The second rule applies to situations where several events can occur in a single trial, but where these events are *independent*. Two or more events are said to be independent if the occurrence of any of them in no way influences the probability of occurrence of any other. For independent events we have the following rule:

Rule 2. If $p_1, p_2, \ldots, p_r$ are the separate probabilities of occurrence of r independent events, the probability P that they will all occur in a single trial is

$$P = p_1 \times p_2 \times \cdots \times p_r \tag{4}$$

Thus in an experiment which consists of drawing a card from a 52-card pack, spinning a coin and shaking a die, the probability of the event: red card, head, six, is

$$P = \tfrac{1}{2} \times \tfrac{1}{2} \times \tfrac{1}{6}, \text{ i.e. } \tfrac{1}{24}$$

Also, having noted in the radiation data that the probability of obtaining a count of 1 is $\tfrac{23}{500}$ and a count of 2 is $\tfrac{54}{500}$, it follows that in two consecutive 10-s exposures to the source, the probability of obtaining first a count of 1 and then a count of 2 will be

$$P = \tfrac{23}{500} \times \tfrac{54}{500} = \tfrac{1242}{25\,000} \quad \text{or} \quad 0.005$$

These two rules are easily proved and we will do this in the worked examples. In remembering the rules it should be borne in mind that a *plus* sign is associated with *or* and a *multiplication* sign with *and*.

15.2.4 Permutations and combinations. Especially when we come to try and estimate probabilities by using the expressions in equations (1) and (2), we find that we have to work out the number of ways in which various events might happen. *Permutations* and *combinations* deal with groupings and arrangements and tend to feature in such calculations. For this reason we need to know about them and what their properties are.

The help focus ideas, consider the situation where a science student has ten experiments to complete in his practical programme and decides one afternoon to complete three of them. Clearly he has a choice of ten for his first experiment. Having completed this there will be nine left so he has a choice of nine for his second experiment and so on. It is easily seen that the total number of ways in which he can carry out three experiments from ten is $10 \times 9 \times 8$ ways. This is just an example of a *permutation of* ten objects taken three at a time and the number is written as $^{10}P_3$. In general ^nP_r is $n(n-1)(n-2)\cdots(n-r+1)$, where there are r terms in the product. Using the factorial notation we see that

$$^nP_r = n!/(n-r)! \tag{5}$$

The important thing about a permutation is that order within an arrangement is important. In the example above let us suppose that what the student actually did was first to complete experiment x, then y, and finally z. In other words he completed the three experiments in the order x, y, z. It could be argued that the order in which these experiments are carried out is irrelevant because at the end of the day the lecturer in charge has to mark the write-ups of x, y and z, which he would still have to do if the particular order had been x, z, y or y, x, z, etc. Many situations arise in science where we have arrangements in which only the objects in the arrangement and not their order within the arrangement are important. Such arrangements are called *combinations*.

To develop a formula for the number of combinations of n objects taken r at a time, we note that for every set of r objects there is just one combination if all r objects are used, whereas there are $r!$ permutations of these objects. Thus there are $r!$ times as many permutations of n objects taken r at a time as there are combinations. Therefore if ^nC_r denotes the number of combinations of n objects taken r at a time, then

$$^nC_r = \frac{1}{r!}\,^nP_r = \frac{n!}{(n-r)!\,r!} \qquad (6)$$

It is important to note that the values of ^nP_r and ^nC_r when worked out will always be whole numbers and that the value of $0!$ is 1 and not zero. ^nP_r and ^nC_r are available as functions on most calculators these days.

15.2.5 General rules. It would be nice if all experiments or trials resulted in outcomes that could be classed as being either mutually exclusive or independent. But in the real world this is not the case. Consider the events: 'has brown eyes' and 'is female'. These are certainly not mutually exclusive events because a person could well be a brown-eyed female. What, then, is the general rule to use to work out the probability that a person selected at random is going to be either brown-eyed or female? On the other hand, suppose we select a playing card from a normal pack and without replacing it we select another one so that we have two playing cards. If we want to work out the probability of the two cards being aces, what general rule could we now use, because without replacing the first card the two events are no longer independent – the probability that the second card is an ace now depends on whether or not the first card was an ace.

Rules of probability, that have more general application than rules 1 and 2 can easily be developed. Suppose an experiment is carried out a total of N times, where N is large, and suppose the set of N outcomes is called E as shown in Fig. 15.1. Let the collection of outcomes where event 1 has occurred be called E_1 and assume

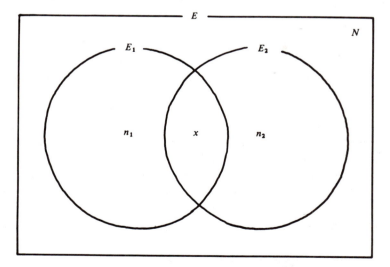

Fig. 15.1. Venn diagram for probabilities.

it contains n_1 outcomes. Let E_2, containing n_2 elements, be the set of outcomes where event 2 has occurred. Let there be x outcomes in the intersection of the two sets where both events 1 and 2 have occurred. If we want the probability of either event 1 or event 2 occurring we see that this is given by

$$P(1 \text{ or } 2) = (n_1 + n_2 - x)/N$$

$$= \frac{n_1}{N} + \frac{n_2}{N} - \frac{x}{N}$$

Since n_1/N is the probability of event 1, n_2/N is the probability of event 2 and x/N is the probability of both event 1 and event 2, then we have the general rule that

$$P(1 \text{ or } 2) = P(1) + P(2) - P(1 \text{ and } 2) \tag{7}$$

Example: What is the probability of a person being either brown-eyed or female?
Solution: Using (7) we have that

$$P(\text{brown or female}) = P(\text{brown}) + P(\text{female}) - P(\text{both})$$

Since having brown eyes and being female are independent events, then $P(\text{both})$ is $P(\text{brown}) \times P(\text{female})$. Also since $P(\text{brown})$ is $\frac{3}{4}$ and $P(\text{female})$ is $\frac{1}{2}$ we have

$$P(\text{brown or female}) = \tfrac{3}{4} + \tfrac{1}{2} - \tfrac{3}{4} \times \tfrac{1}{2} = \tfrac{7}{8}$$

If we now want to derive a more general rule for the probability of both event 1 and event 2 happening, then let us assume, without loss of generality, that event 1 occurred before event 2. With reference to Fig. 15.1 we see that

$$P(1 \text{ and } 2) = \frac{x}{N} = \frac{x}{n_1} \times \frac{n_1}{N}$$

The fraction x/n_1 is just the probability of observing both event 1 and event 2 given that event 1 has already occurred. In other words it is the probability of observing event 2 given that event 1 has happened. Thus we have the general rule that

$$P(1 \text{ and } 2) = P(2 \text{ given } 1) \times P(1) \tag{8}$$

Example: What is the probability of drawing two aces one after the other from a normal 52-card pack of playing cards?
Solution: Using formula (8) we have

$$P(2 \text{ aces}) = P(\text{ace given an ace}) \times P(\text{ace})$$
$$= \tfrac{3}{51} \times \tfrac{4}{52}$$
$$= \tfrac{7}{2652}$$

15.2.6 Probability distributions. A probability, as we have seen, is just a relative frequency. Thus, given a frequency distribution, we can obtain the corresponding probability distribution by dividing each class frequency by the sum of frequencies. *The probability p_i of a reading in a distribution being in the ith class is going to be $f_i / \sum_i f_i$.* For the radiation data the probability distribution is easily worked out and is shown in Table 15.4. A plot of probability against class would have exactly

Table 15.4. *Probability distribution for the radiation decay data*

No. of counts	Frequency	Probability
0	4	4/500
1	23	23/500
2	54	54/500
3	82	82/500
4	93	93/500
5	78	78/500
6	66	66/500
7	53	53/500
8	25	25/500
9	13	13/500
10	5	5/500
11	3	3/500
12	1	1/500
Total:	500	Total: 1

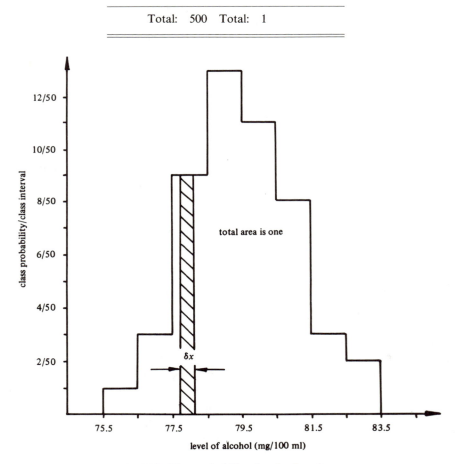

Fig. 15.2. The probability density function.

the same overall shape as a plot of frequency against class and in general frequency and probability distributions convey precisely the same information about the variable of interest.

Now, as we have said, the radiation data concerns a discrete variable. If we are considering a continuous variable, then basically because we are looking for the probability of obtaining a result in a particular range, we find that it becomes more convenient to work not with class probabilities but with *probability densities*. Consider, again, the alcohol concentrations summarized in Table 15.2. We can obtain the class probabilities in the usual way by dividing each frequency by the sum of frequencies. But if we now plot out a histogram of class probability divided by class interval against class we obtain a special histogram whose area will be 1 as shown in Fig. 15.2. The actual profile of this histogram is what we call the *probability density function*. It is so called because it is a plot of probability density against class. Without loss of generality consider the class 77.5–78.5 mg/100 ml. The class probability is $\frac{9}{50}$ and the class interval c is 1 mg/100 ml. The density of probability per unit length of class, in this class, is thus $\frac{9}{50}$ divided by c, i.e. $\frac{9}{50}$ per mg/100 ml. The probability, therefore, of finding a reading, in this class, in a range of width, say δx, will thus be $\frac{9}{50} \times \delta x$. This is the shaded area shown in Fig. 15.2 and we see that it is the *area* under the probability density function that gives us the probability. Thus if we wanted to know what the probability was of obtaining a reading in the range X_1 to X_2, then this would simply be the area under the profile between X_1 and X_2. It is now easy to see why the total area under the profile will be 1. *The notation of an area corresponding to a probability is central in statistics.*

15.2.7 The concept of expectation. We have spoken loosely in this chapter and in the previous chapter of 'what to expect' in future experiments. In statistics the term 'expectation' or 'what to expect' can be given a precise definition in terms of probabilities. The definition looks slightly different depending on whether the variable of interest is discrete or continuous:

Definition of the expected value

1 Discrete variable case: If the possible results in a trial are $x_1, x_2, x_3, \ldots$ and their associated probabilities are $p_1, p_2, p_3, \ldots$ then the *expected value* $E(x)$ of x, is defined to be

$$E(x) = \sum_i x_i p_i \tag{9}$$

2 Continuous variable case: If the particular results lie in the range $a \leqslant x \leqslant b$ and the associated probability density function is $p(x)$, defined over this range, then the expected value of x is defined as

$$E(x) = \int_a^b x p(x)\, \mathrm{d}x \tag{10}$$

Just what the expected value $E(x)$ corresponds to can easily be deduced. If we take the definition given in (9), for example, and note that p_i is really $f_i/\sum_i f_i$, then we see that the expression for $E(x)$ is $\sum_i x_i f_i/\sum_i f_i$, which is just the definition of the arithmetic mean. However, we must take care to note that the mean here is the so-called *population mean* and not the mean of a sample of results. It is implicit in the definition of probability that increasingly larger and larger samples have been taken so that the p_i, here, effectively relate to the totality of readings possible, which we call the population. The value of $E(x)$ in (10) similarly corresponds to the population mean.

Thus, if we regard an experiment or trial as the act of taking a variable, x, at random from some population of readings and noting its value, then the value to expect for x is the mean μ of this population. We should note that the population mean is usually a quantity whose value we can never determine exactly, but as the value to expect it is an extremely useful concept in statistics.

The expected value of $(x - \mu)^2$: Obviously when we measure the value of a quantity and obtain x, this value most likely will be different to μ; so what should we expect as far as this difference is concerned? Well, if we did know the value of μ exactly, we could repeatedly sample from the population and note the value of $(x - \mu)^2$. The expected value of this quantity would turn out to be σ^2, the *variance of the population*.

That the expected value of $(x - \mu)^2$ is the population variance can easily be shown. Assume, without loss of generality, that we have a discrete variable situation, so that by using equation (9) applied to possible $(x - \mu)^2$ values we have

$$E((x - \mu)^2) = \sum_i (x_i - \mu)^2 p_i$$
$$= \sum_i (x_i^2 - 2x_i\mu + \mu^2)p_i$$
$$= \sum_i x_i^2 p_i - 2\mu \sum_i x_i p_i + \mu^2 \sum_i p_i$$
$$= \sum_i x_i^2 p_i - 2\mu^2 + \mu^2$$
$$= \sum_i x_i^2 p_i - \mu^2.$$

Replacing p_i by $f_i/\sum_i f_i$, where the understanding is that $\sum_i f_i$ is a large number, we obtain

$$E((x - \mu)^2) = \sum_i x_i^2 f_i/\sum_i f_i - \mu^2$$

which clearly is the expression for the population variance. Thus for the expected value of $(x - \mu)^2$ we have the following results:

1 Discrete variable case

$$E((x - \mu)^2) = \sum_i (x_i - \mu)^2 p_i = \sigma^2 \tag{11}$$

2 Continuous variable case

$$E((x - \mu)^2) = \int_a^b (x - \mu)^2 p(x) \, dx = \sigma^2 \tag{12}$$

The concept of expectation is very important and we shall use it extensively in the next chapter.

15.3 Worked examples

Example 1. Using the data on radiation counts in Table 15.1, show that the probability of obtaining a count of 4 in a 10-s exposure to the source is some value close to 0.19.

Solution. By inspecting the first 30 readings in the list of data, then the next 50 readings, then the next 80 and so on, and counting the number of 4s in each case, we can obtain the following sequence of relative frequencies:

Number of readings (N)	Number of 4s (n)	n/N
30	5	0.167
50	10	0.200
80	14	0.175
120	23	0.192
220	41	0.186
500	93	0.186

The sequence does appear to be converging to some quantity whose value, to two decimal places, is 0.19. Thus, by working with successively larger and larger samples of readings we can conclude that the probability of obtaining a count of 4 in a 10-s exposure to the source is some value close to 0.19.

Example 2. An electronic device uses ten silicon chips. If the chips are chosen at random from a batch of 100 chips in which there are five defectives, obtain the probability, correct to two decimal places, that the device, when made up, contains just one defective chip.

Solution. In this example we shall work out the number of different ways that the device can be made up to contain just one defective chip. The required probability will be obtained by dividing this number by the total number of different ways of making up the device. We therefore appeal to the definition of probability given in (1).

The device will have one defective chip in it and nine non-defective ones. The number of ways of selecting one defective chip from the five in the batch is $^5C_1$ ways. The number of ways of selecting nine non-defective chips from the 95 non-defective ones is $^{95}C_9$. The total number of ways in which we can obtain ten chips where one is defective is thus $^5C_1 \times {}^{95}C_9$. The total number of ways in which we could choose any ten chips from the batch of 100 is $^{100}C_{10}$. Thus it follows that the probability P of making up the control device with its having just one defective chip is

$$P = \frac{{}^5C_1 \times {}^{95}C_9}{{}^{100}C_{10}}$$

$$= \frac{5 \times 95 \times 94 \times 93 \times 92 \times 91 \times 90 \times 89 \times 88 \times 87 \times 10!}{100 \times 99 \times 98 \times 97 \times 96 \times 95 \times 94 \times 93 \times 92 \times 91 \times 9!}$$

$$= \frac{90 \times 89 \times 88 \times 87}{99 \times 98 \times 97 \times 96} \times \frac{1}{2}$$

$$= 0.34 \text{ correct to two places of decimals.}$$

Example 3. Working with the probability density function shown in Fig. 15.2, obtain the probability of finding a reading in the range 77.30–80.85 mg/100 ml, giving the result correct to two decimal places.

Solution. The required probability is the area under the probability density function profile between 77.30 and 80.85 mg/100 ml. With reference to Fig. 15.2, we see that this area will be

$$0.20 \times \frac{3}{50} + 1.00 \times \frac{9}{50} + 1.00 \times \frac{13}{50} + 1.00 \times \frac{11}{50} + 0.35 \times \frac{8}{50}$$

which is $\frac{36.4}{50}$ or 0.73 correct to two decimal places.

Example 4. In the radiation data we saw that the probability of observing a count of 3 or less was about 0.33. Assuming 0.33 is the value of this probability, how many 10-s exposures to the source would be required for there to be a better than 0.95 chance of observing a count rate of 3 or less on at least one occasion?

Solution. If we consider n exposures to the source, the probability P of there being at least one occasion when there were 3 or less counts is going to be

$$P = p(1) + p(2) + \cdots + p(n)$$

where $p(i)$ is the probability of observing 3 or less counts on i occasions out of n. But the sum of probabilities $p(1) + p(2) + \cdots + p(n)$ will be the same as $1 - p(0)$, so the required probability will be

$$P = 1 - p(0)$$

Now, $p(0)$ is the probability of observing 3 or less counts on no occasions out of n. In other words it is the probability of observing more than three counts on all n occasions. This is easily worked out. The probability of observing more than 3 counts will be $1 - 0.33$, i.e. 0.67. So $p(0)$ is $(0.67)^n$. Thus, requiring that P should be greater than 0.95 leads to the inequality

$$1 - (0.67)^n > 0.95$$

i.e.

$$0.05 > (0.67)^n$$

Solving this equation using logs, or simply by successively multiplying 0.67 by itself, we see that the smallest value of n has to be 8. We thus need at least eight exposures to the source to be 95% sure of observing 3 or less counts on at least one occasion.

15.4 Further developments

The distributions we have met so far have been distributions where the data can be thought of as having been obtained by carrying out some sort of experiment or survey. Not all distributions in statistics are of this type. Model distributions, i.e. *theoretical distributions*, exist which are crucial in the interpretation of experimental data and the rest of this chapter is devoted to studying some of the more important theoretical distributions in science. We have already mentioned that the *Poisson* distribution and the *Normal* distribution are important in analysing data. There is another distribution that is just as important in analysing scientific data, namely the *binomial* distribution, and it is with this distribution that we begin.

15.4.1 The binomial distribution. Suppose we are carrying out an experiment where the outcome is either success or failure. Many experiments can be viewed in this light. For example in the radiation decay situation we can regard a count of 3 or less as success and one of 4 or more as failure. In the alcohol in water experiment we might view the occurrence of a reading below 80 mg/100 ml as success and a reading above this level as failure.

Suppose, therefore, we repeat such an experiment n times and that the n results are as shown:

Experiment no.	1	2	3	$\cdots$	$n-1$	n
result	success	success	failure	$\cdots$	failure	success

If we assume that the probability of success each time we perform the experiment is the same and equal to p, then the probability of failure, q, will be the same and equal to $1 - p$. If, therefore, the outcomes of each experiment are independent of one another, it follows that the probability of obtaining the sequence of results above is

$$p \times p \times q \times \cdots \times q \times p$$

If there are r successes, then this product is $p^r q^{n-r}$, and, provided there are r successes in the n experiments in any order, the probability of the outcome will still be $p^r q^{n-r}$. However, the number of ways of getting r successes in n trials is ^nC_r

and thus the probability of obtaining r successes in n trials (in any order) will be $^nC_r\,p^r q^{n-r}$. The probability distribution given by

$$P(r) = {^nC_r}p^r q^{n-r} \tag{13}$$

for different values of r ranging from 0 to n is the *binomial distribution*.

Example: If $n = 4$ what are the probabilities of 0, 1, 2, 3, 4 successes?
 Answer: Using (13) we have:

$$P(0) = {^4C_0}p^0 q^4 = q^4$$
$$P(1) = {^4C_1}p^1 q^3 = 4pq^3$$
$$P(2) = {^4C_2}p^2 q^2 = 6p^2 q^2$$
$$P(3) = {^4C_3}p^3 q^1 = 4p^3 q$$
$$P(4) = {^4C_4}p^4 q^0 = p^4$$

We should note that the probabilities in the example we have just considered are given, in fact, by the binomial expansion of $(q+p)^4$. In the general case, with n experiments, the probabilities are given by the binomial expansion of $(q+p)^n$ – hence the name of the distribution.

   Like any other distribution, the binomial distribution has a *mean* and a *standard deviation*. These quantities can be worked out by noting that the mean, μ, is the expected value of r and the *variance* is the expected value of $(r-\mu)^2$. It can be shown, therefore, that the mean of the binomial distribution is

$$\mu = np \tag{14}$$

and that the expected value of $(r-\mu)^2$, in other words the expected value of $(r-np)^2$, is equal to the quantity npq. The standard deviation is thus the square root of this quantity and so we have

$$\sigma = \sqrt{(npq)} \tag{15}$$

Thus, for example, if we carried out an experiment, say 72 times, where the probability of success each time was the same and equal to $\frac{1}{3}$, we should expect $72 \times \frac{1}{3}$, i.e. 24 successes and the typical spread about 24 to expect in the actual number of successes would be something of the order of $\sqrt{(72 \times \frac{1}{3} \times \frac{2}{3})}$, i.e. four successes.

   Several problems illustrating the use of the binomial distribution in a scientific context will be considered in the Further Worked Examples.

15.4.2 The Poisson distribution. The binomial distribution assumes a convenient form when the probability of success, p, is very small, but the number of trials n is very large. Under these conditions the binomial distribution is referred to as the *Poisson distribution* after the person who first derived its form.

   Situations where p is small and n is large arise in analysing accidents, or production faults, or events that occur randomly in time and so on. For example the probability p of an accident occurring in any given day in a laboratory is hopefully small, but over a period of many days, such as a year, an accident may well occur and we might typically be interested in the probability of a certain number of accidents occurring during the year. Alternatively a production line may be such that an average 0.5% of goods are defective and we may well want to know what to expect when we analyse a day's production of say 1000 articles. It is in these situations that the Poisson distribution is particularly useful. To derive its form we look at the binomial distribution in the limit of large n and small p.

The binomial distribution gives the probability of r successes in n trials as

$$P(r) = {}^nC_r p^r q^{n-r},$$

which we can write as

$$P(r) = \frac{n(n-1)\cdots(n-r+1)p^r}{r!}(1-p)^{n-r}$$

If we let np be the quantity λ, so that $p = \lambda/n$, we have that

$$P(r) = \frac{n(n-1)\cdots(n-r+1)}{r!}\left(\frac{\lambda}{n}\right)^r\left(1-\frac{\lambda}{n}\right)^{n-r}$$

In the limit of large n and small p the quantity λ, in relative terms, will be neither particularly large nor particularly small, but n will be large enough for the product $n(n-1)\cdots(n-r+1)$ to be regarded as n^r and for $n-r$ to be replaced by n. Thus in this limit the expression for $P(r)$ becomes

$$P(r) = \frac{n^r}{r!}\left(\frac{\lambda}{n}\right)^r\left(1-\frac{\lambda}{n}\right)^n$$

$$= \frac{\lambda^r}{r!}\left(1-\frac{\lambda}{n}\right)^n$$

The expression $(1-(\lambda/n))^n$, for large n, is just $e^{-\lambda}$ and so we have

$$P(r) = \frac{\lambda^r}{r!}e^{-\lambda} \tag{16}$$

This distribution of probabilities, obtained by putting r equal to 0, 1, 2, 3,… is the *Poisson distribution*.

Its *mean* and *standard deviation* are those of the binomial distribution. Thus the mean is

$$\mu = np = \lambda \tag{17}$$

and the standard deviation will be the quantity $\sqrt{(npq)}$, which is the same as $\sqrt{(\lambda q)}$. But since q is $1-p$ and p is very small, it follows that $1-p$, to all intents and purposes, is equal to 1. Thus the standard deviation is given by

$$\sigma = \sqrt{(npq)} = \sqrt{\lambda} \tag{18}$$

One quick check that can be carried out to test if data came from a Poisson distribution is to work out the mean and variance of the data and see if the two are roughly the same. If they are, and the system being studied falls into the general area of application of the Poisson distribution (i.e. the study of accidents and of

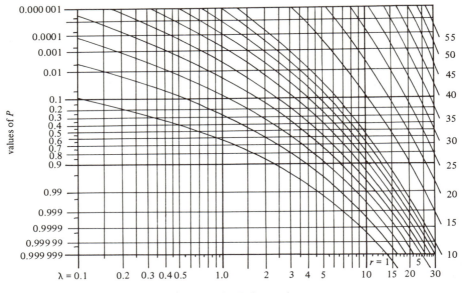

Fig. 15.3. The Poisson chart.

events occurring randomly in time), then the data probably are Poisson distributed.

Several problems illustrating the use of the Poisson distribution will be considered in the Further Worked Examples section. There we shall see it is often the case that what is needed is not the probability of obtaining exactly r successes, but the probability of obtaining r or more successes. Working out the probability of r or more successes can be accomplished by first working out the probability of fewer than r successes and then subtracting this result from 1. However, if r is relatively large, this can be quite tedious and it then becomes easier to use a special chart, called the *Poisson chart*, which is shown in Fig. 15.3. The chart is used in the following way. Suppose we require to know the probability of obtaining five or more successes when we expected 1.5, say. We first find 1.5 on the λ axis. We then move up the $\lambda = 1.5$ line until we cross the 5th curve on the chart. The corresponding value on the P axis is the required probability of obtaining five or more successes. The required value is thus 0.02. The Poisson chart is particularly useful in quality control applications where it is often the case that the probability of r or more defectives or faults is required.

15.4.3 The Normal distribution. The Normal distribution is one of the most important distributions in statistics. Its name stems from the fact that its shape – a nice symmetrical bell shape – is what we might 'normally' expect a distribution to look like. As we mentioned earlier, many quantities are distributed Normally and in particular the distribution of random errors is nearly always a Normal distribution. The Normal distribution is unlike the binomial or the Poisson distribution in that it involves a continuous variable as opposed to a discrete one.

Also its mathematical form is difficult to derive from first principles and for this reason we shall simply state what its form is. Involving as it does a continuous variable it will be characterised by a probability density function. The probability density function of a Normal distribution whose mean is μ, whose standard deviation is σ, and whose variable of interest is x, is given by

$$P(x) = \frac{1}{(2\pi)^{\frac{1}{2}}\sigma} \exp\left[-(x-\mu)^2/2\sigma^2\right] \tag{19}$$

The plot of $P(x)$ against x is shown in Fig. 15.4 and is a bell-shaped curve that is symmetrical about $x = \mu$. The curve is therefore symmetrical about the mean. Probabilities in the Normal distribution are worked out by noting that the probability, $P(x_1, x_2)$, of observing a reading x in the range x_1 to x_2 will be the area under the probability density function between x_1 and x_2. Thus

$$P(x_1, x_2) = \int_{x_1}^{x_2} \frac{1}{(2\pi)^{\frac{1}{2}}\sigma} \exp\left[-(x-\mu)^2/2\sigma^2\right] dx \tag{20}$$

This integral cannot be worked out in terms of functions that we have met so far and requires the use of special tables. At first sight it might appear that separate tables for different values of μ and σ will be needed, but it turns out that just one set of tables relating to the so-called *standard Normal distribution* is all that is needed.

The standard Normal distribution is one whose mean is zero and whose standard deviation is 1. Its probability density function can be written as

$$P(z) = \frac{1}{(2\pi)^{\frac{1}{2}}} \exp\left(-z^2/2\right) \tag{21}$$

where the variable of interest is taken to be z as opposed to x to avoid any confusion

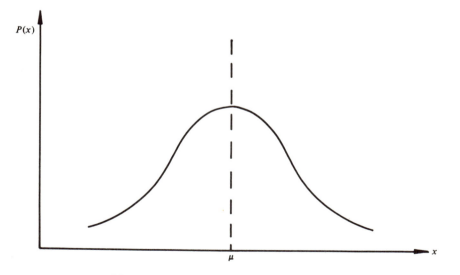

Fig. 15.4. Graph of the Normal distribution.

that might arise when we come to relate z and x later on. The probability of a reading being observed between 0 and z in this standard distribution will be

$$\int_0^z P(z)\,dz$$

and it is this area that is tabulated in the *standard Normal tables* shown in Table 15.5. We see, for example, that the probability of a reading being between $z = 0$ and $z = 1$ is 0.3413, or being between $z = 0$ and $z = 2$ is 0.4772 and so on. In fact by making use of symmetry properties we can actually obtain any area under the standard curve and not just one between 0 and z.

Example: Obtain the area under the standard curve in the following three cases:
  (i) between $z = -1$ and $z = 2$;
  (ii) between $z = -2$ and $z = -1$;
  (iii) above $z = 1.7$.

Solution: In the first case we note that the required area will be the area between -1 and 0 added to the area between 0 and 2. The area between -1 and 0, by symmetry, is the same as that between 0 and 1. Thus the required area is the area between 0 and 1 plus the area between 0 and 2. Consulting Table 15.5 we see that this is $0.3413 + 0.4772$, which is 0.8185.

In case (ii) the required area, by symmetry, will be identical to that between 1 and 2. This area is equal to the area between 0 and 2 minus the area between 0 and 1. The area is thus $0.4772 - 0.3413$, which is 0.1359.

Finally, in case (iii) the area above $z = 1.7$ is obtained by noting that the area between $z = 0$ and ∞ is 0.5. The area is therefore equal to 0.5 minus the area between 0 and 1.7. Thus the area is $0.5 - 0.4554$, which is 0.0446. Since these three cases are typical of all areas that we might require it follows that we can indeed use the one set of tables to work out any area under the standard Normal distribution curve.

One obvious question to ask is how can these tables be used to work out areas under a typical Normal curve between x_1 and x_2, when the distribution has mean μ and standard deviation σ? The answer is simply that the tables can be used if we first convert the normal distribution of interest into the standard Normal distribution and this can always be achieved by a simple change of variable.

The transformation that converts a Normal distribution of mean μ, standard deviation σ and variable x into the standard distribution with variable z is the following:

$$z = (x - \mu)/\sigma \tag{22}$$

Under this transformation the area given by

$$P(x_1, x_2) = \int_{x_1}^{x_2} \frac{1}{(2\pi)^{\frac{1}{2}}\sigma} \exp\left[(x - \mu)^2/2\sigma^2\right] dx$$

becomes equal to

$$P(z_1, z_2) = \int_{z_1}^{z_2} \frac{1}{(2\pi)^{\frac{1}{2}}} \exp\left(-z^2/2\right) dz$$

which is precisely the area under the standard Normal curve between $z_1 = (x_1 - \mu)/\sigma$

Table 15.5. *Areas under the standard Normal curve from 0 to z*

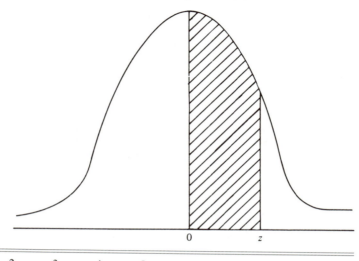

z	0	1	2	3	4	5	6	7	8	9
0.0	0.0000	0.0040	0.0080	0.0120	0.0160	0.0199	0.0239	0.0279	0.0319	0.0359
0.1	0.0398	0.0438	0.0478	0.0517	0.0557	0.0596	0.0636	0.0675	0.0714	0.0754
0.2	0.0793	0.0832	0.0871	0.0910	0.0948	0.0987	0.1026	0.1064	0.1103	0.1141
0.3	0.1179	0.1217	0.1255	0.1293	0.1331	0.1368	0.1406	0.1443	0.1480	0.1517
0.4	0.1554	0.1591	0.1628	0.1664	0.1700	0.1736	0.1772	0.1808	0.1844	0.1879
0.5	0.1915	0.1950	0.1985	0.2019	0.2054	0.2088	0.2123	0.2157	0.2190	0.2224
0.6	0.2258	0.2291	0.2324	0.2357	0.2389	0.2422	0.2454	0.2486	0.2518	0.2549
0.7	0.2580	0.2612	0.2642	0.2673	0.2704	0.2734	0.2764	0.2764	0.2823	0.2852
0.8	0.2881	0.2910	0.2939	0.2967	0.2996	0.3023	0.3051	0.3078	0.3106	0.3133
0.9	0.3159	0.3186	0.3212	0.3238	0.3264	0.3289	0.3315	0.3340	0.3365	0.3389
1.0	0.3413	0.3438	0.3461	0.3485	0.3508	0.3531	0.3554	0.3577	0.3599	0.3621
1.1	0.3643	0.3665	0.3665	0.3708	0.3729	0.3749	0.3770	0.3790	0.3810	0.3830
1.2	0.3849	0.3869	0.3888	0.3907	0.3925	0.3944	0.3962	0.3980	0.3997	0.4015
1.3	0.4032	0.4049	0.4066	0.4082	0.4099	0.4115	0.4131	0.4147	0.4162	0.4177
1.4	0.4192	0.4207	0.4222	0.4236	0.4251	0.4265	0.4279	0.4292	0.4306	0.4319
1.5	0.4332	0.4345	0.4357	0.4370	0.4382	0.4394	0.4406	0.4418	0.4429	0.4441
1.6	0.4452	0.4463	0.4474	0.4484	0.4495	0.4505	0.4515	0.4525	0.4535	0.4545
1.7	0.4554	0.4564	0.4573	0.4582	0.4591	0.4599	0.4608	0.4616	0.4625	0.4633
1.8	0.4641	0.4649	0.4656	0.4664	0.4671	0.4678	0.4686	0.4693	0.4699	0.4706
1.9	0.4713	0.4719	0.4726	0.4732	0.4738	0.4744	0.4750	0.4756	0.4761	0.4767
2.0	0.4772	0.4778	0.4783	0.4788	0.4793	0.4798	0.4803	0.4808	0.4812	0.4817
2.1	0.4821	0.4826	0.4830	0.4834	0.4838	0.4842	0.4846	0.4850	0.4854	0.4857
2.2	0.4861	0.4864	0.4868	0.4871	0.4875	0.4878	0.4881	0.4884	0.4887	0.4890
2.3	0.4893	0.4896	0.4898	0.4901	0.4904	0.4906	0.4909	0.4911	0.4913	0.4916
2.4	0.4918	0.4920	0.4922	0.4925	0.4927	0.4929	0.4931	0.4932	0.4934	0.4936
2.5	0.4938	0.4940	0.4941	0.4943	0.4945	0.4946	0.4948	0.4949	0.4951	0.4952
2.6	0.4953	0.4955	0.4956	0.4957	0.4959	0.4960	0.4961	0.4962	0.4963	0.4964
2.7	0.4965	0.4966	0.4967	0.4968	0.4969	0.4970	0.4971	0.4972	0.4973	0.4974
2.8	0.4974	0.4975	0.4976	0.4977	0.4977	0.4978	0.4979	0.4979	0.4980	0.4981
2.9	0.4981	0.4982	0.4982	0.4983	0.4984	0.4984	0.4985	0.4985	0.4986	0.4986

(Table Contd.)

Table 15.5. (*Contd.*)

z	0	1	2	3	4	5	6	7	8	9
3.0	0.4987	0.4987	0.4987	0.4988	0.4988	0.4989	0.4989	0.4989	0.4990	0.4990
3.1	0.4990	0.4991	0.4991	0.4991	0.4992	0.4992	0.4992	0.4992	0.4993	0.4993
3.2	0.4993	0.4993	0.4994	0.4994	0.4994	0.4994	0.4994	0.4995	0.4995	0.4995
3.3	0.4995	0.4995	0.4995	0.4996	0.4996	0.4996	0.4996	0.4996	0.4996	0.4997
3.4	0.4997	0.4997	0.4997	0.4997	0.4997	0.4997	0.4997	0.4997	0.4997	0.4998
3.5	0.4998	0.4998	0.4998	0.4998	0.4998	0.4998	0.4998	0.4998	0.4998	0.4998
3.6	0.4998	0.4998	0.4999	0.4999	0.4999	0.4999	0.4999	0.4999	0.4999	0.4999
3.7	0.4999	0.4999	0.4999	0.4999	0.4999	0.4999	0.4999	0.4999	0.4999	0.4999
3.8	0.4999	0.4999	0.4999	0.4999	0.4999	0.4999	0.4999	0.4999	0.4999	0.4999
3.9	0.5000	0.5000	0.5000	0.5000	0.5000	0.5000	0.5000	0.5000	0.5000	0.5000

and $z_2 = (x_2 - \mu)/\sigma$. Thus, given a Normal distribution whose mean is 15.2, say, and whose standard deviation is 1.6, the probability of a reading being found between 16.0 and 17.6, for example, would be equal to the area under the standard curve between $z = (16.0 - 15.2)/1.6$ and $z = (17.6 - 15.2)/1.6$, i.e. between 0.5 and 1.5. From Table 15.5 this area is $0.4332 - 0.1915$ which is 0.2417.

Special areas under the Normal curve. Particular areas under the Normal probability curve are used and referred to so often that it is perhaps appropriate to give special mention to them.

Let us consider a typical Normal distribution whose variable x is distributed with mean μ and standard deviation σ. The area under the Normal probability curve from $x = \mu - \sigma$ to $x = \mu + \sigma$, i.e. from $z = -1$ to $z = 1$ under the standard curve, is seen from Table 15.5 to be 0.6826. This means that in any Normal distribution 68.26% (about $\frac{2}{3}$) of the data occur within one standard deviation of the mean. This helps throw light on the fact that in many commonly occurring distributions it is often the case that about $\frac{2}{3}$ of the data occur within this range. If now we consider the area under the normal curve from $x = \mu - 2\sigma$ to $x = \mu + 2\sigma$, i.e. between $z = \pm 2$, this will be found to be 0.9545. Thus in any Normal distribution we can conclude that about 95% of readings occur within two standard deviations of the mean. Again, this proportion checks with what is observed in many commonly occurring distributions. Finally when we look at the area under the Normal curve between $x = \mu + 3\sigma$ and $x = \mu - 3\sigma$, i.e. between $z = \pm 3$, this turns out to be 0.9973, so that in any Normal distribution virtually all the data occur within three standard deviations of the mean. This agrees with the much used rule of thumb that the range is equal to approximately six standard deviations for many distributions.

Significance levels and confidence intervals. If we look at the readings in Table 15.5 we see that whilst 95.45% of the data do indeed occur within two standard deviations of the mean, exactly 95% of readings happen to fall within 1.96 standard deviations of the mean. Any data value, therefore, differing by more than 1.96 standard deviations from the mean could be said to be significantly different from the body of data in that it is not one of the middle 95% of values. In statistical terms we would say that the occurrence of such a reading was *significant at the 5% level*. From the tables we can see that for a reading to be *significant at the 1% level* its deviation from the mean must be at least 2.576 standard deviations. This concept of *level of significance* is fundamental in statistics and has relevance not only in connection with the Normal distribution. In fact, the occurrence of any reading can be said to be significant at the $\alpha\%$ level of significance if the chances of its occurring at random were less than $\alpha/100$. The two most commonly used levels of significance in statistics are the 5% and 1% levels. We shall be meeting these again in the next chapter in connection with the testing of hypotheses.

Levels of significance lead naturally onto the idea of *a confidence interval*. Since 95% of the data in a Normal distribution occur within 1.96σ of μ, then we can be 95% confident that a data value in the distribution, chosen at randon, will occur within the limits $\mu \pm 1.96\sigma$. The interval given by

$$\mu - 1.96\sigma \leqslant x \leqslant \mu + 1.96\sigma \tag{23}$$

is known as the *95% confidence interval* for readings in a Normal distribution. The *99% confidence interval* will thus be the interval

$$\mu - 2.576\sigma \leqslant x \leqslant \mu + 2.576\sigma \tag{24}$$

These confidence intervals can be used in a reverse way so to speak. In many situations it is the case that we have a reading x, obtained from a Normal distribution, and from it we want to deduce something about μ, the mean of the distribution. If we know for certain that the value of the standard deviation is σ, then from the expression for the 95% confidence interval in equation (23) it does not take long to deduce that if we repeatedly obtained readings x, then 95% of the time μ would be included in the range $x \pm 1.96\sigma$. Thus, given a single reading x, we can be 95% sure that μ will lie within the range $x \pm 1.96\sigma$. Similarly, we could be 99% sure that μ would be in the range $x \pm 2.576\sigma$. Thus the 95% and 99% confidence intervals on μ, given a single reading x, are respectively,

$$x - 1.96\sigma \leqslant \mu \leqslant x + 1.96\sigma \tag{25}$$

and

$$x - 2.576\sigma \leqslant \mu \leqslant x + 2.576\sigma \tag{26}$$

Normal probability graph paper. In the previous chapter we saw how most cumulative frequency polygons when plotted out had a typical elongated S-shape. A cumulative plot for data that come from a Normal distribution is no exception. If however, we were to plot out the data on *Normal probability graph paper*, as opposed to ordinary graph paper, we would obtain not an S-shaped plot but a straight-line one. Thus making a cumulative frequency plot on Normal probability paper can be used as a quick test to see whether or not experimental data come from a Normal distribution.

Normal probability graph paper is shown in Fig. 15.5. To cater for all distributions the vertical non-linear scale of the graph paper is arranged for plots

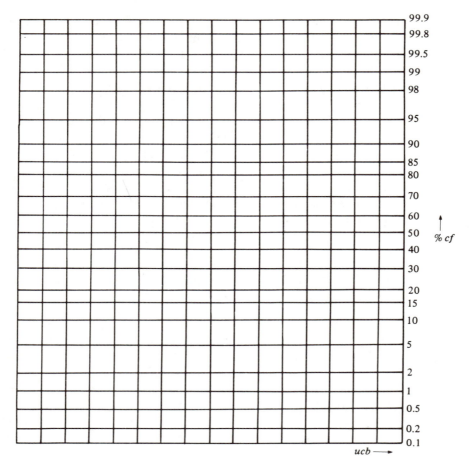

Fig. 15.5. Normal probability graph paper.

of percentage cumulative frequency, whilst the horizontal linear scale is for upper class boundary values.

We should note that the percentage cumulative frequency values of 0 and 100 are not catered for on the vertical scale. This is because in theory the upper class boundary values corresponding to 0 and 100% are the values $-\infty$ and $+\infty$ respectively. Since these cannot be catered for on the linear scale, the percentage cumulative frequency values of 0 and 100% are therefore missing from the vertical scale. This means, for example, that no attempt should be made to plot the last point (of 100%) in any test for Normality in data using this graph paper.

If we are testing some data for Normality, and we do obtain a straight-line plot, then as well as confirming that the data are from a Normal distribution, the plot can also be used to provide us with quick estimates of the mean and standard deviation of the Normal distribution from which the data came. With reference to Fig. 15.6 we see that the mean of the Normal distribution will correspond to a

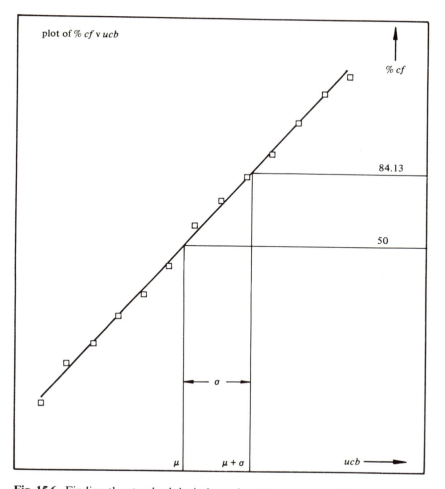

Fig. 15.6. Finding the standard deviation using Normal probability graph paper.

percentage cumulative frequency of 50. This is simply because the Normal distribution is symmetrical and its median is therefore equal to its mean. Also, with regard to the standard deviation, we know that in any Normal distribution 34.13% of readings occur between the mean and one standard deviation above the mean, i.e. between μ and $\mu + \sigma$. Thus if we add 34.13% to the 50% cumulative frequency reading, so that we obtain a percentage cumulative frequency of 84.13%, then this will correspond to the upper class boundary reading of $\mu + \sigma$. The standard deviation σ is then easily read off from the plot as shown in Fig. 15.6.

In the Further Worked Examples we shall show that the alcohol in water data given in Table 15.2 are in fact from a Normal distribution and we will also give estimates of the mean and standard deviation of this normal distribution from the

Normal probability plot. Finally, let us conclude this Further Developments section by looking at the binomial to Normal limit.

The binomial to Normal limit. Suppose we want to predict the probability of obtaining between 45 and 55 heads when we spin a coin 100 times. All we need do is calculate the various probabilities of 45, 46, ... up to 55 heads using the binomial distribution and add these together to obtain the required probability. With a little care, and using a scientific calculator for example, we should obtain 0.7287 as the value of this probability. However, had we wanted to predict the probability of obtaining between say 480 and 520 heads when we spin a coin 1000 times, it would have been virtually impossible to do this calculation in the same way.

In scientific applications situations do arise where we have to calculate such probabilities and fortunately we can make use of the Normal distribution to simplify the calculations involved. It can be shown – but we will not do this here – that whenever the number of trials n becomes large, but at the same time the probability of success p is not too small, then the probability of obtaining r successes in the binomial distribution is well approximated by the probability of finding a reading between $r - \frac{1}{2}$ and $r + \frac{1}{2}$ in the corresponding Normal distribution. By 'the corresponding Normal distribution', we mean the one with mean μ equal to np and standard deviation σ equal to $\sqrt{(npq)}$.

Thus in our example on spinning a coin 100 times, the required probability of observing between 45 and 55 heads is going to be the same as the probability of observing a reading in the range 44.5 to 55.5 in the Normal distribution whose mean is 50 and whose standard deviation is 5. This in turn is equal to the area under the standard curve between $z = (44.5 - 50)/5$ and $z = (55.5 - 50)/5$, i.e. between $z = \pm 1.1$. Consulting the standard normal tables the probability is thus equal to 0.7286 – which is virtually identical to what we calculated initially. In the case of spinning a coin 1000 times, check that the probability of obtaining between 480 and 520 heads is 0.8052. More examples on the binomial to Normal limit are considered in the next section.

15.5 Further worked examples

Example 1.

    (i) Explain why the Poisson distribution is a good model to take to describe the distribution of events in a given time period of interest when the occurrence of events is random in time.

    (ii) With reference to the radiation decay problem use the Poisson distribution to predict the probability of observing 13 or more particles in another 10-s exposure to the source.

Solution. (i) Suppose we are interested in a particular event and the number of times it occurs during a given time period t. Let λ be the average number of occurrences of this event in the time period. Suppose also that we divide t up into a large number n of small time intervals which are so short that the only possibilities are that the event either happens or it does not; multiple occurrences of the event are excluded. If the occurrence of the event is truly random, then the probability of occurrence of the event in each small time interval will be the same – equal to p say. Clearly if λ is not too large in relative terms, p will be very small. Thus the act of observing how many times the event occurs in time t is therefore equivalent to carrying out n experiments, where n is large, and where the probability of occurrence of the event in each experiment, namely p, is very small. We thus have the conditions necessary for the Poisson distribution to be appropriate.

(ii) In the previous chapter we worked out that the mean of the 500 readings making up the radiation decay data was 4.636 counts. In view of what we have just said about the distribution of events whose occurrence is random in time, we could argue that the number of particles entering the geiger counter in a 10-s period will be Poisson distributed. Thus taking 4.636 as the value of λ in the Poisson distribution we can model the probability of r particles by $4.636^r \exp(-4.636)/r!$ The probability of obtaining 13 or more counts is thus going to be $P(13) + P(14) + \cdots$, which can be worked out at $1 - (P(0) + P(1) + \cdots + P(12))$. Thus:

$$P(13 \text{ or more}) = 1 - \exp(-4.636)\left(1 + 4.636 + \frac{4.636^2}{2!} + \cdots + \frac{4.636^{12}}{2!}\right)$$

$$= 1 - 0.998\,95$$

$$= 0.001\,05$$

Correct to three decimal places this probability is 0.001. This small probability explains why 13 or more counts did not appear in the raw data list in Table 15.1.

Instead of carrying out the calculation as we have done, we could have saved time and used the Poisson chart. With a λ value of 4.636, the probability on the left-hand vertical axis, corresponding to the 13th curve, is 0.001, which is in good agreement with the calculated value.

Example 2. The probability of a seed germinating is 0.7. If a batch of 40 000 seeds is planted under proper conditions for germination find:

    (i) the 95% confidence interval for the number of seeds germinating;
    (ii) the lower limit above which we can be 95% sure that the number of seeds germinating will lie.

Solution. (i) The binomial distribution is appropriate here since each seed either germinates or fails to germinate. However, with such a large value of n, namely 40 000, it will be easier to make use of the properties of the binomial to Normal limit.

With $p = 0.7$ and therefore $q = 0.3$ we see that the relevant Normal distribution will have mean μ equal to 28 000 seeds and standard deviation σ equal to $\sqrt{(28\,000)} \times 0.3$, which is 91.652 seeds. The 95% confidence interval for readings x in such a Normal distribution is

$$28\,000 - 1.96 \times 91.652 \leqslant x \leqslant 28\,000 + 1.96 \times 91.652$$

i.e.
$$27\,820.4 \leqslant x \leqslant 28\,179.6$$

Remembering what we said about how the Normal and binomial distributions are related, we see that this interval has to equal
$$r_1 - \tfrac{1}{2} \leqslant x \leqslant r_2 + \tfrac{1}{2}$$
where r_1 and r_2 are integers. Thus we see that $r_1 = 27\,820$ and $r_2 = 28\,180$ if we are to include at least the middle 95% of readings. The 95% confidence interval on the number r of seeds germinating is therefore given by
$$27\,820 \leqslant x \leqslant 28\,180$$

(ii) If we consult Normal distribution tables we see that the upper 95% of results in any Normal distribution occur above the reading $\mu - 1.645\sigma$. Here that reading has the value $28\,000 - 1.645 \times 91.652$, which is $27\,849.2$. In terms of the corresponding binomial distribution this limit corresponds to $r - \tfrac{1}{2}$ where r is an integer. Thus r must equal $28\,849$. We can thus be 95% sure that the number of seeds germinating exceeds $28\,849$.

Example 3.

(i) Using the Poisson model employed in example 1 (ii) in connection with the radiation decay example, estimate the probability of observing fewer than 4 counts in a 10-s exposure to the radioactive source. Compare this result with the figure of 0.33 obtained directly from the raw data.

(ii) Using the probability predicted by the Poisson model, estimate the probability of there being fewer than 4 counts in at least 4 out of the next ten exposures to the radioactive source.

Solution. (i) The Poisson model used in example 1 (ii) gave the probability of observing r counts in a 10-s exposure as $P(r) = 4.636^r \exp(-4.636)/r!$ The probability of observing fewer than 4 counts can be estimated therefore as $P(0) + P(1) + P(2) + P(3)$. This is given by
$$\exp(-4.636)(1 + 4.636 + 4.636^2/2! + 4.636^3/3!)$$
which is 0.320 correct to three decimal places. This is not too different from the figure of 0.33 obtained directly from the raw data. The Poisson model would seem to provide a reasonable description of the radiation decay problem.

(ii) In each 10-s exposure to the radio active source the geiger counter will either record fewer than 4 counts or 4 or more counts. We thus have a situation where the binomial distribution should be applicable. If the probability of fewer than 4 counts is taken as 0.32 then the probability of 4 or more counts will be 0.68. The probability of observing r occasions out of 10 on which fewer than 4 counts occurs is thus given by $P(r) = {}^{10}C_r \, 0.32^r \, 0.68^{10-r}$.

We require the probability of observing fewer than 4 counts on at least four occasions, i.e. we require the sum of probabilities given by $P(4) + P(5) + P(6) + P(7) + P(8) + P(9) + P(10)$. This is easier to work out as $1 - P(0) + P(1) + P(2) + P(3)$.

Noting that
$$P(0) = {}^{10}C_0\, 0.32^0\, 0.68^{10} = 0.021\,14$$
that
$$P(1) = {}^{10}C_1\, 0.32^1\, 0.68^9 = 0.099\,48$$
that
$$P(2) = {}^{10}C_2\, 0.32^2\, 0.68^8 = 0.210\,66$$
and
$$P(3) = {}^{10}C_3\, 0.32^3\, 0.68^7 = 0.264\,36$$

we see that the required probability is $1 - 0.595\,64$, which is 0.404 correct to three decimal places.

Example 4. Show by making a cumulative frequency plot on Normal probability

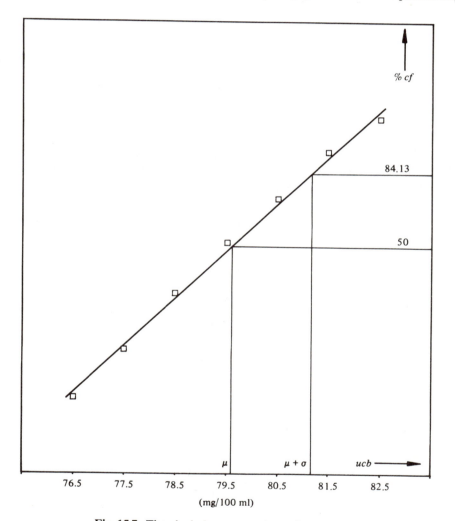

Fig. 15.7. The alcohol concentrations of Table 15.2.

paper that the alcohol concentrations in Table 15.2 are from a Normal distribution
Estimate from the plot the value of the mean and standard deviation of this Normal
distribution.

Solution. Taking the data in Table 15.2 and drawing up a cumulative frequency
table we obtain:

Class (mg/100 ml)	Frequency	ucb (mg/100 ml)	cf	%cf
75.5–76.5	1	76.5	1	2
76.5–77.5	3	77.5	4	8
77.5–78.5	9	78.5	13	26
78.5–79.5	13	79.5	26	52
79.5–80.5	11	80.5	37	74
80.5–81.5	8	81.5	45	90
81.5–82.5	3	82.5	48	96
82.5–83.5	2	83.5	50	100

Plotting out % cumulative frequency against upper class boundary on Normal
probability paper leads to a straight-line plot as shown in Fig. 15.7. It thus follows
that the data are from a Normal distribution. The various alcohol concentrations
are therefore Normally distributed.

From the plot we see that the mean μ of this Normal distribution is
79.60 mg/100 ml and the value of σ is approximately 1.55 mg/100 ml.

15.6 Exercises

1. Standard amounts of five different insecticides are found to kill
   30%, 45%, 65%, 85% and 90% respectively of a fixed size of insect
   population. If one of the insecticides is chosen at random, what is the
   probability that it will kill:
   (i) at least 65% of the insect population?
   (ii) at most 45% of the insect population?
   (iii) between 40% and 80% of the insect population?
   If two of the insecticides are chosen at random, what is the probability
   that any one of the pair chosen will kill at least 85% of the insect
   population?

2. As part of the safety procedure at an oil refinery a fractionating
   column is monitored by three independent warning systems, A, B and C.
   The probabilities that on any given day the warning systems will fail

are $0.1, 0.01$ and 0.05 for A, B and C respectively. If $P(n)$ denotes the probability that n warning systems fail on a given day, obtain $P(0), P(1), P(2)$ and $P(3)$. Hence obtain the expected number of warning systems which fail on any given day.

3. A particular scientific monitoring device is powered by three batteries of the same type. The device is designed so that it will continue to function provided at least two of the three batteries function normally. The three batteries are renewed at a particular time. The probability that a battery will fail within the first 50 hours of operation is 0.25. The probability that a battery will fail within the first 100 hours of operation is 0.70. Determine the probability that:

   (i) the device fails due to battery failure within the first 50 hours of duration;

   (ii) the batteries allow the equipment to operate for longer than 100 hr;

   (iii) the device fails due to battery failure within 50–100 hr of duration.

4. A science class has 45 students in it. Derive a formula for calculating the probability that two or more students in the group have birthdays in common. (Hint: what is the probability that no students share common birthdays?)

   Write a pseudocode algorithm to evaluate the required probability.

5. (a) Show the following:

   (i) If two cards are drawn at random from a well shuffled pack, the probability of drawing two aces is $\frac{1}{221}$;

   (ii) The chance of throwing a 6 at least once in two throws of a die is $\frac{11}{36}$;

   (iii) If eight coins are thrown simultaneously, the chance of obtaining six heads is $\frac{37}{256}$.

   (b) A and B toss a coin alternately on the understanding that the first to obtain 'heads' wins the toss. A tosses first. Show that her probability of winning is $\frac{2}{3}$.

   (c) Three people A, B and C are playing a game for a prize given to the first one who obtains 'heads'. They toss a coin in succession, A tossing first. Show that A's chance of winning is $\frac{4}{7}$. Find the chances of the other two winning.

6. (a) The random variable X can take the values $-10, -20, +30$ with respective probabilities $\frac{1}{5}, \frac{3}{10}, \frac{1}{2}$. Find the expectations $E(X), E(X^2)$, $E(X - \bar{X})^2, E(X^3)$ where $\bar{X} = E(X)$.

   (b) A bag contains two white balls and three black balls. Four persons A, B, C and D in the order named, each draws one ball and does not replace it. The first to draw a white ball receives £1000. Determine the amount each should expect to receive.

7. (a) The lifetime x of a mass produced component is a continuous random variable with a probability density function given by $\rho(x) = k^2 x e^{-kx}$, where k is a positive constant and $x \geq 0$. Show that:

   (i) the expected life-time is $2/k$;

   (ii) the variance is $2/k^2$;

(iii) the probability that a randomly selected component will have a lifetime greater than T is $(1 + kT)e^{-kT}$.

(b) Particles are emitted from a radioactive source. If a collector is switched on at time $t = 0$, then the probability density function for the time of the first emission is given by $\rho(t) = e^{-t/\theta}/\theta$ where θ is a positive constant. Show that:

(i) the probability of a particle being emitted in the interval $[0, T]$, where T is a positive constant, is $1 - e^{-T/\theta}$;

(ii) given that a single particle is emitted in $[0, T]$, the probability density function for the time t of this emission is

$$\frac{e^{-t/\theta}}{\theta(1 - e^{-T/\theta})}$$

where $0 < t < T$.

8. (a) Find the area under the standard normal curve between the following limits:

(i) -1.23 to 2.4

(ii) -1.5 to 0.6

(iii) 0.1 to 2.7

(iv) $-\infty$ to -2.9

(v) 2.31 to ∞

(b) A random variable X is normally distributed with mean 21.82 and standard deviation 5.4. Find the probability that:

(i) $X < 20$

(ii) $19.2 < X < 21.5$

(iii) $20 < X < 22$

(iv) $X > 23.2$

9. A nuclear power station is required to monitor the level of radioactivity of the water it discharges into the sea. The probability that the level will exceed a specified base level on any given day is 0.05. Determine the probability that in a seven day period the base level will be exceeded:

(i) on none of the seven days;

(ii) on at least one of the seven days;

(iii) on exactly three of the seven days;

(iv) on three or more of the seven days.

10. On average 2% of the floppy discs produced by a manufacturer are defective. A customer orders a consignment of 150 discs.

(a) In order to test the validity of the Poisson approximation to the binomial distribution, use both distributions to determine the probability that exactly one of the 150 discs is defective. Compare your answers.

(b) Use the Poisson formula and also the Poisson chart to determine the probability that at least three of the discs are defective. Compare your results.

11. A particular type of casting produced by a foundry has on average 4.5 randomly occurring surface defects.

(a) Calculate the probability that a casting has:
  (i) exactly three defects;
  (ii) more than five defects.

(b) A consignment consists of ten castings: Calculate the probability that three or more of the castings will have more than five surface defects.

12. A laboratory uses on average six bottles of a standard potassium nitrate solution each week. The stock is replenished at the beginning of each week. Calculate the minimum number of bottles which should be in stock at the beginning of each week in order to be at least 95% confident that the stock will be sufficient to meet the week's demand.

13. Tests on the shells of 200 subtidal barnacles yielded the following distribution of breaking strengths:

Breaking pressure $\times 10^{-5}$ (Nm^{-2})	No. of shells
4.2–4.6	8
4.6–5.0	26
5.0–5.4	58
5.4–5.8	60
5.8–6.2	34
6.2–6.6	12
6.6–7.0	2

Show that it is reasonable to assume that the above data are from a population that is Normally distributed. Estimate the mean and standard deviation of this population.

14. (a) In a specially-bred strain of drosophilia fly, the probability of a fly having a certain wing shape is 0.3. If a random sample of ten flies is taken, what is the probability of finding three or more flies with this particular wing shape?

(b) The % nickel in random samples of an alloy is found to be Normally distributed about a mean of 10.05% with a standard deviation of 0.25%. What proportion of alloy samples is expected to contain:
  (i) above 10.50% nickel:
  (ii) below 9.50% nickel:
  (iii) between 9.50% and 10.50% nickel?

(c) Manufacturers of the alloy guarantee the nickel content to be within the range 9.50–10.50%. In a batch of 100 alloy samples what is the expected number of samples failing the guarantee?
Estimate, using the Poisson chart, the probability that ten or more samples fail the guarantee.

15. (a) The percentage of carbon in batches of a mixed powder is known

to be Normally distributed about a mean of 4.56% with a standard deviation of 0.12%. Find the range of percentages within which the middle 50% of readings are expected to lie and estimate the fraction of readings having a percentage carbon content below 4.40%.

(b) The following data refer to readings that are known to be Normally distributed:

class	frequency
below 10	30
10–20	24
20–30	22
30–40	15
above 40	9

By using normal probability paper, estimate the value of the mean and standard deviation of this Normal distribution.

16. (a) The level of lead in water is monitored daily by analysing a sample of water. On average 10% of the daily samples have a lead concentration in excess of a given value.
    Calculate the probability that on five consecutive days
    (i) this given value is not exceeded in any of the five samples,
    (ii) exactly 1 sample has a lead concentration in excess of this value,
    (iii) the level is exceeded in more than 1 sample.

    (b) The amount of iron in samples of a given alloy is Normally distributed about a mean of 28.25% with a standard deviation of 0.05%. If 100 random samples of the alloy are analysed, use the Poisson chart to establish the probability of four or more samples having an iron content below 28.15%.

17. If 500 drosophilia flies, of the type mentioned in Exercise 14(a), are inspected, and their wing shape noted, estimate the probability that the number of flies, having the wing shape referred to, is
    (i) less than 140;
    (ii) more than 170;
    (iii) between 135 and 165.
    Obtain the 95% and 99% confidence limits for the number of flies possessing this wing shape.

18. Small forgings are bought by a firm for machining. The forgings arrive in batches and to check the quality of the batch a sample of 200 forgings from each batch is inspected. If there are nine or more defectives in the sample the entire batch is rejected. If there are eight or less defectives, the batch is accepted. Using the Poisson chart make a plot of the probability of acceptance of the batch against the defective rate in the batch.

If it has been agreed, between the producer of the forgings and the firm, that up to 4% defectives will be acceptable, use your plot to obtain

(i) the risk to the producer of the batch being needlessly rejected when the rate of defectives is 3%,

(ii) the risk to the firm of accepting a batch when the rate of defectives is 5%.

16

Statistics 3: sampling, sampling distributions and hypothesis testing

Contents: 16.1 scientific context – 16.1.1 estimating the value of the acceleration due to gravity; 16.1.2 alcohol in blood analysis; 16.1.3 iron alloy analysis; 16.2 mathematical developments – 16.2.1 definitions; 16.2.2 the distribution of sample means; 16.2.3 the distribution of the difference in two sample means; 16.2.4 best estimates; 16.2.5 sampling from a Normal distribution; 16.2.6 the t-distribution; 16.3 worked examples; 16.4 further developments – 16.4.1 the F-distribution; 16.4.2 elements of hypothesis testing; 16.4.3 the t-test; 16.4.4 the F-test; 16.5 further worked examples; 16.6 exercises.

In Chapter 15 we looked at the basics of probability and at some of the important probability distributions that arise in the handling of scientific data. We did this because, as we stated, a study of probability is first needed if we are properly to address the problem of expressing numerically the inevitable uncertainties that exist in the results of statistical analyses. In this final chapter on statistics we consider the problem of quantifying the uncertainty associated with the results of statistical analyses and look specifically at sampling and at just what can be deduced about a population from only a finite, and often small, sample of results.

16.1 Scientific context

16.1.1 Example 1: estimating the value of the acceleration due to gravity. The time period t of a simple pendulum of length l is given by the well-known expression $t = 2\pi \sqrt{(l/g)}$, where g is the acceleration due to gravity. By observing the time for, say, 50 oscillations of the pendulum, for each of a set of given l values, a plot of t^2 versus l can be drawn up which should be a straight line of slope $4\pi^2/g$. By measuring the gradient it is thus possible to obtain a reasonably accurate estimate of g in a fairly straightforward way.

Suppose, then, that this experiment has been repeated five times and the following five estimates of g (ms^{-2}) are obtained:

$$9.803 \quad 9.807 \quad 9.811 \quad 9.809 \quad 9.806$$

The simple question is what can we deduce about the actual value of the acceleration due to gravity from this sample of five readings?

From what we said in Chapter 15, concerning the Normal distribution, we might

430

argue that the various estimates of g above come from a Normal distribution whose mean is g. Thus, if we knew what the standard deviation of this Normal distribution was, we could choose any of the five readings at random and work out a 95% confidence interval for the actual value of g.

Question: Instead of choosing a single reading at random, would it not be more sensible to incorporate the mean of these five readings into the calculation of a confidence interval?

Answer: Yes it would. The confidence interval would in fact be smaller and we could make a more definite statement about the value of g.

To work out how much smaller the confidence interval would be requires that we know something about *sampling theory* and about how sample means, for example, are distributed. These topics are considered in the next section and later in 16.4.

Now, in working out a confidence interval for g, we must bear in mind the fact that we might not know the standard deviation of the Normal distribution from which these five readings are supposed to have come. If this is the case, then clearly we have to estimate its value from the sample.

Question: What implication does this have for the actual confidence interval for g?

Answer: The confidence interval would obviously widen slightly because of the added uncertainty in the value of the standard deviation.

Just how we would estimate the value of the standard deviation from the sample and by how much the confidence interval would widen are questions whose answers we will find in the next section and in Section 16.4.

16.1.2 Example 2: alcohol in blood analysis. For our second example, let us return to the problem we mentioned right at the beginning of Chapter 14, where a sample of blood was being analysed for alcohol content. Suppose two techniques, namely Conventional Gas Chromatography (CGC) and Head-Space Gas Chromatography (HSGC), are used and the results of the two methods are as shown in Table 16.1.

A little bit of arithmetic tells us that the mean of the six CGC readings is 75.633 mg/100 ml, whereas the mean of the eight HSGC readings is 74.563 mg/100 ml – a difference of 1.070 mg/100 ml. One obvious question that must be answered is whether such a difference actually means anything as far as the two methods of analysis are concerned. In other words, is one method of analysis giving results systematically higher than the other, or is a difference of this magnitude to be expected, given the sizes of the samples we are dealing with and the variability present in the readings of the two samples? Naturally we would like the observed difference to be in no way significant and simply the result of working with small samples because, after all, the same blood sample was analysed by both techniques and the results should be compatible.

Comparing one sample of results with another to see if the difference in sample

Table 16.1. *CGC and HSGC measurements of alcohol content*

CGC mg/100 ml:	76.9, 75.8, 76.1, 75.2, 74.7, 75.1.
HSGC mg/100 ml:	74.6, 74.7, 74.2, 75.2, 74.5, 73.9, 74.6, 74.8.

Table 16.2. *Two laboratory determinations of iron content*

Lab A:	28.91, 28.73, 28.13, 28.35, 28.40, 28.21, 28.36, 28.12, 28.72, 28.62
Lab B:	28.67, 28.52, 28.23, 28.46, 28.42, 28.25, 28.36, 28.71, 28.30, 28.28

means is of any significance requires us to know something about how the differences in two sample means are distributed. This will be considered in the next section and also in 16.4.

16.1.3 Example 3: iron alloy analysis. If it turned out that the two methods of analysis in our second example were giving significantly different results, then one way to decide which technique was superior would be to use each method of analysis to determine the known concentration of a synthetic solution of alcohol which had been previously made up specially for the purpose. In this way various aspects of the two methods of analysis, such as their accuracy and relative precision, could then be compared.

Suppose, then, we have a similar situation in alloy analysis, where we have an alloy that has been made up to contain exactly 28.25% iron, which is now being analysed by two different laboratories skilled in alloy analysis. Assume each laboratory, using its own preferred technique of analysis, makes ten determinations of the iron content and that the various results are as shown in Table 16.2. The mean of Lab A's results can be worked out and is 28.455%, whereas that of Lab B's is 28.420%. Both results, therefore, appear to be slightly in error with Lab A's result seeming to be slightly worse than Lab B's. But is this really the case? Is there really sufficient evidence here for us to say, for example, that Lab A's work is definitely worse than Lab B's from the point of view of accuracy? Also, is there anything significant that we can say about the relative precision of the work of the two labs? It does appear, for example, that the spread in readings associated with Lab A is greater than that associated with Lab B, but again does this difference really mean anything?

Question: What, then, do we need to know to be able to make sensible comments about the accuracy and precision of the labs' work?

Answer: We shall see that the first problem, of comparing the work of the labs from the point of view of accuracy, can easily be answered once we know how the means of samples are distributed. But the problem of looking at the relative precision involves us in understanding how the ratio of two variances is distributed.

The ratio of two variances, under certain conditions, belongs to a distribution called the *F-distribution*. This distribution will be considered in Section 16.4 of this chapter.

16.2 Mathematical developments

16.2.1 Definitions. It is necessary, to begin with, to define several quantities that will constantly be used in this chapter.

> A *population* is the totality of readings or counts obtainable when a quantity of interest is repeatedly measured

We have, of course, already met the concept of a population before in Chapter 15. It is important to realise that in most cases a population is a hypothetical concept, since only rarely can we ever obtain the totality of readings possible in a given experimental situation.

> A *sample* is a set of measurements which constitutes part of a population

The main object of taking a sample is to draw some conclusions about the population from which it is obtained. For example, in our first example, the prime object of looking at the sample of five readings is to obtain an estimate of the mean of the totality of results possible when g is repeatedly measured. In many experiments the mean of the population and the true value we are trying to measure are one and the same thing. Thus from the sample of five readings, the idea is that we should be able to obtain an estimate of the actual value of the acceleration due to gravity.

What we can deduce about a population depends on securing good samples, which is not always easy and usually requires careful design of experiments. A good sample is a random sample.

> A *random sample* is a sample of readings obtained from a population where every individual reading is as likely to be included in the sample as any other

> A *population parameter* is any numerical characteristic of a population, such as its mean or standard deviation

> A *sample statistic* is a similar quantity calculated from a sample

It is important to note that, in contrast to population parameters, which are fixed for a given population, sample statistics obviously differ from sample to sample.

> A *sampling distribution* is the distribution obtained when any particular sample statistic is calculated repeatedly from different samples drawn from the same population

Important sampling distributions that are frequently used in analysing scientific data are the distribution of sample means and the distribution of the difference in two sample means, both of which we now consider.

16.2.2 The distribution of sample means. Imagine we have a population whose mean is μ and whose standard deviation is σ. This population could well be the totality of results possible in a particular experiment for example. Suppose we choose a sample of size n from the population and work out the mean, m_1. In other words imagine we carry out an experiment n times and work out the mean of the results obtained. Suppose we repeatedly choose different samples, all of size n, to obtain further means $m_2, m_3, \ldots$ and so on. The distribution of means $m_1, m_2, m_3, \ldots$ is what we call the *distribution of sample means*. This sampling distribution possesses two important properties:

(1) The mean μ_m of the distribution is equal to the mean of the population from which the samples were drawn:

$$\mu_m = \mu \tag{1}$$

(2) The standard deviation σ_m is equal to the population standard deviation divided by the square root of the sample size n:

$$\sigma_m = \sigma/\sqrt{n} \tag{2}$$

Deriving these properties is easy if we appeal to the concept of *expectation* that we met in the previous chapter. There we only looked at the expected value of x and of $(x - \mu)^2$, but we can of course work out the expected value of other quantities. In particular we find for two independent, randomly distributed variables x and y that

$$E(ax \pm by) = aE(x) \pm bE(y) \tag{3}$$

where a and b are any constants, and

$$E(xy) = E(x)E(y) \tag{4}$$

Using these general results together with the fact that $E(x) = \mu$ and $E((x - \mu)^2) = \sigma^2$ we can now show how the two properties arise:

Derivation of property (1). The mean μ_m of the sampling distribution will be the expected value of m. But

$$E(m) = E\left(\sum_i x_i/n\right)$$

$$= \frac{1}{n}E\left(\sum_i x_i\right)$$

$$= \frac{1}{n}\sum_i E(x_i)$$

$$= \frac{1}{n}\sum_i \mu$$

$$= \frac{1}{n}(n\mu)$$

$$= \mu$$

it follows, therefore, that $\mu_m = \mu$.

Derivation of property (2). The variance σ_m^2 will be the expected value of $(m - \mu_m)^2$, which, in view of equation (1), will be the same as the expected value of $(m - \mu)^2$. Thus we see that

$$E((m - \mu)^2) = E\left(\left(\sum_i x_i/n - \mu\right)^2\right)$$

$$= E\left(\left(\sum_i \frac{(x_i - \mu)}{n}\right)^2\right)$$

$$= E\left(\sum_i \frac{(x_i - \mu)^2}{n^2} + \sum_{i \neq j} \frac{(x_i - \mu)(x_j - \mu)}{n^2}\right)$$

$$= \frac{1}{n^2}\sum_i E((x_i - \mu)^2) + \frac{1}{n^2}\sum_{i \neq j} E((x_i - \mu)(x_j - \mu))$$

Since $(x_i - \mu)$ and $(x_j - \mu)$ in the second term on the right-hand side are independent as $i \neq j$, and since $E(x_i - \mu) = E(x_j - \mu) = 0$, then the second term disappears and it follows that

$$E((m - \mu)^2) = \frac{1}{n^2}\sum_i ((x_i - \mu)^2)$$

$$= \frac{1}{n^2}(n\sigma^2)$$

$$= \frac{\sigma^2}{n}$$

We see therefore that $\sigma_m = \sigma/\sqrt{n}$.

The quantity σ_m is known as the *standard error of the mean*. Essentially, it tells us how good a sample mean is as an estimate of the actual population mean. Thus, for example, if we have a population whose standard deviation is known to be 4.5 units, and a sample of size 25 is taken from this population, then the resulting sample mean can be expected to be accurate to within a quantity of the order of $4.5/\sqrt{(25)}$ or 0.9 units.

16.2.3 The distribution of the difference in two sample means. Suppose we have two different and independently distributed populations whose respective means and standard deviations are μ, σ and μ', σ'. A sample of size n can be taken from the first to give a mean m_1, and a sample of size n' can be taken from the second to give a mean m_1'. We can thus obtain a difference in sample means equal to $m_1 - m_1'$. Similarly, by further sampling, always taking samples of size n and n' respectively, we can get a distribution of differences $m_1 - m'_1$, $m_2 - m'_2$, $m_3 - m'_3, \ldots,$ etc. This sampling distribution is known as the *distribution of the difference in two sample means*. It possesses two important properties:

(1) The *mean* of this distribution, denoted by $\mu_{m-m'}$, is equal to the *difference* in population means:

$$\mu_{m-m'} = \mu - \mu' \tag{5}$$

(2) The standard deviation, denoted by $\sigma_{m-m'}$ is given by

$$\sigma_{m-m'} = (\sigma_m^2 + \sigma_{m'}^2)^{\frac{1}{2}} = \left(\frac{\sigma^2}{n} + \frac{\sigma'^2}{n'}\right)^{\frac{1}{2}} \tag{6}$$

Deriving these properties follows easily using the concept of *expectation*:

Derivation of property (1). The mean $\mu_{m-m'}$, will be the expected value of $(m - m')$. But the expected value of $m - m'$ is given by

$$E(m - m') = E(m) - E(m')$$
$$= \mu - \mu'$$

It therefore follows that $\mu_{m-m'} = \mu - \mu'$.

Derivation of property (2). The quantity $\sigma_{m-m'}^2$, will be the expected value of $(m - m' - \mu_{m-m'})^2$, which in view of the first property is $(m - m' - (\mu - \mu'))^2$. Thus we have

$$E((m - m' - (\mu - \mu'))^2) = E(((m - \mu) - (m' - m))^2)$$
$$= E((m - \mu)^2 + (m' - \mu')^2 - 2(m - \mu)(m' - \mu'))$$
$$= E((m - \mu)^2) + E((m' - \mu')^2) - 2E((m - \mu)(m' - \mu'))$$

But since the two populations are independently distributed, then $2E((m - \mu)(m' - \mu'))$ will be equal to $2E(m - \mu)E(m' - \mu')$, which is zero. Thus

$$E((m - m' - (\mu - \mu'))^2) = E((m - \mu)^2) + E((m' - \mu')^2)$$
$$= \sigma_m^2 + \sigma_{m'}^2$$

It thus follows that

$$\sigma_{m-m'} = \left(\frac{\sigma^2}{n} + \frac{\sigma'^2}{n'}\right)^{\frac{1}{2}}$$

The quantity $\sigma_{m-m'}$ is known as the *standard error of the difference in sample means*. It gives us some indication of how good the difference in two sample means is as an estimate of the actual difference in population mean values.

Example: What is the standard error to associate with a difference in sample means obtained from populations having standard deviations 1 and 1.5 units using samples of size 10 and 12 respectively?

Solution: The standard error is given by $(\sigma^2/n + \sigma'^2/n')^{\frac{1}{2}}$ i.e. by $(1^2/10 + 1.5^2/12)^{\frac{1}{2}}$, which is 0.54 units.

16.2.4 Best estimates. The properties of the sampling distributions listed above make reference to population standard deviations. It would appear, therefore, that in order to work out an appropriate standard error to associate, for example, with a sample mean, we must first of all obtain the value of the population standard deviation. Strictly speaking this is true; but usually, out of necessity, we get by with estimating the value of the population standard deviation from the readings we are working with. It might appear that the natural thing to do in estimating the value of the population standard deviation σ from a sample is simply to use the sample standard deviation s as the best estimate of σ. Unfortunately, the problem is not as straightforward as this, and the best estimate to take depends on sample size, as we now show.

Estimating σ from a sample of n readings. Suppose we have a sample of n readings taken from a population whose mean is μ and whose standard deviation is σ. Let the sample mean and standard deviation be m and s respectively. If we look at the expected value of s^2 we see that it is not the same as σ^2. Explicitly we have:

$$E(s^2) = E\left(\sum_i (x_i - m)^2/n \right)$$

$$= E\left(\sum_i x_i^2/n - m^2 \right)$$

Now because the value of s^2 is unaltered if the same amount (μ in this case) is subtracted from each reading in the sample (including m), then

$$E(s^2) = E\left(\sum_i (x_i - \mu)^2/n - (m - \mu)^2 \right)$$

$$= \frac{1}{n}\sum_i E((x_i - \mu)^2) - E((m - \mu)^2)$$

$$= \frac{1}{n}\sum_i \sigma^2 - \frac{\sigma^2}{n}$$

$$= \sigma^2 - \frac{\sigma^2}{n}$$

$$= \left(\frac{n-1}{n} \right)\sigma^2$$

We see, therefore, that the expected value of s^2 is not σ^2 but $(n-1)\sigma^2/n$. In other words, on average, s^2 is too small as an estimate of σ^2 by a factor $(n-1)/n$. Turning this relationship round, we see that the best estimate of σ^2 in terms of s^2 is therefore $s^2 [n/(n-1)]$. If we call $\hat{\sigma}^2$ the best estimate of σ^2, then from a single sample of size n we have the very important result that

$$\hat{\sigma}^2 = s^2 \left(\frac{n}{n-1} \right) \tag{7}$$

By taking the square root of this equation we obtain an expression for the quantity

$\hat{\sigma}$ that is used as the best estimate of σ, namely

$$\hat{\sigma} = s\left(\frac{n}{n-1}\right)^{\frac{1}{2}} \tag{8}$$

The factor $(n/n-1)^{\frac{1}{2}}$ is known as *Bessel's correction* and the quantity $\hat{\sigma}$ is what is labelled as σ_{n-1} on many electronic calculators. The key that is labelled as σ_n is what we have called the sample standard deviation, s. Perhaps a word of caution is needed at this point. Just because the expression in (7) is the best estimate of σ^2, it does not follow that the square root of the expression will automatically be the best estimate of σ. The expected value of s, in fact, is not equal to $\sigma[(n-1)/n]^{\frac{1}{2}}$. However, the difference between $E(s)$ and this expression is small enough for us not to worry about the need for small correction terms. Consequently, in estimating the value of the standard error of the mean from a sample, we use the quantity given by

$$\hat{\sigma}_m = \hat{\sigma}/\sqrt{n} = s/\sqrt{(n-1)} \tag{9}$$

Estimating σ by pooling readings from two samples. Many situations arise where we have two samples that come from populations having different means but having the same standard deviation. In our third example we could envisage such a situation arising if one of the labs was sent two alloys to analyse for iron content – one with 28.25% iron in it and the other one with 30.00% iron in it. If the same technique of analysis is used to determine the iron content in each case then we would expect the respective estimates of iron content to come from populations that had the same standard deviation. What we are saying is that we would expect the level of precision to be the same when each set of iron determinations is carried out.

When we have a situation like this it is obviously good practice to *pool* the results of both samples for the purposes of estimating this common standard deviation, for by pooling the results we have effectively a larger set of readings to work with.

Suppose, therefore, we have two samples, one of size n and the other of size n', with respective means m and m', that come from populations having means μ and μ', but having the same standard deviation σ. The mean sum of squares of deviations from the respective means m and m' of the $n + n'$ readings in both samples is

$$s_p^2 = \frac{\sum_i (x_i - m)^2 + \sum_i (y_j - m')^2}{n + n'}$$

where x_i and y_j are the ith and jth readings, respectively, in the two samples and the subscript p on s_p^2 refers to the fact that we have pooled the results of the two samples. In terms of the individual sample variances s^2 and s'^2, s_p^2 can be expressed as

$$s_p^2 = \frac{ns^2 + n's'^2}{n + n'}$$

To see how this quantity relates to σ^2 we look at the expected value of s_p^2 and make use of the result $E(s^2) = [(n-1)/n]\sigma^2$. Explicitly we have

$$E(s_p^2) = E\left(\frac{ns^2 + n's'^2}{n + n'}\right)$$

$$= \frac{n}{n + n'} E(s^2) + \frac{n'}{n + n'} E(s'^2)$$

$$= \frac{n}{n + n'}\left(\frac{n - 1}{n}\right)\sigma^2 + \frac{n'}{n + n'}\left(\frac{n' - 1}{n'}\right)\sigma^2$$

$$= \left(\frac{n + n' - 2}{n + n'}\right)\sigma^2$$

Thus, if we call $\hat{\sigma}_p^2$ the best estimate of σ^2 obtained by pooling the results of the two samples, it follows that

$$\frac{ns^2 + n's'^2}{n + n'} = \left(\frac{n + n' - 2}{n + n'}\right)\hat{\sigma}_p^2$$

from which we obtain the important result that

$$\hat{\sigma}_p^2 = \frac{ns^2 + n's'^2}{n + n' - 2} \tag{10}$$

The quantity, therefore, that is used as the best estimate of the population standard deviation, when results have been pooled, is the square root of this expression, namely

$$\hat{\sigma}_p = \left(\frac{ns^2 + n's'^2}{n + n' - 2}\right)^{\frac{1}{2}} \tag{11}$$

In working out the best estimate of the standard error to associate with a difference in two sample means, we should note that there are now two choices. If we choose not to pool the results, then the best estimate, $\hat{\sigma}_{m-m'}$, of the standard error will be given by

$$\hat{\sigma}_{m-m'} = (\hat{\sigma}_m^2 + \hat{\sigma}_{m'}^2)^{\frac{1}{2}} = \left(\frac{s^2}{n - 1} + \frac{s'^2}{n' - 1}\right)^{\frac{1}{2}} \tag{12}$$

If, on the other hand, we choose to pool the results, for the reasons given above, then the expression for the standard error will now be

$$\hat{\sigma}_{m-m'} = (\hat{\sigma}_m^2 + \hat{\sigma}_{m'}^2)^{\frac{1}{2}} = \left(\frac{\hat{\sigma}_p^2}{n} + \frac{\hat{\sigma}_p^2}{n'}\right)^{\frac{1}{2}}$$

Taking $\hat{\sigma}_p$ out as a factor and replacing it by the expression involving s and s' we have

$$\hat{\sigma}_{m-m'} = \left(\frac{ns^2 + n's'^2}{n + n' - 2}\right)\left(\frac{1}{n} + \frac{1}{n'}\right)^{\frac{1}{2}} \tag{13}$$

16.2.5 Sampling from a Normal distribution. The nature of sampling distributions, such as the distribution of sample means and the distribution of the difference in two sample means, is fortunately well understood when the populations from which the samples come happen to be Normal distributions. Many populations, it can be argued, are Normal distributions especially when the only source of variation is random factors of one kind or another. Consequently, the results of many measuring processes in science will form samples from Normally distributed populations. What follows, therefore, in this section on sampling from a Normal distribution, will be applicable to many situations where data are collected in science.

The nature of the distribution of sample means. Suppose we have a Normally distributed population with mean μ and standard deviation σ. We know that if samples of size n are repeatedly taken from the population then the resulting distribution of sample means will have mean equal to μ and standard deviation equal to $\sigma/\sqrt{n}$. This result is true irrespective of the nature of the parent population. What is important is the fact that if the parent population is Normally distributed, then the distribution of sample means will also be Normally distributed. In other words if a sample mean m is chosen at random from the distribution of sample means the quantity given by

$$z = \frac{m - \mu_m}{\sigma_m} = \frac{m - \mu}{\sigma/\sqrt{n}} \tag{14}$$

will belong to the standard Normal distribution that we spoke about in the previous chapter.

The proof of this fact is outside the scope of this book and cannot therefore be given here. The result does mean, however, that we can now work out a confidence interval for the population mean. We can do this by noting that since the distribution of sample means is Normally distributed about μ with a standard deviation of $\sigma/\sqrt{n}$, then 95% of the time a sample mean will be found in the range $\mu \pm 1.96\sigma/\sqrt{n}$. Thus given a sample whose mean is m, 95% of the time the interval $m \pm 1.96\sigma/\sqrt{n}$ will include μ. A 95% confidence interval for the population mean is therefore given by

$$m - 1.96\sigma/\sqrt{n} \leqslant \mu \leqslant m + 1.96\sigma/\sqrt{n} \tag{15}$$

A 99% confidence interval for μ is therefore

$$m - 2.576\sigma/\sqrt{n} \leqslant \mu \leqslant m + 2.576\sigma/\sqrt{n} \tag{16}$$

The nature of the distribution of the difference in two sample means. Suppose now we have two Normally distributed populations, the first having mean μ and standard deviation σ, and the second having mean μ' and standard deviation σ'. The distribution of the difference in sample means is also a Normal distribution with mean $\mu_{m-m'}$ equal to $\mu - \mu'$ and standard deviation $\sigma_{m-m'}$ equal to $(\sigma^2/n + \sigma'^2/n')^{\frac{1}{2}}$. Thus any difference $m - m'$, chosen at random from the distribution of differences, will be such that the quantity

$$z = \frac{(m - m') - \mu_{m-m'}}{\sigma_{m-m'}} = \frac{(m - m') - (\mu - \mu')}{(\sigma^2/n + \sigma^2/n')^{\frac{1}{2}}} \tag{17}$$

belongs to the standard Normal distribution. Therefore, given two samples of size n and n', having means m and m', and coming from respective populations having means μ and μ', a 95% confidence interval for $\mu - \mu'$ is given by:

$$(m - m') - 1.96\sigma_{m-m'} \leqslant \mu - \mu' \leqslant (m - m') + 1.96\sigma_{m-m'} \qquad (18)$$

A 99% confidence interval is thus

$$(m - m') - 2.576\sigma_{m-m'} \leqslant \mu - \mu' \leqslant (m - m') + 2.576\sigma_{m-m'} \qquad (19)$$

16.2.6 The *t*-distribution. When sampling from a Normal distribution it is clear, therefore, that if we know the value of the population standard deviation, we should know exactly what to expect from the sample we are working with. Unfortunately, in most situations where data are being collected, we do not know the value of σ exactly, and have to work with $\hat{\sigma}$, the best estimate of σ from the sample. This now involves us working not with the Normal distribution, but with a new distribution called the *t*-distribution.

The t-value associated with a sample mean. Consider, then, the situation where we have a sample of size n obtained by sampling fom a Normally distributed population. We know that the quantity $z = (m - \mu)/\sigma_m$ would be Normally distributed, but in the absence of knowing the exact value of σ_m, we have to be content to work with a so-called *t*-value given by

$$t = \frac{m - \mu}{\hat{\sigma}_m} = \frac{m - \mu}{s/\sqrt{(n - 1)}} \qquad (20)$$

These *t*-values possess the very important property that if samples of size n are repeatedly taken from the population, then the *t*-values obtained belong to a *t*-distribution having $n - 1$ degrees of freedom. The term 'degrees of freedom' is one that is used extensively in statistics. For our purposes it is best to regard the term simply as a device for classifying the different *t*-distributions that arise. As indicated in Fig. 16.1, *t*-distributions are bell-shaped and symmetrical like the standard Normal distribution, but tend to be 'flatter'. With reference to the shape of the *t*-distribution it is a fact that the smaller the value of the degrees of freedom, the flatter the distribution compared to the standard Normal curve. To all intents and purposes, when the degrees of freedom number is large, the *t*-distribution is indistinguishable from the Normal.

To enable us to use the distribution, *t*-tables exist which give us *percentage points* in the distribution. In the same way that Normal tables enable us to know that the middle 95% of readings in a Normal distribution have associated z values in the range $z = \pm 1.96$, or the middle 99% in the range $z = \pm 2.576$, then *t*-tables can also be used to provide us with similar information about *t*-values. Thus, for example, with reference to Table 16.3, we see that for a distribution with ten degrees of freedom, the middle 95% of *t*-values occur within $t = \pm 2.228$, or for 16 degrees of freedom, the middle 99% of *t*-values occur within $t = \pm 2.921$. The quantity α in the table gives the required tail area in the distribution and v is the associated

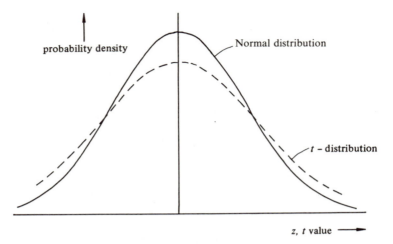

Fig. 16.1. t-distribution compared with the Normal distribution.

degrees of freedom value. Percentage points for given values of α and v are often written t_α^v. So, for example, $t_{0.025}^{10}$ would be 2.228 and $t_{0.005}^{16}$ would be 2.921. Using these percentage points we can quote 95% and 99% confidence intervals for the population mean μ. If the sample is of size n, has mean m and standard deviation s, then a 95% confidence interval for μ is given by

$$m - t_{0.025}^{n-1}\hat{\sigma}_m \leqslant \mu \leqslant m + t_{0.025}^{n-1}\hat{\sigma}_m$$

i.e.

$$m - t_{0.025}^{n-1}s/\sqrt{(n-1)} \leqslant \mu \leqslant m + t_{0.025}^{n-1}s/\sqrt{(n-1)} \tag{21}$$

and a 99% one by

$$m - t_{0.005}^{n-1}s/\sqrt{(n-1)} \leqslant \mu \leqslant m + t_{0.005}^{n-1}s/\sqrt{(n-1)} \tag{22}$$

It will be seen that the corresponding percentage points in the t-distribution are larger than those in the Normal distribution, so that on average, confidence intervals for the population mean are larger when worked out using the t-distribution, than when they can be worked out using the Normal distribution. This simply reflects the fact that uncertainty in the value of σ inevitably widens the confidence interval for μ.

The t-value associated with a difference in sample means. When we have two sample means m and m', obtained by taking samples of size n and n' from two Normally distributed populations, then the t-value to associate with the difference $m - m'$ is

$$t = \frac{(m - m') - (\mu - \mu')}{\hat{\sigma}_{m-m'}} \tag{23}$$

Strictly speaking, this expression for t only belongs to a t-distribution if $\hat{\sigma}_{m-m'}$ has been worked out by pooling the results of the two samples involved, that is if the

Table 16.3 *Percentage points of the t-distribution*

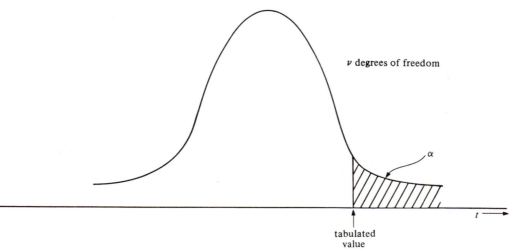

ν degrees of freedom

α

tabulated
value

ν	0.25	0.20	0.15	0.10	0.05	0.025	0.01	0.005	0.0005
1	1.000	1.376	1.963	3.078	6.314	12.706	31.821	63.657	636.619
2	0.816	1.061	1.386	1.886	2.920	4.303	6.965	9.925	31.598
3	0.765	0.978	1.250	1.638	2.353	3.182	4.541	5.841	12.941
4	0.741	0.941	1.190	1.533	2.132	2.776	3.747	4.604	8.610
5	0.727	0.920	1.156	1.476	2.015	2.571	3.365	4.032	6.859
6	0.718	0.906	1.134	1.440	1.943	2.447	3.143	3.707	5.959
7	0.711	0.896	1.119	1.415	1.895	2.365	2.998	3.499	5.405
8	0.706	0.889	1.108	1.397	1.860	2.306	2.896	3.355	5.041
9	0.703	0.883	1.100	1.383	1.833	2.262	2.821	3.250	4.781
10	0.700	0.879	1.093	1.372	1.812	2.228	2.764	3.169	4.587
11	0.697	0.876	1.088	1.363	1.796	2.201	2.718	3.106	4.437
12	0.695	0.873	1.083	1.356	1.782	2.179	2.681	3.055	4.318
13	0.694	0.870	1.079	1.350	1.771	2.160	2.650	3.012	4.221
14	0.692	0.868	1.076	1.345	1.761	2.145	2.624	2.977	4.140
15	0.691	0.866	1.074	1.341	1.753	2.131	2.602	2.947	4.073
16	0.690	0.865	1.071	1.337	1.746	2.120	2.583	2.921	4.015
17	0.689	0.863	1.069	1.333	1.740	2.110	2.567	2.898	3.965
18	0.688	0.862	1.067	1.330	1.734	2.101	2.552	2.878	3.922
19	0.688	0.861	1.066	1.328	1.729	2.093	2.539	2.861	3.883
20	0.687	0.860	1.064	1.325	1.725	2.086	2.528	2.845	3.850
21	0.686	0.859	1.063	1.323	1.721	2.080	2.518	2.831	3.819
22	0.686	0.858	1.061	1.321	1.717	2.074	2.508	2.819	3.792
23	0.685	0.858	1.060	1.319	1.714	2.069	2.500	2.807	3.767
24	0.685	0.857	1.059	1.318	1.711	2.064	2.492	2.797	3.745
25	0.684	0.856	1.058	1.316	1.708	2.060	2.485	2.787	3.725
26	0.684	0.856	1.058	1.315	1.706	2.056	2.479	2.779	3.707
27	0.684	0.855	1.057	1.314	1.703	2.052	2.473	2.771	3.690
28	0.683	0.855	1.056	1.313	1.701	2.048	2.467	2.763	3.674

(Table Contd.)

Table 16.3. (*Contd.*)

v \ α	0.25	0.20	0.15	0.10	0.05	0.025	0.01	0.005	0.0005
29	0.683	0.854	1.055	1.311	1.699	2.045	2.462	2.756	3.659
30	0.683	0.854	1.055	1.310	1.697	2.042	2.457	2.750	3.646
40	0.681	0.851	1.050	1.303	1.684	2.021	2.423	2.704	3.551
60	0.679	0.848	1.046	1.296	1.671	2.000	2.390	2.660	3.460
120	0.677	0.845	1.041	1.289	1.658	1.980	2.358	2.617	2.373
∞	0.674	0.842	1.036	1.282	1.645	1.960	2.326	2.576	3.291

Note: v is degrees of freedom.
$\quad\quad$ α is area.

two populations have the same standard deviation. If pooling has occurred, then the degrees of freedom of the resulting t-distribution is $n + n' - 2$.

As we have mentioned before, pooling of results must only take place if we have strong grounds for suspecting that the two populations have equal standard deviations. Obviously, there are many situations where this will not be the case. What we can do then is to work out the t-value given above, using the unpooled expression for $\hat{\sigma}_{m-m'}$ in equation (12) and note that the t-value approximately belongs to a t-distribution having degrees of freedom equal to the smaller of n and n' minus one, i.e. equal to $\min(n, n') - 1$. This result is useful in providing rough and ready confidence intervals for the difference in two population means when it is clear that the two populations have different standard deviations.

95% and 99% confidence limits on $\mu - \mu'$ are thus given respectively by:

$$(m - m') - t_{0.025}^v \hat{\sigma}_{m-m'} \leqslant \mu - \mu' \leqslant (m - m') + t_{0.025}^v \hat{\sigma}_{m-m'} \tag{24}$$

and

$$(m - m') - t_{0.005}^v \hat{\sigma}_{m-m'} \leqslant \mu - \mu' \leqslant (m - m') + t_{0.005}^v \hat{\sigma}_{m-m'} \tag{25}$$

The degrees of freedom, v, in each case takes the value $n + n' - 2$ or $\min(n, n') - 1$ depending on whether pooling of results has taken place or not as mentioned above.

16.3 Worked examples

Example 1. Obtain a 95% confidence interval for the value of the acceleration due to gravity from the sample of five readings given in 16.1.1 assuming that

$\quad\quad$ (i) the value of the population standard deviation is 0.003;
$\quad\quad$ (ii) the value of the population standard deviation is unknown.

State the assumptions being made in each case.

Solution. (i) Taking the arithmetic mean of the five readings given, we see that this is $m = 9.8072 \, \mathrm{m\,s^{-2}}$. Since we are told that the population standard deviation is $0.003 \, \mathrm{m \, s^{-2}}$, then it follows that sample means will be distributed about the population mean with a standard deviation σ_m equal to $0.003/\sqrt{5}$, i.e. $0.001\,342 \, \mathrm{m \, s^{-2}}$.

Assuming, therefore, that the parent population from which the five readings come is Normally distributed, so that the distribution of sample means is also Normally distributed, a 95% confidence interval for the value of the population mean, i.e. g, is thus given by:

$$9.8072 - 1.96 \times 0.001342 \leqslant g \leqslant 9.8072 + 1.96 \times 0.001342$$

i.e. by

$$9.8046 \leqslant g \leqslant 9.8098$$

where the units are in $\mathrm{m \, s^{-2}}$.

(ii) If the population standard deviation is unknown then we must estimate it from the sample value. The sample standard deviation works out to be $0.002\,71 \, \mathrm{m \, s^{-2}}$, so, using Bessel's correction, the value of $\hat{\sigma}$ is thus $0.002\,71 \times \sqrt{(\tfrac{5}{4})}$, i.e. $0.003\,03 \, \mathrm{m \, s^{-2}}$ and the best estimate of σ_m, namely $\hat{\sigma}/\sqrt{5}$ is $0.001\,36 \, \mathrm{m \, s^{-2}}$. The required confidence interval will now be given in terms of t-values as

$$9.8072 - t^4_{0.025} \times 0.00136 \leqslant g \leqslant 9.8072 + t^4_{0.025} \times 0.001\,36$$

Since $t^4_{0.025}$ is equal to 2.776, then the confidence interval, in $\mathrm{m \, s^{-2}}$, is

$$9.8034 \leqslant g \leqslant 9.8110$$

Again, the assumption we must make for the analysis to be valid is that the parent population from which the five estimates of g came is normally distributed about a mean equal to the actual value of g.

Example 2. Obtain a 95% confidence interval for the difference in blood alcohol levels as measured by conventional gas chromatography and head-space gas chromatography, using the data given in Table 16.1, assuming
  (i) the data values have been pooled;
  (ii) the data values have not been pooled.
What does the difference in sample means indicate if anything?

Solution. For the conventional gas chromatography readings the sample standard deviation is $0.730 \, \mathrm{mg/100 \, ml}$ and the mean is $75.633 \, \mathrm{mg/100 \, ml}$. For the head space results the standard deviation is $0.364 \, \mathrm{mg/100 \, ml}$ and the mean is $74.563 \, \mathrm{mg/100 \, ml}$.

(i) If the results are pooled then the best estimate of the common standard deviation is

$$
\begin{aligned}
\hat{\sigma}_p &= \left(\frac{ns^2 + n'm'^2}{n + n' - 2} \right)^{\frac{1}{2}} \\
&= \left(\frac{6 \times 0.730^2 + 8 \times 0.364^2}{6 + 8 - 2} \right)^{\frac{1}{2}} \\
&= 0.596 \, \mathrm{mg/100 \, ml}
\end{aligned}
$$

The value of $\hat{\sigma}_{m-m'}$ is thus given by

$$\hat{\sigma}_{m-m'} = \hat{\sigma}_p \left(\frac{1}{n} + \frac{1}{n'} \right)^{\frac{1}{2}}$$

$$= 0.596(\tfrac{1}{6} + \tfrac{1}{8})^{\frac{1}{2}}$$

$$= 0.322 \, \text{mg}/100 \, \text{ml}.$$

The observed difference, $m - m'$, in sample means is $(75.633 - 74.563) \, \text{mg}/100 \, \text{ml}$, i.e. $1.070 \, \text{mg}/100 \, \text{ml}$. A 95% confidence interval for the difference, $\mu - \mu'$, in blood alcohol levels, assuming Normally distributed populations, is thus given by

$$1.070 - t_{0.025}^{6+8-2} \times 0.322 \leqslant \mu - \mu' \leqslant 1.070 + t_{0.025}^{6+8-2} \times 0.322$$

Since $t_{0.025}^{12}$ is 2.179, then this confidence interval is

$$0.368 \leqslant \mu - \mu' \leqslant 1.772 \, \text{mg}/100 \, \text{ml}$$

(ii) If the results have not been pooled, then the best estimate of $\hat{\sigma}_{m-m'}$ is given by

$$\hat{\sigma}_{m-m'} = \left(\frac{s^2}{(n-1)} + \frac{s'^2}{(n'-1)} \right)^{\frac{1}{2}}$$

$$= (0.730^2/5 + 0.364^2/7)^{\frac{1}{2}}$$

$$= 0.354 \, \text{mg}/100 \, \text{ml}$$

A 95% confidence interval for the difference in blood alcohol levels is now given by

$$1.070 - t_{0.025}^{\min(6,8)-1} \times 0.354 \leqslant \mu - \mu' \leqslant 1.070 + t_{0.025}^{\min(6,8)-1} \times 0.354$$

Since $t_{0.025}^5$ is 2.571, then this confidence interval is

$$0.160 \leqslant \mu - \mu' \leqslant 1.980 \, \text{mg}/100 \, \text{ml}$$

With regard to the problem of deciding what, if anything, the difference in sample means is indicating, we should note that from analysing the data we can now be 95% sure that the mean difference in measured levels of alcohol lies (at worst) somewhere between 0.160 and 1.980 mg/100 ml. It therefore seems unlikely that the difference in blood alcohol levels as measured by the two methods could be zero. Thus, since the same blood sample is being analysed we are forced to conclude that it is more than likely that the two methods of analysis are giving significantly different results.

Example 3. Using the data given in Table 16.2 work out the t-value to associate with the mean of the iron content readings of both Lab A and Lab B. What can we conclude about the accuracy of each of the laboratory's work?

Solution. Lab A's results have a sample mean of 28.455% with a standard deviation of 0.261%. The associated t-value, knowing that the population mean is 28.25%, is thus going to be

$$t = \frac{m - \mu}{s/\sqrt{(n-1)}}$$

$$= \frac{28.455 - 28.25}{0.261/\sqrt{9}}$$

$$= 2.356$$

For Lab B's work, the mean is 28.420% and the standard deviation is 0.161%. The associated t-value is thus

$$t = \frac{28.420 - 28.25}{0.161/\sqrt{9}}$$

$$= 3.168$$

To interpret these two results we have to note that if the work of the two labs is accurate, then means of samples of size 10 should in each case be Normally distributed about a mean iron content of 28.25%. This assumes, of course, that each sample comes from a Normally distributed population – which we will take to be the case. In other words associated t-values should, in each case, come from t-distributions having nine degrees of freedom. Thus, 95% of the time, the t-values should occur between ± 2.262. Since the two calculated t-values are not in this range, it is therefore more than likely that the work of both laboratories is inaccurate. It would appear that the work in each case is susceptible to systematic sources of variation that give rise to measurements that are slightly too large on average.

16.4 Further developments

16.4.1 The *F*-distribution. In our third example at the beginning of this chapter we considered how the relative precision of the work of the two laboratories involved in the alloy analysis could be compared and we stated that this could be achieved by comparing sample variances. Drawing conclusions from such comparisons involves making use of the *F-distribution*.

Suppose we have two Normally distributed populations possessing the same standard deviation. If a sample of size n is drawn from the first population then from the sample we can obtain an estimate $\hat{\sigma}$ of this population standard deviation. Similarly, if a sample of size n' is drawn from the second population, an estimate $\hat{\sigma}'$ of this same standard deviation can be obtained. We would expect $\hat{\sigma}$ and $\hat{\sigma}'$ to be roughly equal and in fact it can be shown that the quantity $F(v, v')$, given by

$$F(v, v') = \hat{\sigma}^2/\hat{\sigma}'^2 \tag{26}$$

belongs to a so-called *F-distribution* that peaks about $F = 1$. There are two separate degrees of freedom associated with an F-value: a numerator degrees of freedom v and a denominator degrees of freedom v'. The numerator degrees of freedom is usually one less than the size of the sample used to obtain the variance estimate in the numerator; so $v = n - 1$. The denominator degrees of freedom, by analogy, is given by $v' = n' - 1$.

A typical F-distribution is illustrated in Fig. 16.2. It is seen that the F-distribution

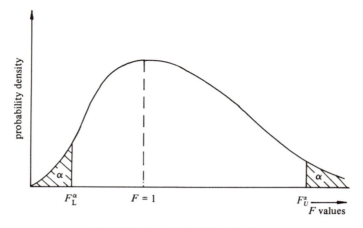

Fig. 16.2. A typical F-distribution.

is not symmetrical and the upper and lower percentage points, denoted by F_U^α and F_L^α, respectively, for a given tail area α, are not symmetrically spaced about any centre of symmetry. F-tables giving percentage points for both the upper and lower tails of the distribution do exist, but usually it is possible to get by with only the upper percentage points being given. A table of upper percentage points, F_U^α, can be used to furnish a corresponding set of F_L^α values by noting that since an F-value associated with two variance estimates $\hat{\sigma}^2$ and $\hat{\sigma}'^2$ can be worked out either as $\hat{\sigma}^2/\hat{\sigma}'^2$, to give $F(v, v')$, or as $\hat{\sigma}'^2/\hat{\sigma}^2$ to give $F(v', v)$, then as far as percentage points are concerned, we must have

$$F_L^\alpha(v, v') = 1/F_U^\alpha(v', v) \tag{27}$$

Thus, using this relation, lower percentage points can be obtained from upper percentage points by appropriately switching the numerator and denominator degrees of freedom. In Table 16.4 percentage points in the F-distribution for upper tail areas of 1% and 5% are given for a range of numerator and denominator degrees of freedom.

It should be noted that the F-distribution is used primarily to test whether or not two population variances are the same and in this role we seldom need to know lower percentage points. In fact most F-values are deliberately calculated to be greater than one, and hence to lie in the upper half of the distribution, by always dividing the larger variance estimate by the smaller one. Using the F-distribution in F-tests is considered in 16.4.4 after we have looked at the elements of *hypothesis testing*.

16.4.2 Elements of hypothesis testing. Hypothesis testing is the last topic to be considered in these three chapters on statistics and in some ways, therefore, it is the culmination of all the material that has been covered so far. Hypothesis testing is a very important part of statistics and is primarily concerned with looking at what can be deduced about a population from only a finite and often small sample of results. To carry out a hypothesis test requires nothing that we do not essentially know already in the way of theory. However, it tends to be couched in a language, some of whose terms we have not yet met, so in this section we spend some time defining and discussing these terms.

A *hypothesis*, then, is any statement we might care to make about a population parameter. For example, in the acceleration due to gravity experiment, before ever measuring the value of g, we might postulate that its value is $9.810 \, \text{m s}^{-2}$; in other words we might state that μ, the population mean of results to expect when g is repeatedly measured, will be $9.810 \, \text{m s}^{-2}$.

Hypotheses in statistics that are easiest to test are those known as *null hypotheses*, and the statement $\mu = 9.810 \, \text{m s}^{-2}$ is an example of such a hypothesis. Why the statement is called a null hypothesis can be appreciated as follows. Anybody making the above statement about μ would probably do so because experience or the results of similar previous experiments would tell them that this was the value to expect for μ. Expecting μ to be $9.810 \, \text{m s}^{-2}$, when g is measured again, is tantamount, therefore, to expecting nothing new in the experiment, and for this reason the hypothesis is called a null hypothesis.

As we have said, hypothesis testing is concerned with looking at what can be deduced about a population, in particular about a population parameter, from a finite and often small sample of results. The idea, therefore, is to look at the sample and on the basis of calculations and analysis decide whether or not to accept the null hypothesis. Obviously, because a sample is only part of a population, errors are bound to arise when we decide to accept or reject the null hypothesis. It is easily seen that two types of error can be made. If, in a particular problem, the null hypothesis (usually referred to as H_0) is true, but based on the sample we decide to reject it, then we have made a *type-1 error*. If on the other hand the null hypothesis is false, but we decide to accept it, we have then made a *type-2 error*. These two types of errors are summarized in Table 16.5.

If a hypothesis test is to be useful, it must be designed to minimise the chances of making these two types of error. For a given sample size this is not always easy because it is generally the case that if we attempt to reduce the chances of making one type of error, we invariably increase the chances of making the other type. The only sure way of reducing the chances of making both types of error is to increase the sample size – but for all kinds of reasons this might not always be possible. So what we do in practice when our sample size is fixed is to decide which type of error it is more important we do not commit in the test and then seek to minimise the probability of making such an error.

The probability of making a type-1 error is known as the *significance level* of the test and commonly used values of this probability are 0.05 and 0.01 i.e. 5% and 1% significance levels. If it is important that we do not commit a type-1 error we choose a low significance level – say 1%. On the other hand, if it is important that we do not commit a type-2 error we choose a higher significance level such as 5%. In view of what we have said about how the two types of error are related,

16.4. *Percentage points of the F-distribution*

5% points	Degrees of freedom of numerator (v)									
	1	2	3	4	5	6	8	12	24	∞
1	161.4	199.5	215.7	224.6	230.2	234.0	238.9	243.9	249.0	254.3
2	18.51	19.00	19.16	19.25	19.30	19.33	19.37	19.41	19.45	19.50
3	10.13	9.55	9.28	9.12	9.01	8.94	8.84	8.74	8.64	8.53
4	7.71	6.94	6.59	6.39	6.26	6.16	6.04	5.91	5.77	5.63
5	6.61	5.79	5.41	5.19	5.05	4.95	4.82	4.68	4.53	4.36
6	5.99	5.14	4.76	4.53	4.39	4.28	4.15	4.00	3.84	3.67
7	5.59	4.74	4.35	4.12	3.97	3.87	3.73	3.57	3.41	3.23
8	5.32	4.46	4.07	3.84	3.69	3.58	3.44	3.28	3.12	2.93
9	5.12	4.26	3.86	3.63	3.48	3.37	3.23	3.07	2.90	2.71
10	4.96	4.10	3.71	3.48	3.33	3.22	3.07	2.91	2.74	2.54
11	4.84	3.98	3.59	3.36	3.20	3.09	2.95	2.79	2.61	2.40
12	4.75	3.83	3.49	3.26	3.11	3.00	2.85	2.69	2.50	2.30
14	4.60	3.74	3.34	3.11	2.96	2.85	2.70	2.53	2.35	2.13
16	4.49	3.63	3.24	3.01	2.85	2.74	2.59	2.42	2.24	2.01
18	4.41	3.55	3.16	2.93	2.77	2.66	2.51	2.34	2.15	1.92
20	4.35	3.49	3.10	2.87	2.71	2.60	2.45	2.28	2.08	1.84
25	4.24	3.38	2.99	2.76	2.60	2.49	2.34	2.16	1.96	1.71
30	4.17	3.32	2.92	2.69	2.53	2.42	2.27	2.09	1.89	1.62
40	4.08	3.23	2.84	2.61	2.45	2.34	2.18	2.00	1.79	1.51
60	4.00	3.15	2.76	2.53	2.37	2.25	2.10	1.92	1.70	1.39
∞	3.84	3.00	2.60	2.37	2.21	2.10	1.94	1.75	1.52	1.00

Degrees of freedom of denominator (v')

Degrees of freedom of numerator (ν)

1% points	1	2	3	4	5	6	8	12	24	∞
1	4052.	4999.	5403.	5625.	5764.	5859.	5982.	6106.	6234.	6366.
2	98.50	99.00	99.17	99.25	99.30	99.33	99.37	99.42	99.46	99.50
3	34.12	30.82	29.46	28.71	28.24	27.91	27.49	27.05	26.60	26.10
4	21.20	18.00	16.69	15.90	15.52	15.21	14.80	14.37	13.93	13.50
5	16.26	13.27	12.06	11.39	10.97	10.67	9.29	9.89	9.47	9.02
6	13.74	10.92	9.78	9.15	8.75	8.47	8.10	7.72	7.31	6.88
7	12.25	9.55	8.45	7.85	7.46	7.19	6.84	6.47	6.07	5.65
8	11.26	8.65	7.59	7.01	6.63	6.37	6.03	5.67	5.28	4.86
9	10.56	8.02	6.99	6.42	6.06	5.80	5.47	5.11	4.73	4.31
10	10.04	7.56	6.55	5.99	5.64	5.39	5.06	4.71	4.33	3.91
11	9.65	7.20	6.22	5.67	5.32	5.07	4.74	4.40	4.02	3.60
12	9.33	6.93	5.95	5.41	5.06	4.82	4.50	4.16	3.78	3.36
14	8.86	6.51	5.56	5.04	4.69	4.46	4.14	3.80	3.43	3.00
16	8.53	6.23	5.29	4.77	4.44	4.20	3.89	3.55	3.18	2.75
18	8.28	6.01	5.09	4.58	4.25	4.01	3.71	3.37	3.00	2.57
20	8.10	5.85	4.94	4.43	4.10	3.87	3.56	3.23	2.86	2.42
25	7.77	5.57	4.68	4.18	3.86	3.63	3.32	2.99	2.62	2.17
30	7.56	5.39	4.51	4.02	3.70	3.47	3.17	2.84	2.47	2.01
40	7.31	5.10	4.31	3.83	3.51	3.29	2.99	2.66	2.29	1.80
60	7.08	4.98	4.13	3.63	3.34	3.12	2.82	2.50	2.12	1.60
∞	6.63	4.60	3.78	3.32	3.02	2.80	2.51	2.18	1.79	1.00

Degrees of freedom of denominator (ν')

Note: The table gives for various degrees of freedom the values of F such that $P/100 = \int_F^\infty p(u)\,du$, where $p(u)$ is the probability density of the F-distribution.

451

Table 16.5. *Types of error in hypothesis testing*

	H_0 is true	H_0 is false
accept H_0	decision correct	type-2 error
reject H_0	type-1 error	decision correct

a higher significance level will ensure a smaller probability of making a type-2 error. Once the significance level of a test has been fixed, then we can go on to work out the actual value of the probability of a type-2 error if we have in mind some definite alternative to the null hypothesis. The quantity which has the value 1 – prob (type-2 error) is known as the *power* of a test. A hypothesis test is only deemed to be powerful, i.e. useful as a means of discerning between the null hypothesis and the alternative hypothesis, if its power is close to its maximum value of 1. How all these ideas come together in a hypothesis test will be illustrated in problems considered in the Further Worked Examples section. But before going on to look at specific tests we need to consider *one- and two-tail tests*.

Without loss of generality, suppose we are testing the null hypothesis that some population mean μ is equal to a, say. To test this hypothesis we might take a sample of size n from the population and obtain the t-value to associate with the sample mean assuming $\mu = a$. What we will expect of this t-value, if the null hypothesis is true, is that it will belong to a t-distribution of $n-1$ degrees of freedom. We would therefore only suspect the null hypothesis of not being true if we obtained a t-value significantly far from the centre of the t-distribution, i.e. a t-value far out in one of the tails. Now if, for any reason, we could predict before we did the test which of the two tails the calculated t-value was more likely to lie in, if the null hypothesis were false, we would be performing what is called a *one-tail test*. A situation where this could arise might be the following: we might know, for example, that the mean hardness reading of a particular metal is a, say, and we may well have then subjected the metal to some heat treatment that we know can only increase the hardness of the metal if, indeed, it is going to affect it at all. So, in measuring the new hardness reading to see if it was significantly different to a, we would be expecting the calculated t-value to be found, with enhanced probability, in the upper tail of the t-distribution if the hardness reading was affected by the treatment. If the level of significance of our hypothesis test was 5%, then we would be tempted to reject the null hypothesis, that the mean hardness reading was a, if the calculated t-value exceeded the critical value $t_{0.05}^{n-1}$.

Many tests, of course, are such that we have no way of knowing in which tail of the distribution the calculated t-value is likely to occur when the null hypothesis happens to be untrue and they are called *two-tail tests*. Therefore, for a 5% level of significance, we would be tempted to reject the null hypothesis if the calculated t-value either exceeded $t_{0.025}^{n-1}$ or was less than $-t_{0.025}^{n-1}$. One- and two-tail tests are summarised in general in the diagrams in Fig. 16.3 for a 5% level of significance.

One-tail tests

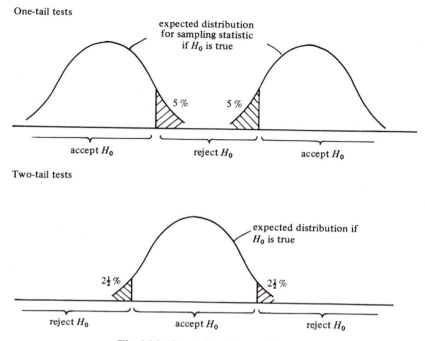

Fig. 16.3. One-tail and two-tail tests.

(*N.B.* What we have said about one- and two-tail tests in connection with the *t*-distribution applies equally well to hypothesis tests where other sampling distributions (besides the *t*) are used.)

16.4.3 The *t*-test. The *t*-test is usually used to compare either a sample mean with a known or assumed known value, or to compare two sample means one with another. To illustrate how a *t*-test is carried out we will consider three examples. More *t*-tests are carried out in the Further Worked Examples section.

Example: *Comparing a sample mean with a known value*
A sample of 16 readings is found to have a mean of 28.0 and a standard deviation of 3.0. On the basis of this evidence is there any reason to reject the claim that the population mean μ is 30.0?

Solution: The null hypothesis to take is clearly $H_0 : \mu = 30.0$. But before carrying out any test we should first decide what the level of significance is to be – in other words we should decide how important it is that we do not make a type-1 error. If it is crucial that we do not make a type-1 error then the significance level would probably be 1% or less. However, a significance level this low means that we may well be accepting the null hypothesis as true when it is in fact false. For completeness, therefore, we will also see what happens in the test when it is carried out at the 5% level (where, in relative terms, the probability of a type-2 error will in general be less). We will further assume that we are carrying out a two-tail test.

The *t*-value associated with the sample mean of 28.0, for the null hypothesis

$\mu = 30.0$, is

$$t = \frac{m - \mu}{s/\sqrt{(n-1)}}$$

$$= \frac{28.0 - 30.0}{3.0/\sqrt{(16-1)}}$$

$$= -2.583$$

At the 1% level of significance we therefore accept the null hypothesis as true because the calculated t-value of -2.582 lies between $\pm t_{0.005}^{16-1}$, i.e. between ± 2.947. We would claim that there was not sufficient evidence in our findings to enable us to reject the null hypothesis.

On the other hand, at the 5% level of significance, we would reject the null hypothesis because -2.582 is outside the range $\pm t_{0.025}^{16-1}$, i.e. outside ± 2.131. Here we would claim that there was sufficient evidence for us to doubt the validity of the null hypothesis.

Clearly it is of fundamental importance in any hypothesis test to choose an appropriate level of significance because our conclusions, and particularly the way we choose to defend them, depend crucially on this choice.

Table 16.6. *Experiments showing percentage of chloride in potassium chloride*

1st analyst	2nd analyst
(%)	(%)
47.59	47.63
47.65	47.59
47.57	47.62
47.55	47.57
47.61	46.66
47.58	
47.57	

Example: *Comparing two sample means*

Suppose the results of repeated estimates of the percentage of Cl^- in KCl, obtained by two analysts using the same automatic analyser, are as shown in Table 16.6. Is the difference in mean readings of these two sets of data of any significance?

Solution: The appropriate null hypothesis here is that the two populations, from which the samples are drawn, have means μ and μ' that are equal. So H_0 is $\mu - \mu' = 0$. We shall work to a 5% level of significance and choose to pool the results of the samples because, using the same automatic analyser, we would expect the scatter in both samples to be more or less the same.

The mean m of the first set of readings is 47.589% and the standard deviation s is 0.030 44%. The mean m' of the second set is 47.614% with a standard deviation s' of 0.031 37%. The pooled estimate of σ will thus be given by

$$\hat{\sigma}_p = \left(\frac{ns^2 + n's'^2}{n + n' - 2} \right)^{\frac{1}{2}}$$

$$= 0.033\,77\%$$

so that the *t*-value to associate with the difference $m - m'$ will be

$$t = \frac{(m - m') - (\mu - \mu')}{\hat{\sigma}_p(1/n + 1/n')^{\frac{1}{2}}}$$

$$= \frac{(47.589 - 47.614) - 0}{0.033\,77(1/7 + 1/5)^{\frac{1}{2}}}$$

$$= -1.264$$

Assuming, also, that the test is a two-tail test then we will accept the null hypothesis if this *t*-value occurs in the range $\pm t_{0.025}^{n+n'-2}$. Checking the value of $t_{0.025}^{10}$ in tables we see that $t_{0.025}^{10}$ is 2.228 so that we have to accept the null hypothesis. Our conclusion, on the basis of the *t*-test, is that there is not sufficient evidence to reject the null hypothesis. The difference in sample means is probably, therefore, of no significance.

Example: *The paired t-test*

Ten specimens of the same alloy are tested for hardness and after heat treatment their hardness is measured again with the following results that are set out below in Table 16.7. Test whether or not heat treatment gives rise to a significant change in hardness.

Table 16.7. *Hardness tests on specimens of alloy*

Specimen no.	1	2	3	4	5	6	7	8	9	10
1st reading	20.7	18.4	19.8	19.9	18.8	23.4	23.7	20.8	20.0	22.0
2nd reading	21.9	20.8	21.1	19.9	20.1	24.4	25.5	21.6	24.6	23.4
Difference	1.2	2.4	1.3	0.0	1.3	1.0	1.8	0.8	4.6	1.4

Before considering the solution to this problem it is worth commenting on the information we have here. We could simply perform a *t*-test on the difference in sample means, as in the previous example, but if we did this we would implicitly be assuming that each sample contained readings from the same Normally distributed population. This assumption might be false because even though each specimen is a specimen of the same alloy there might well be systematic differences in the make up of the alloy in each specimen that affect the hardness. In trying to assess if heat treatment alone has affected the hardness it is better to work with differences in hardness for individual specimens. The systematic factors affecting the hardness should be the same before and after heat treatment, so the *paired* differences should be a better indication of the effect of heat treatment alone. Also, any scatter in the sample of ten paired differences should now only be due to random factors of one kind or another and there is thus every possibility that similar differences would be Normally distributed. The conditions needed to perform a *t*-test are thus likely to be satisfied by the paired difference, whereas they need not necessarily be satisfied by the hardness readings themselves. A so-called *paired* *t*-test is thus the best *t*-test to use here.

In the test we will work to a 5% level of significance, but use a one-tail test, assuming that if the heat treatment is going to have any affect then it can only increase the hardness.

Solution: Working solely with differences, then, we find that the mean difference m is 1.58 and the standard deviation s is 1.699. The null hypothesis is clearly that the effect of heat treatment is negligible, so that μ, the population mean difference, is zero. Thus we have for H_0 the statement $\mu = 0$. The associated t-value is therefore

$$t = \frac{m - \mu}{s/\sqrt{(n-1)}}$$

$$= \frac{1.58 - 0.0}{1.1669/\sqrt{9}}$$

$$= 4.062$$

In this one-tail test we will reject the null hypothesis if the calculated t-value is greater than $t_{0.05}^{10-1}$, i.e. greater than 1.833. We thus reject H_0 and claim that heat treatment significantly increases the alloy's hardness.

One important question is at what point in a hypothesis test should we decide whether or not to carry out a paired t-test? The answer has to be: right at the beginning, before we ever collect any data. Whether or not a paired t-test is appropriate is a fundamental decision that must be taken at the *experimental design* stage. Taking the decision after the data have been collected is often too late. Careful design of experiments is crucial in science and usually an experiment is carried out in a way that suits the analysis that follows; not vice versa.

16.4.4 The F-test. The F-test makes use of the F-distribution and is carried out when conclusions have to be drawn from a comparison of two variances. The F-test is particularly useful as an objective way of deciding whether or not to pool the readings of two samples, and it can also be used to decide which of two experimental techniques is better from the point of view of relative precision. Here we consider two situations that illustrate how F-tests are carried out; further examples are given in Section 16.5.

(1) *Checking two samples for homogeneity*. Samples from populations possessing the same standard deviation are said to be *homogeneous*, and checking two samples for homogeneity is useful for several reasons. One good reason, as we have just mentioned, is if we want to decide whether or not to pool the results for the purposes of estimating the common standard deviation.

Suppose, therefore, that we have two samples, one of size 9 and one of size 6 with respective standard deviations equal to 3.01 and 1.80, and suppose further that we have good reason for suspecting that the two samples are homogeneous, but we want some objective means of testing if this is so. Our null hypothesis will be that the two samples are from Normally distributed populations having the same standard deviation. Thus H_0 is the statement $\sigma = \sigma'$. The best estimate $\hat{\sigma}$ of this standard deviation from the first sample will be $\hat{\sigma} = 3.01 \times \sqrt{(\frac{9}{8})} = 3.193$, and from the second sample the best estimate $\hat{\sigma}'$ will be $1.80 \times \sqrt{(\frac{6}{5})}$, which is 1.972. The F-value to associate with these two estimates is therefore

$$F = 3.193^2/1.9972^2$$
$$= 2.622$$

Note that we have deliberately chosen to work the F-ratio out as the larger variance estimate divided by the smaller one to give an F-value larger than one. This is so that we can use the percentage points given in Table 4, which refer only to the upper tail of the F-distribution.

In the absence of any knowledge about which sample is likely to come from the population with the larger variance, should H_0 be false, we shall choose to regard the test as being two-tailed. We shall also choose to work at the 10% level of significance, using percentage points in the F-distribution for upper tail areas of 5%, and reject the null hypothesis if the calculated F-value exceeds the appropriate upper percentage point $F_U^{0.05}(v, v')$. In our problem the degrees of freedom v and v' are, respectively, 8 and 5, so since $F_U^{0.05}(8, 5)$ is 4.82 we choose to accept the null hypothesis in view of our F-value only being 2.622. Our conclusion must be that there is not sufficient evidence for us to regard the samples as being significantly inhomogeneous. We accept that the samples are probably from populations possessing the same standard deviation and we therefore would pool the results for the purposes of estimating the common standard deviation.

We might note that here a 10% level is appropriate because this reduces our chances of making a type-2 error. We are after all bound to reject the null hypothesis more often with a 10% level than with a smaller one, and consequently the probability of accepting the null hypothesis when it is false, must also be lower. When we are wondering whether or not to pool results it is crucial to bear in mind the fact that it is undesirable to pool results that come from inhomogeneous samples, so it is important to work with a significance level that reflects this fact.

(2) *Comparing the relative precision of two experimental techniques.* The relative precision of two experimental techniques can be compared by simply comparing the variances of the results obtained by the two techniques.

Suppose, then, that the value of a physical quantity is being determined by two different experimental techniques and suppose, using the first technique that six measurements of the physical quantity are obtained which have a standard deviation of 12.31. Assume also that when the second technique is used, 11 measurements are taken which have a standard deviation of 4.32. The simple question is what can we deduce about the relative precision of the two techniques?

The null hypothesis to test is that the two sets of measurements come from populations having the same standard deviation, so H_0 is again the statement $\sigma = \sigma'$, and we shall assume that a two-tail test is appropriate. Also, since it is probably important here not to claim that one technique is better than the other, without having strong grounds on which to make this claim, then we shall choose to work at a 2% level of significance and use the $F_U^{0.01}$ percentage points of Table 16.4.

The F-value to associate with the two samples, ensuring that its value is greater than one, is

$$F = \hat{\sigma}^2 / \hat{\sigma}'^2$$

$$= \frac{(12.31^2) \times (6/7)}{(4.32^2) \times (11/10)}$$

$$= 6.327$$

The degrees of freedom are 5 and 10, respectively, for the numerator and denominator, so the appropriate upper percentage point against which to compare this value of 6.327 will be $F_U^{0.01}$ (5, 10), which is 5.64. Since our F-value is greater than 5.64 we thus reject the null hypothesis and claim that it is unlikely that the two samples come from populations having the same standard deviation. It is unlikely therefore that the two techniques have comparable precision and taking the sample evidence at face value it would seem that the first technique is worse compared to the second in this respect.

16.5 Further worked examples

Example 1. Using the five estimates of the value of the acceleration due to gravity given in 16.1.1, in the first example, test the hypothesis that $g = 9.810 \, \text{m s}^{-2}$.

Solution. If μ is the mean of all possible estimates of g in the experiment, then the appropriate null hypothesis here is $\mu = 9.810 \, \text{m s}^{-2}$.

We shall work (quite arbitrarily) to a 5% level of significance and assume that we have no *a priori* knowledge about the value of g in the event of its value not being $9.810 \, \text{m s}^{-2}$. We shall therefore carry out a two-tail test.

Our sample of five readings has a mean value m equal to $9.807 \, \text{m s}^{-2}$ and a standard deviation s of $0.0027 \, \text{m s}^{-2}$, so that the t-value to associate with this mean is

$$t = \frac{m - \mu}{s/\sqrt{(n-1)}}$$

$$= \frac{9.807 - 9.810}{0.0027/\sqrt{(5-1)}}$$

$$= -2.222$$

We reject the null hypothesis only if this calculated t-value is outside the range $\pm t_{0.025}^4$, i.e. outside the range ± 2.776. Since -2.222 lies inside this range we accept the null hypothesis, claiming insufficient evidence to cause us to believe that the value of g is not $9.810 \, \text{m s}^{-2}$.

Example 2. By carrying out an appropriate t-test at the 5% level decide if the difference in blood alcohol levels as measured by the two gas chromatography methods, described in the alcohol in blood analysis example of 16.1.2, is of any significance.

Solution. If μ is the mean of all possible estimates of the blood alcohol level using conventional gas chromatography, and μ' is the corresponding population mean for head space results, then the null hypothesis to test is $H_0: \mu - \mu' = 0$. We shall assume that the test is two-tailed and carry out a *t*-test on the observed difference in sample means.

First of all, though, we will test the two samples for homogeneity, using an *F*-test (at the 10% level of significance) to decide whether or not to pool the results.

Since the standard deviation of the six conventional gas chromatography readings is 0.730 mg/100 ml and that of the eight head space results is 0.364 mg/100 ml, the *F*-value to associate with the results of the two samples is

$$F = \frac{0.730^2 \times 6 \times 7}{0.364^2 \times 5 \times 8}$$

$$= 4.223$$

The two degrees of freedom associated with this *F* value are $v = 5$ and $v' = 7$. The critical value $F_U^{0.05}(5, 7)$ in tables is 3.97, so we have to reject the null hypothesis that the two samples come from populations having the same standard deviation. We therefore do not pool the results of the two samples.

The appropriate *t*-test to carry out is thus the one where the *t*-value is given by the expression

$$t = ((m - m') - (\mu - \mu'))/(s^2/(n-1) + s'^2/(n'-1)^{\frac{1}{2}})$$

Here, therefore, we have that

$$t = \frac{(75.633 - 74.563) - 0}{(0.730^2/5 + 0.364^2/7)^{\frac{1}{2}}}$$

$$= 3.020$$

The degrees of freedom associated with this value are $\min(n, n') - 1$, i.e. 5. The value of $t_{0.025}^4$ is 2.571, so we reject H_0. There would appear, therefore, to be sufficient evidence to suggest that the two gas chromatography methods of estimating the blood alcohol level are giving significantly different results. This, of course, was the conclusion we arrived at previously in 16.3, example 2, where essentially the same question was considered using confidence intervals.

Example 3. Using the iron alloy analysis data given in Table 16.2 test whether the precision of the work of the two laboratories involved is different.

Solution. The null hypothesis we shall test is that which assumes that the two sets of results given in Table 16.2 come from populations having the same standard deviation. Thus H_0 is the statement $\sigma = \sigma'$.

We shall use an *F*-test and will only reject H_0 if the *F*-value associated with the two samples exceeds the appropriate $F_U^{0.01}$ value given in Table 16.4.

The standard deviation of Lab A's result is 0.261% and that of Lab B's is 0.161%.

The associated F-value is thus

$$F = \frac{0.261^2 \times 10 \times 9}{0.161^2 \times 9 \times 10}$$

$$= 2.628$$

The degrees of freedom here are 9 and 9, respectively, for the numerator and denominator, and since $F_U^{0.01}(9,9)$ is 5.38 (using interpolation) we have to accept the null hypothesis. The observed difference in the spread of results within each sample is thus not sufficient for us to claim that the relative precision of the work of the two laboratories is significantly different at the 2% level of significance.

Example 4. A sample of size n is taken from a population whose mean is 30.5 and whose standard deviation is known from experience to be 0.4. The null hypothesis being tested is that the population mean is 30.0. If the test is being carried out at the 5% level of significance and is one-tailed (where if μ is not 30.0, then it must be larger than this value), obtain expressions in terms of sample size n for

　　　(i) the probability of making a type-2 error;
and
　　　(ii) the power of the test.

What is the power of the test when $n = 10$?

Solution. The situation is best considered with reference to Fig. 16.4 showing the distribution of sample means to expect when samples of size n are drawn from populations having the same standard deviation of 0.4, but respective means of 30.0 and 30.5. If the null hypothesis were true we would expect the distribution of sample means to be a normal distribution having mean $\mu = 30.0$ and standard deviation σ_m equal to $0.4/\sqrt{n}$. We would only reject the null hypothesis if the sample mean reading exceeded the critical value m_c shown such that the right tail area of this distribution is 0.05. The value of m_c is therefore

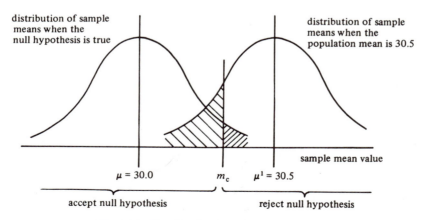

Fig. 16.4. The distribution of sample means.

$$30.0 + \frac{0.4}{\sqrt{n}} \times z(0.05)$$

where $z(0.05)$ is the z-value in the standard normal distribution that gives us a right tail area of 0.05. The value of $z(0.05)$ from standard normal tables is 1.645, so m_c is

$$30.0 + \frac{0.4}{\sqrt{n}} \times 1.645$$

Now, a type-2 error is committed when we accept the null hypothesis when in fact it is false. The probability of a type-2 error is thus the left tail area, below m_c, in the distribution of sample means that is centred about the value $\mu' = 30.5$, which is the true population mean. The z-value associated with m_c in the distribution is given by:

$$z = \frac{m_c - \mu'}{0.4/\sqrt{n}}$$
$$= \frac{30.0 + 0.4/\sqrt{n} \times 1.645 - 30.5}{0.4/\sqrt{n}}$$
$$= 1.645 - 0.5\sqrt{n}/0.4$$
$$= 1.645\sqrt{n}$$

Assuming, therefore, that n is larger than 1, the probability of a type-2 error is thus the area below $z = 1.645 - 1.25\sqrt{n}$ in the standard normal distribution. In other words it is 0.5 – the area tabulated for $z = (1.25\sqrt{n} - 1.645)$. Call the tabulated area $\emptyset(z)$, the expression for the probability of a type-2 error is thus

$$\text{Prob (type-2 error)} = 0.5 - \emptyset(1.25\sqrt{n} - 1.645)$$

The power of a test we defined as $1 - \text{Prob (type-2 error)}$ and so the power, P say, is given by

$$P = 0.5 + \emptyset(1.25\sqrt{n} - 1.645)$$

If therefore n is equal to 10, the power will be

$$P = 0.5 + \emptyset(1.25\sqrt{10} - 1.645)$$
$$= 0.5 + \emptyset(2.308)$$
$$= 0.5 + 0.4895$$
$$= 0.9895$$

The test should therefore be powerful enough to reject the null hypothesis with very little chance of making a type-2 error in the case of n being equal to 10.

16.6 Exercises

1. A chemicals company produces barium sulphate for medical purposes. 50 estimates of the % purity of a batch of the product are found to have

a mean of 95.60%. The standard deviation of purity readings is known, from experience, to be 0.20%. Obtain both a 95% and a 99% confidence interval for the mean % purity of the batch, stating any assumptions that have been made.

2. The lead levels $(mg\,kg^{-1})$ in ten specimens of drinking water taken from a reservoir were as follows:

$$19, 11, 18, 19, 14, 17, 11, 15, 17, 10$$

   (a) Using your calculator, obtain:
       (i) the sample mean;
       (ii) the sample standard deviation;
       (iii) an estimate of the population standard deviation.
   (b) Stating any assumptions you make, obtain a 95% confidence interval for the true mean lead level of the resovoir.
       for the true mean lead level of the reservoir.

3. It is known from observations over a long period that repeated measurements of the same resistance with a particular measuring technique will vary slightly, and that the standard deviation of such measurements is 0.24 M ohm. How many measurements of a particular resistance must be carried out in order to be 95% confident that the mean of the measurements will lie within 0.013 M ohm of the true

4. Lead level readings (in $mg\,kg^{-1}$), obtained from random samples of earth in a kilometre square of land near a motorway, are as follows:

$$65, 70, 85, 62, 91, 75, 69, 83,$$
$$88, 70, 85, 78, 76, 68, 81, 75.$$

   (i) Using the first six readings obtain a 95% confidence interval for the mean lead level in the kilometre square of land.
   (ii) By using the first six (i.e. $n = 6$), the first seven ($n = 7$), the first eight ($n = 8$), etc., readings in the list, make a plot of $t^{n-1}_{0.025}\hat{\sigma}/\sqrt{n}$ against n, and hence determine how many readings are required to be 95% sure of obtaining an estimate of the mean lead level accurate to $8\,mg\,kg^{-1}$.
   (iii) Estimate roughly the sample size needed to achieve an accuracy of $\pm 5\,mg\,kg^{-1}$ for the lead level.

5. Material from a spoil heap is being used for land fill purposes and is being tested for harmful chemicals, particularly dioxin. The following readings of dioxin levels are made:
   dioxin level $mg\,kg^{-1}$ 0.492, 0.487, 0.513, 0.502, 0.496

$$0.486, 0.499, 0.508, 0.501$$

   How many of these readings need to be included (in the order written) to be 95% sure of obtaining an estimate for the true mean dioxin level accurate to within $0.01\,mg\,kg^{-1}$?

6. In the design of a new heating system, the thermal capacity of a particular fluid must be known accurately. Six measurements are made with the

following results (all in Kcals $kg^{-1}K^{-1}$):

$$0.982, 0.963, 0.941, 0.972, 0.963, 0.953$$

Test the hypothesis that the thermal capacity is 0.970, against the hypothesis that it is different to this value. A 5% level of significance should be used.

7. For digesters of a certain shape and size in paper making, the mean weight of fuel used per day is 20.10 units. A new shape of digester of the same capacity, when tested over ten days, used the following weights of fuel:

$$20.15, 19.84, 18.32, 17.50, 20.32$$
$$19.41, 19.57, 18.54, 22.43, 19.82$$

Design experts claim that the new digester can only be more efficient. Do the figures indicate that there is a significant decrease in the fuel used by the new shape of digester compared with the old shape? A 5% level of significance may be used.

8. A metallurgist carried out four determinations of the melting point of manganese, obtaining (in °C) the results

$$1269, 1271, 1263, 1265$$

Test at the 1% significance level the hypothesis that the true melting point is greater than 1260 °C.

9. The following are eight measurements of the wavelength of a particular spectral line (in nanometres):

$$503.29, 503.31, 503.37, 503.33, 503.42, 503.39, 503.36, 503.41$$

Test at the 5% significance level whether the true wavelength is different from 503.30.

10. Ten samples of material from a spoil heap are tested for a hazardous chemical. The ten results gave a mean level for the chemical of $0.42\,mg\,kg^{-1}$ with a standard deviation of $0.10\,mg\,kg^{-1}$. The permitted mean level for the spoil heap is $0.35\,mg\,kg^{-1}$. If you were an inspector whose job it was to pick up any potential environmental hazards, what would you decide on the basis of this evidence? If you were the firm operating the spoil heap, how would you look at the evidence?

11. Material for land fill from two sources is being tested for polychlorinated biphenyls. 12 estimates from the first source had a mean reading (in coded units) of 53.6 with a standard deviation of 7.26. Eight estimates from the second source had a mean of 45.3 with a standard deviation of 9.81.
   (i) Using an F-test decide if the results of the two samples may be pooled.
   (ii) Carrying out an appropriate t-test decide if there is any significant difference in PCB levels from the two sources.

12. Helium gas, produced commercially, is known to contain unwanted oxygen. Gas chromatography was used to estimate the concentration of oxygen in helium produced by two processes. Seven readings were made

on the helium produced by process A and five on the helium produced by process B. The results are as shown:

A/ppm 7.3, 5.5, 7.4, 8.1, 6.9, 4.3, 8.8
B/ppm 4.2, 5.5, 5.6, 7.1, 6.1

Test the samples for homogeneity and decide if the difference in sample means is significant at the 5% level.

13. Tensile strength tests of two types of thin metal wire gave the following results (in $N\,mm^{-2}$):

Type 1: 1380, 1270, 1340, 1250
Type 2: 1340, 1370, 1350, 1400, 1300, 1340

Using a t-test decide whether the mean strength of the two types of wire differ significantly.

14. The accepted level of silicon in standard alloy type LM2 is 10.22%. Two new methods for determining this level are being compared. Six determinations of the level are made using method A and eight are made using method B. The results are as follows:

Method A(%): 10.18, 10.20, 10.23, 10.23, 10.25, 10.29
Method B(%): 10.18, 10.20, 10.22, 10.20, 10.21, 10.19, 10.23, 10.17

Compare the two methods from the point of view of their accuracy and relative precision.

15. The amount of iron(III) in iron(III) ammonium citrate was determined by treating it with potassium iodide and titrating the iodine liberated with thiosulphate solution, using starch as an indicator. The results of a series of determinations A, in which titration was done immediately after the mixing, and a series B, in which the mixture was titrated after 30 min, are as follows for eight samples:

sample number	A (Fe^{III}, %)	B (Fe^{III}, %)
1	13.29	13.86
2	13.36	13.99
3	13.32	13.88
4	13.53	13.91
5	13.56	13.89
6	13.43	13.94
7	13.30	13.80
8	13.43	13.89

Using a paired t-test decide, on the basis of these results, whether the delay before titration is of significance.

16. A production process is considered to be under control if the machine parts by it have a mean length, l, of 35.50 mm, with a standard deviation

of 0.45 mm. If the process 'wanders' in any way the mean length of machine parts will increase, but the standard deviation should remain constant at 0.45 mm. A quality control engineer measures 10 machine parts at random and halts production (i.e. he rejects the null hypothesis that $l = 35.50$ mm) if the result for the sample mean is significant at the 1% level of significance:

(i) Plot a graph showing the probability of making a type-2 error as a function of the actual mean length of machine parts produced by the process.

(ii) Hence determine the power of the test when the mean is running at 35.90 mm.

17

Partial differentiation 1: introduction

Contents: 17.1 Scientific context – 17.1.1 van der Waal's equation of state; 17.1.2 a vibrating string; 17.1.3 potential fields; 17.2 mathematical developments – 17.2.1 functions of several variables; 17.2.2 definition of a partial derivative; 17.2.3 higher partial derivatives; 17.2.4 notation; 17.2.5 rules of partial differentiation; 17.3 worked examples; 17.4 further developments – 17.4.1 total changes; 17.4.2 error analysis; 17.4.3 differentials; 17.4.4 the chain rule; 17.5 further worked examples; 17.6 exercises.

17.1 Scientific context

In earlier chapters we made a study of many of the functions commonly used in science and we showed how to obtain their derivatives. The functions we have considered so far have been functions of a single variable. In science there are many situations where some quantity of interest is related to not just a single variable but to several. In such situations these relationships would be modelled mathematically by functions of several variables.

In order to understand the properties of functions of one variable, we needed to differentiate and sketch out the graph of the functions. Similarly, we need to carry out the same operations in the several variable case. To differentiate functions of several variables involves knowing how to work out what are called *partial derivatives* and obtaining partial derivatives is the central theme of this first chapter on partial differentiation.

As we shall see, the partial derivative of a function of several variables is a straightforward extension of ordinary differentiation just as functions of several variables are a straight forward extension of the single variable case.

17.1.1 Example 1: van der Waal's equation of state. Pure substances can exist in various phases, namely gas, liquid and solid, and whether the phase is a gas, a liquid or a solid is determined essentially by the values of pressure, volume and temperature of the substance.

A pVT plot for a pure substance is what we obtain if we draw a diagram showing the pressure for all possible values of the volume and temperature. The result would be a complicated *surface* such as the plot in Fig. 17.1, which represents the pVT plot of a substance that contracts on freezing. It is clear from the diagram that pressure depends in some special way on volume and temperature. Indeed we can state, just by looking at the diagram, that pressure will be a *function* of V

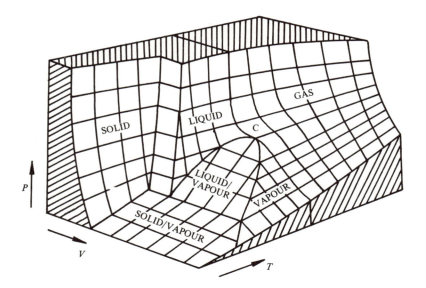

Fig. 17.1. The pVT plot of a substance.

and T because given any value of V and T it is possible to find only one corresponding value of the pressure (this idea of there being only one corresponding value of the dependent variable for any value of the independent one, that we first met in Chapter 1, extends to functions of more than one variable). It is extremely difficult, however, to obtain a single mathematical expression representing this functional relationship between p, V and T, that models the entire surface and all its phases. Many approximate formulae have been put forward and one of the most famous of these, which gives a good description in the liquid–vapour–gas region is van der Waal's equation of state, namely

$$\left(p + \frac{a}{V^2}\right)(V - b) = RT$$

where a and b are constants, and R is the usual gas constant. In this liquid–vapour–gas region it is very important, especially in the liquefaction of gases, to know where these three phases all come together at the so-called critical point of the surface. This critical point is the point C of the pVT plot in Fig. 17.1. It is extremely useful to have an equation, such as van der Waal's equation, to give us the critical point.

Referring again to Fig. 17.1, if we take a slice through the pVT plot along the isothermal (i.e. the line of constant temperature) that passes through the critical point C, then the resulting plot of p against V is shown in Fig. 17.2. The plot is no longer a surface but a curve and the critical point now manifests itself as a point of inflexion. At a point of inflexion we know that both the first and second derivatives are zero (see Chapter 7), so with reference to van der Waal's equation,

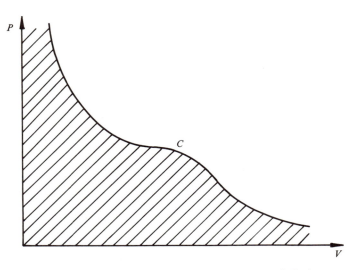

Fig. 17.2. Section showing the critical point as an inflexion.

if we find both the first and second derivatives of p with respect to V to be zero along an isothermal, we should be able to obtain mathematically where the critical point is.

Now the pressure p is a function of two variables V and T, and rearranging van der Waal's equation we have

$$p = f(V, T) = \frac{RT}{(V - b)} - \frac{a}{V^2}$$

Thus, when forming the derivative of p with respect to V along an isothermal, what we are really doing is differentiating p with respect to V in the ordinary way, but keeping T constant. The resulting derivative of p with respect to V is called the *partial derivative* of p with respect to V.

To obtain the critical point mathematically, we require both the first and second partial derivative of p with respect to V to be zero. The subject of thermodynamics deals with the investigation of properties of substances in this way.

17.1.2 Example 2: a vibrating string. Suppose we have a string held taut at both ends so that it is under tension. If the string is plucked, or somehow set in motion, it will vibrate. One way of studying these vibrations is to obtain the equation of motion of the string. This equation as we shall now show is one that involves partial derivatives, it is an example of a *partial differential equation*.

Consider the motion of a small element AB of the string shown in Fig. 17.3. Assume that the equilibrium position of AB is A_0B_0, so that the mass of AB is ρA_0B_0, i.e. $\rho \delta x$ where ρ is the equilibrium mass per unit length of the string. As the string vibrates its overall profile will clearly change shape, so that the displacement y of the element AB from its equilibrium position will be a *function*

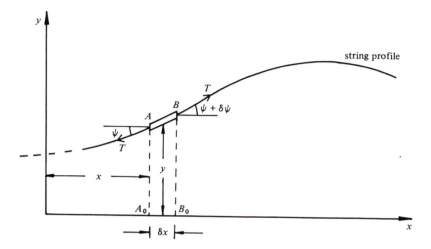

Fig. 17.3. The forces on an element of the string.

of both x and t (time), as will ψ, the angle that the string makes with the horizontal at A. The dependence of y (and ψ) on x and t will be a functional dependence because unless the string happens to break, its displacement y (and hence ψ) will be uniquely specified at any given x and t value. The problem in studying the motion mathematically is to obtain the precise nature of the functional relationship that connects y, x and t.

Consider Fig. 17.3. The force in the y direction causing the element AB to move is the difference in string tension in that direction. Resolving vertically we see that the difference is

$$T \sin (\psi + \delta\psi) - T \sin (\psi)$$

which is $T\delta \sin (\psi)$.

Now according to Newton's second law of motion, this force must cause an acceleration given by the second derivative of y with respect to t, which for the moment we write as $\ddot{y}$.

Equating the force to the mass times acceleration, we have

$$T\delta \sin (\psi) = \rho \delta x \ddot{y}$$

This is essentially the equation of motion of the element of the string AB. To obtain the equation of motion of any *point* on the string we divide both sides by δx and take the limit as $\delta x \to 0$ (in other words we shrink AB into a point by making δx vanishing small). Dividing also by the tension T the motion of any point on the string is therefore described by the equation

$$\lim_{x \to 0} \frac{\delta \sin (\psi)}{\delta x} = \frac{\rho \ddot{y}}{T}$$

This is called the wave equation for a string. It is clearly a differential equation since both sides contain derivatives. However, on the right-hand side, the derivative of y with respect to t is worked out at an x value (see Fig. 17.3) which does not vary during the motion, since the motion is in a vertical plane only. The derivative $\ddot{y}$ is thus a partial derivative. Similarly the derivative on the left-hand side of $\sin(\psi)$ with respect to x, is evaluated by treating t as a constant, for we are interested in how $\sin(\psi)$ varies along the string profile at a particular instant in time. It too is thus a partial derivative.

In 17.3 we shall write down explicitly the form of this wave equation.

17.1.3 Example 3: potential fields. In this example we consider forces in electrostatics and we give another illustration of the application of partial differentiation. Suppose we have a charge q located at the origin of a three-dimensional Cartesian coordinate system (Fig. 17.4). The electrostatic potential ϕ at an arbitrary point P due to the charge q is defined in terms of (x, y, z) as

$$\phi = \frac{q}{4\pi\varepsilon_0\sqrt{(x^2 + y^2 + z^2)}}$$

where ε_0 is the permeability of free space. The potential ϕ is thus a function of these three variables. This functional relationship defines the so-called *potential field* due to q at any arbitrary point P in space.

The component of the force experienced by a unit positive charge at P (in

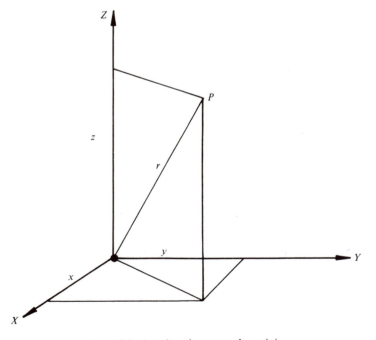

Fig. 17.4. A point charge at the origin.

other words the electric field at *P*), in the *x* direction, say, is defined to be minus the derivative of ϕ with respect to *x* keeping *y* and *z* fixed. This is called the *partial derivative* of ϕ with respect to *x*. In a similar way the component of the force in the *y* direction is minus the partial derivative of ϕ with respect to *y*, and in the *z* direction it is the partial derivative of ϕ with respect to *z*. This definition of force involving the partial derivatives of a potential function is quite general in science. Here in electrostatics it gives an expression for the force that is exactly in accord with Coulomb's law for two point charges. This approach generalises in a straightforward manner when we have a charge distribution as opposed to a single point charge. The calculation of the resulting electrostatic field by differentiating the potential function is usually easier than working with Coulomb's law directly.

17.2 Mathematical developments

17.2.1 Functions of several variables. Before we proceed with the central topic of this chapter, namely partial differentiation, we consider briefly the essential features of functions of several variables. As we mentioned in the introduction, functions of several variables are really nothing more than extensions of ideas we have already met concerning functions of a single variable. We begin by looking at functions of two variables.

Suppose that we have a quantity *z* that is a function of two variables *x* and *y*, then by analogy with the one-dimensional case we write the functional relationship connecting *z* to the two independent variables *x* and *y* as

$$z = f(x, y)$$

The set of *z* values corresponding to all possible *x* and *y* values in the domain makes up the range of the function, just as in the one-variable case. To illustrate such a functional relation we plot the dependent variable *z* on the vertical axis and *x* and *y* as the horizontal axes as in the example in Fig. 17.5.

It is important to note that we now obtain a *surface* in three-dimensional space as opposed to a curve, which we obtain when plotting out a function of one variable. The shape of the surface depends of course on the nature of the functional relationship.

In the first example in 17.1.1 we have that the pressure is a function of volume and temperature, i.e. $p = f(V, T)$ and the *pVT* plot in Fig. 17.1 is nothing more than a plot of the functional relationship

$$p = f(V, T)$$

which can be approximated by van der Waal's equation.

In the second example (17.1.2) where the displacement *y* is a function of *x* and *t*, i.e. $y = f(x, t)$, a plot of this relationship would be made with *y* on the vertical axis and *x* and *t* as horizontal axes.

The generalisation to *n* variables is now straightforward. Suppose that the

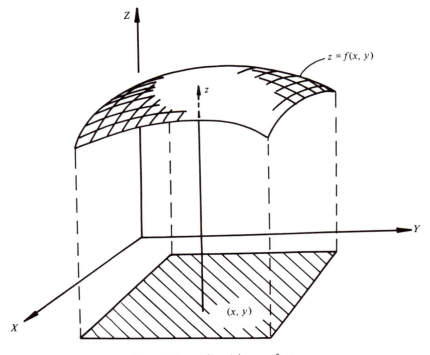

Fig. 17.5. $z = f(x, y)$ is a surface.

dependent variable is the quantity, w say, and the independent variables are $x_1, x_2, x_3, \ldots, x_n$. Then we write

$$w = f(x_1, x_2, \ldots, x_n).$$

The f notation means exactly what it has always meant, namely that given any particular values of $x_1, x_2, \ldots, x_n$ within the domain of the function the corresponding value of w is then uniquely specified by the functional relationship. However, now we would not be able to illustrate the function by drawing a graph because space is only three dimensional.

In this chapter we restrict our discussion mainly to functions of two variables x and y.

17.2.2 Definition of a partial derivative. Suppose that z is a function of two variables x and y given by

$$z = f(x, y)$$

Consider a change in the variable x by a small amount δx from x to $x + \delta x$ and no change in the variable y. Then the small partial change in the variable z, denoted by δz, will be given by

$$\delta z = f(x + \delta x, y) - f(x, y)$$

We have called δz a *partial* change because it arises through a small change occurring in one and not both of the independent variables. Dividing this partial change by δx, we obtain the ratio $\delta f/\delta x$ given by

$$\frac{\delta f}{\delta x} = \frac{f(x + \delta x, y) - f(x, y)}{\delta x} \tag{1}$$

The limit of this ratio, as δx tends to zero is defined as the *partial derivative* of the function f with respect to x, and is written as

$$\frac{\partial f}{\partial x}$$

We have

$$\frac{\partial f}{\partial x} = \lim_{\delta x \to 0} \frac{f(x + \delta x, y) - f(x, y)}{\delta x} \tag{2}$$

This partial derivative would normally be worked out not by formally carrying out the limiting process in this equation, but by regarding z as a function of just *one* variable x and using the normal rules of *ordinary* differentiation. Note that in the limiting process the variable y is not changing in the differentiation process and can be treated as a constant.

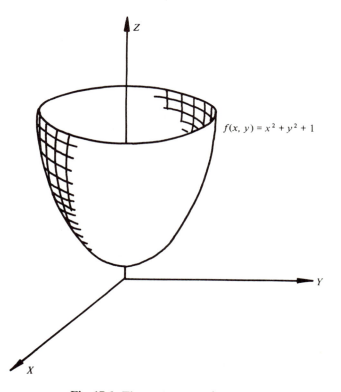

Fig. 17.6. The surface $z = x^2 + y^2 + 1$.

The other partial derivative with respect to y, $\partial f/\partial y$ (keeping x constant) can be defined in a similar way as

$$\frac{\partial f}{\partial y} = \lim_{\delta y \to 0} \frac{f(x, y + \delta y) - f(x, y)}{\delta y} \qquad (3)$$

To illustrate the process geometrically consider the function $f(x, y) = x^2 + y^2 + 1$; the surface $z = f(x, y)$ is shown in Fig. 17.6.

Now if y is held constant at some value 2 say, the function f takes the form $f(x, 2) = g(x) = x^2 + 5$. This function of one variable can be differentiated in the 'ordinary' sense to give $dg/dx = 2x$. The plane $y = 2$ gives a section through the surface and a graph of the section function $f(x, 2) = g(x) = x^2 + 5$ is just a curve (see Fig. 17.7). The derivative dg/dx at some point x represents the slope of the tangent to this curve. In terms of the surface $z = f(x, y)$ the derivative dg/dx is simply the slope of the tangent to the surface in the x direction.

Similarly, if x is held constant, $x = 1$ say, then $f(1, y) = h(y) = y^2 + 2$ and $dh/dy = 2y$ represents the direction of the gradient of the tangent to the surface

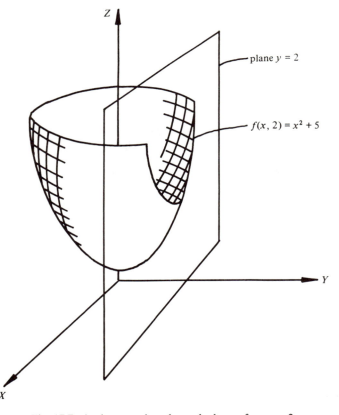

Fig. 17.7. A plane section through the surface $y = 2$.

in the y direction. Clearly, at the point $(1, 2)$ the derivatives take different values; $(dg/dx) = 2$ and $(dh/dy) = 4$ there.

More generally if y is a constant with value y_0 then $f(x, y_0) = g(x) = x^2 + y_0^2 - 1$ and so

$$\frac{dg}{dx} = 2x = \frac{\partial f}{\partial x}$$

The partial derivative of a function $f(x, y)$ with respect to x is just the gradient of the tangent to the section function. This section function is simply a function of one variable which is represented geometrically by a curve. To evaluate partial derivatives we treat one variable as a constant and differentiate with respect to the other variable in the usual way. The following examples illustrate how we find the partial derivatives $\partial f/\partial x$ and $\partial f/\partial y$.

Example: Determine the partial derivative of each function with respect to x

(i) $f(x, y) = x^2 + 2y^2 + xy + 7$;
(ii) $f(x, y) = x^3 - y^2 + 2xy^2$.

Solution: To calculate $\partial f/\partial x$ we consider y fixed at say y_0 and differentiate the resulting function.

(i) $f(x, y_0) = x^2 + 2y_0^2 + xy_0 + 7$

Differentiating with respect to x in the usual way we have

$$\frac{\partial f}{\partial x} = 2x + y_0$$

Hence in general

$$\frac{\partial f}{\partial x} = 2x + y$$

(ii) $f(x, y_0) = x^3 - y_0^2 + 2xy_0^2$

Differentiating with respect to x in the usual way we have

$$\frac{\partial f}{\partial x} = 3x^2 + 2y_0^2$$

Hence in general

$$\frac{\partial f}{\partial x} = 3x^2 + 2y^2$$

Example: Determine the partial derivative of each function with respect to y:

(i) $f(x, y) = x^2 + 2y^2 + xy + 7$;
(ii) $f(x, y) = x^3 - y^2 + 2xy^2$.

Solution: To calculate $(\partial f/\partial y)$ we consider x fixed at say x_0 and differentiate the resulting function.

(i) $f(x_0, y) = x_0^2 + 2y^2 + x_0 y + 7$

Differentiating with respect to y in the usual way we have

$$\frac{\partial f}{\partial y} = 4y + x_0$$

Hence in general

$$\frac{\partial f}{\partial y} = 4y + x$$

(ii) $f(x_0, y) = x_0^3 - y^2 + 2x_0 y^2$

Differentiating with respect to y in the usual way we have

$$\frac{\partial f}{\partial y} = -2y + 4x_0 y$$

Hence in general

$$\frac{\partial f}{\partial y} = -2y + 4xy$$

We can see from these examples that the two partial derivatives $\partial f/\partial x$ and $\partial f/\partial y$ do not necessarily give the same value. So we cannot call either *the* derivative of f, instead we call them *partial* derivatives.

17.2.3 Higher partial derivatives. The expressions for $\partial f/\partial x$ and $\partial f/\partial y$ are themselves functions of x and y. They can therefore be differentiated again to form second-order partial derivatives. Similarly, these second-order derivatives can also be differentiated to form third-order partial derivatives and so on.

Example: Determine the second-order partial derivatives of the function

$$f(x, y) = x^2 + xy^2 - y^3 + 4x$$

Solution: Differentiating f with respect to x and y

$$\frac{\partial f}{\partial x} = 2x + y^2 + 4 \quad \text{and} \quad \frac{\partial f}{\partial y} = 2xy - 3y^2$$

Differentiating $\partial f/\partial x$ with respect to x and y in turn

$$\frac{\partial}{\partial x}\left(\frac{\partial f}{\partial x}\right) = 2 \quad \text{and} \quad \frac{\partial}{\partial y}\left(\frac{\partial f}{\partial x}\right) = 2y$$

Differentiating $\partial f/\partial y$ with respect to x and y in turn

$$\frac{\partial}{\partial x}\left(\frac{\partial f}{\partial y}\right) = 2y \quad \text{and} \quad \frac{\partial}{\partial y}\left(\frac{\partial f}{\partial y}\right) = 2x - 6y$$

To denote the second-order partial derivatives with respect to x and y we adopt the notation $\partial^2 f/\partial x^2$ and $\partial^2 f/\partial y^2$ respectively. (This is similar to the ordinary differentiation case $\mathrm{d}^2 f/\mathrm{d}x^2$.) Thus for the last example we would write

$$\frac{\partial}{\partial x}\left(\frac{\partial f}{\partial x}\right) = \frac{\partial^2 f}{\partial x^2} = 2 \quad \text{and} \quad \frac{\partial}{\partial y}\left(\frac{\partial f}{\partial y}\right) = \frac{\partial^2 f}{\partial y^2} = 2x - 6y$$

For $(\partial/\partial x)(\partial f/\partial y)$ and $(\partial/\partial y)(\partial f/\partial x)$ we write $\partial^2 f/\partial x\partial y$ and $\partial^2 f/\partial y\partial x$ respectively. The functions $\partial^2 f/\partial x\partial y$ and $\partial^2 f/\partial y\partial x$ are called *mixed partial derivatives* since we are differentiating f with respect to more than one of the variables involved. In principle we should be careful to differentiate the function f with respect to the variables in the correct order. However, for any well-behaved function (and we shall only consider such functions) the two mixed partial derivatives $\partial^2 f/\partial x \partial y$ and $\partial^2 f/\partial y \partial x$ are equal as in the last example.

17.2.4 Notation. The most usual notation for the partial derivative of f with respect to x is the expression $\partial f/\partial x$. However, other notations are sometimes used. One alternative to writing $\partial f/\partial x$ is to write f_x. Similarly, f_y would mean $\partial f/\partial y$. Another often used notation, especially in thermodynamics, consists of writing the partial derivative inside brackets and indicating by a subscript, outside the brackets, which variables are being held constant. Thus $(\partial p/\partial V)_T$ in thermodynamics would mean the partial derivative of the pressure with respect to the volume keeping the temperature constant. This notation might at first sight appear redundant because, as we saw in 17.1.1, p is normally a function of V and T. However, the importance of the notation in thermodynamics lies in the fact that it specifies the variables on which the function being differentiated actually depends. Pressure is not always explicitly treated as a function of volume and temperature, it could just as easily be regarded as a function of volume and total energy, E say. Also E is related to V and T.

Seeing $(\partial p/\partial V)_E$ would indicate that pressure was being treated as a function of volume and energy, as opposed to volume and temperature and that in the differentiation process, E was being held constant. The notation is even more important when considering partial derivatives of other thermodynamic quantities such as entropy and enthalpy, for example. The expression, $\partial S/\partial p$, for the partial derivative of the entropy with respect to pressure is difficult to interpret unless we know the two variables on which S depends. $(\partial S/\partial p)_V$ and $(\partial S/\partial p)_T$, for example, in the case of a gas, are quite different expressions (see Section 17.3).

For higher-order derivatives the second derivative $\partial^2 f/\partial x^2$ may often be seen written as f_{xx} and the mixed derivative $(\partial^2 f/\partial x \partial y)$ as f_{xy}.

17.2.5 Rules of partial differentiation. We have said that partial differentiation is essentially a straightforward extension of ordinary differentiation, and this is true. All the simple rules of ordinary differentiation that we considered in Chapter 6 carry over into partial differentiation. For example, the product and quotient rules apply: if u and v are each functions of the variables x and y then

Product rule

$$\frac{\partial}{\partial x}(uv) = u\frac{\partial v}{\partial x} + v\frac{\partial u}{\partial x}$$

Quotient rule

$$\frac{\partial}{\partial x}(u/v) = \frac{[v(\partial u/\partial x)] - [u(\partial v/\partial x)]}{v^2}$$

The same rules apply for partial differentiation with respect to y, we replace x by y in the above formulas.

17.3 Worked examples

Example 1. For a pure substance the critical point will occur as a point of inflexion in the plot of p against V. Now at a point of inflexion both partial derivatives $\partial p/\partial V$ and $\partial^2 p/\partial V^2$ are zero. Determine the critical point of a pure substance whose equation of state is given by van der Waal's equation

$$\left(p + \frac{a}{V^2}\right)(V - b) = RT$$

expressing the critical volume, pressure and temperature in terms of a, b and R.

Solution. Rearranging van der Waal's equation to give an expression for pressure we have

$$p = \frac{RT}{(V - b)} - \frac{a}{V^2}$$

Thus, regarding T as a constant, and differentiating p partially with respect to V we get that

$$\frac{\partial p}{\partial V} = \frac{-RT}{(V - b)^2} + \frac{2a}{V^3}$$

Differentiating $\partial p/\partial V$ again partially with respect to V keeping T constant we get

$$\frac{\partial^2 p}{\partial V^2} = \frac{2RT}{(V - b)^3} - \frac{6a}{V^4}$$

Equating $\partial p/\partial V$ and $\partial^2 p/\partial V^2$ each to zero yields the following equations:

$$\frac{RT}{(V - b)^2} = \frac{2a}{V^3}$$

and

$$\frac{2RT}{(V - b)^3} = \frac{6a}{V^4}$$

Dividing the first equation by the second and simplifying gives

$$\frac{(V - b)}{2} = \frac{V}{3}$$

from which

$$V = 3b$$

In other words, at the critical point, the critical volume assumes the value $3b$. Substituting back in the first equation we obtain

$$\frac{RT}{(V - b)^2} = \frac{2a}{V^3}$$

$$\frac{RT}{4b^2} = \frac{2a}{27b^3}$$

Solving for T we get

$$T = 8a/27Rb$$

The critical temperature thus takes the value $8a/27Rb$.

Finally, substituting these values of T and V back into the expression for the pressure, we see that the critical pressure is given by

$$p = a/27b^2$$

To summarise then, the critical point occurs on the pVT surface at the point where

$$p = \frac{a}{27b^2}, \quad V = 3b \quad \text{and} \quad T = \frac{8a}{27Rb}$$

Example 2. Show that provided the amplitude of the oscillations is everywhere relatively small, the wave equation for a string (see 17.1.2) can be written as

$$\frac{\partial^2 y}{\partial x^2} = \frac{1}{c^2} \frac{\partial^2 y}{\partial t^2}$$

where ρ/T is the quantity $1/c^2$.

Solution. The wave equation derived in the first section is given by

$$\lim_{\delta x \to 0} \frac{\delta(\sin(\psi))}{\delta x} = \frac{\rho}{T} \ddot{y}$$

where y is the displacement of the string from its equilibrium position. Considering the right-hand side first, we know that the displacement y is a function of both x and t, and hence $\ddot{y}$, the second derivative of y with respect to time, can therefore be written as $\partial^2 y/\partial t^2$. Thus, noting that ρ/T is the quantity $1/c^2$, the right-hand side of the wave equation is $(1/c^2)(\partial^2 y/\partial t^2)$.

Consider now the left-hand side. The angle ψ measures the slope of the string's profile at any point along it and at any time. It, too, is a function of both x and t. Thus since $\tan(\psi)$ is the gradient, it follows that $\tan(\psi) = \partial y/\partial x$, i.e. that

$$\frac{\sin(\psi)}{\cos(\psi)} = \frac{\partial y}{\partial x}$$

Now, if ψ is small everywhere, which will be the case if the amplitude of the oscillations is everywhere relatively small, then $\cos(\psi)$ is approximately 1 so that

$$\frac{\sin(\psi)}{\cos(\psi)} \simeq \sin(\psi) = \frac{\partial y}{\partial x}$$

The left-hand side of the wave equation is thus

$$\lim_{\delta x \to 0} \frac{\delta(\partial y/\partial x)}{\delta x}$$

This limiting process gives the second-order partial derivative of y with respect to x,

$\partial^2 y/\partial x^2$. Thus the wave equation for a string may be written as

$$\frac{\partial^2 y}{\partial x^2} = \frac{1}{c^2}\frac{\partial^2 y}{\partial t^2}$$

Example 3. The electrostatic potential ϕ at an arbitrary point P, (x, y, z), due to a charge q placed at the origin is given by

$$\phi = \frac{q}{4\pi\varepsilon_0\sqrt{(x^2 + y^2 + z^2)}}$$

Using partial differentiation determine the components of the electrostatic force experienced by a unit positive charge at P. Show that the magnitude of this force is given by

$$F = \frac{q}{4\pi\varepsilon_0(x^2 + y^2 + z^2)}$$

Solution. In 17.1.3 we saw that the x component of the force on the unit charge at P is $-\partial\phi/\partial x$. So differentiating ϕ partially with respect to x we get that

$$F_1 = -\frac{\partial\phi}{\partial x}$$

$$= \frac{1}{2}\cdot\frac{q}{4\pi\varepsilon_0}\cdot\frac{2x}{(x^2 + y^2 + z^2)^{\frac{3}{2}}}$$

$$= \frac{qx}{4\pi\varepsilon_0(x^2 + y^2 + z^2)^{\frac{3}{2}}}$$

Similarly the y component of the force F_2 is

$$F_2 = -\frac{\partial\phi}{\partial y} = \frac{qy}{4\pi\varepsilon_0(x^2 + y^2 + z^2)^{\frac{3}{2}}}$$

The z component F_3 is minus the partial derivative of ϕ with respect to z, i.e.

$$F_3 = -\frac{\partial\phi}{\partial z} = \frac{qz}{4\pi\varepsilon_0(x^2 + y^2 + z^2)^{\frac{3}{2}}}$$

Finally, the magnitude of the force F is given by $\sqrt{(F_1^2 + F_2^2 + F_3^2)}$. Thus

$$F = \frac{q(x^2 + y^2 + z^2)^{\frac{1}{2}}}{4\pi\varepsilon_0(x^2 + y^2 + z^2)^{\frac{3}{2}}}$$

$$= \frac{q}{4\pi\varepsilon_0(x^2 + y^2 + z^2)}$$

which is the expression we would obtain using Coulomb's law.

Example 4. The entropy S of a gas, whose equation of state is $pV = RT$, is given by $S = c_v \ln(T) + R\ln(V) + a$ where c_v and a are constants and R is the gas constant.

    (i) Obtain expressions for the partial derivatives $(\partial S/\partial V)_T$, $(\partial S/\partial p)_T$ and $(\partial S/\partial p)_V$.

(ii) Show that

$$\left(\frac{\partial S}{\partial V}\right)_T = \left(\frac{\partial p}{\partial T}\right)_V$$

and

$$\left(\frac{\partial S}{\partial p}\right)_T = -\left(\frac{\partial V}{\partial T}\right)_p.$$

Solution. (i) The partial derivative $(\partial S/\partial V)_T$ is the partial derivative of S with respect to V, regarding the entropy S as a function of the variables V and T. (It might be useful to refer back to 17.2.4 at this stage.)

As a function of V and T the entropy S is the expression

$$S = c_v \ln(T) + R \ln(V) + a$$

Thus

$$\left(\frac{\partial S}{\partial V}\right)_T = R \cdot \frac{1}{V} = \frac{R}{V}$$

The partial derivative $(\partial S/\partial p)_T$ has to be obtained from the expression for S as a function of the variables p and T. The given expression for S involves V and T, and we must therefore substitute for V in terms of p and T using the equation of state. From the equation of state we see that $V = RT/p$. Thus

$$S = c_v \ln(T) + R \ln(RT/p) + a$$
$$= c_v \ln(T) + R \ln(R) + R \ln(T) - R \ln(p) + a$$

The partial derivative $(\partial S/\partial p)_T$ is therefore given by

$$\left(\frac{\partial S}{\partial p}\right)_T = -R \cdot \frac{1}{p} = -\frac{R}{p}$$

For $(\partial S/\partial p)_V$ we now need the entropy S expressed as a function of p and V. This expression is obtained from the expression given in the question by substituting for T in terms of p and V. Thus

$$S = c_v \ln(pV/R) + R \ln(V) + a$$
$$= c_v \ln(p) + c_v \ln(V) - c_v \ln(R) + R \ln(V) + a$$

$(\partial S/\partial p)_V$ is therefore given by

$$\left(\frac{\partial S}{\partial p}\right)_V = c_v \cdot \frac{1}{p} = \frac{c_v}{p}$$

(ii) The quantities $(\partial p/\partial T)_V$ and $(\partial V/\partial T)_p$ can be obtained directly from the equation of state. As a function of T and V the pressure is given as

$$p = RT/V$$

Thus

$$\left(\frac{\partial p}{\partial T}\right)_V = \frac{R}{V}$$

which is the same expression as for $(\partial S/\partial V)_T$. Thus we have the property that

$$\left(\frac{\partial S}{\partial V}\right)_T = \left(\frac{\partial p}{\partial T}\right)_V$$

In a similar way we observe that from the equation of state the volume in terms of p and T is given as

$$V = RT/p$$

Hence $(\partial V/\partial T)_p = R/p$ which is the same as $-(\partial S/\partial p)_T$. Thus we also have the result that

$$\left(\frac{\partial S}{\partial p}\right)_T = -\left(\frac{\partial V}{\partial T}\right)_p$$

17.4 Further developments

17.4.1 Total changes. Suppose that we have a function of two variables x and y, $f(x, y)$, but instead of just one of the variables changing giving rise to a partial change, suppose that both variables change by small amounts δx and δy. The total change in the function is given by

$$\delta f = f(x + \delta x, y + \delta y) - f(x, y) \tag{4}$$

By adding and subtracting a pair of identical terms we can write this expression in terms of partial changes in the following way:

$$\delta f = f(x + \delta x, y + \delta y) - f(x, y + \delta y) + f(x, y + \delta y) - f(x, y)$$

Dividing and multiplying the first pair of terms by δx and the second pair by δy we have

$$\delta f = \frac{[f(x + \delta x, y + \delta y) - f(x, y + \delta y)]}{\delta x} \delta x + \frac{[f(x, y + \delta y) - f(x, y)]}{\delta y} \delta y \tag{5}$$

If we now make the changes δx and δy arbitrarily small, each of the square brackets become first-order partial derivatives. We can write

$$\delta f = \frac{\partial f}{\partial x} \delta x + \frac{\partial f}{\partial y} \delta y \tag{6}$$

In other words, if the values of x and y change by small amounts δx and δy, then the subsequent change in f, which we have denoted by δf, is approximately equal to the right-hand side of the above equation. This formula is often called *Taylor's approximation formula*. This formula is of fundamental importance in studying small changes of all kinds that can occur in functions of several variables and is the starting point for a whole range of applications, some of which we now consider.

17.4.2 Error analysis. In Chapter 8 we discussed the type of errors that can occur in science, for example, there are rounding errors due to the approximation of decimal numbers and there are experimental errors due to the inaccuracy of reading instruments. In Chapter 8 we used the chain rule of ordinary differentiation to evaluate the error in a function due to an error in one of the variables. Now we investigate the error when more than one variable is in error. We use Taylor's approximation formula and an important result called *the triangle inequality*:

For any real numbers a and b,

$$|a + b| \leqslant |a| + |b| \tag{7}$$

Suppose that w is some physical quantity of interest that depends on the measurable quantities x and y via a relationship $w = f(x, y)$, and suppose that w is to be obtained by first measuring x and y and then using the functional relationship to work out its value. Errors made in measuring x and y will clearly propagate into an overall error in the value of w.

To see what this overall error is, we can assume that $\delta x = e_1$ and $\delta y = e_2$ are the small errors made in measuring x and y respectively and these errors could be positive or negative depending on the nature of the errors. Using Taylor's approximation formula (6), the resulting overall error in w, e say, will be approximately

$$e \simeq \frac{\partial f}{\partial x} e_1 + \frac{\partial f}{\partial y} e_2$$

It is important to realise that we will never know exactly what the values of e_1 and e_2 are, however we can usually estimate the range in which they occur. Let us assume that

$$-r_1 \leqslant e_1 \leqslant r_1$$
$$-r_2 \leqslant e_2 \leqslant r_2$$

We can now obtain a bound on the value of e. If we look at the worst that can happen, namely that each e_i is either $\pm r_i$ and that all the errors made are such that their signs cause them to combine to give the maximum possible error in w, then we see that w will be some quantity in error by e that is bounded by $+ e_{max}$, where

$$e_{max} = \left| \frac{\partial f}{\partial x} e_1 \right| + \left| \frac{\partial f}{\partial y} e_2 \right| \tag{8}$$

The modulus signs ensure that all the terms in the summation are given the same sign to cause the maximum possible error to arise. This is essentially an application of the triangle inequality. We have

$$e = \left| \frac{\partial f}{\partial x} e_1 + \frac{\partial f}{\partial y} e_2 \right| \leqslant \left| \frac{\partial f}{\partial x} e_1 \right| + \left| \frac{\partial f}{\partial y} e_2 \right| = e_{max}$$

Example: The quantity w is calculated from variables x and y by using the formula

$$w = f(x, y) = 2 + 3x^2 + 3y^2 - 3x^2 y - y^3$$

In an experiment the variable x is measured to be 1.0 and the variable y is measured to be 0.0. If each measurement is subject to a maximum experimental error of modulus 0.05, estimate the maximum possible error in the calculated value of w.

Solution: Using Taylor's approximation formula (6) we have

$$e \simeq \delta w \simeq \frac{\partial f}{\partial x}\delta x + \frac{\partial f}{\partial y}\delta y$$

Now $\partial f/\partial x = 6x - 6xy$ and $\partial f/\partial y = 6y - 3x^2 - 3y^2$. At the point $x = 1$, $y = 0$ the values of these partial derivatives are

$$\frac{\partial f}{\partial x}(1,0) = 6 \quad \text{and} \quad \frac{\partial f}{\partial y}(1,0) = -3$$

So from (8)

$$e_{max} = |6e_1| + |-3e_2|$$

But

$$|e_1| \leqslant 0.05 \quad \text{and} \quad |e_2| \leqslant 0.05$$

Hence

$$e_{max} = 9 \times 0.05 = 0.45$$

Hence the maximum error in the calculated value of w is 0.45.

17.4.3 Differentials. If we refer back to the first example in Section 17.1, we see that the equation of state for a pure substance can be regarded as a relationship that connects the three variables p, V and T. What this means is that any one of the variables p, V and T can be expressed as a function of the other two. This dependence on two of the variables permeates all the way through thermodynamics and in fact any thermodynamic quantity, such as entropy or enthalpy for example, can be expressed in terms of just two thermodynamic variables. Properties of functions of two variables are therefore important in thermodynamics and what is especially important is being able to tell whether or not an expression involving infinitesimal changes in two thermodynamic variables corresponds to the total change in any given thermodynamic quantity. This has deep physical significance as far as the study of reversible and irreversible processes is concerned and it is here that the theory of differentials is important. Suppose that f is a function of one variable, x say, and x increases by an amount δx, then the Taylor polynomial in δx for f about x (see Chapter 8) gives

$$\delta f = f(x + \delta x) - f(x) = \frac{df}{dx}\delta x + \frac{1}{2!}\frac{d^2 f}{dx^2}\delta x^2 + \cdots \tag{9}$$

The sum on the right-hand side contains an infinite number of terms for equality to hold. Suppose we call the first term δf_t, i.e.

$$\delta f_t = \frac{df}{dx}\delta x$$

Now δf and δf_t are not equal in general since we have

$$\delta f = \delta f_t + \text{second- and higher-order terms in } \delta x$$

The difference between δf and δf_t is easy to visualise geometrically. In Fig. 17.8 the point $(x + \delta x, f + \delta f)$ lies on the graph of $y = f(x)$. However the point $(x + \delta x, f + \delta f_t)$ lies on the tangent at $(x, f(x))$ to the graph. The difference between the lengths δf and δf_t is now clear. The change δx is not restricted in

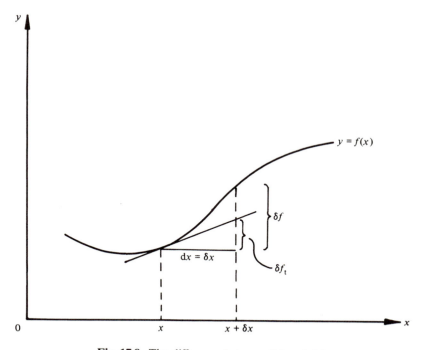

Fig. 17.8. The difference between δf_t and δf.

size, the larger δx is then the more terms we need to take in (9) to achieve a good approximation to δf, so that in general the greater is the difference between δf and δf_t. However, the smaller we take δx then the closer is the tangent approximation of f to the function values of f. If x increases by an infinitesimally small amount we write such a change as dx and define *the differential of f* as

$$df = \frac{df}{dx} dx$$

Differentials are infinitesimally small changes and are not to be confused with actual changes δf and the approximation to such a change δf_t.

Consider now a function $f(x, y)$ of two variables x and y. If x and y change by small amounts δx and δy, then according to Taylor's approximation formula, the change in f is given by

$$\delta f = \frac{\partial f}{\partial x} \delta x + \frac{\partial f}{\partial y} \delta y + \text{second and higher-order terms}$$

By a similar argument we define the *total or perfect differential of a function of two variables* df to be

$$df = \frac{\partial f}{\partial x} dx + \frac{\partial f}{\partial y} dy$$

Now in general the partial derivatives $\partial f/\partial x$ and $\partial f/\partial y$ will be functions of x and y, so that we can write df as

$$df = M(x, y)\,dx + N(x, y)\,dy$$

where M and N are expressions for $\partial f/\partial x$ and $\partial f/\partial y$ respectively.

One question we might now ask is whether all expressions of the form $P(x, y)\,dx + Q(x, y)\,dy$, where P and Q are arbitrary functions of x and y, are in fact perfect differentials. The answer is no, unless both $\partial P/\partial y$ and $\partial Q/\partial x$ happen to be equal. Why this should be the case can easily be appreciated if we consider the perfect differential. Here we have $\partial f/\partial x = M(x, y)$ and $\partial f/\partial y = N(x, y)$. Now since the mixed derivatives $\partial^2 f/\partial x\partial y$ and $\partial^2 f/\partial y\partial x$ are in general equal, it follows that

$$\frac{\partial}{\partial y}\left(\frac{\partial f}{\partial x}\right) = \frac{\partial M}{\partial y} = \frac{\partial}{\partial x}\left(\frac{\partial f}{\partial y}\right) = \frac{\partial N}{\partial x}$$

i.e.

$$\frac{\partial M}{\partial y} = \frac{\partial N}{\partial x}$$

Thus for $P(x, y)\,dx + Q(x, y)\,dy$ to be a perfect differential, we must therefore have that

$$\frac{\partial P}{\partial y} = \frac{\partial Q}{\partial x}$$

Example: Which of the following expressions are perfect differentials?

(i) $2(x - y - 1)dx - (2x - 2y - 1)dy$;

(ii) $\left(\dfrac{RT}{(V - b)^2} - \dfrac{2a}{V^3}\right)dV - \left(\dfrac{R}{(V - b)}\right)dT$;

(iii) $(\cos(x + y) - 2\sin(x - y))dx - (\cos(x + y) + 2\sin(x - y))dy$.

Solution:

(i) $P(x, y) = 2(x - y - 1)$ and $Q(x, y) = -(2x - 2y - 1)$

For a perfect differential we must have $\partial P/\partial y = \partial Q/\partial x$. So $\partial P/\partial y = -2$ and $\partial Q/\partial x = -2$. We conclude that the expression is a perfect differential.

(ii) With the expression in (ii) we see that

$$P(V, T) = \frac{RT}{(V - b)^2} - \frac{2a}{V^3} \quad \text{and} \quad Q(V, T) = -\frac{R}{(V - b)}$$

Finding the partial derivatives we have

$$\frac{\partial P}{\partial T} = \frac{R}{(V - b)^2} \quad \text{and} \quad \frac{\partial Q}{\partial V} = \frac{R}{(V - b)^2}$$

and since these two partial derivatives are the same we conclude that the expression is a perfect differential.

(iii) Here we have

$$P(x, y) = \cos(x + y) - 2\sin(x - y) \text{ and}$$
$$Q(x, y) = -\sin(x + y) + 2\cos(x - y)$$

So

$$\frac{\partial P}{\partial y} = -\sin(x+y) + 2\cos(x-y)$$

and

$$\frac{\partial Q}{\partial x} = \cos(x+y) - 2\sin(x-y)$$

These two partial derivatives are not the same and we do not, therefore, have a perfect differential.

17.4.4 The chain rule. In 6.4.1 we introduced the chain rule for functions of one variable. This says that if $y = y(u)$ and $u = u(x)$ then

$$\frac{dy}{dx} = \frac{dy}{du}\frac{du}{dx}$$

We shall now derive a similar expression for functions of two variables $f(x, y)$. Suppose that x and y are functions of a variable t, so that $x = x(t)$ and $y = y(t)$. Suppose that t increases by a small amount δt so that we can write the small changes in x and y as

$$\delta x = x(t + \delta t) - x(t) \quad \text{and} \quad \delta y = y(t + \delta t) - y(t)$$

Then using Taylor's approximation formula we have

$$\delta f = f(x + \delta x, y + \delta y) - f(x, y)$$

Now adding and subtracting $f(x, y + \delta y)$ and multiplying and dividing by δx and δy (as in 17.4.1 above) we obtain

$$\delta f = \left[\frac{f(x + \delta x, y + \delta y) - f(x, y + \delta y)}{\delta x} \right] \delta x + \left[\frac{f(x, y + \delta y) - f(x, y)}{\delta y} \right] \delta y$$

Dividing each side by δt we get

$$\frac{\delta f}{\delta t} = \left\{ \frac{f(x + \delta x, y + \delta y) - f(x, y + \delta y)}{\delta x} \right\} \frac{\delta x}{\delta t} + \left\{ \frac{f(x, y + \delta y) - f(x, y)}{\delta y} \right\} \frac{\delta y}{\delta t}$$

The next step is to take the limit as δt tends to zero. The left-hand side becomes df/dt. Also $\delta x/\delta t$ and $\delta y/\delta t$ become dx/dt and dy/dt respectively. Now as δt tends to zero so do δx and δy. The expressions in parentheses then give the first-order partial derivatives of $f(x, y)$. So we have

$$\frac{df}{dt} = \frac{\partial f}{\partial x}\frac{dx}{dt} + \frac{\partial f}{\partial y}\frac{dy}{dt} \tag{10}$$

This is called *the chain rule* for a function of two variables.

For a function of three variables $f(x, y, z)$, where $x = x(t)$, $y = y(t)$ and $z = z(t)$ the corresponding chain rule is

$$\frac{df}{dt} = \frac{\partial f}{\partial x}\frac{dx}{dt} + \frac{\partial f}{\partial y}\frac{dy}{dt} + \frac{\partial f}{\partial z}\frac{dz}{dt}$$

Example: If $f(x, y) = 1 - x^2 - y^2$, $x = 2t$ and $y = 1 - 2t$, evaluate df/dt directly and using the chain rule.

Solution: For the direct method we start with $f(x, y)$ and substitute for x and y as functions of t,

$$f = 1 - x^2 - y^2$$
$$= 4t - 8t^2$$

Now ordinary differentiation with respect to t gives

$$\frac{df}{dt} = 4 - 16t$$

Using the chain rule (10), first we find the partial derivatives,

$$\frac{\partial f}{\partial x} = -2x \quad \text{and} \quad \frac{\partial f}{\partial y} = -2y$$

Also

$$\frac{dx}{dt} = 2 \quad \text{and} \quad \frac{dy}{dt} = -2$$

So on substitution into (10)

$$\frac{df}{dt} = \frac{\partial f}{\partial x}\frac{dx}{dt} + \frac{\partial f}{\partial y}\frac{dy}{dt}$$
$$= (-2x)(2) + (-2y)(-2)$$
$$= -4x + 4y$$

and replacing x and y in terms of t

$$\frac{df}{dt} = -4(2t) + 4(1 - 2t)$$
$$= 4 - 16t$$

The results agree as they should.

Very often it is not the case that the variables $x, y, \ldots$ in the chain rule formula are functions of just one variable t. Suppose without loss of generality that we have a function $f(x, y)$ where $x = x(u, w)$ and $y = y(u, w)$. The chain rule in (10) generalises now to a form involving only partial derivatives, namely

$$\frac{\partial f}{\partial u} = \frac{\partial f}{\partial x}\frac{\partial x}{\partial u} + \frac{\partial f}{\partial y}\frac{\partial y}{\partial u} \tag{11}$$

and

$$\frac{\partial f}{\partial w} = \frac{\partial f}{\partial x}\frac{\partial x}{\partial w} + \frac{\partial f}{\partial y}\frac{\partial y}{\partial w} \tag{12}$$

17.5 Further worked examples

Example 1. The focal length f of a convex lens is measured by observing the image distance v and the object distance u, on an optical bench, and using the formula

$$\frac{1}{u} + \frac{1}{v} = \frac{1}{f}$$

If u and v can each be measured to an accuracy of 1% and the image is a real one, obtain the maximum possible percentage error in f.

Solution. If $(1/u) + (1/v) = 1/f$ then the expression for the focal length is

$$f = \frac{uv}{(u+v)}$$

If δu and δv are small errors made in the measurements of u and v, then the corresponding error δf in f will be given to a good approximation by

$$\delta f = \frac{\partial f}{\partial u}\,\delta u + \frac{\partial f}{\partial v}\,\delta v$$

$$= \left(\frac{(u+v)v - uv}{(u+v)^2}\right)\delta u + \left(\frac{(u+v)u - uv}{(u+v)^2}\right)\delta v$$

$$= \frac{v^2\,\delta u}{(u+v)^2} + \frac{u^2\,\delta v}{(u+v)^2}$$

The maximum possible error in f will occur when each term in this expression contributes with the same positive sign to the overall error and each error made is the worst one possible. Thus we set δu to $+u/100$ and δv to $+v/100$ and obtain for e_{max} the expression

$$e_{max} = \frac{v^2 u}{100(u+v)^2} + \frac{u^2 v}{100(u+v)^2}$$

$$= \frac{uv}{100(u+v)}$$

$$= \frac{f}{100}$$

Thus dividing both sides by the focal length f we see that

$$\frac{e_{max}}{f} = \frac{1}{100}$$

The maximum possible error in the focal length calculation is thus 1%.

Example 2.

(i) If $f = x^\alpha y^\beta z^\gamma$, where α, β and γ are constants, show that when x, y and z change by small amounts $\delta x, \delta y$ and δz, then the relative change in f is equal to the corresponding total change in the quantity $\ln(f)$.

(ii) The rate of flow of a liquid through a tube is given by Poiseuille's equation as

$$Q = p\pi a^4/8\eta l$$

where Q is the rate of flow, p is the pressure difference between the ends of the tube, a is the radius of the tube and η is the coefficient of viscosity of the liquid. If η is obtained by measuring Q, p, a and l and if Q can be measured accurate to 0.5%, p accurate to 2%, a accurate to 2% and l accurate to 0.5%, calculate the maximum possible % error in the value of η.

Solution.

(i) If $f = x^{\alpha}y^{\beta}z^{\gamma}$ then to a good approximation we can write for the small change in f the following:

$$\delta f = \frac{\partial f}{\partial x}\delta x + \frac{\partial f}{\partial y}\delta y + \frac{\partial f}{\partial z}\delta z$$

$$= \alpha x^{\alpha-1}y^{\beta}z^{\gamma}\delta x + \beta x^{\alpha}y^{\beta-1}z^{\gamma}\delta y + \gamma x^{\alpha}y^{\beta}z^{\gamma-1}\delta z$$

Thus dividing both sides by f to obtain the relative change in f we have

$$\frac{\delta f}{f} = \alpha\frac{\delta x}{x} + \beta\frac{\delta y}{y} + \gamma\frac{\delta z}{z}$$

Consider now the function $\ln(f)$. Using the properties of logarithms, $\ln(f)$ is given by

$$\ln(f) = \alpha\ln(x) + \beta\ln(y) + \gamma\ln(z)$$

The total change in $\ln(f)$ will be given by

$$\delta\ln(f) = \frac{\partial(\ln(f))}{\partial x}\delta x + \frac{\partial(\ln(f))}{\partial y}\delta y + \frac{(\partial(\ln(f)))}{\partial z}\delta z$$

$$= \frac{\alpha\delta x}{x} + \frac{\beta\delta y}{y} + \frac{\gamma\delta z}{z}$$

which is clearly the same as the expression for the relative change $\delta f/f$ in the function f.

(ii) From Poiseuille's equation we can express η in the form

$$\eta = \frac{p\pi a^{4}}{8Ql}$$

Using the result in part (i) above, we can express the relative error in η as the error in $\ln(\eta)$. Similarly, by the same token, the maximum possible % error in η will be 100 times the maximum possible error in $\ln(\eta)$. Now, the expression for $\ln(\eta)$ may be written as

$$\ln(\eta) = \ln(p) + \ln(\pi/8) + 4\ln(a) - \ln(Q) - \ln(l)$$

For which

$$\delta(\ln(\eta)) = \frac{\delta p}{p} + \frac{4\delta a}{a} - \frac{\delta Q}{Q} - \frac{\delta l}{l}$$

The maximum possible error, $e_{\max}$, in $\ln(\eta)$ will thus be given by

$$e_{\max} = \left|\frac{\delta p}{p}\right| + 4\left|\frac{\delta a}{a}\right| + \left|\frac{\delta Q}{Q}\right| + \left|\frac{\delta L}{L}\right|$$

where δp, δa, δQ and δl are the worst errors that can arise in measurements of p, *a*, *Q* and *l*. Using the data given, we see that

$$e_{max} = \frac{2}{100} + \frac{(4)(2)}{100} + \frac{0.5}{100} + \frac{0.5}{100}$$

$$= \frac{11}{100}$$

This maximum possible error in $\ln(\eta)$ is the maximum fractional error in η since $\delta \ln(\eta) = \delta\eta/\eta$. The maximum possible percentage error in η is thus 100 times this quantity, and is therefore 11%.

Example 3. If the thermodynamic expression $C\,\mathrm{d}T + p\,\mathrm{d}V$, where C is a constant, can be made into a perfect differential by multiplying by a function of T only, deduce the form of this function given that $pV = RT$.

Solution. Let this function of T be $I(T)$, so that the perfect differential is $I(T)C\,\mathrm{d}T + I(T)p\,\mathrm{d}V$. Since this is a perfect differential we must necessarily have that

$$\frac{\partial(CI)}{\partial V} = \frac{\partial(pI)}{\partial T}$$

Now $\partial(CI)/\partial V$ is zero, since CI is independent of V, and so $\partial(pI)/\partial T$ must also be zero.

Expressing p in terms of V and T (the two relevant variables here) we see that $pI = IRT/V$. If this variable quantity has to be such that its partial derivative with respect to T vanishes, then it must be independent of T. Thus, given that I is a function of T only, I must take the form of a constant divided by T. Thus

$$I(T) = A/T$$

where A is any arbitrary constant.

Example 4. The internal energy U of a gas for which $pV = RT$ is such that

$$\mathrm{d}U = T\mathrm{d}S + p\,\mathrm{d}V$$

where S is the entropy of the gas. By noting that $\mathrm{d}S$ is a perfect differential, deduce that U is a function of temperature only.

Solution. The expression for $\mathrm{d}S$ is $[(1/T)\mathrm{d}U] - [(p/T)\mathrm{d}V]$. Letting the variables of interest in the problem be V and T, we can express $\mathrm{d}U$ as $[(\partial U/\partial V)\mathrm{d}V] + (\partial U/\partial T)\mathrm{d}T$. Thus the expression for $\mathrm{d}S$ is now given by

$$\mathrm{d}S = \left(\frac{1}{T}\frac{\partial U}{\partial T}\right)\mathrm{d}T + \left(\frac{1}{T}\frac{\partial U}{\partial V} - \frac{p}{T}\right)\mathrm{d}V$$

If $\mathrm{d}S$ is to be a perfect differential, then we require

$$\frac{\partial}{\partial V}\left(\frac{1}{T}\frac{\partial U}{\partial T}\right) = \frac{\partial}{\partial T}\left(\frac{1}{T}\frac{\partial U}{\partial V} - \frac{p}{T}\right)$$

Considering first of all the partial derivative with respect to V, we see that

$$\frac{\partial}{\partial V}\left(\frac{1}{T}\frac{\partial U}{\partial T}\right) = \frac{1}{T}\frac{\partial^2 U}{\partial V \partial T}$$

Replacing p by RT/V and considering the partial derivative with respect to T, we see that

$$\frac{\partial}{\partial T}\left(\frac{1}{T}\frac{\partial U}{\partial V} - \frac{p}{T}\right) = \frac{\partial}{\partial T}\left(\frac{1}{T}\frac{\partial U}{\partial V} - \frac{R}{V}\right)$$

$$= \frac{1}{T}\frac{\partial^2 U}{\partial T \partial V} - \frac{1}{T^2}\frac{\partial U}{\partial V}$$

If, therefore, these two partial derivatives have to be equal, as dS is a perfect differential, then we must have that $(1/T^2)(\partial U/\partial V) = 0$. The partial derivative $\partial U/\partial V$ is therefore zero, making U at best a function of temperature only.

Example 5. The wave equation in two-dimensions can be written as

$$\frac{\partial^2 z}{\partial x^2} + \frac{\partial^2 z}{\partial y^2} = \frac{1}{c^2}\frac{\partial^2 z}{\partial t^2}$$

where $z(x, y, t)$ measures the displacement from equilibrium at the point (x, y), at time t, of the wave disturbance. Express this wave equation in terms of plane polar coordinates, for which:

$$x = r\cos(\theta) \quad \text{and} \quad y = r\sin(\theta)$$

Solution. The quantities that need transforming to plane polar coordinates are $\partial^2 z/\partial x^2$ and $\partial^2 z/\partial y^2$. These derivatives can only be transformed by first transforming $\partial z/\partial x$ and $\partial z/\partial y$. Consider first $\partial z/\partial x$; using the chain rule ·we can write

$$\frac{\partial z}{\partial x} = \frac{\partial z}{\partial r}\frac{\partial r}{\partial x} + \frac{\partial z}{\partial \theta}\frac{\partial \theta}{\partial x}$$

Since $r^2 = x^2 + y^2$, we have

$$\frac{\partial r}{\partial x} = \frac{x}{r} = \cos(\theta)$$

Also, since $\tan(\theta) = y/x$, we have $\sec^2(\theta)\partial\theta/\partial x = -y/x^2$. Writing this in terms of r and θ we have

$$\frac{\partial \theta}{\partial x} = \frac{-r\sin(\theta)}{r^2\cos^2(\theta)\sec^2(\theta)} = \frac{-\sin(\theta)}{r}$$

Thus

$$\frac{\partial z}{\partial x} = \frac{\partial z}{\partial r}\cos(\theta) - \frac{\partial z}{\partial \theta}\frac{\sin(\theta)}{r}$$

Similarly we have

$$\frac{\partial z}{\partial y} = \frac{\partial z}{\partial r}\frac{\partial r}{\partial y} + \frac{\partial z}{\partial \theta}\frac{\partial \theta}{\partial y}$$

$$= \frac{\partial z}{\partial r} \sin(\theta) + \frac{\partial z}{\partial \theta} \frac{\cos(\theta)}{r}$$

To obtain $\partial^2 z / \partial x^2$ we differentiate $\partial z / \partial x$ partially with respect to x using the chain rule:

$$\frac{\partial^2 z}{\partial x^2} = \frac{\partial}{\partial r}\left(\frac{\partial z}{\partial x}\right)\frac{\partial r}{\partial x} + \frac{\partial}{\partial \theta}\left(\frac{\partial z}{\partial x}\right)\frac{\partial \theta}{\partial x}$$

$$= \left(\frac{\partial^2 z}{\partial r^2}\cos\theta + \frac{\partial z}{\partial \theta}\frac{\sin\theta}{r^2} - \frac{\partial^2 z}{\partial r \partial \theta}\frac{\sin\theta}{r}\right)\frac{\partial r}{\partial x}$$

$$+ \left(-\frac{\partial z}{\partial r}\sin\theta + \frac{\partial^2 z}{\partial r \partial \theta}\cos\theta - \frac{\partial^2 z}{\partial \theta^2}\frac{\sin\theta}{r} - \frac{\partial z}{\partial \theta}\frac{\cos\theta}{r}\right)\frac{\partial \theta}{\partial x}$$

$$= \frac{\partial^2 z}{\partial r^2}\cos^2\theta - 2\frac{\partial^2 z}{\partial r \partial \theta}\frac{\cos\theta\sin\theta}{r} + 2\frac{\partial z}{\partial \theta}\frac{\cos\theta\sin\theta}{r^2} + \frac{\partial z}{\partial r}\frac{\sin^2\theta}{r} + \frac{\partial^2 z}{\partial \theta^2}\frac{\sin^2\theta}{r^2}$$

Similarly, it can be shown that $\partial^2 z / \partial y^2$ is given by

$$\frac{\partial^2 z}{\partial r^2}\sin^2\theta + \frac{2\partial^2 z}{\partial r \partial \theta}\frac{\cos\theta\sin\theta}{r} - 2\frac{\partial z}{\partial \theta}\frac{\cos\theta\sin\theta}{r^2} + \frac{\partial z}{\partial r}\frac{\cos^2\theta}{r} + \frac{\partial^2 z}{\partial \theta^2}\frac{\cos^2\theta}{r^2}$$

Adding together these last two expressions, and collecting terms as appropriate, and noting that $\cos^2(\theta) + \sin^2(\theta) = 1$, we get

$$\frac{\partial^2 z}{\partial x^2} + \frac{\partial^2 z}{\partial y^2} = \frac{\partial^2 z}{\partial r^2} + \frac{1}{r}\frac{\partial z}{\partial r} + \frac{1}{r^2}\frac{\partial^2 z}{\partial \theta^2}$$

The wave equation in plane polar coordinates is thus expressible as

$$\frac{\partial^2 z}{\partial r^2} + \frac{1}{r}\frac{\partial z}{\partial r} + \frac{1}{r^2}\frac{\partial^2 z}{\partial \theta^2} = \frac{1}{c^2}\frac{\partial^2 z}{\partial t^2}$$

17.6 Exercises

1. Find $\partial f/\partial x$, $\partial f/\partial y$, $\partial^2 f/\partial x^2$, $\partial^2 f/\partial x \partial y$ and $\partial^2 f/\partial y^2$ for the following functions:

   (i) $f = xy^2 - x^2 + y^2 - 3$ (ii) $f = (x - y)^4$
   (iii) $f = \sin(xy)$ (iv) $f = xe^{x-y}$
   (v) $f = xy - \ln(xy)$ (vi) $f = (x/y) - (y/x)$
   (vii) $f = 4x^3 - 5xy^2 + 3y^3$ (viii) $f = \cos(2x + 3y)$

2. For the following functions verify that $\partial^2 f/\partial x \partial y = \partial^2 f/\partial y \partial x$:

   (i) $f(x, y) = x^2 \sin(y) + y^2 \cos(x)$
   (ii) $f(x, y) = y \ln(x)/x$
   (iii) $f(x, y) = \sin(y/x)$

3. If $z = g(y/x)$ where g is an arbitrary function prove that:

$$x\frac{\partial z}{\partial x} + y\frac{\partial z}{\partial y} = 0$$

4. If $z = xy\, e^{y/x}$ show that:

(i) $x\dfrac{\partial z}{\partial x} + y\dfrac{\partial z}{\partial y} = 2z$

(ii) $x^2\dfrac{\partial^2 z}{\partial x^2} + 2xy\dfrac{\partial^2 z}{\partial x \partial y} + y^2\dfrac{\partial^2 z}{\partial y^2} = 2z$

5. Find dz/dt in the following cases:
   (i) $z = x^n y^n$, where $x = \cos(at)$ and $y = \sin(at)$;
   (ii) $z = \ln(x^2 + y^2)$, where $x = a(1 - \cos(t))$ and $y = a\sin(t)$;
   (iii) $z = (ax - by)/(cx + dy)$, where $x = e^t\sin(t)$ and $y = e^t\cos(t)$.

6. In a calculation of g from $T = 2\pi\sqrt{(1/g)}$ what is the maximum possible error if $1 = 100\,\text{cm}$ with a possible error of $0.15\,\text{cm}$ and $T = 2\,\text{s}$ with a possible error of $0.01\,\text{sec}$?

7. The amplitude I of a current flowing in a certain resonant circuit is given by the expression

$$I^2 = V^2\left\{ C^2 w^2 + \frac{1 - 2LCw^2}{R^2 + L^2 w^2} \right\}$$

If R and C are subject to small variations δR and δC while L, w and V are kept constant, find an appropriate expression for the resulting variation δI of I.

Verify that this variation is negligible if $C = 1/2Lw^2$ and $R = Lw$.

8. A formula for turbulent gas flow is

$$v = 5.3\, L^{5/4} D^{1/4} / (e_p^{5/4} T^{5/6})$$

If the measurements of D and T are subject to errors of $\pm 1.5\%$ and $\pm 1\%$ respectively, show that the maximum error in the value of v will be 1.2% approximately.

9. If $u(x, y) = x^2 - y^2$, find a function $v(x, y)$ such that $\partial v/\partial x = -\partial u/\partial y$ and $\partial v/\partial y = \partial u/\partial x$ for all x and y.

10. If $f(x, y) = xe^{xy}$, and the values of x and y are slightly changed from 1 and 0 to $1 + \delta x$ and δy, respectively, so that δf, the change in f is very nearly $3\delta x$, show that δy must be very nearly $2\delta x$.

11. The volume of a segment of a sphere is $\pi h(h^2 + 3R^2)/6$, where h is the height of the segment and R is the radius of its base. If the measurement of h is too large by a small amount α, and that of R is too small by an equal amount, show that the calculated volume is too large by an amount

$$\tfrac{1}{2}\pi\alpha(h - R)^2$$

approximately.

12. If g is a function of the independent variables x, y, z which are changed to independent variables u, v, w by the transformation

$$x = vw/u, \quad y = wu/v, \quad z = uv/w$$

show that

$$u\frac{\partial g}{\partial u} + v\frac{\partial g}{\partial v} + w\frac{\partial g}{\partial w} = x\frac{\partial g}{\partial x} + y\frac{\partial g}{\partial y} + z\frac{\partial g}{\partial z}$$

13. If $x = e^r \cos(\theta)$ and $y = e^r \sin(\theta)$ show that

$$\frac{\partial^2 v}{\partial x^2} + \frac{\partial^2 v}{\partial y^2} = e^{-2r} \left(\frac{\partial^2 v}{\partial r^2} + \frac{\partial^2 v}{\partial \theta^2} \right)$$

14. In a Wheatstone bridge $R_1 R_4 = R_2 R_3$ at balance. If R_2, R_3 and R_4 have tolerances of $\pm a\%$, $\pm b\%$ and $\pm c\%$ respectively, show, using the method of partial differentiation, that the maximum possible percentage error in R_1 is approximately $\pm (a + b + c)$.

15. The natural frequency of oscillation of a series LRC circuit is given by

$$f = \frac{1}{2\pi} \sqrt{\frac{1}{LC} - \frac{R^2}{4L^2}}$$

If L is *increased* by 1% and C is *decreased* by 1% show that the percentage increase in f is $R^2 C/(4L - R^2 C)$.

16. Show that the following equations are exact:

   (i) $x\,dy + y\,dx = 0$

   (ii) $1/x\,dy - y/x^2\,dx = 0$

   (iii) $(ax + hy + g)\,dx + (hx + by + f)\,dy = 0$

   (iv) $[1 - \cos(2x)]\,dy + 2y \sin(2x)\,dx = 0$

   (v) $(3x^2 + 2y + 1)\,dx + (2x + 6y^2 + 2)\,dy = 0$

17. The entropy S of a gas for which $pV = RT$ may be written as

$$S = c_V \ln(p) + c_p \ln(V) + \text{constant}$$

where $c_p - c_V = R$. Show that:

(i) $\dfrac{\partial T}{\partial p} \dfrac{\partial S}{\partial V} - \dfrac{\partial T}{\partial V} \dfrac{\partial S}{\partial p} = 1$

(ii) $\dfrac{1}{T^2} \dfrac{\partial^2 T}{\partial p \partial V} = \dfrac{1}{V^2} \dfrac{\partial^2 S}{\partial p^2} - \dfrac{1}{p^2} \dfrac{\partial^2 S}{\partial V^2}$

18. Given that $c_p - c_V = T(\partial p/\partial T)_V (\partial V/\partial T)_p$, obtain an explicit expression for $c_p - c_V$ assuming the equation of state to be $(p + a/V^2)(V - b) = RT$.

19. (a) Verify that $\partial^2 f/\partial x \partial y = \partial^2 f/\partial y \partial x$ when $f = \cos(y) \exp(\sin(x))$.

   (b) If $y = A e^{i(x - ct)} + B e^{i(x + ct)}$, where $i^2 = -1$ and A and B are constants, show that y satisfies the wave equation given by

$$\frac{\partial^2 y}{\partial x^2} = \frac{1}{c^2} \frac{\partial^2 y}{\partial t^2}$$

20. The propagation of stress waves along a conical bar is described by the relation

$$u = \frac{1}{r} \sin(\lambda[r - ct])$$

where λ is the wave number $c^2 = E/\rho$, E is the Young's modulus and ρ is the density of the cone. Find $\partial u/\partial r$ and $\partial u/\partial t$ and show that the displacement u satisfies the equation for spherical waves

$$\rho r \frac{\partial^2 u}{\partial t^2} = E r \frac{\partial^2 u}{\partial r^2} + 2E \frac{\partial u}{\partial r}$$

21. In considering the flow of heat in one dimension the following equation arises:

$$\frac{\partial^2 u}{\partial x^2} = 2\frac{g^2}{n}\frac{\partial u}{\partial t}$$

where g and n are constants. Show that

$$u = e^{-gx}\sin(nt - gx)$$

satisfies this equation.

22. (a) For the gas whose equation of state is given by $pV = RT$ show that

$$\left(\frac{\partial p}{\partial V}\right)_T \left(\frac{\partial V}{\partial T}\right)_p \left(\frac{\partial T}{\partial p}\right)_V = -1$$

(b) For a substance whose thermodynamic equation of state is given by

$$pV = RT\left(1 + \frac{a}{V} + \frac{b}{V^2}\right)$$

where a and b are constants, obtain expressions for the partial derivatives $\partial p/\partial V$ and $\partial V/\partial T$.

23. (a) The density, ρ of a body can be calculated from the formula

$$\rho = W/(W - w)$$

where W is the weight in air, and w is the weight in water. Show that the relative error in ρ due to small errors δW, δw in W and w respectively is approximately

$$\frac{\delta\rho}{\rho} = \left(\frac{w}{(W - w)}\right)\left(\frac{\delta w}{w} - \frac{\delta W}{W}\right)$$

(b) Raoult's law for dilute solutions can be expressed in the form

$$(p_0 - p)/p_0 = w_2 M_1/(w_1 M_2)$$

where p is the vapour pressure of the solution, w_1 and w_2 are the masses of solvent and solute, and p_0, M_1 and M_2 are constants for the system. Calculate the relative error in p due to small errors δw_1 and δw_2 in w_1 and w_2 respectively.

24. (a) The coefficient of diffusion D of a gas is given in terms of the pressure p and temperature T in the form

$$D = ApT^n$$

where A and n are constants. Find an expression for the relative change $\delta D/D$ in D corresponding to small changes δT and δp in the temperature and pressure respectively.

(b) The resistance R of a wire is proportional to its length and inversely proportional to the square of its radius r, i.e. $R = kl/r^2$ where k is a constant. If the relative error in the measurement of the length and radius are 5% and 2.5% respectively, find the relative error in R in the worst possible case.

25. (a) The variation of the molar entropy S of an ideal monatomic gas with temperature T and volume V is given by

$$S = (3R/2)\ln(T) + R\ln(V) + a$$

where R and a are constants. Using partial differentiation find the approximate increase in entropy ΔS when the temperature changes from $T = 320\,\text{K}$ to $324\,\text{K}$ and the volume increases by 2%.

(b) The magnitude of the impedance of a certain electrical circuit measured in ohms, is given by

$$Z = \sqrt{[(R^2 + 10^9 L^2)]}$$

where the resistance R is measured in ohms and the inductance L in henrys. Using partial differentiation, obtain an expression for the approximate change in Z, in terms of R, L, ΔR and ΔL, when the resistance changes from R to $R + \Delta R$ and the inductance changes from L to $L + \Delta L$.

Using the expression, calculate the nominal value and approximate percentage tolerance in Z when the resistance has a nominal value of $100\,\text{ohm}$ and a tolerance of 1% and the inductance is nominally $0.001\,\text{H}$ with a tolerance of 3%.

26. The following equation is used by geophysicists to calculate shear velocity V_s in minerals

$$V_s = \frac{\theta}{280}\left(\frac{A}{\rho}\right)^{\frac{1}{3}}$$

where θ, A and ρ are respectively the thermal specific heat, the mean atomic mass of the mineral and the mean density of the mineral.

(i) Find $\partial V_s/\partial\theta$, $\partial V_s/\partial A$ and $\partial V_s/\partial\rho$. Hence or otherwise determine the approximate relationships between $\%$ errors in θ, A, and ρ and the resulting $\%$ error in V_s.

(ii) What is the estimated maximum $\%$ error that could result in V_s from simultaneous errors of $\pm 0.1\%$ in θ, $\pm 0.2\%$ in A and $\pm 0.4\%$ in ρ?

18

Partial differentiation 2: stationary points

Contents: 17.1 Scientific context – 18.1.1 positions of equilibrium; 18.1.2 optimum conditions; 18.1.3 making a clock 18.2 mathematical developments – 18.2.1 Taylor polynomials of second order; 18.2.2 stationary points; 18.3 worked examples; 18.4 further developments – 18.4.1 least squares analysis; 18.5 further worked examples; 18.6 exercises.

18.1 Scientific context

In Chapter 7 we saw how ordinary differentiation could be used to determine the stationary points of a function of the form $y = f(x)$. Functions of several variables also have stationary points and in this chapter we consider how partial differentiation may be used to determine where these stationary points occur and also their exact nature. Stationary points in functions of several variables are important in many areas of science and we shall be considering several applications to demonstrate this. One notable application is in curve fitting using the criterion of least squares analysis.

18.1.1 Example 1: positions of equilibrium. Imagine we have n fixed point charges of strengths $q_1, q_2, \ldots, q_n$ located at arbitrary points on the x–y plane. Assume that the signs of the charges are quite arbitrary and suppose, too, that the ith point charge is located at the point (x_i, y_i) as shown in Fig. 18.1. One question we might ask is where could we place another arbitrary point charge, that was free to move, so that it remained at rest on the plane? From Section 17.1.3, we know that the fixed point charges will give rise to an electrostatic potential field defined at any arbitrary point (x, y) on the plane by the potential function

$$\phi(x, y) = \sum_{i=1}^{n} \frac{q_i}{4\pi\varepsilon_0 r_i}$$

where distance r_i (see Fig. 18.1) is given by

$$r_i^2 = (x - x_i)^2 + (y - y_i)^2$$

This potential function corresponds to the potential energy of a unit positive charge placed at the point (x, y). An arbitrary point charge of strength q, say (i.e. the one that we are going to add to the others), placed at this point, will thus have electrostatic potential energy $V(x, y)$ given by

$$V(x, y) = q\phi(x, y)$$

It will remain in equilibrium only if the forces acting on it due to the other fixed

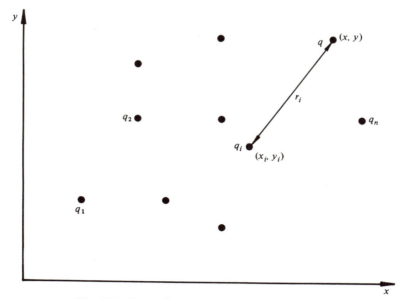

Fig. 18.1. Point charges located in the $x-y$ plane.

charges cancel one another out, in other words if the individual components add up to zero in any given direction. Now from what we know about the electric field (i.e. the force on a unit positive charge) being equal to minus the derivative of the potential function $\phi(x, y)$, it follows that the forces on q, for example in the x and y directions, will only add up to zero if both $(\partial q\phi/\partial x)$ and $(\partial q\phi/\partial y)$, i.e. both $\partial V/\partial x$ and $\partial V/\partial y$, are zero. These conditions, as we shall see in the next section, are just the conditions necessary to locate the stationary point of the function $V(x, y)$. The simpler answer to our question, therefore, is that for the arbitrary charge to be at rest it should be placed at one of the stationary points of the potential energy function $V(x, y)$.

Question: Not all positions of equilibrium are stable ones. How can we ensure that we choose a stationary point of $V(x, y)$ that corresponds to a position of stable equilibrium?

Answer: In general, states of unstable equilibrium are relatively short-lived, so we should certainly be seeking to place the charge in a position of stable equilibrium if it is to remain at rest on the plane for any period of time. A state of stable equilibrium corresponds to a minimum in the potential energy function, so the answer has to be that we place the charge at a minimum point in the potential energy function $V(x, y)$.

Thus, as well as being able to locate where stationary points are, it is crucial that we can also identify their nature. This very important aspect of stationary points is covered in 18.2.2.

18.1.2 Example 2: optimum conditions. In this section we consider a simple optimisation problem. Suppose that a certain chemical product is being made industrially by what is essentially a two-stage process, which is shown in Fig. 18.2.

In process 1 the input materials are fed into a reaction vessel where various chemical reactions take place under the effects of pressure, temperature and catalysts. The important factor in this process is the pressure, which can be varied: the individual reactions go better at higher pressures, but the cost then becomes higher as a result.

In process 2, the products of process 1 are fed into a separator where the final chemical product is removed along with waste and reusable input materials. These reusable input materials are recycled as shown in Fig. 18.2. The longer the products from process 1 remain in the separator the better as far as the amounts of final product and reusable inputs are concerned, but, if they remain too long in the separator, process 1 is likely to be adversely affected and might even be unable to operate continuously. In order to meet the demand that exists for the chemical product a range of operating pressures and separation times are possible but the obvious question is what are the optimum values of pressure and separation time going to be?

One way of looking at this problem is to look at the monthly cost of making the product. This overall cost will depend on quantities such as the size of the demand, the cost of raw materials, labour costs, plant overheads and so on, but it will also depend on the operating pressure P and the separation time T which are quantities that can readily be varied by the operators of the process. In operational research terms, we say that these quantities are not part of the environment (as are the other factors mentioned like labour costs, for example). The monthly cost C of producing the product can thus be expressed as a function of several variables in the form

$$C = C(P, T, \text{environmental factors})$$

Now, if the environment is not going to suddenly change during the space of a month, then C is essentially a function of the two controllable variables P and T. Finding the values of P and T that make C a minimum will thus give us an optimum solution to the question posed. The problem, therefore, is simply one of

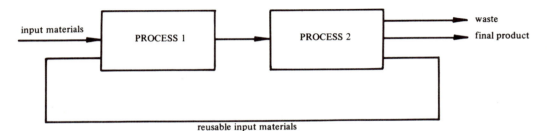

Fig. 18.2. A simple model for the production of a chemical product.

finding and identifying the stationary values of a function of several variables, C in this case.

In Section 18.3 we shall look at this problem again and consider a specific mathematical form for the cost function C. We shall therefore arrive at optimum operating conditions of pressure and separation time for the overall chemical process.

18.1.3 Example 3: making a clock. For our last example, we consider some of the problems involved in making a measuring device. Suppose that we want to make a clock using our heart beat as a simple oscillator. What we must do first is establish some relationship between heart beat and time. This can be done by using our pulse and recording the number of heart beats in several known time intervals with a digital watch for example. Suppose we have done this as accurately as we can and obtained the following results:

Time interval (s)	Number of recorded heart beats
20	19
30	28
40	36
50	44
60	57
70	68

These data are plotted in Fig. 18.3. The graph clearly suggests that the number of heart beats Y and the time interval X are linearly related. In other words, allowing for a certain amount of scatter, which we would expect anyway, the relationship between Y and X is of the form

$$Y = mX + c$$

where m and c are constants to be determined from the data.

Question: How do we determine m and c from the data?
Answer: Clearly we want the line to be a good fit so that it would seem reasonable to minimise some quantity that measures the deviation of the data points from the line.

One method of approach is to draw what looks like the 'best' line by eye and then calculate m and c from this line. *Least squares analysis* does this by minimising the sum of the squares of deviations of each data point from the line. We shall see that the sum of squares of deviations is simply a function of the quantities m and c, so the problem becomes one of finding where a function of two variables has a minimum. So we are back to partial differentiation and least squares analysis is another application of the theory of stationary points.

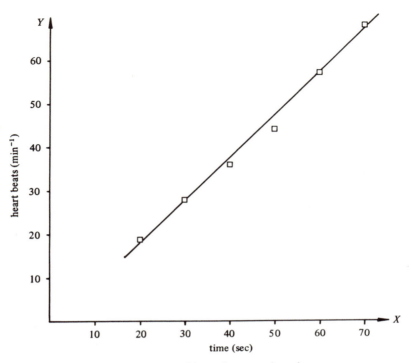

Fig. 18.3. Graph of heart beats against time.

18.2 Mathematical developments

The approach to this chapter will be to develop the theory of stationary points by concentrating primarily on functions of two variables. Where the two-dimensional theory easily generalises to many dimensions we shall state, without proof, the appropriate results.

The conditions necessary to locate and identify stationary points of a function are usually obtained by looking at the Taylor polynomial expansion of the function. We begin by considering the Taylor polynomial expansion of a function of two variables. We have already met Taylor polynomials in Chapter 8 for functions of one variable, and we shall make use of some of the results we established there.

18.2.1 Taylor polynomials of second order. Suppose that $f(x, y)$ is a function of two variables x and y. The idea of a Taylor polynomial is to express the value of the function at some general point in terms of its behaviour at an arbitrary fixed point.

For a function of one variable $f(x)$ the Taylor polynomial of second order is

$$p_2(x) = f(a) + \frac{\mathrm{d}f}{\mathrm{d}x}(a)(x - a) + \frac{\mathrm{d}^2 f}{\mathrm{d}x^2}(a)\frac{(x - a)^2}{2!} \qquad (1)$$

With this polynomial we can find a good approximation to $f(x)$ close to the point $x = a$.

Suppose now that we want the value of the function of two variables $f(x, y)$ close to some point (a, b). The coordinates of a point close to (a, b) can be written as

$$x = a + h \quad \text{and} \quad y = b + k$$

where h and k are small. To express the value of $f(x, y)$ in terms of its behaviour at (a, b) we introduce the function of one variable $g(r)$ given by

$$g(r) = f(a + rh, b + rk) \tag{2}$$

It is clear that

$$g(0) = f(a, b)$$

and

$$g(1) = f(a + h, b + k) \tag{3}$$

It is also clear that $g(r)$ can be written as $f(x, y)$ where x and y are now no longer independent but are connected via the equations

$$x = a + rh \quad \text{and} \quad y = b + rk \tag{4}$$

Provided the partial derivatives of $f(x, y)$ exist, we can use the chain rule to differentiate $g(r)$ to obtain

$$\frac{\mathrm{d}g}{\mathrm{d}r} = \frac{\partial f}{\partial x}\frac{\mathrm{d}x}{\mathrm{d}r} + \frac{\partial f}{\partial y}\frac{\mathrm{d}y}{\mathrm{d}r} \tag{5}$$

and differentiating x and y in (4) with respect to r we have

$$\frac{\mathrm{d}x}{\mathrm{d}r} = h \quad \text{and} \quad \frac{\mathrm{d}y}{\mathrm{d}r} = k \tag{6}$$

Substituting from (6) in (5) gives

$$\frac{\mathrm{d}g}{\mathrm{d}r} = h\frac{\partial f}{\partial x} + k\frac{\partial f}{\partial y} \tag{7}$$

Using the chain rule again to obtain the second derivative of $g(r)$ we get

$$\frac{\mathrm{d}^2 g}{\mathrm{d}r^2} = \frac{\partial}{\partial x}\left(h\frac{\partial f}{\partial x} + k\frac{\partial f}{\partial y}\right)\frac{\mathrm{d}x}{\mathrm{d}r} + \frac{\partial}{\partial y}\left(h\frac{\partial f}{\partial x} + k\frac{\partial f}{\partial y}\right)\frac{\mathrm{d}y}{\mathrm{d}r}$$

$$= h\frac{\partial}{\partial x}\left(h\frac{\partial f}{\partial x} + k\frac{\partial f}{\partial y}\right) + k\frac{\partial}{\partial y}\left(h\frac{\partial f}{\partial x} + k\frac{\partial f}{\partial y}\right)$$

$$= h^2\frac{\partial^2 f}{\partial x^2} + 2hk\frac{\partial^2 f}{\partial x \partial y} + k^2\frac{\partial^2 f}{\partial y^2} \tag{8}$$

We now use (1) to approximate $g(r)$ by its second-order Taylor polynomial about $r = 0$

$$p_2(r) = g(0) + r\frac{\mathrm{d}g}{\mathrm{d}r}(0) + \frac{r^2}{2!}\frac{\mathrm{d}^2 g}{\mathrm{d}r^2}(0)$$

Substituting for $g(0)$ and the derivatives $(dg/dr)(0)$ and $(d^2g/dr^2)(0)$ from (3), (7) and (8) we have

$$p_2(r) = f(a, b) + r\left(h\frac{\partial f}{\partial x}(a, b) + k\frac{\partial f}{\partial y}(a, b)\right)$$

$$+ \frac{r^2}{2!}\left(h^2\frac{\partial^2 f}{\partial x^2}(a, b) + 2hk\frac{\partial^2 f}{\partial x\partial y}(a, b) + k^2\frac{\partial^2 f}{\partial y^2}(a, b)\right)$$

In particular, replacing h by $(x - a)$ and k by $(y - b)$, and putting $r = 1$ we have the following result:

For values of x and y close to a and b

$$f(x, y) = f(a, b) + \frac{\partial f}{\partial x}(a, b)(x - a) + \frac{\partial f}{\partial y}(a, b)(y - b)$$

$$+ \frac{1}{2!}\frac{\partial^2 f}{\partial x^2}(a, b)(x - a)^2 + \frac{1}{2!}\frac{\partial^2 f}{\partial y^2}(a, b)(y - b)^2 \qquad (9)$$

$$+ \frac{\partial^2 f}{\partial x\partial y}(a, b)(x - a)(g - b)$$

The second degree polynomial on the right-hand side of this expression is called *the Taylor polynomial of second order for $f(x, y)$ at (a, b)* and is often denoted by $p_2(x, y)$.

This Taylor polynomial provides a very good approximation to $f(x, y)$ for points sufficiently close to (a, b). Further away from (a, b) we would need to add the third- and higher-degree terms. For example the Taylor polynomial of third order would be

$$p_3(x, y) = p_2(x, y) + \frac{1}{3!}\left(\frac{\partial^3 f}{\partial x^3}(a, b)(x - a)^3 + 3\frac{\partial^2 f}{\partial x^2\partial y}(a, b)(x - a)^2(y - b)\right.$$

$$\left. + 3\frac{\partial^2 f}{\partial x\partial y^2}(a, b)(x - a)(y - b)^2 + \frac{\partial^3 f}{\partial y^3}(a, b)(y - b)^3\right)$$

Example: Determine the Taylor polynomial of second order for the function

$$f(x, y) = x^3 + xy - 2y^2$$

at the point $(1, 2)$

Solution: Finding the first and second partial derivatives of f we have

$$\frac{\partial f}{\partial x} = 3x^2 + y \quad \text{and} \quad \frac{\partial f}{\partial y} = x - 4y$$

$$\frac{\partial^2 f}{\partial x^2} = 6x, \quad \frac{\partial^2 f}{\partial x\partial y} = 1 \quad \text{and} \quad \frac{\partial^2 f}{\partial y^2} = -4$$

It follows that

$$f(1, 2) = -5, \quad \frac{\partial f}{\partial x}(1, 2) = 5, \quad \frac{\partial f}{\partial y}(1, 2) = -7$$

$$\frac{\partial^2 f}{\partial x^2}(1, 2) = 6, \quad \frac{\partial^2 f}{\partial x\partial y}(1, 2) = 1 \quad \text{and} \quad \frac{\partial^2 f}{\partial y^2}(1, 2) = -4$$

Substituting into the formula for $p_2(x, y)$ we conclude that the Taylor polynomial for $f(x, y)$ at the point $(1, 2)$ is

$$p_2(x, y) = -5 + 5(x - 1) - 7(y - 2) + 3(x - 1)^2 + (x - 1)(y - 2) - 2(y - 2)^2$$

18.2.2 Stationary points. One of the most common applications of the derivative of functions of one variable is to the problem of finding when the function has a local maximum or minimum value. Finding such points is often the first step in sketching the graph of the function. Similarly, one of the most useful features of the derivative of functions of two variables is in finding the position of maxima and minima.

We begin by giving a mathematical meaning to the terms maximum and minimum of a surface.

Let $f(x, y)$ be a function of two variables.

We say that the function f has a *local maximum* at the point (a, b) if for all points (x, y) sufficiently close to (a, b) (Fig. 18.4),

$$f(x, y) \leqslant f(a, b)$$

Similarly we say that f has a *local minimum* at (a, b) if for all points (x, y) sufficiently close to (a, b) (Fig. 18.5),

$$f(x, y) \geqslant f(a, b)$$

We often omit the word local in our discussion; we call a point which is either a maximum or a minimum, *an extremum*.

Now consider the value of the first partial derivative $\partial f / \partial x$ at the position of an extremum. When we introduced partial derivatives in Chapter 17 we interpreted this partial derivative as the change in f in the x direction. In particular if we slice through the surface $z = f(x, y)$ with the plane $y = b$ we get a curve of the function $f(x, b)$ (see Fig. 18.6). If the surface has a maximum or minimum at the point (a, b) then this curve has a maximum or minimum (in the function of one variable sense)

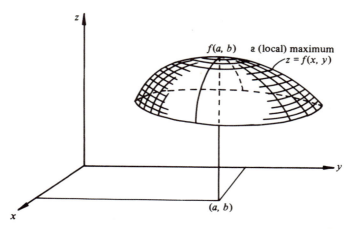

Fig. 18.4. A local maximum of a surface.

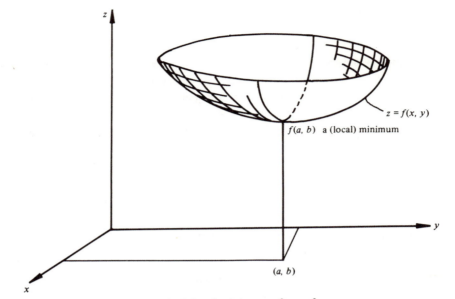

Fig. 18.5. A local minimum of a surface.

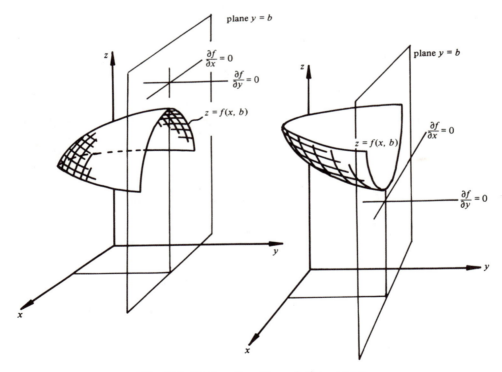

Fig. 18.6. Taking slices through the extrema.

at $x = a$. Thus the derivative of the function $f(x, b)$ with respect to x, namely $(\partial f / \partial x)(x, b)$ must be zero when $x = a$.

Similarly by slicing through the surface with the plane $x = a$, we can deduce that $(\partial f / \partial y)(a, y)$ must also be zero if (a, b) is an extremum of f. We call a point (a, b) where $(\partial f / \partial x)(a, b) = (\partial f / \partial y)(a, b) = 0$ a *stationary point* of f. Thus we can find the position of a (local) maximum or minimum by finding where the first derivatives of $f(x, y)$ are zero. In summary the *stationary point criterion* is stated as:

If $f(x, y)$ has an extremum at the point (a, b) then

$$\frac{\partial f}{\partial x}(a, b) = 0 \quad \text{and} \quad \frac{\partial f}{\partial y}(a, b) = 0 \tag{10}$$

Example: Find the position of the stationary points of the function

$$f(x, y) = 4x^5 + 5y^5 - 10x^2 - 25y$$

Solution: The first-order partial derivatives of f are

$$\frac{\partial f}{\partial x} = 20x^4 - 20x \quad \text{and} \quad \frac{\partial f}{\partial y} = 25y^4 - 25$$

The coordinates of the stationary points of f are determined from the pair of equations (10) which in this case become

$$20x^4 - 20x = 0 \quad \text{and} \quad 25y^4 - 25 = 0$$

Solving for x we have $x = 0$ and $x = 1$. Solving for y we have $y = 1$ and $y = -1$. For either choice of x, either choice of y is possible. Hence the stationary points of f are at the points

$$(0, -1), \quad (1, 1), \quad (0, 1), \quad (1, -1)$$

We now have a mathematical statement for finding the positions of the stationary points. However it is not clear from the function statement whether each stationary point is a maximum or a minimum (or in fact any other type). For functions of one variable we used the sign of the second derivative as a test for deciding what type of stationary point we had. We can derive a similar test for functions of two variables, but now there are three second partial derivatives to take into account. As a starting point we consider the Taylor polynomial of second order for the position of the stationary point. Suppose that the function $f(x, y)$ has a stationary point at (a, b), then we know that

$$\frac{\partial f}{\partial x}(a, b) = 0 \quad \text{and} \quad \frac{\partial f}{\partial y}(a, b) = 0$$

Then if h and k are small, the Taylor polynomial of second-order (9) becomes

$$f(a + h, b + k) = f(a, b) + \frac{1}{2}\left(h^2 \frac{\partial^2 f}{\partial x^2}(a, b) + 2hk \frac{\partial^2 f}{\partial x \partial y}(a, b) + k^2 \frac{\partial^2 f}{\partial y^2}(a, b) \right)$$

since the first derivative terms vanish. For convenience let

$$A = \frac{\partial^2 f}{\partial x^2}(a, b), \quad B = \frac{\partial^2 f}{\partial x \partial y}(a, b), \quad C = \frac{\partial^2 f}{\partial y^2}(a, b) \tag{11}$$

Then we can write (11) as

$$f(a+h, b+k) - f(a,b) = \tfrac{1}{2}(Ah^2 + 2Bhk + Ck^2) \tag{12}$$

According to our definition of a maximum and minimum, the sign of $f(a+h, b+k) - f(a,b)$ will provide a means of classification for any stationary points.

If $f(a+h, b+k) - f(a,b)$ is *positive* for all choices of h and k then the stationary point is a *minimum*, and if $f(a+h, b+k) - f(a,b)$ is *negative* for all choices of h and k then the stationary point is a *maximum*.

From (12) it follows that the sign of $f(a+h, b+k) - f(a,b)$ is the same as the sign of

$$L = Ah^2 + 2Bhk + Ck^2$$

We rewrite this as

$$L = A\left(h^2 + \frac{2Bhk}{A} + \frac{Ck^2}{A} \right)$$

Completing the square we have

$$L = A\left[\left(h + \frac{Bk}{A} \right)^2 + \left(\frac{C}{A} - \frac{B^2}{A^2} \right)k^2 \right]$$

and simplifying

$$L = A\left[\left(h + \frac{Bk}{A} \right)^2 + \frac{(AC - B^2)}{A^2}k^2 \right] \tag{13}$$

The first term in the parenthesis in (13) is always positive, so that the sign of L depends on (i) the sign of A, (ii) the sign of $AC - B^2$ and (iii) the relative sizes of the terms in the parenthesis.

Case 1: $AC - B^2 > 0$. The condition $AC > B^2$ implies that $A \neq 0$ and each term in the parenthesis is positive. Hence the sign of L depends only on the sign of A. We have from (12)

$$f(a+h, b+k) - f(a,b) = L/2$$

so it follows that

if $A > 0$ then $L > 0$ and (a,b) is a minimum
if $A < 0$ then $L < 0$ and (a,b) is a maximum

Example: For the function

$$f(x, y) = 4x^5 + 5y^5 - 10x^2 - 25y$$

show that the stationary point $(0, -1)$ is a maximum and the stationary point $(1, 1)$ is a minimum.

Solution: For

$$f(x, y) = 4x^5 + 5y^5 - 10x^2 - 25y$$

the first derivatives are

$$\frac{\partial f}{\partial x} = 20x^4 - 20x \quad \text{and} \quad \frac{\partial f}{\partial y} = 25y^4 - 25$$

Differentiating again we have

$$A = \frac{\partial^2 f}{\partial x^2} = 80x^3 - 20$$

$$B = \frac{\partial^2 f}{\partial x \partial y} = 0$$

$$C = \frac{\partial^2 f}{\partial y^2} = 100y^3$$

For $(0, -1)$ we have $A = -20$, $B = 0$ and $C = -100$. For $(1, 1)$ we have $A = 60$, $B = 0$ and $C = 100$. In each case $AC - B^2 > 0$. For $(0, -1)$, $A < 0$ so the point is a (local) maximum. For $(1, 1)$, $A > 0$ so the point is a (local) minimum.

The classification of two of the stationary points as a maximum and a minimum in this example is quite straightforward.

But what about the other two stationary points $(0, 1)$ and $(1, -1)$? In each case $AC - B^2$ is negative so that the sign of L depends on the choice of h and k, the sign of A and the relative sizes of the terms in the parenthesis in (13). A sketch of the surface near to the stationary point $(1, -1)$, for instance, shows that the point is neither a (local) maximum nor a (local) minimum (Fig. 18.7).

Case 2: $AC - B^2 < 0$. When the value of $AC - B^2$ is negative we have a special kind of stationary point called *a saddle-point*. (Sometimes such a point is called *a col*.) Fig. 18.7 shows the typical shape of the function at a saddle point. At such

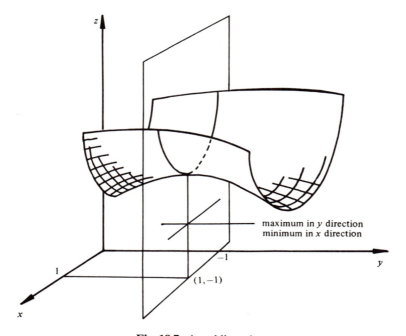

maximum in y direction
minimum in x direction

Fig. 18.7. A saddle-point.

points $f(a+h, b+k) - f(a, b)$ takes positive and negative values in any small region close to a stationary point. We define a saddle-point of a function f to be a stationary point which is neither a maximum nor a minimum. We summarise the criterion for classifying stationary points in the following way:

Classification of stationary points
Let (a, b) be a stationary point of $f(x, y)$ and

$$A = \frac{\partial^2 f}{\partial x^2}(a, b) \quad B = \frac{\partial^2 f}{\partial x \partial y}(a, b) \quad C = \frac{\partial^2 f}{\partial y^2}(a, b)$$

Case 1: If $AC - B^2 > 0$ then
(a, b) is a maximum if $A < 0$
(a, b) is a minimum if $A > 0$
Case 2: if $AC - B^2 < 0$ then
(a, b) is a saddle-point

Example: Locate and classify the stationary points of the function
$$f(x, y) = 2x^3 - 6x^2 - 8y^2 + 2$$

Solution: The stationary points occur where

$$\frac{\partial f}{\partial x} = 0 \quad \text{and} \quad \frac{\partial f}{\partial y} = 0$$

Differentiating the function partially with respect to x and y gives

$$\frac{\partial f}{\partial x} = 6x^2 - 12x \quad \text{and} \quad \frac{\partial f}{\partial y} = -16y$$

Equating both to zero gives $x = 0$ or 2 and $y = 0$. There are stationary points at $(0, 0)$ and $(2, 0)$. To classify the stationary points we look at the second partial derivatives. We have

$$A = \frac{\partial^2 f}{\partial x^2} = 12x - 12, \quad B = \frac{\partial^2 f}{\partial x \partial y} = 0 \quad \text{and} \quad C = \frac{\partial^2 f}{\partial y^2} = -16$$

At the point $(0, 0)$, $AC - B^2 = (-12)(-16) - 0^2$, which is positive. We therefore have a maximum or a minimum. Since $A(= -12)$ is negative the point $(0, 0)$ is a local maximum. For the point $(2, 0)$ the value of $AC - B^2$ is $(12)(-16) - 0^2$ which is negative. The stationary point at $(2, 0)$ is therefore a saddle-point.

18.3. Worked examples

Example 1. Equal point charges q are located at the vertices of an equilateral triangle. If the vertices are at the points $(0, 0)$, $(2, 0)$ and $(1, \sqrt{3})$ in the x–y plane, show that a point charge, q', will be in equilibrium at the point $(1, 1/\sqrt{3})$. Investigate the stability of the point charge q' at this position of equilibrium assuming it is constrained to move in the x–y plane.

Solution. Let the charge q' be located at an arbitrary point (x, y) as shown in Fig. 18.8. The potential energy of the charge q', $V(x, y)$ will be given by

$$V = \frac{qq'}{4\pi\varepsilon_0} \left(\frac{1}{r_1} + \frac{1}{r_2} + \frac{1}{r_3} \right)$$

$$= \frac{qq'}{4\pi\varepsilon_0} \left(\frac{1}{(x^2 + y^2)^{\frac{1}{2}}} + \frac{1}{((2-x)^2 + y^2)^{\frac{1}{2}}} + \frac{1}{((x-1)^2 + (\sqrt{3}-y)^2)^{\frac{1}{2}}} \right)$$

Differentiating this function partially with respect to x and y gives

$$\frac{\partial V}{\partial x} = \frac{qq'}{4\pi\varepsilon_0} \left(\frac{x}{r_1^3} - \frac{(2-x)}{r_2^3} + \frac{(x-1)}{r_3^3} \right)$$

$$\frac{\partial V}{\partial y} = \frac{qq'}{4\pi\varepsilon_0} \left(\frac{y}{r_1^3} + \frac{y}{r_2^3} - \frac{(\sqrt{3}-y)}{r_3^3} \right)$$

The charge will be in equilibrium when $V(x, y)$ has a stationary point. Clearly both $\partial V/\partial x$ and $\partial V/\partial y$ are zero at the point $(1, 1/\sqrt{3})$, for which $r_1 = r_2 = r_3$, showing that we have a stationary point of V, i.e. an equilibrium point for q'. In order to investigate the stability of q' at this point we must look at the second partial derivatives. Differentiating partially again we obtain

$$A = \frac{\partial^2 V}{\partial x^2} = \frac{qq'}{4\pi\varepsilon_0} \left(\frac{1}{r_1^3} - \frac{3x^2}{r_1^5} + \frac{1}{r_2^3} - \frac{3(2-x)^2}{r_2^5} + \frac{1}{r_3^3} - \frac{3(x-1)^2}{r_3^5} \right)$$

$$B = \frac{\partial^2 V}{\partial x \partial y} = \frac{qq'}{4\pi\varepsilon_0} \left(-\frac{3xy}{r_1^5} + \frac{3(2-x)y}{r_2^5} - \frac{3(x-1)(\sqrt{3}-y)}{r_3^5} \right)$$

$$C = \frac{\partial^2 V}{\partial y^2} = \frac{qq'}{4\pi\varepsilon_0} \left(\frac{1}{r_1^3} - \frac{3y^2}{r_1^5} + \frac{1}{r_2^3} - \frac{3y^2}{r_2^5} + \frac{1}{r_3^3} - \frac{3(\sqrt{3}-y)^2}{r_3^5} \right)$$

The value of $AC - B^2 = (\partial^2 V/\partial x^2)(\partial^2 V/\partial y^2) - [(\partial^2 V)/\partial x \partial y]^2$ evaluated at $(1, 1/\sqrt{3})$, is $\frac{243}{256}(qq')^2/(4\pi\varepsilon_0)$, which is positive. The position of equilibrium thus

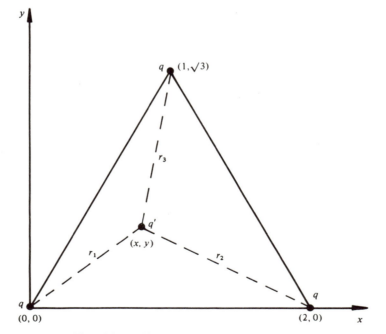

Fig. 18.8. Stability due to equal point charges.

corresponds to a maximum or minimum in potential energy. The values of both $\partial^2 V/\partial x^2$ and $\partial^2 V/\partial y^2$ at $(1, 1/\sqrt{3})$ are $(-9\sqrt{3}qq'/(64\pi\varepsilon_0))$.

If qq' is positive, then the second derivatives are negative implying that V has a maximum value. Thus if q and q' are of the same sign, the equilibrium position at $(1, 1/\sqrt{3})$ is a position of unstable equilibrium.

If qq' is negative, then $\partial^2 V/\partial x^2$ and $\partial^2 V/\partial y^2$ are both positive and the potential energy is a minimum. The position of equilibrium, if q and q' are of opposite sign, is thus one of stable equilibrium.

Example 2. In the two-stage chemical process considered in 18.1.2, the monthly operating cost C (£) is related to the pressure p (Pa) and separator time T (hr) by the expression

$$C = 0.01p + \frac{4 \times 10^{12}}{pT} + 2.5 \times 10^4 T$$

Obtain the values of p and T that make the monthly operating cost a minimum.

Solution. At the minimum we require both $\partial C/\partial p$ and $\partial C/\partial T$ to be zero. Thus, working out the two partial derivatives we obtain

$$\frac{\partial C}{\partial p} = 0.01 - \frac{4 \times 10^{12}}{p^2 T}$$

$$\frac{\partial C}{\partial T} = \frac{-4 \times 10^{12}}{pT^2} + 2.5 \times 10^4$$

Equating each partial derivative to zero we have the two equations

$$\frac{4 \times 10^{12}}{p^2 T} = 0.01; \quad \frac{4 \times 10^{12}}{pT^2} = 2.5 \times 10^4$$

Dividing one equation by the other gives $p = 2.5 \times 10^6 T$. Substituting back into the second of these two equations and solving for T we have $T = 4$. The separation time is therefore 4 hr and the corresponding pressure $(2.5 \times 10^6 T)$ is thus 10^7 Pa.

We now have to show that these values of p and T give rise to a minimum in C.

Looking at the second partial derivatives we have

$$\frac{\partial^2 C}{\partial p^2} = \frac{8 \times 10^{12}}{p^3 T}$$

$$\frac{\partial^2 C}{\partial p \partial T} = \frac{4 \times 10^{12}}{p^2 T^2}$$

$$\frac{\partial^2 C}{\partial T^2} = \frac{8 \times 10^{12}}{pT^3}$$

At the stationary point, where $p = 10^7$ and $T = 4$, the value of

$$A = \frac{\partial^2 C}{\partial p^2} \text{ is } 2 \times 10^{-9} \text{ (which is positive)}$$

$$C = \frac{\partial^2 C}{\partial T^2} \text{ is } 12\,500 \text{ (which is also positive)}$$

and

$$B = \frac{\partial^2 C}{\partial p \partial T} \text{ is } 0.0025$$

The value of $AC - B^2$ is thus 1.875×10^{-5} (which is again positive). Using the classification criterion the stationary point of the cost function is a minimum and the monthly operating cost will be least when the pressure is 10^7 Pascals and the separation time is 4 hr.

18.4 Further developments

In this section we illustrate another use of the stationary point analysis for a function of two variables. This is associated with the problem of fitting the 'best' straight line to experimental data points.

18.4.1 Least squares analysis. Suppose that we have a situation similar to the problem described in 18.1.3, where we have some experimental data points which, apart from a certain amount of scatter, clearly show a linear trend. Fitting a linear polynomial to experimental data was discussed in Chapter 1 where we drew a 'best' straight line 'by eye'. Now we introduce some theory which gives a 'mathematical best' straight line.

Fig. 18.9 shows a plot of data, where for convenience, the variables of interest are x and y and there are N points. What we require is the equation that best

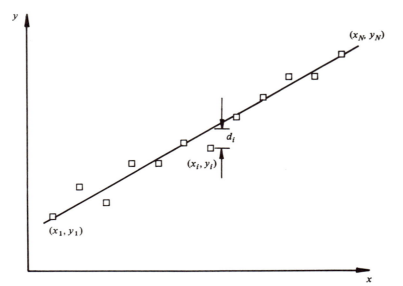

Fig. 18.9. A linear model for a set of data.

describes the linear relationship connecting x and y. The scatter that does exist about the line we have drawn can be measured by first noting the various deviations d_i of each of the points from the line. Taking the equation of the line as $y = mx + c$, then if the point happens to be below the line the deviation d_i will be given by

$$d_i = mx_i + c - y_i$$

or by

$$d_i = y_i - mx_i - c$$

when the point is above the line. In either case we can write

$$d_i^2 = (mx_i + c - y_i)^2$$

A measure of the total scatter can now conveniently be taken as the sum of these deviations squared. Calling the sum S we have

$$S = \sum_i (mx_i + c - y_i)^2$$

In *least squares analysis* we define as '*best*' the line that minimises this quantity S. Now, as far as S is concerned it is important to remember that the quantities x_i and y_i are constants, so S is simply a function of the two variables m and c. We have therefore to vary m and c to make S a minimum, and consequently what we must do first is solve the equations

$$\frac{\partial S}{\partial m} = \frac{\partial S}{\partial c} = 0$$

Differentiating S partially with respect to m and c leads to the following expressions:

$$\frac{\partial S}{\partial m} = \sum_i 2x_i(mx_i + c - y_i)$$

$$\frac{\partial S}{\partial c} = \sum_i 2(mx_i + c - y_i)$$

Equating both these partial derivatives to zero gives rise to the following pair of simultaneous equations in the variables m and c:

$$m\sum_i x_i^2 + c\sum_i x_i = \sum_i x_i y_i$$

$$m\sum_i x_i + cN = \sum_i y_i.$$

These equations are known as the *normal equations of linear regression analysis*, and solving them, first for m, gives

$$m = \frac{N\sum x_i y_i - \sum x_i \sum y_i}{N\sum x_i^2 - (\sum x_i)^2} \tag{14}$$

The value of c is then found by substituting back into the second of the normal equations, using the value of m already found, to give

$$c = \frac{\sum y_i}{N} - \frac{m\sum x_i}{N} \tag{15}$$

Note that this expression for c implies that the mean of the x_i values and the mean of the y_i values lie on the line.

Having located where the stationary value of S occurs we have now to show that this point is a minimum. We need therefore to consider the classification criterion developed in Section 18.2.

Working out the second derivatives of S we obtain

$$\frac{\partial^2 S}{\partial m^2} = 2\sum_i x_i$$

$$\frac{\partial^2 S}{\partial c^2} = 2N$$

and

$$\frac{\partial^2 S}{\partial c \partial m} = 2\sum_i x_i$$

We note that $\partial^2 S/\partial m^2$ and $\partial^2 S/\partial c^2$ are both positive quantities, as required, and that

$$AC - B^2 = \frac{\partial^2 S}{\partial m^2}\frac{\partial^2 S}{\partial c^2} - \left(\frac{\partial^2 S}{\partial m \partial c}\right)^2 = 4N\sum_i x_i^2 - 4\left(\sum_i x_i\right)^2.$$

This latter quantity is required also to be positive. Without too much difficulty it is possible to show that

$$N\sum_i x_i^2 - \left(\sum_i x_i\right)^2 = (x_1 - x_2)^2 + (x_1 - x_3)^2 + \cdots + (x_{N-1} - x_N)^2$$

$$= \frac{1}{2}\sum_{j \neq i}^{N}\sum_{i=1}^{N}(x_i - x_j)^2$$

which because of the squared terms will always be positive. Thus, all the sufficient conditions in the classification criterion are satisfied and the stationary point is therefore a minimum.

To summarise then, the best straight line fitted to N data points, using least squares analysis, is given by $y = mx + c$ where

$$m = \frac{N\sum_i x_i y_i - \sum_i x_i \sum_i y_i}{N\sum_i x_i^2 - (\sum_i x_i)^2}$$

$$c = \frac{\sum_i y_i}{N} - \frac{m\sum_i x_i}{N}$$

It is worth noting that when carrying out least squares analysis, the problem is simply one of taking the data values and forming the sums $\sum_i x_i$, $\sum_i y_i$, $\sum_i x_i^2$ and $\sum_i x_i y_i$. With a calculator this is easily accomplished and, indeed, many calculators can do this automatically once the data have been entered. With respect to computers, linear regression using least squares analysis is a standard feature of most statistical software packages.

Having obtained the regression line it can then, of course, be used for predictive purposes to furnish y values from given x values: the predicted y value being given simply by $mx + c$. We illustrate the least squares analysis in the next section.

Example: A manufacturer of optical equipment has the following data on the unit cost (Y) of certain custom made lenses and number of units (X) in each order:

number of units (X) 1   3   5   10   12

cost per unit (Y) 58   55   40   37   22

Assuming a linear law between X and Y, $Y = mX + c$, calculate using the method of least squares the values of the constants m and c.

Solution: To obtain the linear equation we need the sums X, Y, X^2 and XY. The calculations are laid out in tabular form in the following way:

X	Y	X^2	XY	
1	58	1	58	
3	55	9	165	
5	40	25	200	
10	37	100	370	
12	22	144	264	
Totals: 31	212	279	1057	$N = 5$

The expressions for m can be found from (14)

$$m = \frac{(N\sum XY - \sum X \sum Y)}{(N\sum X^2 - (\sum X)^2)}$$

On substitution from the table we get

$$m = \frac{5 \times 1057 - 32 \times 212}{5 \times 279 - (31)^2}$$

$$= -2.965$$

To find c we use (15),

$$c = (\sum Y - m\sum X)/N$$

$$= (212 - (-2.965)(31))/5$$

$$= 60.79$$

The equation relating X and Y is therefore

$$Y = 60.79 - 2.965X$$

The principle of least squares is not confined to the fitting of straight lines to data points. If f is any function of one variable relating two quantities x and y by $y = f(x)$, then least squares analysis applied to the deviation of a data point (x_i, y_i) from the graph of $f(x)$ minimises the sum

$$(y_i - f(x_i))^2$$

For example, suppose that f is a quadratic function of the form

$$f(x) = ax^2 + bx + c$$

we can minimise the expression

$$S = \sum_i (y_i - (ax_i^2 + bx_i + c)^2)$$

by setting the three first partial derivatives with respect to a, b and c equal to zero, i.e.

$$\frac{\partial S}{\partial a} = \frac{\partial S}{\partial b} = \frac{\partial S}{\partial c} = 0$$

The resulting values of a, b and c will give the quadratic function which best fits the data points in the least squares sense.

18.5 Further worked examples

Example 1. In this example we use the data on heart beats introduced in 18.1.3 and repeated below.

Time interval X	Number of heart beats Y
20	19
30	28
40	36
50	44
60	57
70	68

Obtain the regression line that relates Y to X. How many beats would be recorded for a time interval of 45 s?

Solution. In this example the calculations will be simplified if we work not with X, but with a new variable x where

$$x = (X - 20)/10$$

Thus for the data given we have

x	y	x^2	$x \cdot Y$
0	19	0	0
1	28	1	28
2	36	4	72
3	44	9	132
4	57	16	228
5	68	25	340
Totals: 15	252	55	800 and $N = 6$

The values of $\sum x, \sum Y, \sum x^2$ and $\sum x \cdot Y$ are therefore 15, 252, 55 and 800 respectively. The regression line relating Y and x will be $Y = mx + c$ where

$$m = \frac{n \sum x \cdot Y - \sum x \sum Y}{N \sum x^2 - (\sum x)^2}$$

$$= \frac{6 \times 800 - 15 \times 252}{6 \times 55 - 15 \times 15}$$

$$= 9.7143$$

and where

$$c = (\sum Y - m \cdot \sum X)/N$$

$$= (252 - 9.7143 \times 15)/6$$

$$= 17.7143$$

The equation relating Y and X is therefore

$$Y = 9.7143(X - 20)/10 + 17.7143$$

$$= 0.97143X - 1.7143.$$

Thus the number of beats recorded for an interval of 45 s using this equation as a model is

$$Y = 0.97143 \times 45 - 1.7143$$

which is roughly 42 beats.

Example 2. The shear strengths of electric welds of different thickness are given in the table below:

Thickness of weld (mm) t	Shear strength (kg) s
0.2	102
0.3	129
0.4	201
0.5	342
0.6	420
0.7	591
0.8	694
0.9	825
1.0	1014
1.1	1143
1.2	1219

It is proposed to fit a linear model between the variables s and t such that $s = at + b$. Calculate, using the method of least squares, the values of the constants a and b.

Solution. First we find the sums $\sum t$, $\sum s$, $\sum t^2$ and $\sum ts$. These are carried in the following table:

t	s	t^2	$t \cdot s$
0.2	102	0.04	20.4
0.3	129	0.09	38.7
0.4	201	0.16	80.4
0.5	342	0.25	171.0
0.6	420	0.36	252.0
0.7	591	0.49	413.7
0.8	694	0.64	555.2
0.9	825	0.81	742.5
1.0	1014	1.00	1014.0
1.1	1143	1.21	1256.2
1.2	1219	1.44	1462.8

Total: 7.7   6680   6.49   6006.9   and $N = 11$

Substituting for these sums in the appropriate formulas we get that

$$a = \frac{N\sum t \cdot s - \sum t \sum s}{N\sum t^2 - (\sum t)^2}$$

$$= \frac{11 \times 6006.9 - 7.7 \times 6680}{11 \times 6.49 - (7.7)^2}$$

$$= 1209.9$$

and b is given by

$$b = \frac{\sum s}{N} - \frac{a\sum t}{N}$$

$$= (6680 - 7.7a)/11$$

$$= -239.7$$

The equation relating s and t is therefore

$$s = 1209.9t - 239.7$$

Example 3. In an experiment the flow of a liquid along a straight channel is observed and measurements are made of the speed of flow u_i at a distance y_i from the wall of the channel. The results of the experiment are shown in the following table:

y_i (cm)	1	3	5	7	9	11	13	15	17
u_i (cm s^{-1})	7.2	13.1	16.8	19.5	21.3	22.4	23.2	23.65	23.9

It is proposed to fit a quadratic relationship between u and y of the form

$$u = ay^2 + by + c$$

(i) Use the method of least squares to find the three normal equations for a, b and c.

(ii) Hence find the 'best' (in the least squares sense) quadratic relationship between these data.

Solution. (i) The sum of the squares of the deviations is given by

$$S = \sum_{i=1}^{N} (u_i - (ay_i^2 + by_i + c))^2$$

Now we seek to minimize this expression by setting the partial derivatives with respect to a, b and c equal to zero. Differentiating, we have

$$\frac{\partial S}{\partial a} = \sum_{i=1}^{N} -2y_i^2(u_i - (ay_i^2 + by_i + c))a$$

$$\frac{\partial S}{\partial b} = \sum_{i=1}^{N} -2y_i(u_i - (ay_i^2 + by_i + c))$$

$$\frac{\partial S}{\partial c} = \sum_{i=1}^{N} -2(u_i - (ay_i^2 + by_i + c))$$

Using the rules for manipulating the summation symbol and setting each partial derivative to zero, we have the following three equations for a, b and c:

$$\sum u_i y_i^2 - a\sum y_i^4 - b\sum y_i^3 - c\sum y_i^2 = 0$$
$$\sum u_i y_i - a\sum y_i^3 - b\sum y_i^2 - c\sum y_i = 0$$
$$\sum u_i - a\sum y_i^2 - b\sum y_i - cN = 0$$

These are the normal equations.

(ii) For the data given in the question we have

y_i	u_i	y_i^2	y_i^3	y_i^4	$y_i u_i$	$u_i y_i^2$
1	7.2	1	1	1	7.2	7.2
3	13.1	9	27	81	39.3	117.9
5	16.8	25	125	225	84.0	420.0
7	19.5	49	343	2401	136.5	955.5
9	21.3	81	729	6561	191.7	1725.3
11	22.4	121	1331	14 641	246.4	2710.4
13	23.2	169	2197	28 561	301.6	3930.8
15	23.65	225	3375	50 625	354.75	5321.25
17	23.9	289	4913	83 521	406.3	6907.01

Totals: 81   171.05   969   13041   186 617   1767.75   22 085.45 and $N = 9$

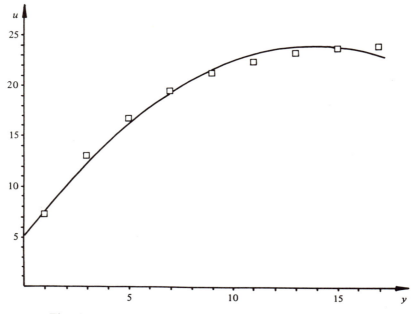

Fig. 18.10. A quadratic function as a model for the flow data.

Substituting these sums into the normal equations, we have

$$186\,617a + 13041b + 969c = 22\,085.45$$
$$13\,041a + 969b + 81c = 1767.75$$
$$969a + 81b + 9c = 171.05$$

Solving these equations for a, b and c we get that

$$a = -0.097, \quad b = 2.70 \quad \text{and} \quad c = 5.15$$

The quadratic function of closest fit (in the least squares sense) to the data is

$$u = -0.097y^2 + 2.70y + 5.15$$

Fig. 18.10 shows the graph of this function along with the experimental data.

18.6 Exercises

1. Find and classify the stationary points of the following functions:
   (i) $f(x, y) = xy - x - y - (x - 1)^2 - (y - 2)^2$
   (ii) $f(x, y) = x^4 + y^4 - 2(x - y)^2$
   (iii) $f(x, y) = x^3/3 + y^3 + x^2 - 3(x + y)$
   (iv) $f(x, y) = 2x^2y - 2xy^2 + x^2 + y^2 + 2xy + 6(x - y)$
   (v) $f(x, y) = x^3 + ay^2 - 6axy$, where $a > 0$
   (vi) $f(x, y) = (x^2 + 4xy + 2y^2 + 4x - 10)e^{-x-y}$

2. Show that the rectangular closed box, of a given surface area, which has maximum volume is a cube.

3. A rectangular block with edges x, y and z is cut from a solid sphere of radius b so that the volume has a maximum value. Find x, y and z and show that the maximum value is $8b^3/3\sqrt{3}$.

4. Positive numbers x, y and z are subject to the condition $x + y + z =$ constant. Prove that their product is greatest when they are equal.

5. The area A of a triangle with sides a, b and c can be written as

$$A = \sqrt{[s(s-a)(s-b)(s-c)]}$$

where $s = (a + b + c)/2$. Show that, for a given perimeter, the area A will be a maximum when $a = b = c$.

6. An open rectangular box is to be made out of thin plastic of thickness t. A fixed amount of plastic is available for each box. If the dimensions of the base of the box are x and y, and the height is z, show that the volume of the box is a maximum when $x = y = 2z$. (You may assume that t is very much smaller than x.)

7. The ground state energy E of a particle in a rectangular box is

$$E = \frac{h^2}{8m}\left(\frac{1}{a^2} + \frac{1}{b^2} + \frac{1}{c^2}\right)$$

where h is Planck's constant, m is the mass of the particle and a, b and c are dimensions of the box. Find the minimum energy of the particle if the box is constrained to have fixed volume k.

8. An elastic string of unstretched length a has n equal masses, each of weight mg, attached to it at equal intervals of a/n, leaving one end free and the other with a mass m on it. The string, with masses, is suspended from the free end and allowed to hang vertically. If the top piece of the string is stretched by an amount x_1 from its unstretched length of a/n, the next to the top piece by x_2, and so on, so that the bottom piece of the string is stretched by an amount x_n, then the potential energy V of the system when hanging is given by

$$V(x_1, x_2, \ldots x_n) = \frac{n\lambda x_1^2}{2a} - mg(a + x_1) + \frac{n\lambda x_2^2}{2a} - mg(a + x_1 + x_2)$$

$$+ \frac{n\lambda x_3^2}{2a} - mg(a + x_1 + x_2 + x_3)$$

$$+ \cdots$$

$$+ \frac{n\lambda x_n^2}{2a} - mg(a + x_1 + x_2 + \cdots + x_n)$$

where λ is the modulus of elasticity. Assuming that the stationary value of V is a minimum, find the total length of the string when it is hanging in equilibrium. Find the single mass required to stretch the string by the same amount if it is hung at the end of the string.

9. The entropy S of a collection of N distinguishable particles, distributed

amongst states of different energy, is given by

$$S = \left[N \ln(N) - N - \sum_i (n_i \ln(n_i) - n_i) \right] \Big/ k$$

where k is Boltzmann's constant and n_i is the number of particles in the ith energy state. The most likely distribution of particles to states, when the total energy of N particles is E, can be found by requiring that the following function is stationary

$$L(n_1, n_2, \ldots n_i, \ldots) = Sk + \beta\left(E - \sum_i n_i e_i \right) + \lambda\left(N - \sum_i n_i \right)$$

where e_i is the energy of a particle in state i, and β and λ are constants. Show that the most likely distribution of particles to states is given by

$$n_i = C \exp(-\beta e_i)$$

where the constant C is equal to $e^{-\lambda}$.

10. The molar volume V of a series of iodo hydrocarbons varies with carbon chain length x as follows:

V(ml)	64.1	85.6	106.8	128.3	149.6
x	1	2	3	4	5

Show that these results are described by a straight-line relationship. Using least squares analysis find the equation of this line.

11. The solubility, s of a compound at various temperatures T is as follows:

T(°C)	10	20	30	40	50	60	70	80
s(gl^{-1})	0.290	0.296	0.312	0.318	0.327	0.340	0.342	0.355

The value of s and T are known to be connected via a relationship given by $s = aT + b$ where a and b are constants. Using least squares analysis obtain values for a and b.

12. The contraction ratio σ associated with lateral tension in a certain glass-fibre-reinforced material was measured at different volume ratios V of the glass fibre. The readings were as follows:

V	0.05	0.10	0.15	0.20	0.25	0.30	0.35	0.40
σ	0.63	0.75	0.85	0.91	0.94	0.95	0.92	0.87

It is thought that the relationship between σ and V is of the form

$$\sigma = 0.48 + aV + bV^2$$

where a and b are constants. Reduce this expression to a straight line form and obtain values of a and b by linear regression. Using your results, predict the value of σ when $V = 0.12$.

13. The following table shows the percentage y of initial potassium content in human red blood cells remaining after suspension in glucose solution for varying lengths of time t under certain specified conditions:

time t(min)	40	60	80	100	120
% potassium remaining	68.7	65.9	64.1	62.1	61.0

To a reasonable approximation, y and t are connected by the relation

$$y = R_0 e^{\alpha t} + 55$$

where R_0 and α are constants.

(i) Transform the above equation to give a linear relationship.

(ii) Find the least squares regression line connecting the transformed data and hence determine R_0 and α.

(iii) Use your regression line to estimate the length of time for which the cells must be suspended in the glucose solution before the potassium content drops to 60% of its initial value.

14. The following results were obtained for the amount of acetone y (millimole g^{-1}) absorbed on charcoal from an aqueous solution of concentration C (millimole dm^{-3}) at $18°C$:

y	0.60	0.75	1.05	1.50	2.15	3.50	5.10
C	15.0	23.0	42.0	84.0	165	390	800

Show that y and C are connected by an equation of the form $y = kC^{1/n}$ where k and n are constants, and determine values for k and n.

15. Write a pseudocode algorithm that will

(i) read in n sets of (x, y) values;

(ii) use least squares to fit a best straight line of the form $y = mx + c$ to these data;

(iii) print out the values of m and c.

ANSWERS TO THE EXERCISES

1.6 Exercises

1. (1.10, 8.94), (2.85, 37.64)

2. (i) $y = 3.4x + 0.3$ (ii) $y = 2x - 8$
   (iii) $y = -0.5x + 4$ (iv) $y = -0.4407x + 4.3$

3. Roots are
   (i) $x = 0$
   (ii) $x = 0$ (repeated)
   (iii) $x = 0$ (repeated)
   (iv) $x = 0$ (repeated)
   (v) $x = 0$ (repeated)
   (vi) $x = 1$ (repeated)
   (vii) $x = 1, x = -1; f(x) = (x-1)(x+1)$
   (viii) $x = 1, x = 2; f(x) = (x-1)(x-2)$
   (ix) $x = \frac{1}{2}, x = 1; f(x) = (x - \frac{1}{2})(x - 1)$
   (x) $x = 0, x = 1/\sqrt{2}, x = -1/\sqrt{2}; f(x) = 2x^2(x - 1/\sqrt{2})(x + 1/\sqrt{2})$

4. (i) $x = 0.3158$
   (ii) $x = 6$ or $x = 10$
   (iii) $x = -1, x = 1$ or $x = -2$
   (iv) $x = 0, x = 1, x = -1$ (repeated), $x = 2$

5. (i) $x = 3$ or $x = 2$
   (ii) $x = -1$ or $x = -7$
   (iii) $x = -1$ or $x = -3$
   (iv) $x = 2$ or $x = -3$
   (v) $x = 5$ or $x = -5$
   (vii) $x = \frac{1}{2}$ or $x = -1$
   (viii) $x = \frac{2}{3}$ or $x = -\frac{1}{2}$
   (ix) $x = (-3 + \sqrt{37})/2$ or $x = (-3 - \sqrt{37})/2$
   (x) $x = (-7 + \sqrt{41})/4$ or $x = (-7 - \sqrt{41})/4$

6. (i) 540.4361   (ii) 383.0988   (iii) 17.2941

7. (i) $-16 + 20i$ (ii) $19 - 4i$ (iii) $0.6 - 0.2i$ (iv) $-0.5 + 0.5i$

8. $x = -1.462$ and $y = -3.308$

9. (i) $p = 8 \pm 6i$ (ii) $x = 0.42$ or $x = -1.33$
   (iii) $y = -0.50 \pm 0.87i$ (iv) $z = 0, z = 0.16$ or $z = -6.16$

10. (i) $y = 43.02$ (linear interpolation)
    (ii) $y = 39.21$ (linear extrapolation)
    (iii) $y = 46.38$ (linear extrapolation)

11. $x = 0.74$

12. $x = 0.6566, x = -0.3283 \pm 2.112i$

13. Approximately 10 bisections

14. $m = 2.8t$; 7 grammes after $2\frac{1}{2}$ hours

15. £4.94

16. $\theta = 1.5f + 5; f = 13.3\,\text{kHz}$ when $\theta = 25$

17. $L = T$; scale markings 10 mm for 10°C, 60 mm for 60°C

18. $R = 190 - 20.8t + t^2$; minimum value is 81.84 and occurs when $t = 10.4$ min

19. (i) $R = 84.5$ beats/min (linear interpolation)
     (ii) $R = 83.75$ beats/min (linear extrapolation)

21. $R = R_1 R_2 R_3 w^2 c^2 / (1 + w^2 c^2 R_3{}^2)$
     $L = R_1 R_2 c / (1 + w^2 c^2 R_3^2)$

22. **begin**
     input $x_1, x_2,$ e
     while absolute value of $(1 - x_1/x_2) >$ e **do**
         set x **to** $x_1 - (x_2 - x_1) * f(x_1)/(f(x_2) - f(x_1))$
         if $f(x) < 0$ **then**
             set x_1 **to** x
         else
             set x_2 **to** x
         endif
     endwhile
     print x
     stop

2.6 Exercises

1. (i) 3^6 (ii) x^7 (iii) x^{-5} (iv) $1/10^2$
    (v) $1/a^2$ (vi) $1/x^3$ (vii) b^5 (viii) 10^{15}
    (ix) $1/x^8$ (x) b^8 (xi) $64y^3$ (xii) a^3/b^3
    (xiii) $5b^4c/2a^2$

2. (i) $a^{3.04}$ (ii) 1   (iii) $0.8409b$

3. (i) $1 + 7x + 21x^2 + 35x^3 + 35x^4 + 21x^5 + 7x^6 + x^7$
    (ii) $1 - 7x + 21x^2 - 35x^3 + 35x^4 - 21x^5 + 7x^6 - x^7$
    (iii) $1 - 10p + 40p^2 - 80p^3 + 80p^4 - 32p^5$
    (iv) $x^n + nx^{n-1}y + (n(n-1)/2!)x^{n-2}y^2 + \cdots + y^n$

4. 2.698703×10^{-38} joules

5. (i)   7.39   (ii) 4.48   (iii) 1.00
    (iv) 0.368   (v)   1.00   (vi) 1000
    (vii) 0.001   (viii) 1.65   (ix) 0.100

6. (i) 7   (ii) 5

10. (i) 0.0954   (ii) 0.2254   (iii) ± 0.5477

12. $x = 2$

13. (i) $x = 0.567$ (ii) $x = -1.10$

14. (i) 0.84%, 5.13%, 10.52%   (ii) 1.68%, 10.52%, 22.14%

15. (i) 7 years   (ii) 3 years   (iii) 2 years

16. $T = 120 \times 0.8^{t/10.5}$, 63°C

17. 80.25 days, 8.637×10^{-3} day^{-1}

18. 4.56 hr, 91.20 hr

19.   101.04 m, 6.28 m

20.   7.51 m

3.6 Exercises

1.   (i) $\sin(50°)$ (ii) $\cos(40°)$ (iii) $\sin(270°)$ (iv) $\tan(0°)$ (v) $\cos(120°)$
       (vi) $\sin(240°)$ (vii) $\cos(0°)$ (viii) $\tan(60°)$

2.   (i) $-\sin(84°)$ (ii) $\sin(18°)$ (iii) $-\sin(88°)$ (iv) $-\sin(34°)$

3.   (i) $\cos(30°)$ (ii) $\cos(80°)$ (iii) $-\cos(57°)$ (iv) $-\cos(80°)$

4.   (i) $\cos(A) = \sqrt{3/8}$, $\tan(A) = 1/\sqrt{8}$
       (ii) $\sin(B) = 4/5$, $\tan(B) = 4/3$
       (iii) $\sin(C) = 4/5$, $\cos(C) = 3/5$
       (iv) $\cos(x) = 2ab/(a^2 + b^2)$, $\tan(x) = (a^2 - b^2)/2ab$

5.   (i) $60°, 120°$ (ii) $30°, 150°, 210°, 330°$
       (iii) $30°, 150°, 270°$ (iv) $60°, 150°, 240°, 330°$

6.   (i) $150° \pm 360\,n$, $210° \pm 360n$ (ii) $15° + 180n$, $165° \pm 180n$ (iii) $15° \pm 60n$

8. $5\sin(2x + 36.87°)$
       $5\cos(2x - 53.13°)$

9. $A = 90°$ or $A = 298°$

10.   (i) 0.271 (ii) 1.211 (iii) 5.089 (iv) 10.414

11.   (i) $49.8°$ (ii) $154.7°$ (iii) $300.2°$ (iv) $14.3°$

12.   (i) $2e^{2\pi ni}$ (ii) $3e^{(\pi/2 + 2\pi n)i}$
       (iii) $2e^{(\pi + 2\pi n)i}$ (iv) $2e^{(3\pi/2 + 2\pi n)i}$
       (v) $5e^{(0.295\pi + 2\pi n)i}$ (vi) $2\sqrt{3}e^{(\pi/6 + 2\pi n)i}$
       (vii) $2\sqrt{2}e^{(\pi/4 + 2\pi n)i}$ (viii) $2e^{(11\pi/6 + 2\pi n)i}$

13.   (i) $-1.99 + 14.00i$ (ii) $4\sqrt{3}$
       (ii) $\dfrac{\sqrt{3}}{2} + \dfrac{1}{2}i$ (iv) $0.480 + 0.360i$
       (v) $-0.184 - 1.29i$

14.   (i) $a = 3\cos(A)$, $b = -3\sin(A)$
       (ii) $a = 1 + \dfrac{\sin(A)}{2(1 + \cos(A))}$, $b = \dfrac{\sin(A)}{(1 + \cos(A))} - \dfrac{1}{2}$

15.

	amplitude	period	phase
(i)	2.3	$2\pi/7$	$\pi/2$
(ii)	5	π	1.5
(iii)	5	π	0.93
(iv)	2.25	$\pi/3$	0

16.   (i) $0,$ $\pi/4,$ $3\pi/4,$ $\pi,$ $5\pi/4,$ $7\pi/4,$ 2π
       (ii) $\pi/6,$ $\pi/4,$ $\pi/2,$ $5\pi/6,$ $7\pi/6,$ $5\pi/4,$ $3\pi/2,$ $11\pi/6$
       (iii) $\pi/6,$ $5\pi/6,$ $3\pi/2$

17.   for $x = 0.1$; $\sin(0.1) \simeq 0.0998$; $3 \times 10^{-7}\%$
                $\cos(0.1) \simeq 0.9950$; $3 \times 10^{-7}\%$
                $\tan(0.1) \simeq 0.1003$; $3.4 \times 10^{-6}\%$

      for $x = 0.2$; $\sin(0.2) \simeq 0.1987$; $1.7 \times 10^{-6}\%$
                $\cos(0.2) \simeq 0.9801$; $8.9 \times 10^{-6}\%$
                $\tan(0.2) \simeq 0.2027$; $3.5 \times 10^{-4}\%$

18.   (i) $x = 0.0125$
      (ii) $x = \pm 0.316$

20. $x = 0.74$

21. $x = 2.310$

22. $x = 4.49$

23. $x = 0.8\cos(2t - 2.2143)$
      amplitude $= 0.8$ m, phase $= -2.2143$, period $= \pi$
      distance travelled $= 0.9523$ m

24.   Temperature $= 58.5 + 13.5\cos(\pi(t + 4)/6)$
      June 65.25°C, November 58.5°C

26. $w = \pi/6.2$, $A = 6.5$, $\delta = \pi/2$
      5.97 m, 4.48 m, 2.26 m

4.6 Exercises

1.   (i)   1.95   (ii)   1.61   (iii) 5.49
      (iv) -0.357 (v)   0.301   (vi)   0.903
      (vii) 1.5   (viii) 3   (ix)   1.5
      (x)   0.75   (xi)   1   (xii) 1
      (xiii) 1   (xiv) 2   (xv) -2

4.   (i) $x = \ln(2) = 0.6931$ (ii) $x = \pm\frac{1}{2}\ln(3) = \pm 0.5493$
      (iii) $x = 0$, $x = \frac{1}{2}\ln(2) = 0.3466$

5.   (i) $\pi/4$ (ii) $-\pi/4$ (iii) $\pi/4$
      (v) $3\pi/4$ (vi) $-\pi/4$ (vii) $-\pi/2$
      (viii) $5\pi/6$ (ix) $2\pi/3$ (x) 0.131π
      (xi) 0.436π (xii) -0.205π (xiii) 0.212π

7.   (i) $x = \pm 0.830 + 2\pi n$
      (ii) $x = \pi/4 + n\pi$, $x = \pm\pi/6 + n\pi$
      (iii) $x = \pm\pi/3 + 2\pi n$, $x = \pm 0.392\pi + 2\pi n$
      (iv) $x = \pm 0.196\pi + 2\pi n$, $x = 0.804\pi + 2\pi n$, $x = 1.196\pi + 2\pi n$

8.   (i) $x = 0.5\pi + 2\pi n$, $x = 1.20\pi + 2\pi n$ (ii) $x = 0.205\pi + 2\pi n$
      (iii) $x = 2\pi n$, $x = \pi/2 + 2\pi n$

10.   30°($\pi/6$), 150°($5\pi/6$), 270°($3\pi/2$)

11.   (i) $(1 \pm \sqrt{7})/3$ (ii) 1.107, 4.249, 2.356, 5.498

12.   amplitude $= \sqrt{5}$, phase $= 0.4636$, $t = 0.136$ s

13. $f^{-1}(x) = \sqrt{(1 - x)/(2x - 1)}$

14. $f^{-1}(x) = \frac{1}{2}\ln(x)$;   domain $0 < x < \infty$,   range $-\infty < f^{-1}(x) \leqslant \infty$
    $g^{-1}(x) = -\frac{1}{2}\sqrt{1-x^2}$;   domain $0 \leqslant x \leqslant 1$,   range $-\frac{1}{2} \leqslant g^{-1} \leqslant 0$

15. range $1 \leqslant f < \infty$; restrict f to $\{x : x \geqslant 0\}$

17. 44 days

18. $n = 0.924$, $k = 16.284$, $H = 604.8$

19. $t = 39.5°C$

20. $k = 0.07696$

21. $t = 7.1$ hours

22. $\lambda = 1.5753 \times 10^{-11}$

23. $A = 20.14\,\text{mol}.l^{-1}$, $a = 161.2$, $\text{sol} = 32.96\,\text{mol}.l^{-1}$

24. $A = 2.94 \times 10^4 R$

5.6 Exercises

1. (i)   0   (ii) 0   (iii) $\frac{1}{2}$
    (iv) 3   (v)   0   (vi) $\frac{1}{4}$
    (vii) 5/7   (vii) 1   (ix) 1
    (x)   divergent

2. (i) -3 (ii) $3, -3$ (iii) 1
    (iv) $\pi/2 + 2\pi n$ (v) None   (vi) πn
    (vii) $\pm\sqrt{12}, \pm\sqrt{3}$

3. (i) $\dfrac{2/3}{(x-1)} - \dfrac{2/3}{(x+2)}$ (ii) $\dfrac{1/5}{(x-2)} - \dfrac{1/5}{(x+3)}$

    (iii) $\dfrac{3}{(x-2)} - \dfrac{2}{(x-1)}$ (iv) $\dfrac{-3/2}{(3x-1)} - \dfrac{1/2}{(x-1)}$

    (v) $\dfrac{3/8}{x} - \dfrac{5/4}{(x-2)} + \dfrac{7/8}{(x-4)}$ (vi) $1 + \dfrac{1/3}{(x-1)} - \dfrac{4/3}{(x+2)}$

    (vii) $1 + \dfrac{4}{(x+1)^2} - \dfrac{4}{(x+1)}$ (viii) $(x-3) + \dfrac{14}{(x+2)} - \dfrac{4}{(x+1)}$

    (ix) $\dfrac{1}{(x-3)^2} - \dfrac{1}{(x-3)}$ (x) $\dfrac{1/27}{(x-2)} - \dfrac{1/3}{(x+1)^3} - \dfrac{1/9}{(x+1)^2} - \dfrac{1/27}{(x+1)}$

4. (i) 1500 m   (ii) 1056 kHz

5. 2.4

6. (i) $x = -3$, $y = 0$ (ii) $x = 1$, $x = -5$, $y = 0$
    (iii) $x = 1$, $y = 1$ (iv) $y = 1$
    (v) $x = 1/5$, $y = 3/5$ (vi) $x = 0$, $y = 2$

7. $V_T = mg/k$

8. $N_B = 0$

9. $T(0, t) = T_0(1 - 2\exp(-\pi^2 kt/R^2))$

10. $\lim_{x \to \infty} k = 0$

11. $t = L/(\beta - \alpha)$; the tank is empty
     $\lim_{t \to \infty} c(t) = \alpha/[(\beta + 1)(\alpha - \beta)]$

12. At equilibrium $P = a/b$

6.6 Exercises

1.   0.3

2.   0.3

3.   6

4.   (i) -23 (ii) $2t_0 - 25$

5.   (i) 7   (ii) 4   (iii) $32x$ (iv) $9x^2 + 4x - 1$
     (v) -1 (vi) $28x^6$ (vii) $38x^{18}$ (viii) $5x^4$
     (ix) 0   (x) $63x^8 - 12x^3$

6.   (i) (a) $20x^3$ (b) $-\dfrac{2}{x^3}$ (c) 0

     (d) $2.4(x + 2)^{1.4}$ (e) $\dfrac{1}{2\sqrt{x + 1}}$ (f) $3\cos(x) + 3x^2$

     (g) $-\sin(x) + \dfrac{1}{x}$ (h) $-\dfrac{1}{3}x^{-4/3}$ (i) $3e^x$

     (j) e^{x+17} (k) $2x + 1$ (l) $-\dfrac{(\sin(x) + \cos(x))}{2}$

     (m) $\ln(2)x^{\ln(2) - 1}$ (n) $\dfrac{2}{(x + 1)}$ (o) $-\dfrac{1}{2} \cdot x^{-3/2}$

     (ii) (a) $x^3 \cos(x) + 3x^2 \sin(x)$ (b) $e^x \cos(x) - e^x \sin(x)$
          (c) $1 + \ln(x)$ (d) $\cos^2(x) - \sin^2(x)$

          (e) $\dfrac{x^{-1/2}(1 + x)^{1/2}(4x + 1)}{2}$ (f) $x\cos(x) + \sin(x)$

          (g) $xe^x + x^2 e^x \ln(x) + 2xe^x \ln(x)$ (h) $x^{-1}\cos(x) - x^{-2}\sin(x)$

     (iii) (a) $x^{-1}\cos(x) - x^{-2}\sin(x)$ (b) $\dfrac{x(2 + x)}{(1 + x)^2}$ (c) $\dfrac{(\cos(x) - \sin(x))}{e^x}$

          (d) $\dfrac{4x}{(1 - x^2)^2}$ (e) $\dfrac{1 - 3\ln(x)}{x^4}$ (f) $-\dfrac{x\sin(x)}{(1 - \cos(x))^2}$

          (g) $\dfrac{x^2}{\cos^2(x)} + \dfrac{2x\sin(x)}{\cos(x)}$ (h) $\dfrac{x - x^2}{e^x}$

     (iv) (a) $3e^{3x}$ (b) $4x\exp(x^2)$ (c) $3\cos(3x + \pi/2)$

(c) $\dfrac{2x}{1+x^2}$ (d) $e^x \cos(e^x)$ (f) $\cos x (e^{\sin x})$

(g) $\dfrac{1}{(1+x)} + \dfrac{1}{(1-x)}$ (h) $\dfrac{x}{\sqrt{1+x^2}}$

(v) (a) $\dfrac{1}{1+x^2}$ (b) $\dfrac{2x}{\sqrt{1-x^4}}$ (c) $\dfrac{x\cos(x)}{2+\sin x} + \ln(1 + \tfrac{1}{2}\sin(x))$

(d) $\dfrac{e^x \cos(x) + x(e^x \cos(x) + e^x \sin(x))}{(1+x)^2}$ (e) $\exp(x^2)(1/x + 2x\ln(x))$

(f) $-\dfrac{1}{2}\sqrt{\left(\dfrac{1-\cos(x)}{x+\sin(x)}\right)} \cdot \dfrac{x\sin(x)}{(1-\cos(x))^2}$

7. (i) 30   (ii) 10

8. 9.72

9. $\dfrac{3\,ba^4}{16k}$

10. $\dfrac{\pi}{\sqrt{g}}$

12. $\dfrac{2n^2h^2}{8\pi^2 m\, z_0^2}$

13. $-200\,\text{cm sec}^{-2}$; the velocity is a decreasing function of r

14. (i) $(3t^2 + 8t + 9)e^t$ (ii) $\cos^2(t) - \sin^2(t)$
    (iii) $(2t^3 + 15t^2 + 30t + 7)\cos(t) + (6t^2 + 30t + 30)\sin(t)$
    (iv) $-\dfrac{(6t^5 + 6t + 2)}{(t^6 + 3t^2 + 2t)^2}$ (v) $-\operatorname{cosec}(t)\cot(t)$

    (vi) $\dfrac{t^2 - 2t}{(t-1)^2}$

15. (i) $\cos(x)\exp(\sin(x))$ (ii) $21(3x+4)^6$
    (iii) $(4x+4)\cos(2x^2 + 4x - 1)$ (iv) $\tfrac{1}{3}\cosh(x/3) + 15\sinh(5x)$

    (v) $3\tanh(3x)$ (vi) $-\dfrac{x}{\sqrt{1-x^2}}$

16. (i) $\dfrac{1}{2\sqrt{x}}$; $-\dfrac{1}{4}x^{-3/2}$ (vi) $-\dfrac{x}{\sqrt{1-x^2}}$

    (ii) $-\dfrac{7}{2}\dfrac{(3x^2 + 4x + 7)}{(x^3 + 2x^2 + 7x)^{9/2}}$;

    $-\dfrac{7}{2}\left\{(x^3 + 2x^2 + 7x)^{-9/2}(6x+4) - \dfrac{9}{2}(x^3 + 2x^2 + 7x)^{-11/2}(3x^2 + 4x + 7)^2\right\}$

    (iii) $\dfrac{4x+1}{(2x^2 + x - 5)}$; $-(4x+1)^2(2x^2 + x - 5)^{-2} + 4(2x^2 + x - 5)^{-1}$

17. (i) $36t^2 - 5\cos(t); \ 72t + 5\sin(t)$
     (ii) $24t + 4; \ 24$
     (iii) $-4\sin(t) - \exp(t); \ -4\cos(t) - \exp(t)$

20. (i) $(-1)^{n/2}\cos(x)$ if n is even
           $(-1)^{(n+1)/2}\sin(x)$ if n is odd

     (ii) $-\dfrac{(n-1)!}{(1-x)^n} + \dfrac{(-1)^n(n-1)!}{(1+x)^n} \quad (n \geqslant 1)$

7.6 Exercises

1. $y = x + 1$
     minimum at $(0, 0)$
     maximum at $(-2/3, \ 4/27)$

3.   Stationary values at $x = 0, \ 1, \ 3/5$
     minimum at $x = 1$, maximum at $x = 3/5$

4.   (i) $(3, -1)$, minimum
     (ii) $(-1, 19)$, maximum
     (iii) $(2, -11)$, minimum; $(-2, 21)$, maximum
     (iv)   no stationary points exist
     (v) $(1, 0)$, point of inflexion
     (vi) $(0, 40)$, minimum; $(1, 43)$, maximum; $(5, -85)$, minimum
     (vii) $(1, 0)$, maximum; $(5/3, -4/27)$, minimum
     (viii) $(1, 0)$, point of inflexion; $(2, 0)$, minimum; $(\frac{8}{5}, \frac{108}{3125})$, maximum
     (ix) $(1, 0)$, minimum; $(5, 2/3)$, maximum
     (x) $(2, 1/3)$, minimum; $(-2, \ 3)$, maximum
     (xi) $(-4, \frac{1}{2})$, minimum; $(4, \ 1/18)$, maximum
     (xii) $\ldots(-225°, \sqrt{2})$, maximum; $(-45°, -\sqrt{2})$, minimum; $(135°, \sqrt{2})$, maximum;$\ldots$
     (xiii) $\ldots(-135°, -0.067)$, minimum; $(45°, 1.551)$, maximum;
            $(225°, -35.89)$, minimum;$\ldots$
     (xiv) $(0, 0)$, minimum
     (xv) $\ldots(-\pi, -\pi), (\pi, \pi), (3\pi, 3\pi), \ldots$ all points of inflexion
     (xvi) $\ldots(-\pi, 0), (0, 0), (\pi, \pi), \ (2\pi, 0), \ldots$ all points of inflexion

     $$\ldots\left(\frac{-\pi}{3}, \frac{-3\sqrt{3}}{16}\right), \left(\frac{2\pi}{3}, \frac{-3\sqrt{3}}{16}\right), \left(\frac{5\pi}{3}, \frac{-3\sqrt{3}}{16}\right), \ldots \text{all minima}$$

     $$\ldots\left(-\frac{2\pi}{3}, \frac{3\sqrt{3}}{16}\right), \left(\frac{\pi}{3}, \frac{3\sqrt{3}}{16}\right), \left(\frac{4\pi}{3}, \frac{3\sqrt{3}}{16}\right), \ldots \text{all maxima}$$

9. $R/\sqrt{2}$

10. $\sqrt{2aL}, \ 16.7 \ \text{mil hr}^{-1}$

12. $0.33\,\text{s}$

13. $\sqrt{T/3a}, \quad a > 0$

16. $x = y, \ 10^{-7} \ \text{mol dm}^{-3}$

17. $300, \ £120$

8.6 Exercises

1. (i) $x + \dfrac{2x^2}{2!} + \dfrac{2x^3}{3!}$

    (ii) $1 + x + \dfrac{x^2}{2!} + \dfrac{0x^3}{3!}$

    (iii) $\log_e 2 + \dfrac{x}{2} + \dfrac{1}{4}\dfrac{x^2}{2!} + \dfrac{0.x^3}{3!}$

    (iv) $x - \dfrac{x^2}{2!} + \dfrac{x^3}{3!}$

2. $1 + \dfrac{x^2}{2!} + \dfrac{5x^4}{4!}$

4. $\sin(\alpha) + (x - \alpha)\cos(\alpha) - \dfrac{(x - \alpha)^2}{2!}\sin(\alpha) - \dfrac{(x - \alpha)^2}{3!}\cos(\alpha) + \cdots$

    $\tan(\alpha) + (x - \alpha)\sec^2(\alpha) + \dfrac{(x - \alpha)^2}{2!}2\sec^2\alpha\tan(\alpha) + \dfrac{(x - \alpha)^3}{3!}2\sec^2\alpha(1 + 3\tan^2\alpha) + \cdots$

    1.0538

5. $e^1 + (x - 1)e^1 + \dfrac{(x - 1)^2}{2!}e^1 + \dfrac{(x - 1)^3}{3!}e^1 + \cdots$

    $1 + (x - 1).\dfrac{1}{2} - \dfrac{(x - 1)^2}{2!}\cdot\dfrac{1}{4} + \dfrac{(x - 1)^3}{3!}\cdot\dfrac{3}{8} + \cdots$

    $\ln 2 + \dfrac{(x - 1)}{2} - \dfrac{(x - 1)^2}{2!4} + \dfrac{(x - 1)^3}{3!4} + \cdots$

6. (i) $-\dfrac{1}{6}$,   (ii) $-\dfrac{1}{2}$,   (iii) $\dfrac{1}{4}$,   (iv) 0

7. $\dfrac{Va}{27b^2}, \dfrac{Va}{27b^2} - \dfrac{(V - 3b)^3}{3!}\dfrac{a}{81b^5}$

8. $0.0025\left(\dfrac{kt}{125} - \dfrac{3}{4}\left(\dfrac{kt}{125}\right)^2\right)$

9. $\dfrac{2M}{d^3}\left\{1 - \dfrac{3}{2}\left(\dfrac{1}{d}\right) + 2\left(\dfrac{1}{d}\right)^2 - \left(\dfrac{5}{2}\right)\left(\dfrac{1}{d}\right)^3 + \cdots\right\}$

10. (i) 0.735   (ii) $-1.895, 1.895$ (iii) -1.478 (iv) 0.112, 3.577

11. (i) 1.5121
    (ii) $-0.5045, -3.9372, -8.5582$
    (iii) $0, \pm 4.4934, \pm 7.7253, \pm 10.9041, \ldots$
    (iv) 2.2214
    (v) 0.46392, 5.3567
    (vi) $\pm 0.7408, \pm 2.9726. \pm 6.3619, \pm 9.3714, \ldots$

12. (i) 0.6566   (ii) 1.8342   (iii) -2.4909

13. $-9, -6.0711, -4.1336$

14. $x_1 = x_0 - (x_0^2 - A)/(2x_0)$

15. $c_1 = c_0 - (A(2c_0 - 1) - \ln(c_0/(1 - c_0)))/(2A - 1/c_0(1 - c_0))$

9.6 Exercises

1. (i) $\dfrac{x^4}{4} + C$ (ii) $\frac{2}{5}x^{10} + C$ (iii) $-\dfrac{1}{2x^2} + C$

   (iv) $16x + \dfrac{x^3}{3} + C$ (v) $\dfrac{x^2}{2} + 2\ln x + C$ (vi) $-\frac{1}{3}\cos(3x) + C$

   (vii) $2\sin\left(\dfrac{x}{2}\right) + C$ (viii) $\tan(x) + C$ (ix) $[\sin(x) + \cos(x)]/2 + C$

   (x) $\frac{1}{3}e^{3x} + C$ (xi) $e^{(x-1)} + C$ (xii) $\frac{6}{7}x^{7/2} + \frac{4}{3}x^3 + \frac{1}{2}e^{2x} + C$

2. (i) $\dfrac{\sin(2x)}{4} + \dfrac{x}{2} + C$ (ii) $\dfrac{x}{2} - \dfrac{\sin(2x)}{4} + C$

   (iii) $\dfrac{\sin 4x}{32} + \dfrac{\sin 2x}{4} + \dfrac{3x}{8} + C$ (iv) $\dfrac{\sin 4x}{32} - \dfrac{\sin 2x}{4} + \dfrac{3x}{8} + C$

3. (i) 33 (ii) $\frac{15}{2}$ (iii) $-\frac{29}{30}$ (iv) $\frac{7}{2}$ (v) $3 - \dfrac{1}{\sqrt{2}}$

   (vi) 2232.97 (vii) 2 (viii) 1.09861 (ix) 25.9735

4. (i) 2 (ii) 0.69315 (iii) 13.5 (iv) 4.5 (v) 16

5. (i) $v_0 - \dfrac{eV_1(1 - \cos\omega T)}{dm\omega}$ (ii) $v_0 T - \dfrac{eV_1(T\omega - \sin(\omega T))}{dm\omega^2}$

7. 36.74

8. $\left[\dfrac{L^2 A^4 \omega^2}{2} + \dfrac{3}{8} R^2 A^4\right]^{1/2}, \dfrac{RA^2}{2}$

10. $\dfrac{I_0}{2L\omega^2}, \left(\dfrac{A^2}{2} + \dfrac{I_0^2 \pi^2}{L\omega^3} + \dfrac{I_0^2}{8L\omega} + \dfrac{AI_0 \cos\varphi}{2L\omega}\right)$

11. $\displaystyle\int_{V_1}^{V_2} RT\left(\dfrac{1}{V} + \dfrac{a}{V^2} + \dfrac{b}{V^3}\right)dV, \quad RT\ln\left(\dfrac{V_2}{V_1}\right) - a\left(\dfrac{1}{V_2} - \dfrac{1}{V_1}\right) - \dfrac{b}{2}\left(\dfrac{1}{V_2^2} - \dfrac{1}{V_1^2}\right)$

12. (i) $300\,\text{kg}\,\text{m}^{-3}$ (ii) $640\,\text{kg}\,\text{m}^2$

13. (ii) $\dfrac{1}{\lambda} - \dfrac{e^{-\lambda T}}{\lambda} - Te^{-\lambda T}$

16. 0.9770, area should be 1.0

17. (i) 23.99, 24.23, 25.18 (ii) 23.91, 23.92, 23.97

19. 33.27

20. (i) 0.7434, (ii) 0.7457, ' (iii) 0.7463

21. 0.845

22.   0.0997, 0.1974, 0.2912, 0.3797, 0.4613,
     0.5352, 0.6007, 0.6577, 0.7062, 0.7468

23.   (i) 6356.0×10^3 gallons ph
     (ii) 6345.3×10^3 gallons ph, negligible difference

24.   (i) $49.03°C$ (ii) $49.05°C$, average is $49.45°C$

25. $\frac{1}{2}e^{-0.25T} - \frac{1}{2}e^{-0.89T}$

10.6 Exercises

1.   (i) $\frac{2}{3}(x+1)^{3/2} + C$ (ii) $\dfrac{(x+3)^3}{3} + C$

    (iii) $1.4\ln(x+1.4) + C$ (iv) $\ln(x^2+1) + C$

    (v) $\sqrt{(x^2+2x+33)} + C$ (vi) $\dfrac{\exp(x^2)}{2} + C$

    (vii) $\dfrac{-(\cos(x)+\sin(x))^4}{4} + C$ (viii) $-\ln(2\cos(x)+3\sin(x)) + C$

    (ix) $\dfrac{\sin^2 x}{2} + C$ (x) $-\ln(\cos(x)) + C$

    (xi) $x+1-4\sqrt{x+1}+8\ln(2+\sqrt{x+1}) + C$

2.   (i) $-\dfrac{1}{4x} + C$ (ii) $\dfrac{x}{2}\sqrt{(1-x^2)} + \dfrac{\sin^{-1}x}{2} + C$

    (iii) $\frac{1}{3}\tan^{-1}\left(\dfrac{x+2}{3}\right)$ (iv) $\dfrac{x}{8(x^2+4)} + \dfrac{\tan^{-1}(x/2)}{16} + C$

    (v) $\dfrac{x\sqrt{(1+9x^2)}}{2} + \dfrac{\sinh^{-1}3x}{6} + C$ (vi) $\dfrac{x}{\sqrt{1+x^2}} + C$

    (vii) $\frac{1}{4}\ln\left(\dfrac{2+\tan(x/2)}{2-\tan(x/2)}\right) + C$ (viii) $-x - \dfrac{2}{(1-\tan(x/2))} + C$

3.   (i) $\frac{1}{10}\ln\left(\dfrac{5+x}{5-x}\right) + C$ (ii) $\ln\left(\dfrac{x-3}{x-2}\right) + C$

    (iii) $\frac{1}{9}\ln\left(\dfrac{2x-1}{x+4}\right) + C$ (iv) $\ln(x^2-x-1) - 2\ln(2x-3) + C$

    (v) $-\ln(x) - 1/x + \ln(x+1) + C$
    (vi) $\ln(x) - \frac{3}{4}\ln(x-1) - \frac{1}{2}(x-1)^{-1} - \frac{1}{4}\ln(x+1) + C$

4.   (i) $\dfrac{x^2}{2}\ln(x) - \dfrac{x^2}{4} + C$ (ii) $x\tan(x) + \ln(\cos(x)) + C$

    (iii) $-\dfrac{x}{a}e^{-ax} - \dfrac{e^{-ax}}{a^2} + C$ (iv) $x^2e^x - 2xe^x + 2e^x + C$

    (v) $x\sin^{-1}(x) + \sqrt{1-x^2} + C$ (vi) $x\tan^{-1}(x) - \frac{1}{2}\ln(1+x^2) + C$
    (vii) $x^2\sin(x) + 2x\cos(x) - 2\sin(x) + C$
    (viii) $\frac{3}{8}e^x\sinh(3x) - \frac{1}{8}e^x\cosh(3x) + C$
    (ix) $x^n\cosh(x) - nx^{n-1}\sinh(x) + n(n-1)x^{n-2}\cosh(x) - \cdots + C$

5.　(i)　$2\sqrt{2}-2$　　(ii)　0.28768　　(iii)　0.32175
　　(iv)　2/3　　　　　(v)　$e-2$　　　　(vi)　$a^3/3$
　　(vii) 0.08961　　(viii) 0.21450　　(ix)　0.21232
　　(x)　0.01841　　(xi)　0.07619　　(xiii) $3\pi/2$

6.　$x-\dfrac{x^3}{3}+\dfrac{x^5}{5}-\dfrac{x^7}{7}$,　$x+\dfrac{x^3}{6}+\dfrac{3x^5}{40}+\dfrac{5x^7}{112}$,　$\dfrac{7}{20}$

7.　0.08127

8.　$I_4 = 6.2515$,　$I_5 = 5.5709$

9.　(i)　$i = \dfrac{1}{e^{Rt/L}}\left\{\dfrac{V_0}{L}\left(\dfrac{R}{Lp^2}e^{Rt/L}\sin(pt)-\dfrac{e^{Rt/L}}{p}\cos(pt)\right)\middle/\left(1+\dfrac{R^2}{L^2p^2}\right)+C\right\}$

　　(ii)　$C = \dfrac{V_0 Lp}{(L^2 p^2 + R^2)}$

11.　$\dfrac{1}{k}\ln 2$,　$\dfrac{(2^{n-1}-1)}{ka^{n-1}(n-1)}$

12.　$3(2-x)^2 - 4(3-x)(1-x)e^{2kt} = 0$

13.　(i)　$t = \ln\left(\dfrac{100(1+kV)}{V(1+100k)}\right)$　　　　(ii)　$\ln\left(\dfrac{2+100k}{1+100k}\right)$

　　(iii)　$V = 100e^{-t}/(1 + 100k - 100ke^{-t})$　　　(iv)　0

15.　$I_n = -\dfrac{x^n}{m}\cos(mx) + \dfrac{n}{m^2}x^{n-1}\sin(mx) - \dfrac{n(n-1)}{m^2}I_{n-2}$

　　$I_n = \dfrac{x^n}{m}\sin(mx) + \dfrac{n}{m^2}x^{n-1}\cos(mx) - \dfrac{n(n-1)}{m^2}I_{n-2}$

11.6 Exercises

1.　(i) $\frac{1}{2}$　　(ii) 1　　(v) $\pi/12$　　(vi) $-\frac{1}{4}$　　(vii) $\frac{1}{2}$

2.　(i) 3　　(ii) a　　(vi) 0.54931

4.　$2.1 \times 10^{12}\,\text{s}^{-1}$

5.　$RT\ln\left(\dfrac{V_2-b}{V_1-b}\right)+\dfrac{a}{V_2}-\dfrac{a}{V_1}$

6.　$b\pi/R$

8.　(a)　$aq(1+q^2)^{1/2} + a/2\sinh^{-1}(q)$

　　(b)　$\dfrac{u^2}{2g}\left(\sin(\alpha)+\dfrac{\cos^2(\alpha)}{2}\sinh^{-1}(\tan(\alpha))\right)$

9.　(iii)　5.67233

10.　$\frac{2}{5}MR^2$

12.6 Exercises

1. (i) $y^2 = \frac{2}{3}x^3 + C$ (ii) $\ln(\sin(y)) = e^x + C$

    (iii) $y = C(1 + x)$ (iv) $y = Ce^{-\cos(x)} - 1$

    (v) $\tan^{-1}\left(\dfrac{y}{2}\right) = 2\tan^{-1}(x) + C$

    (vi) $\dfrac{e^{-2y}}{2}(y + \frac{1}{2}) + \dfrac{e^x}{2}(\sin(x) + \cos(x)) + C = 0$

    (vii) $\dfrac{1}{y} = \dfrac{1}{4x^2} + \dfrac{3}{2x} - \dfrac{3}{4}$

    (viii) $e^y = \tan^{-1}(x) + 1 - \dfrac{\pi}{4}$

2. (i) $y = \sin^2(x)(e^x + C)$ (ii) $y = \dfrac{1}{2x}(\ln(x))^2 + \dfrac{C}{x}$

    (iii) $y = \dfrac{x^4}{(1 + x)} + \dfrac{Cx}{(1 + x)}$

    (iv) $y = x\operatorname{cosec}(x) + \operatorname{cosec}(x)\cot(x) + C\operatorname{cosec}^2(x)$

    (v) $y = 10e^{x-1} - x^2 - 2x - 2$ (vi) $y = \dfrac{\sin(x)\tan(x)}{2} + \frac{1}{2}\sec(x)$

3. (i) $y = -\dfrac{2}{x} - \dfrac{3}{2x^2} + C$ (ii) $y = -x^2 - 2 + Ce^{x^2/2}$

    (iii) $y = \dfrac{1}{a}\ln(ax + b) + C$ (iv) $y = C\cot(x)$

    (v) $y = \left(\dfrac{x}{2} + C\right)^2 - 1$ (vi) $y = \dfrac{x^2}{2(2x + 1)} + \dfrac{C}{(2x + 1)}$

4. (i) first order, non-linear
    (ii) second order, linear
    (iii) first order, linear
    (iv) second order, non-linear
    (v) fourth order, linear
    (vi) first order, separable, non-linear

5. (i) $i = i_0 e^{-Rt/L}$ (ii) $y = -x^2 + 2 - 2e^{x^2/2}$
    (iii) $p = p_0(T_0/(T_0 - 0.0061z))^{g/(0.0061R)}$

6. 3

7. 8727 yr

8. 6.93 month

9. (i) $T = \dfrac{10}{k}(1 - e^{-kt})$ (ii) $0.1594\,\text{min}^{-1}$

10. 4.2, 4.4254, 4.6779, 4.9590, 5.2708

    $y = \dfrac{1}{4}\left(\dfrac{x^2}{2} + 3.5\right)^2$,   % error is 1.44%

11. (i) 1.08   (ii) 1.25   (iii) 0.602

12. $x = \dfrac{a(1 - e^{-2akt})}{(1 + e^{-2akt})}$

13. $\dfrac{ka}{(k + k_1)}$, $\dfrac{1}{k + k_1} \ln(2)$

14. $t = \dfrac{1}{k}\left\{\dfrac{1}{(2 - x)} - \dfrac{1}{2} + \ln\left(\dfrac{3(2 - x)}{2(3 - x)}\right)\right\}$, $\dfrac{0.2123}{k}$ time units

15. $x = \dfrac{M}{(1 + e^{-Mkt})}$

16. (i) $C = C_0 - C_0 e^{-vt/V}$
    (ii) $C = C_0 - C_0 e^{-vt/V}$ $(0 < t \leqslant 2)$,
    $C = C_0 e^{-v(t-2)/V} - C_0 e^{-vt/V}$ $(2 < t < \infty)$

17. $V = 100 e^{-t}/(1 + 100k - 100k e^{-t})$, 0

18. $k = 1.405$

19. $T = T_S - (T_S - T_A)e^{-(2\pi Rk/vs)z}$

$v = \dfrac{2\pi RkL}{s} \bigg/ \ln\left(\dfrac{T_S - T_A}{T_S - T_0}\right)$

20.

t	$P(r = 1.5)$	$P(r = 3)$	$P(r = 4)$	$P(r = 2)$
0	1.0	1.0	1.0	2
0.25	2.0	2.0	2.0	3
0.5	3.58	1.0	−3.0	3
0.75	5.57	2.0	−48	”
1.0	7.35	1.0	−265438	”
1.25	8.41	2.0		”
1.5	8.83	1.0	becomes	”
1.75	8.96	2.0	unstable	”
2.0	8.99	1.0		”
2.25	9.00	2.0		”
2.5	”	1.0		”
2.75	”	2.0		”
3.0	”	1.0		”
3.25	”	2.0		”
3.5	”	1.0		”
3.75	”	2.0		”
4.0	”	1.0		”
4.25	”	2.0		”
4.5	”	1.0		”
4.75	”	2.0		”
5.0	”	1.0		”

21. 4.90, 2.21

13.6 Exercises

1. (i) $s = \dfrac{t^5}{20} + \dfrac{7t^3}{12} - \dfrac{7}{4}t$ (ii) $y = -x \sin(x) - 2\cos(x) - x + 3$

2. (i) $y = Ae^{-5t} + Be^{3t}$
   (ii) $y = (At + B)e^{3t}$
   (iii) $y = e^{2t}(A\cos(3t) + B\sin(3t))$
   (iv) $y = e^{-t/4}(A\cos(\sqrt{7}t/4) + B\sin(\sqrt{7}t/4))$

3. (i) $y = Ae^{-2x} + Be^{-x} + e^x$
   (ii) $y = Ae^{2x} + Be^{3x} + \frac{95}{108} + \frac{25}{18}x + \frac{5}{6}x^2$
   (iii) $y = e^{\sin(a)x}(A\cos(\cos(a)x) + B\sin(\cos(a)x)) + \cot(a)\cos(x)$
   (iv) $y = Ae^{-5/2x} + Be^{7x} - \frac{43}{2173}\sin(2) + \frac{18}{2173}\cos(2x)$
   (v) $y = Ae^{4x} + Be^x - \dfrac{4x}{3}e^x$

5. $x = d(wt + 1)e^{-wt}$

6. $\theta = \frac{2}{9} - [\frac{2}{3}t + \frac{2}{9}]e^{-3t}$

7. $v = \dfrac{mg}{6\pi\eta a}\left(1 - \exp\left(-\dfrac{6\pi\eta at}{m}\right)\right)$

8. (i) $y = 2t - 1 - 5e^{0.2t}\sin(0.4t)$
   (ii) $y = \frac{1}{3}e^{6t} - \frac{3}{4}e^{4t} + t + \frac{5}{12}$
   (iii) $y = e^{-x}(1 - \cos(2x))$
   (iv) $y = (\frac{3}{4}x + \frac{1}{8})e^{-2x} + \dfrac{x^2}{4} - \dfrac{x}{2} + \dfrac{3}{8} + \dfrac{x^2e^{-2x}}{2}$
   (v) $x = \frac{7}{24}\sin(3t) + \frac{1}{8}\cos(3t) - \frac{1}{8}e^{-t}$

9. (i) $x = e^{-3t/2}\left(A\sin\left(\dfrac{\sqrt{31}t}{2}\right) + B\cos\left(\dfrac{\sqrt{31}t}{2}\right)\right)$

   (ii) $A = 0.3/\sqrt{31}$, $B = 0.1$
   (iii) 0.0702 time units

10. (i) $\psi(x) = A\sin\left(\dfrac{2\pi}{h}(2Em)^{1/2}x\right) + B\cos\left(\dfrac{2\pi}{h}(2Em)^{1/2}x\right)$

    (ii) $\psi_1(x) = A\sin\left(\dfrac{n\pi x}{L}\right)$

    (iii) $\psi_1(x) = \sqrt{\dfrac{2}{L}}\sin\left(\dfrac{n\pi x}{L}\right)$

11. $y = 4te^{-3/2t}$

12. $y = \dfrac{k_1 a}{k_2 - k_1}(e^{-k_1 t} - e^{-k_2 t})$

13. (i) $Q_0 = E_0/(-Lw^2 + 1/C + Riw)$

14. (ii) $y = \left(1 - \dfrac{F}{w^2 - p^2}\right)\cos(wt) + \dfrac{F\cos(pt)}{w^2 - p^2}$

16. $T = T_A + (T_S - T_A)\sinh(\alpha(L - z))\operatorname{cosech}(\alpha L)$, where $\alpha = \sqrt{\dfrac{4h}{kR}}$

14.6 Exercises

1. (i) 9.12, 1.66   (ii) 9.18, 1.73

2. (ii) 10.5 − 12.0, 11.25   (iii) 11.87, 2.86   (iv) 11.86, 2.84

3. 11.87, 2.75

4. 0.416

5. (ii) 3.251, 0.0548   (iii) 3.251, 0.039   (iv) 32.5%

6. (i) 51.02 kg   2.458 kg   (ii) (a) 22%   (b) 44%

7. (i) 68%   (ii) 95.9%   (iii) 99.9%

8. (i) 46.9%   (ii) 15.95 kg   (iii) 0.149

9. (i) 0.0936, 0.0866 − 0.0988   (ii) − 0.085

10. (i)   6.448, 0.3673
    (ii) 6.62, 6.32, 6.22, 6.38, 6.94
          6.84, 7.10, 6.64, 6.82, 6.34
          6.26, 6.44, 6.56, 6.38, 6.46
          6.60, 5.94, 6.06, 5.90, 6.14
    (iii) 5, 6, 7, 9, 17, 18, 19

15.6 Exercises

1. (i) 3/5   (ii) 2/5   (iii) 2/5, 16/25

2. 0.84645, 0.14715, 6.35×10^{-3}, 5×10^{-5}, 0.16

3. (i) 0.15625   (ii) 0.216   (iii) 0.744

4. $1 - {}^{365}P_{45}/365^{45}$

5. (c) $P(B) = \frac{2}{7}$, $P(C) = \frac{1}{7}$

6. (a) 7, 590, 23.259, 11600   (b) £400, £300, £200, £100

8. (a) (i) 0.9,   (ii) 0.659,   (iii) 0.4567,   (iv) 0.0019,   (v) 0.0104
    (b) (i) 0.368,   (ii) 0.163,   (iii) 0.145,   (iv) 0.399,

9. (i) 0.698   (ii) 0.302   (iii) 0.00356   (iv) 0.0041

10. (a) 0.148 (binomial), 0.149 (Poisson)   (b) 0.647

11. (a) (i) 0.169   (ii) 0.297   (b) 0.609

12. 11 bottles

13. $5.46 \times 10^{-5} \text{Nm}^{-2}$, $0.50 \times 10^{-5} \text{Nm}^{-2}$

14. (a) 0.617   (b) (i) 0.0359   (ii) 0.0139   (iii) 0.9502   (c) 4.98, 0.03

15. (a) $4.56 \pm 0.081\%$, 9.13%   (b) 19, 15.5

16. (a) (i) 0.590   (ii) 0.328   (iii) 0.082   (b) 0.2

17. (i) 0.153   (ii) 0.0228   (iii) 0.130
    150 ± 20, 150 ± 27

18. (i) 0.15   (ii) 0.34

16.6 Exercises

1. $95.60 \pm 0.0554\%$, $95.60 \pm 0.0729\%$

2. (a) (i) $15.1 \, \text{mg} \, \text{kg}^{-1}$ (ii) $3.270 \, \text{mg} \, \text{kg}^{-1}$ (iii) $3.446 \, \text{mg} \, \text{kg}^{-1}$
   (b) $15.1 \pm 2.465 \, \text{mg} \, \text{kg}^{-1}$

3. 1310

4. (i) $74.67 \pm 11.961 \, \text{mg} \, \text{kg}^{-1}$ (ii) 10 readings   (iii) 15 readings

5. 7 readings

6. $t = 1.313$, accept null hypothesis

7. $t = 1.204$, no significant decrease

8. $t = 3.834$, not significant

9. $t = 3.618$, significant difference

10. $t = 2.1$

11. (i) $F = 1.913$, pool results   (ii) $t = 2.060$, not significant at 5% level

12. $F = 2.139$, pool results; $t = 1.503$, not significant

13. $F = 3.274$, pool results; $t = 1.360$, not significant

14. B is inaccurate at 5% level; both of comparable precision

15. $t = 13.93$, delay of significance

16. (ii) 0.687

17.6 Exercises

1. (i) $y - 2x$, $2xy + 2y$, -2, $2y$, $2x + 1$
   (ii) $4(x - y)^3$, $-4(x - y)^3$, $12(x - y)^2$, $-12(x - y)^2$, $12(x - y)^2$
   (iii) $y\cos(xy)$, $x\cos(xy)$, $-y^2\sin(xy)$, $-xy\sin(xy) + \cos(xy)$, $-x^2\sin(xy)$
   (iv) $xe^{x-y} + e^{x-y}$, $-xe^{x-y}$, $xe^{x-y} + 2e^{x-y}$, $-xe^{x-y} - e^{x-y}$, xe^{x-y}
   (v) $y - 1/x$, $x - 1/y$, $1/x^2$, 1, $1/y^2$

   (vi) $\dfrac{1}{y} + \dfrac{y}{x^2}$, $-\dfrac{x}{y^2} - \dfrac{1}{x}$, $-\dfrac{2y}{x^3}$, $-\dfrac{1}{y^2} + \dfrac{1}{x^2}$, $\dfrac{2x}{y^3}$

   (vii) $12x^2 - 5y^2$, $-10xy + 6y^2$, $24x$, $-10y$, $-10x + 12y$
   (viii) $-2\sin(2x + 3y)$, $-3\sin(2x + 3y)$, $-4\cos(2x + 3y)$, $-6\cos(2x + 3y)$,
        $-9\cos(2x + 3y)$

5. (i) $na(xy)^{n-1}(x^2 - y^2)$ (ii) $2ay/(x^2 + y^2)$
   (iii) $(y^2(ad - bc) - 2yxbc + x^2(ad + cb))/((cx + dy))^2$

6. $0.113 \, \text{ms}^{-2}$

7. $V\left(C^2w^2 + \dfrac{1 - 2LCw^2}{R^2 + L^2w^2}\right)^{-1/2}\left\{\dfrac{R(2LCw^2 - 1)}{(R^2 + L^2w^2)}\delta R + \dfrac{Cw^2 - Lw^2}{(R^2 + L^2w^2)}\delta C\right\}$

9. $x^2 + y^2 + C$

18. $$\frac{R^2}{(V-b)\left(p-\dfrac{a}{V^2}+\dfrac{2ab}{V^3}\right)}$$

20. $\dfrac{\lambda}{r}\cos\left(\lambda[r-ct]\right)-\dfrac{1}{r^2}\sin\left(\lambda[r-ct]\right),\dfrac{\lambda c}{r}\cos\left(\lambda[r-ct]\right)$

22. (b) $-RT\left(\dfrac{1}{V^2}+\dfrac{2a}{V^3}+\dfrac{3b}{V^4}\right),\ \dfrac{R}{p}\dfrac{(V^3+aV^2+bV)}{(V^2+2aV+3b)}$

23. (b) $\dfrac{\delta_p}{p}=\left(\dfrac{w_2m_1}{w_1m_2-w_2m_1}\right)\cdot\left(\dfrac{\delta w_1}{w_1}-\dfrac{\delta w_2}{w_2}\right)$

24. (a) $\dfrac{\delta D}{D}=\dfrac{\delta p}{p}+\dfrac{n\delta T}{T}$ (b) 10%

25. (a) $\Delta S=0.03875R$ (b) $\dfrac{R\delta S+10^9L\delta L}{\sqrt{[R^2+10^9L^2]}}$, 104.88 ohm, 1.24%

26. (i) $\dfrac{\partial V_s}{\partial\theta}=\dfrac{1}{280}\left(\dfrac{A}{\rho}\right)^{1/3},\ \dfrac{\partial V_s}{\partial A}=\dfrac{\theta}{840}\left(\dfrac{A}{\rho}\right)^{-2/3},$

$$\dfrac{\partial V_s}{\partial\rho}=-\dfrac{\theta}{840\rho^2}\left(\dfrac{A}{\rho}\right)^{-2/3}$$

% error in V_s = % error in $\theta+\frac{1}{3}$% error in $A-\frac{1}{3}$% error in ρ
(ii) 0.267%

18.6 Exercises

1. (i) $(1,1)$, maximum
   (ii) $(0,0)$, double saddle
   (iii) $(1,0)$, saddle; $(1,1)$; minimum; $(-3,0)$, maximum; $(-3,1)$, saddle
   (iv) $(1,-1)$, saddle; $(-1,1)$, minimum
   (v) $(0,0)$, saddle; $(6a,18a)$, minimum
   (vi) $(2,-3)$, saddle; $(2,1)$, maximum

7. $E=\dfrac{h^2}{8m}\cdot\dfrac{3}{k^{2/3}}$

8. $a+\dfrac{mga(n+1)}{2\lambda n},\ mg\left(\dfrac{n+1}{2n}\right)$

10. $V=21.37x+42.77$

11. $S=9.2619\times10^{-4}T+0.280821$

12. $\sigma=0.48+3.09V-4.88V^2$, 0.78

13. (i) $\ln(y-55)=\alpha t+\ln R_0$ (ii) $R_0=20.599,\alpha=-0.0104$ (iii) 136.1 min

14. $k=0.13861,n=1.8539$

INDEX

Alternating current, 94
Amplitude, 84, 135, 345
Anti-derivative, 262
Approximation
  of functions, 42, 211–32
  of roots, 23–6
Arbitrary constants, 311, 315, 340
Arc length, 82
Area
  as a definite integral, 239
Argand diagram, 88
Arithmetic mean, 367, 380
Asymptote, 57, 146–9, 202–4
Atomic structure, 280
Auxiliary equation, 340
Average, 381
  rate of decay, 49
  rate of growth, 49
  speed, 159

Bacterial growth, 35, 101
Bar chart, 370, 377
Base of logarithm, 115–6
Bessel's correction, 438
Best estimates 434–5, 466–7
Binary code, 35
Binomial distribution, 409–10, 420, 421
  mean, 410
  Normal approximation to, 420
  Poisson approximation to, 411
  standard deviation, 410
  variance, 410
Binomial expansion, 40, 219
Bohr model, 197
Bohr radius, 280
Boundary conditions, 349
Boyle's law, 133

Catenary, 52–4, 300
Cell growth, 162–3
Centre of mass, 259–61, 298–300
Centre, measures of, 380–3
Centroid, 259, 269
Chain rule, 179, 262, 487–9, 503
Chemical reactions, 261, 288, 305, 310
Circular measure, 81
Class
  boundaries, 369
  contiguous, 369
  end marks, 369
  frequency, 366, 386
  mid points, 369
Coding formulae, 386

Combinations, 401
Complementary function, 352
Complex numbers, 12–14, 88–92
  addition, 12
  argument, 90
  Cartesian form, 89
  equality, 13
  Euler's formula, 88
  exponential form, 89
  imaginary part, 12
  modulus, 90
  product, 12
  real part, 12
  roots, 12
Composite function, 108–12
Computer program, 15, 26, 42, 107, 115, 226, 253, 384
Confidence interval, 417, 421, 440, 441
Continuity, 139–47
Continuous function, 23, 141
Convex lens, 142, 220
Cosecant, 73
Cosine, 69, 75
Cotangent, 73
Coulomb's law, 279, 480
Critical damping, 347
Critical point, 467
Crude mode, 383, 387
Cumulative frequency, 367, 372
Curve sketching, 200–4
Curves
  frequency, 373–5
  length of, 293–6

Damped harmonic motion, 281, 348
Damping, 346–8
  critical, 347
Data
  continuous, 369
  discrete, 369
  organisation of, 367–70
  presentation of, 370–3
Decay function, 49–51
Definite integral, 238, 242–8
Degrees of freedom, 441–4, 447
Degree of polynomial, 6
Dependent variable, 4
Derivative, 168, 239
  chain rule, 179, 262
  geometric interpretation, 190–1
  higher, 181
  of basic functions, 169–72
  of product, 177

Derivative (*Contd.*)
  of quotient, 179
  partial, 468, 471, 473–6
  second, 181
Deviation
  mean, 385
  standard, 383–4, 388, 410
Differential equations, 261, 305–63
  arbitrary constants, 311, 315, 340
  auxiliary equation, 340
  boundary conditions, 349
  classification of, 310–11
  complementary function, 352
  exact methods of solution, 320
  first order, 305–33
  general solution, 312, 340
  initial conditions, 345, 347
  integrating factor, 314, 326, 339
  linear first order, 306, 310, 313–15
  non-linear first order, 307, 310, 327
  numerical solution, 320–6
    accuracy of, 324
    rounding errors, 326
  order, 310, 337
  ordinary, 305
  particular solution, 315–6, 345, 351
  second order, 334–63
    complex solution, 343
    homogeneous, 338–9
    linear, 338
    non-homogeneous, 351
    real solution, 342
  solution, 311
  tangent method, 321
  trial solutions, 352–4
  variables separable, 311–13, 327
Differentials, 484–6
  perfect, 486
Differentiation, 159–232
  higher partial, 476–7
  of basic functions, 169–74
  parametric, 181
  partial, 466–524
  product rule, 177
  quotient rule, 179
  rules of, 176–81
Distribution
  binomial, 409–10, 420, 421
  cumulative frequency, 367, 372
  difference of sample means, 435–9
  F, 447–8
  frequency, 403
  Normal, 398, 409, 412–20
  Poisson, 410–12
  probability, 403–5
  sample means, 434–5
  sampling, 433
  t, 441–4
Domain, 4, 106, 107

e, 39
Eckart potential, 183

Electric currents, 350, 355
Electric field, 291
Electric potential, 286
Enthalpy, 480
Entropy, 480, 491
Equilibrium
  position of, 190
  stable, 190
  unstable, 190
Error analysis, 452–84
Errors in measurement, 211–12, 364
Errors, rounding, 326
Estimates, 433, 437, 439
  best, 434–5, 436–7
Euler's formula, 88
Euler's method, 248–50
  for differential equations, 320–5, 328
Event, 398
  mutually exclusive, 400
  statistically independent, 400
Expectation, 405–9, 434
Expected value, 405
Experimental design, 456
Exponential function, 34–46
Exponential series, 40–2
Extrapolation, 19–23

Factor
  linear, 15
  repeated, 17
F distribution, 447–78
Force–distance graph, 236–8, 245–6
Fourier's law, 336
Fraction
  partial, 149–54, 262, 327
Frequency
  classes, 367, 386
  cumulative, 372
  curves, 373–5
  distribution, 367
  of vibrations, 135
  polygon, 370, 380, 386
  relative, 374, 398
  table, 367–70
Functions, basic, 1–30
  composite, 108–12
  continuous, 23, 141
  decay, 49–51
  discontinuous, 134–8
  exponential, 34–46
  hyperbolic, 51–62
  inverse, 101–28
  limit of, 134
  logarithmic, 102, 112–6
  of a function, 108–9
  one-to-one, 110
  polynomial, 5
  singularity of, 289
  trigonometric, 66–95
Functional notation, 4
Functional relation, 1, 467
Fundamental Theorem of Calculus, 240

Gradient
  of a curve, 121
  of a line, 6
Graph, 1
Graphical methods, 23
Gravity, 2, 11, 430

Half life, 50, 119, 309
Harmonic motion, 281, 345
  damped, 281, 348
Heat
  conduction, 308, 338, 349
  convection, 308, 336
  radiation, 228, 271, 293
Higher derivatives
  ordinary, 181
  partial, 476–7
Histogram, 370, 378
Homogeneity, 456–8
Hooke's law, 1, 17, 103
Hyperbola, 148
Hyperbolic functions, 51–62
  inverse, 123
Hypothesis
  null, 449
  testing, 448–53
Hysteresis, 57–8

Ideal gas, 237
Imaginary number, 12
Improper integral, 287
Indefinite integral, 243–4
Independent event, 400
Independent variable, 4
Index laws, 43–5
Infinite integral, 288–9
Inflexion, point of, 467
Initial condition, 345, 347
Input-output process, 305, 334
Instantaneous
  rate of change, 166
  speed, 159
Integral
  definite, 238
  evaluation of, 242–8
  improper, 287
  indefinite, 243–4
  infinite, 288–9
Integrand, 239
Integration, 233–304
  applications of, 295–302
  as a summation, 238–44
  by indirect substitution, 265–8
  by partial fractions, 271–3
  by parts, 273–7
  by quadrature, 248
  by substitution, 262–9
  by the substitution $t = \tan(x/2)$, 268–9
  change of variable, 266
  constant of, 244
  definite, 239
  Euler's method, 249–50

limits of, 239, 288
  numerical, 248–52
  reduction formulae, 277–8
  Simpson's method, 251–2
  techniques of, 259–85, 286–396
  Trapezoidal method, 250–1
Interpolation
  linear, 19–22
Inter-quartile range, 385
Interval bisection, 24
Inverse functions, 101–28
Isothermal, 468
Iteration, 25

Kepler, 128

Length of plane curve, 293–6
  linear relationship, 1
Limit, 161, 166
Limiting process, 161
Lower quartile, 385

Magnetic field, 246–7
Maximum of a function
  global, 192
  local, 192
Mean
  arithmetic, 367, 380
  assumed, 386
  binomial distribution, 410
  deviation, 385
  Poisson distribution, 411
  population, 434
  sampling distribution, 434, 436
  standard error of, 435
Measures of centre, 367, 380–3
Measures of scatter, 367, 383–6
Median, 381
  class, 382
Minimum of function (local), 192
Modal class, 383
Mode, 383
Model, 5, 38, 46, 213, 305
Modulus of complex number, 90

Natural logarithm, 112–16
Nesting of polynomials, 14
Newton-Raphson method, 211, 223–7
Newton's
  law of cooling, 307–8
  laws of motion, 11, 104, 334, 337, 469
Normal distribution, 398, 409, 412–20
  approximation to binomial, 420
  probability graph paper, 417–20
  special areas of, 416
  standard, 413–16
Null hypothesis, 449
Numerical integration, 248–52

Ogive, 379–80, 390
Ohm's law, 81
One-tail test, 452

Orders of magnitude, 34
Ordinary differential equation, 305
Oscillations
    natural 354
    forced, 354
Overdamping, 347

Paired *t*-test, 455
Parabola, 4, 8
Parameters
    population, 433
Partial derivative, 466–524
Partial differentiation, 466–524
Partial fractions, 149–54, 262, 327
Particular solution, 351
Parts, integration by, 273–7
Pendulum, 66, 94, 213, 221, 337, 345–8
Perfect differential, 486
Periodic function, 66
Periodic systems, 102–3
Permutations, 401–2
Phase, 84
Phasor, 91, 95
Point of inflexion, 194
Poiseuille's equation, 489, 490
Poisson
    approximation to binomial, 411
    chart, 412
    distribution, 396, 409, 410–12, 420
    mean, 411
    standard deviation, 411
    variance, 411
Polar co-ordinates, 492–3
Polynomial, 4
    cubic, 14
    factors of, 14
    function, 5
    nested form of, 14
    of degree n, 14
    quartic, 14
    roots of, 14, 23
    Taylor, 211, 214–23
Pooled data, 439, 442
Population, 433
    mean, 406
    parameter, 433
    variance, 406
Potential
    energy, 164, 190, 499
    field, 470–1
    function, 498
Power
    law, 121
    of statistical test, 452, 460–1
    series, 41, 87
Pressure
    atmospheric, 123, 175
    diastolic, 92
    systolic, 92
Primitive function, 240, 262
Probability, 395, 398–400
    addition rule, 400

densities, 397
density function, 397, 405
distribution, 403–5
product rule, 400
scale, 400

Quadratic
    approximation, 251
    formula, 10
    polynomial, 8
    rule, 179

Radian, 81
Radioactive decay, 50, 118, 308, 310, 326, 365, 395
Random sample, 433
Range, 4, 385
Rate of change, 159, 162–9
Raw data, 369
Real number, 12
Reciprocal rule, 178
Reduction formulae, 277–8
Relationship, 1, 467
Relative frequency, 375
    curves, 374–5, 377
Relative precision, 457
Resonance, 136, 357–9
Roots
    of a linear function, 7
    of a quadratic function, 8
    of polynomials, 15–17, 23–7
    repeated, 17

Saddle point, 509
Sample, 433
    estimation from, 433, 437, 439
    mean, 406, 441
    random, 433
    size
    statistic, 433
Sampling
    distributions, 433
    from a Normal distribution, 439–44
    theory, 431
Scatter, measures of, 367, 383–6
Secant, 73
Semi-interquartile range, 385
Series
    exponential, 40–2
    logarithmic, 114
    power, 41, 87
    trigonometric, 87
Shot-putter's dilemma, 281
Siderial period, 126
Significance level, 416, 449, 454
Simple harmonic motion, 345
Simpson's method, 251–2
Sine function, 69, 75
Singularity, 141
Singular point, 141, 202
Solid of revolution, 295–300
Solubility, 28

Speed
  average, 159
  instantaneous, 159
Speed time graph, 233–5, 245
Stability, 189–190
Standard deviation, 383–4, 388, 410
  of binomial, 410
  of Normal, 413
  of Poisson, 410
Standard error of mean, 433, 436
Stationary points, 187–210, 502–10
Statistical
  hypothesis, 448–53
  tests, 452
Steady-state
  response, 135
  solution, 356
Substitution, integration by, 262–9

Tangent function, 78
Taylor
  approximation formula, 482
  polynomials, 211, 214–23, 484, 502–5
t-distribution, 453–6
Test
  F, 447–8
  one tail, 452
  paired *t*, 455
  t, 453–6
  two tail, 452
Theoretical distributions, 409
Thermocouple, 18
Thermodynamics, 484, 491
Transient, 356
  response, 135

Trapezoidal method, 250–1
Trial, 398
Triangle inequality, 483
Trigonometric
  functions, 66–95
  formulae, 99–110
  inverse functions, 116–18
  series, 87
Turning points
  of a quadratic function, 11

Upper quartile, 385

Van der Waal's equation, 135, 145, 149, 154, 466
Variability, 364
Variable
  coded, 387
  dependent, 4
  independent, 4
Variance, 384, 388
  binomial distribution, 410
  Normal distribution, 413
  Poisson distribution, 410
Velocity
  of chemical reactions, 33, 278, 319
  of light, 34, 366
Venn diagram, 402
Vibrating string, 468–70
Vibrations, 68, 135–6, 337, 345, 350
Volume of revolution, 295–8
Von Bertalonffy growth, 60

Wave equation, 470, 479, 492
Work done, 237, 286